Comparative Brain Research in Mammals
Volume 1

Heinz Stephan Georg Baron Heiko D. Frahm

Insectivora

With a Stereotaxic Atlas of the Hedgehog Brain

With 129 Figures

Springer-Verlag
New York Berlin Heidelberg London
Paris Tokyo Hong Kong Barcelona

Heinz Stephan
Max-Planck-Institut für Hirnforschung
6000 Frankfurt a.M. 71 (Niederrad)
Federal Republic of Germany

Georg Baron
Département de Sciences Biologiques
Université de Montréal
Montréal, Québec H3C 3J7
Canada

Heiko D. Frahm
Anatomisches Institut der Universität
5000 Köln 41
Federal Republic of Germany

Cover art: *Top*, Figure 16, page 37, *Tenrec ecaudatus*; *middle*, Figure 19, page 37, *Potamogale velox*; *bottom*, Figure 20, page 38, *Chrysochloris stuhlmanni*.

Library of Congress Cataloging-in-Publication Data
Insectivora/ Heinz Stephan, Georg Baron, Heiko D. Frahm
p. cm. – (Comparative brain research in mammals: v. 1)
Includes bibliographical references and index.

ISBN-13: 978-1-4613-9126-5 e-ISBN-13: 978-1-4613-9124-1
DOI: 10.1007/978-1-4613-9124-1

1. Brain – Physiology. 2. Physiology, Comparative. 3. Mammals – Physiology. I. Stephan, H. (Heinz), 1924– . II. Baron, G. (Georg), 1930– . III. Frahm, Heiko D.
QP376.C627 1991 90-25353
599'.0188 – dc20

Printed on acid-free paper.

Softcover reprint of the hardcover 1st edition 1991

Camera-ready copy prepared by Arts & Graphic, Hanau, Federal Republic of Germany, and provided by the authors.

9 8 7 6 5 4 3 2 1

Editors' Preface

This first volume in the series on "Comparative Brain Research in Mammals" deals with the brains of Insectivora. The importance of Insectivora lies above all in the fact that, (1) insectivore-like ancestors are regarded as the initial group for the evolution of higher mammals, and (2) the insectivore brains retained many conservative traits, though the animals have adapted to different living environments. Therefore, the study of Insectivora brains enables an insight into the size and composition of the brain structure of earlier ancestors of the higher mammals including primates and humans; in addition, it illuminates the various evolutionary trends which made the successful adaptation to different biotopes possible.

The alterations which the brain has experienced in the course of the phylogeny and the related adaptive radiation will be examined in the succeeding volumes on the brains of other Eutheria, e.g., bats (Volume 2) and primates (Volume 4). In Volume 3 the brains of marsupials will be compared with those of conservative and evolved Eutheria.

Important topics of this series are the quantitative comparison of the brain and the discussion of the brain differences with regard to variations in the biology of the animals. Since all aspects of the life of an animal are directly or indirectly controlled by the brain, an intimate connection between the structure of the brain and the behavioral habits of the animal is to be expected. Differences will be discussed in detail, both in view of the brain structures as well as the taxonomic groups of the animals. Quantitative comparisons are only of limited worth without presenting the raw data. Otherwise they allow neither a critical discussion of the original material nor a continuation of the work by other scientists. The advantage of our data-collection consists above all in the fact that all histological work follows the same methods and all measurements follow the same criteria. For this reason the tables are important contents of the prospective volumes. A large portion of the data presented here is still unpublished. The collection of data will be supplemented by a listing of all the neuroanatomical work on the brains of Insectivora which is known to us. These studies will be critically evaluated when appropriate. Therefore, the book contains (as far as possible) a complete survey of the present knowledge regarding Insectivora brains.

The series addresses not only the interests of the neuroanatomists or neurobiologists but also those of scientists interested in ecology, ethology, and evolution. In the long run, it will be an indispensable source of data for morphometric studies. The data, which has been collected and worked on for decades, should retain its validity for a long time.

Contents

1 Introduction

The value of the comparative approach as a research strategy lies in the possibility it offers of examining various aspects of the diversity in nature. For example, it makes it possible to compare closely related species living in different ecological conditions as well as different species occupying similar ecological niches.

Adopting a dynamic approach to morphological studies, in other words, analyzing the structure of an organism with respect to its ecoethological conditions, is not new. However, despite several investigations which tried to relate the structural organization of particular animals to their living conditions and activities (e.g., Chapman 1919), consideration of species' natural environments was long neglected by students of morphology. Böker (1924, 1935, 1937), who sums up his approach under the term "biological anatomy" (biologische Anatomie), was perhaps the first to present an integrated view of anatomy as an ecoethological adaptation. His very promising first attempts unfortunately had little success because of his adherence to the Lamarckian position (Starck 1978). In the last few decades, however, the assumption that morphological traits of an organism reflect its ecological niche and that therefore even slight differences in structure should reflect at least slight differences in function is gaining acceptance by ecologists. More and more studies have been relating anatomical features, such as form and size of skeletal structures (e.g., Reed 1951) and shape of hairs (e.g., Hutterer and Hürter 1981), to ecological aspects and life style. The ecoethological approach has been particularly fruitful in the study of chiropteran morphology, as is shown in the summary by Findley and Wilson (1982). More recently, Hutterer (1985) has given a detailed review of anatomical adaptations of shrews.

Comparative studies have proven to be particularly fruitful in quantitative brain research, especially when brain composition is considered. Total brain size alone may not be sufficient since the same size found in different species may be the result of a differential growth of different brain parts, e.g., reduction of olfactory structures and simultaneous neocortical enlargement (Stephan and Andy 1964) or compensatory effects in sensory structures (Jolicoeur and Baron 1980). Therefore, whenever possible, brain composition should be investigated. As a rule, the size and differentiation of a brain structure do not vary independently. Increased size is almost always accompanied by progressive differentiation and vice versa (e.g., Stephan and Andy 1970; Stephan 1975; for entorhinal cortex and olfactory bulb). The characters of structural differentiation can vary in many ways (e.g., in construction, arrangement and connections of the cells, layers, larger units, etc.), whereas there are only two possibilities for variations in size: to become larger or smaller. Thus, an interspecific comparison of size may serve as a relatively simple first step in determining the functional importance of a cerebral structure in a given species.

Furthermore, since an animal's interactions with its environment are controlled largely by the nervous system and the size of a given brain structure i.e., the amount of brain tissue, is highly interrelated with the functional requirements of a species' habits, quantitative brain organization should reflect major evolutionary trends. The brain and its structures have to meet all the external and internal requirements if the animals and species are to survive. Higher and more complex behavioral patterns need larger and more differentiated cerebral structures.

Therefore, analyzing differences in brain structure may not only reveal trends in brain evolution, but it may ultimately lead to a deeper understanding of the complex relationship between brain and behavior. Numerous studies in mammals and birds have shown that variations in the relative size of the entire brain, of its major regions and, particularly, of specific sensory and motor nuclei, are paralleled by ecoethological characteristics. For example, in Insectivora, the form and size of the brain may change in relation to a particular substrate adaptation (Spatz and Stephan 1961; Stephan and Spatz 1962). Other studies have shown that the most reduced olfactory bulbs are found in semi-aquatic forms regardless of their taxonomic affiliation (Stephan and Bauchot 1959; Bauchot and Stephan 1968; Stephan and Kuhn 1982; Baron et al. 1983). Since, in mammals, olfaction is linked with breathing, it is of no use in detecting prey under water. Similarly, Bang and Cobb (1968) suggested a relationship between olfactory ability in various birds and the development of their olfactory bulbs. Pirlot and Stephan (1970), Stephan and Pirlot (1970), Pirlot and Pottier (1977) and Stephan et al. (1981b) were able to establish a relationship between feeding habits and the size of the total brain and of its major components in bats. Similarly, Baron (1972, 1974) suggested a relationship between the differential development of acoustic nuclei, feeding behavior, and echolocative abilities in bats and Boire (1989) showed a close correlation between feeding strategies and the size of the trigeminal and visual brain centers in birds.

An advantage of quantitative comparisons is that they allow us to rank data. However, in the absence of other evidence it should not be assumed that such data rankings represent an evolutionary lineage or historical progression. Nevertheless, although great caution is required, data rankings may allow inferences about conservative and derived characters of the brain. The validity of comparative studies and rank arrangements depends largely on the composition and the size of the sample investigated. The more species studied within a given taxon, the greater the likelihood of revealing the whole extent of adaptive radiation as well as all cases of convergent evolution. This, in turn, will facilitate differentiation between ancestral and derived characters.

Based on paleontological studies on fossils, the Insectivora form geologically the oldest eutherian group. In fact, many, if not all, recent orders of mammals originated phylogenetically in insectivore-like ancestors (see Thenius 1969a, b, 1972; Thenius and Hofer 1960). Although recent forms exhibit a great variety of evolved (apomorphic, derived, advanced) characters, they possess numerous conservative (plesiomorphic, primitive, ancestral) traits which identify them as the basic eutherian group. This is also true of the brain. The brain and nearly all its components have been found to be smallest in certain groups of Insectivora compared with other mammalian orders. The most conservative species of the extant Insectivora probably have undergone little evolutionary change since their first appearance (permanent types). Hence, they can be expected to be similar to the early forerunners of other mammalian orders, as e.g., Primates, and to represent a good reference base for evaluating brain evolution. Therefore, in previous investigations (Andy, Baron, Bauchot, Frahm, Pirlot, Stephan, since 1956), Insectivora (or subgroups of them) were used as the reference base for brain comparisons with other orders (Scandentia, Chiroptera, Primates, Macroscelidea).

Insectivora have not only retained a great number of primitive characters, they also show an enormous variety of ecotypes. The Insectivora are in fact more diverse anatomically than the members of any other order (Butler 1972). Comparative study of insectivoran brain structure may therefore contribute to a better understanding of the trends in brain evolution in relation to ecological constraints.

The aims of this book are (1) to bring out major trends in brain evolution among Insectivora and (2) to determine the extent to which brain structure and composition relate to ecoethological considerations. This review does not, of course, pretend to be complete in either its coverage of taxonomic and structural diversity or in its treatment of the ecological and ethological facts. It should, rather, be considered a first attempt to show how brain structure may reflect an organism's ecological role and to gain some insight into the problem of brain evolution.

In order to reflect the present state of knowledge about insectivoran brain structure, the book also gives, in addition to many new data, a review of all pertinent descriptive and experimental studies of which we were aware.

The raw data on brain structure volumes in a great number of tables are presented in order (1) to make possible comparisons with similar data obtained in other laboratories and from other taxonomic groups, (2) to encourage further analysis with appropriate mathematical methods (e.g., multivariate analyses) and (3) to stimulate further thought about possible interrelations with more and new ecological and behavioral observations as knowledge in these fields progresses.

Acknowledgements

Thanks are due to many colleagues and friends in various parts of the world, who helped to collect the material. Only some of them could be mentioned in the key to the source of body and brain weight data given at the end of Table 5. For careful histological work, which was done in the Max-Planck-Institut für Hirnforschung in Frankfurt, formerly Giessen, we are profoundly grateful to Hermine Keiner, Cläre Roberg, Monika Martin, Sigrid Wadle, Christina Ronnacker and Anja Niehuis. Excellent assistance was given by Helga Grobecker (photography), Hensy Rietveld-Fernandez (measurements), Michael Stephan (computer programs and works), Volker Stephan (drawings and graphics) and Helma Lehmann (secretarial work). Heartly thanks are due to John Nelson for stimulating discussions and to Jessica Pottier for checking the English text.

List of the Available Species

TENRECIDAE

TENRECINAE

Tenrec ecaudatus, Setifer setosus, Hemicentetes semispinosus, Echinops telfairi

GEOGALINAE

Geogale aurita

ORYZORICTINAE

Oryzorictes talpoides, Microgale dobsoni, M. talazaci, Limnogale mergulus

POTAMOGALINAE

Potamogale velox, Micropotamogale lamottei, M. ruwenzorii

CHRYSOCHLORIDAE

Chrysochloris asiatica, C. stuhlmanni

SOLENODONTIDAE

Solenodon paradoxus

ERINACEIDAE

ECHINOSORICINAE

Echinosorex gymnurus, Hylomys suillus

ERINACEINAE

Atelerix algirus, Erinaceus europaeus, Hemiechinus auritus

TALPIDAE

DESMANINAE

Desmana moschata, Galemys pyrenaicus

TALPINAE

Talpa europaea, T. micrura, Parascalops breweri, Scalopus aquaticus

SORICIDAE

SORICINAE

Sorex alpinus, S. araneus, S. cinereus, S. fumeus, S. minutus, Microsorex hoyi, Neomys anomalus, N. fodiens, Blarina brevicauda, Cryptotis parva, Anourosorex squamipes

CROCIDURINAE

Crocidura attenuata, C. flavescens, C. hildegardeae, C. jacksoni, C. russula, C. suaveolens, Suncus etruscus, S. murinus, Scutisorex somereni, Sylvisorex granti, S. megalura, Ruwenzorisorex suncoides, Myosorex babaulti

Abbreviations

AA	anterior amygdaloid area
AD	nucleus anterodors. thalami
AM	aqueductus mesencephali
AMY	amygdala
ANT	nuclei anteriores thalami
AOB	bulbus olfact. accessorius
AP	area postrema
APT	area pretectalis
ARCH	archicortex
AS	area striata
ASG	area striata grey matter
AUD	auditory nuclei
AvCg	average of *Crocidura*-group
AvEr	average of Erinaceinae
AvIF	average of Insectivora families and subfamilies
AvIS	average of Insectivora species
AvOr	average of Oryzorictinae
AvPo	average of Potamogalinae
AvSg	average of *Sylvisorex*-group
AvSn	average of Soricinae
AvTa	average of Talpinae
AvTn	average of Tenrecinae
AVT	area ventralis tegmenti
BA	basal amygdaloid nucleus
BCI	brachium colliculi inferioris
BH	brain height
BL	brain length
BOL	olfactory bulbs (MOB+AOB)
BoW	body weight
BrV	brain volume
BrW	brain weight
BW	brain width
CAI	capsula interna
CAM	cortical amygdaloid nucleus
CA 1 - CA 4	areas of hippocampus retrocommissuralis
CC	corpus callosum
CEA	central amygdaloid nucleus
CER	cerebellum
CGL	corpus geniculatum laterale
CGM	corpus geniculatum mediale
CHO	chiasma opticum
CLT	ncl. centrolateralis thalami
CM	corpus mamillare
CMT	ncl. centromedialis thalami
COA	commissura anterior
COC	commissura posterior

COE	locus coeruleus
COH	commissura habenulare
CON	cochlear nuclei
CR	cubic root of brain volume
CU	nucleus cuneiformis
CV	coefficient of variation
DCO	dorsal cochlear nucleus
DIAG	diagonal band of Broca
DIE	diencephalon
DMH	nucleus dorsomedialis hypothalami
DOT	nuclei dorsales thalami
DR	nucleus dorsalis raphe
EI	encephalization index
ENT	regio entorhinalis
EPI	epiphysis, pineal body
ETH	epithalamus
EXT	pars externa of RB
F	fimbria-fornix complex
FCE	nucleus fascicularis cuneatus externus
FCM	nucleus fascicularis cuneatus medialis
FD	fascia dentata
FGR	nucleus fascicularis gracilis
FR	formatio reticularis
FRTM	formatio reticularis tegmenti mesencephali
FUN	funicular nuclei
GC	griseum centrale
GLD	dorsal nucleus of CGL
GLV	ventral nucleus of CGL
GNF	genu nervi facialis
HA	hipp. anterior (HP + HS)
HAL	nucleus habenularis lateralis
HAM	nucleus habenularis medialis
HemL	length of cerebral hemisphere
HIP	hippocampus
HP	hipp. precommissuralis
HR	hipp. retrocommissuralis
HS	hipp. supracommissuralis
HSE	half surface of the eye
HTA	area hypothalamica anterior
HTD	area hypothalamica dorsalis
HTH	hypothalamus
HTL	hypothalamus lateralis
HTP	area hypothalamica posterior
HYP	hypophysis
ICN	ncl. interpositus cerebelli
IMD	ncl. intermediodors. thalami
INC	inferior colliculus
INF	nucleus infundibularis
INO	inferior olive
INT	intercalated cell masses
IP	nucleus interpeduncularis
LA	lateral amygdaloid nucleus
LAM	lateral amygdaloid division
LCN	nucleus lateralis cerebelli
LD	ncl. laterodorsalis thalami
LP	ncl. lateralis posterior thalami
LT	lamina terminalis
LTC	lamina tectoria
LUY	corpus subthalamicum Luysi
L 1-6	layers in allocortex
MA	medial amygdaloid nucleus
MAM	medial amygdaloid division
Max	maximum
MCB	magnocellular part of basal nucleus of AMY
MCN	nucleus medialis cerebelli
MD	ncl. medialis thalami dorsalis
MES	mesencephalon
MHT	nucleus medianus hypothalami
Min	minimum
MLT	nuclei mediales thalami
MNT	nuclei mediani thalami
MOB	main olfactory bulb
MOT	sum of measured motor nuclei
MTC	mesencephalic tectum
MTG	mesencephalic tegmentum

N	number of species
n	number of individuals
NEO	neocortex (isocortex)
NET	net brain volume
NG	neocortical grey matter
NG 1	molecular layer of NG
NG 2-6	cell layers 2-6 of NG
NLL	nuclei of lateral lemniscus
NN	number of subfamilies and families
NTO	bed nucleus of lateral olfactory tract
NW	neocortical white matter
N V	nervus trigemini
N VII	nervus facialis
OBL	medulla oblongata
OLC	olfactory cortices (RB+PRPI+TOL)
OLS	superior olive
PAL	paleocortex
PALL	pallidum, globus pallidus
PAM	regio periamygdalaris
PAR	area parasubicularis
PC	pedunculus cerebri
PH	nucleus parahypothalamicus
PI	proportion index
PO	area preoptica
POL	area preoptica lateralis
POM	area preoptica medialis
PRH	regio perirhinalis
PRP	nucleus prepositus hypoglossi
PRPI	regio prepiriformis
PRS	regio presubicularis
PT	nucleus parataenialis thalami
PTE	pretectal region
PVO	periventricular organs
PvZ	periventricular zone of MOB
RA	nucleus raphes
RB	regio retrobulbaris (anterior olfactory nucleus)
RBrS	relative brain size
RC	nucleus residualis cornus ventralis
RE	nucleus reuniens thalami
RED	nucleus reticularis dorsalis
REL	nucleus reticularis lateralis
REV	nucleus reticularis ventralis
RH	nucleus rhomboides thalami
RLN	relay nuclei
RMO	nucleus reticularis medullae oblongatae
RP	ncl. reticularis paramediani
RTH	nuclei reticulares thalami
RUB	nucleus ruber
SAN	sulcus semiannularis
SC	nucleus suprachiasmaticus
SCB	subcommissural body
SCH	schizocortex
SEA	ncl. of anterior commissure
SED	nucleus dorsalis septi
SEF	nucleus fimbrialis septi
SEL	nucleus lateralis septi
SEM	nucleus medialis septi
SEP	septum telencephali
SES	nuclei of stria terminalis
SET	nucleus triangularis septi
SFB	subfornical body
SFF	fimbria-fornix complex
SIN	substantia innominata
SN	substantia nigra
SOC	supraoptic commissures
SON	supraoptic nucleus
STH	subthalamus
STR	striatum
SUB	subiculum (hippocampi)
SUC	superior colliculus
TCN	total cerebellar nuclei
TEL	telencephalon
THA	thalamus
THP	thalamus posterior
TOL	tuberculum olfactorium
TR	complexus sensorius nervi trigemini

TRL lateral olfactory tract
TRM nucleus mesencephalicus nervi trigemini
TRO optic tract
TRP nucleus sensorius principalis nervi trigemini
TRS ncl. spinalis nervi trigemini
TSO nucleus tractus solitarii

V ventricle
VC vestibular nuclei complex
VCO nucleus cochlearis ventralis
VET nuclei ventrales thalami
VI nucleus vestibularis inferior
VL nucleus vestibularis lateralis
VLT ncl. ventrolateralis thalami
VM nucleus vestibularis medialis
VMH ncl. ventromed. hypothalami
VMT ncl. ventromedialis thalami
VNS vomeronasal system
VP thalamus ventroposterior
VPL ncl. ventroposterior lateralis thalami
VPM ncl. ventroposterior medialis thalami
VPO ventral pons
VS nucleus vestibularis superior
VT nucleus ventralis tegmenti
V III third ventricle
V IV fourth ventricle

ZI zona incerta

Motor nuclei

III ncl. oculomotorius communis
V ncl. motorius nervi trigemini
VI nucleus nervi abducentis
VII nucleus nervi facialis
X nucleus dorsalis motorius nervi vagi
XII nucleus nervi hypoglossi

Arbor vitae cerebelli

I lingula
II lobulus centralis ventralis
III lobulus centralis dorsalis
IV lobulus ventralis culminis
V lobulus dorsalis culminis
VI declive
VII tuber vermis
VIII pyramis
IX uvula
X nodulus

1 fissura preculminata
2 fissura prima
3 fissura secunda
4 fissura postpyramidalis (f. posterolateralis)

2 Material and Methods

Methods of Fixation and Preparation

The animals were weighed immediately after trapping in order to avoid errors due to changes in body weight (BoW) caused by captivity and thus provide a more accurate measure of "normal" BoW. Dead animals were measured immediately (BoW, lengths of head and body, tail, ear, hind foot), perfused and prepared. Live animals were weighed a second time after being sacrificed with Nembutal to determine changes in BoW, and then were measured. The bodies were perfused transcardially with Bouin's fluid after the blood had been washed out with physiological saline. After fixation with Bouin's fluid, brains harden immediately and, unlike other adjacent tissues, they become uniformly yellow. The brains were prepared by breaking small pieces of the skull away from the hardened brain, starting at the foramen magnum and proceeding rostrally at the base of the brain case and then dorsally to expose the lobulus petrosus and the cerebellum. Further forward movement exposed the hypophysis, then the optic chiasma and olfactory lobes. Finally, the dorsal surface of the cerebrum was exposed. The vault of the cranium formed a bowl in which the brain rested during preparation. The upper jaw and snout formed a kind of handle with which the vault could be held. The cranial nerves were cut close to the brain, except for one of the trigeminal nerves, which was cut at the level of the optic chiasma. Very thin optic nerves were often broken off during the preparation. When the optic nerves were thicker and could be preserved, they were cut some distance from the chiasma. The meninges were removed from the brain, and the spinal cord was cut so that the part attached to the brain formed a square (seen from below). The prepared brain was weighed on a Mettler PE 360 Delta Range balance having an accuracy of 1 mg, and some linear dimensions of the brain were measured (see next page). Then the brain was stored in Bouin's fluid for four days before being placed in 70% alcohol, which was changed several times during the first weeks.

The reason for using the relatively complicated method of fixation and preparation was that in very small animals, such as shrews, generally only a pulp of brain tissue is obtainable from fresh preparations and that, in brains of all sizes, the natural proportions are much better preserved. In situ fixation with Bouin's fluid makes it possible to prepare uninjured and complete brains and thus meets the requirements for exact measurements.

Determination of Fresh Brain Weights

The question arises as to what extent brain weight changes by fixation with Bouin's fluid and/or during the time of preparation. According to investigations in guinea pigs (Stephan 1960b) and rats (Stephan et al. 1981a), the weight of brains fixed in Bouin's fluid corresponds to the weight of fresh brains when prepared within about the first five hours. There is not even a slight tendency towards enlargement or reduction. Later on, the weight of brains stored in Bouin's fluid slowly decreases.

Brain and Body Weight Collection

Pairs of brain and body weights were derived from at least 466 individuals representing 53 species from the six Insectivora families. The exact number of individuals cannot be given for those species for which data taken from the literature were included, because reports do not always specify the number of individuals used. In such cases, we counted the data as if they were obtained from one individual. In all, 441 specimens from 52 species were collected by the senior author in cooperation with friends and colleagues. Pairs of data on 25 specimens from five species were taken from the literature. Only one of these species was not represented in our own material. For three of the 53 species, pairs of data came from either two or more independent sources.

In some cases, either body weights (BoWs) or brain weights (BrWs), i.e., no pairs of data, were available from our own or other sources. These data were also taken into account because of their interest, especially when they came from rare species. For several common species, such supplementary data came from numerous sources and at times produced large numbers of individuals, e.g., up to more than 700 for the body weight and 200 for the brain weight of *Sorex araneus* (Table 5). In Table 5, only species averages and standards (see p. 16) are given; data on individuals are available on request.

Comparison of Linear Brain Measurements

Linear brain measurements were made of (1) the length of the brain from the rostral extremity of the olfactory bulb to the caudal extremity of the cerebellar uvula, (2) the length of the cerebral hemispheres without the olfactory bulb, (3) the width of the hemispheres and (4) their height (Figs. 1 and 2). (5) The length of the corpus callosum (Fig. 3) was determined in reconstructions of the midsagittal plane from serially sectioned brains; the measures were taken at the midline of the corpus callosum. Linear shrinkage was corrected to fresh brain dimensions.

Brain shapes were usually compared by using ratios of two measures, e.g., brain width to brain length. In the case of differences it is, however, impossible to say which of the two measures has really changed. Such difficulties can be avoided if the measurements are referred to a base independent of brain shape but reliably reflecting brain size. Such a measure is the cube root of brain weight (Herre and Stephan 1955) or, even better, brain volume (Stephan 1977; Stephan et al. 1977; Stephan and Nelson 1981). All linear measurements can be referred to this cube root to give informative proportion indices (PI). A PI of 1.7 signifies that the linear measurement is 1.7 times the edge of a cube having the same volume as the brain. With such indices, all measures (e.g., lengths, widths and heights) can be compared independently of one another.

Brain Sectioning

A total of 215 brains from 50 species were embedded in paraffin and serially sectioned (10, 15 or 20 μm thick) in the three standard planes: 152 brains from 50 species transversally, 51 brains from 33 species sagittally and 12 brains from 9 species horizontally. Sections of all brains were used in the histologic investigations. The midsagittal planes were reconstructed from the sagittal sections (Figs. 40 - 51). All volume determinations were based on transversally sectioned brains.

In addition, three brains of *Atelerix algirus* were sectioned in stereotaxic levels: one hemisphere sagittally, one hemisphere horizontally, and two brains transversally. The stereotaxic atlas (see pp. 503 - 553) was prepared from one of the transversally sectioned brains.

Determination of Volumes

In all, 129 transversally sectioned brains from 50 species met all the requirements for exact volume measurements. As a rule, 250 sections of each of these brains taken at equal distances were mounted and stained with either cresyl violet or gallocyanine. Normally, every fourth mounted section, i.e., 60 to 80 equidistant sections, was used for the volume estimations. The sections were projected directly onto photographic paper (18 x 24 cm), the borders delineated, and the components of each structure either measured by planimetry or cut out and weighed. In the second case, the photographic paper had been weighed to determine the number of square millimeters per milligram, and care was taken to compensate for changes in paper weight due to varying temperature and humidity. Distances between the histologic sections were calculated, and the volumes for each division were derived by applying the formula:

$$V = \frac{AP \times WS \times D}{M^2}$$

where V = volume in mm^3; AP = average unit weight of the photographic paper in mm^2/mg; WS = weight of the cut-out structure in mg; D = distance of measured sections in mm; M^2 = square of linear magnification.

The sum of the volumes of all measured structures (= total brain) is considerably smaller than the volume of the fresh brain. This difference is due to shrinkage from fixation and embedding. The extent of shrinkage is different in each brain, even if all brains undergo the same procedures. To obtain comparable values, it is therefore necessary to correct the figures to the fresh brain volume. This value can be obtained by dividing the weight of the fresh brain by the specific brain weight. The specific gravity of fresh brains has been found to be close to 1.036 g/cm^3 (Stephan, 1960b). Thus,

$$\text{volume of fresh brain} = \frac{\text{weight of fresh brain}}{1.036}$$

To obtain the conversion factor for an individual brain (C_{ind}), this volume was divided by the serial section volume:

$$C_{ind} = \frac{\text{volume of fresh brain}}{\text{serial section volume}}$$

The size of this conversion factor depends on the type of fixation. The following means were found:

Fixation fluid	n	Conversion factor		Volume loss in % of fresh brain	
		mean	range	mean	range
Bouin's fluid	648	2.04 ± 10.0%	1.66-2.70	50.9	37.0-60.2
Formol-alcohol	5	1.54 ± 7.6%	1.41-1.73	35.0	29.0-42.0
Formol 10%	21	2.40 ± 9.3%	2.02-2.86	58.4	50.6-65.1

Number of individuals (= n) includes Chiroptera, Scandentia, Macroscelidea, and Primates

Since volume loss varies widely, as indicated by the table, it is inappropriate to compare uncorrected figures from material with unknown or different types of fixation, and even if the same fixation and embedding procedures are applied.

The conversion factor calculated for the total brain was then used to correct all the parts. While it could be argued that the correction is not ideal since different brain parts may shrink to different extents, such differences are difficult to ascertain and, since they are likely to be small, they were not taken into account.

Introduction of Standards

To obtain brain structure volumes as representative as possible for a given species, we performed a second conversion (C_{sta}) that takes into account the difference between the individual weight of the measured brain and the standard brain weight for each species as given in Tables 5 and 6. Such differences are due to intraspecific variation in brain weights.

$$C_{sta} = \frac{\text{weight of standard fresh brain}}{\text{weight of individual fresh brain}}$$

We introduced these standards since, in general, our information about average brain size within a species is much more extensive than our knowledge about the average volume size of the various brain structures for the following reasons: (1) Even when there is ample brain material from a given species, it would be cumbersome and of little value to utilize all the material for comparative studies merely to have a better base on which to determine the average. The number of brains we have sectioned serially is limited since such work is very time-consuming and expensive. Furthermore, for our purpose, consideration of as many species as possible is more informative than investigation of many specimens of a single species. (2) By introducing standards, serial-sectioned brains for which the individual brain weight is not known can be used for interspecific volume comparisons. (3) Standards generally permit inclusion of data taken from the literature that otherwise would have to be rejected.

This second conversion (C_{sta}), determined by comparing total brain, was utilized to adjust all the parts of the brain. This is possible since, from our experience, extremely small or extremely large brains of adult specimens within the same species generally have the same composition. Intraspecific comparisons of small and large brains do not reveal any directional change in brain composition (e.g., Kruska and Stephan 1973).

We believe that by applying standard brain weights for each species, we can obtain a good approximation to the unknown mean from a few values or even only one specimen.

In addition to standard brain weights, we introduced standard body weights (Tables 5, 6). The body weight standards are the best substantiated values derived from all available material after exclusion of juvenile, thin and sick, as well as excessively fat animals, and after subtracting the uterus weights from the body weights of pregnant females. Thus, the resulting standard BoWs sometimes differ widely from the arithmetic means of all known data, which in part are from the literature and may be from juveniles.

Survey of Structures and Complexes

Volumes and size indices of structures and complexes are given in Tables 10 - 70. An outline of the tables is given in Table 1. Several measurements were made primarily for earlier publications to obtain a reference base for comparisons with other mammalian orders, e.g., Primates and/or Chiroptera. They were based on fewer species, most of which belong to a group classified as "Basal Insectivora" (see p. 21). Some of the data on diencephalic structures were taken from Bauchot (1979a, b) (Tables 65 - 70).

Delimitations or at least topography of the structures are given in the stereotaxic atlas of the hedgehog brain (see pp. 503 - 553).

Variability

Most of the volume data in the tables contain at least two real numbers, e.g., 0.0011, 0.011, 0.11. In cases of very large figures, the last places are, of course, fairly unimportant since a variability of up to 10% may be expected. In most species (39 of 50), the main structures and/or complexes were measured in more than one brain. In these cases, only the average volumes are given in the tables. Volume data on individuals will be provided on request.

The coefficients of variation (CV = standard error of the mean in percentage of the mean) varied from 0.1 to 79.3. The degree of variability is obviously not the same in all structures. In some structures, high values predominate; in others, low values. When the CVs were averaged, the highest variability was found for the ventricles, a complex that does not consist of true brain tissue (av. 23.6, max. 60.4). The ventricles account for an average of 1.2% of the total brain. They were included in the measurements because they are normally filled with liquid when the brain is weighed and thus influence brain weight. During embedding, which results in considerable shrinkage, the ventricles may behave differently from brain to brain, thereby accounting for the relatively high variability shown here.

A similar high variability was obtained for a complex (= "rest") comprising meninges, nerves, etc. (av. 21.2, max. 79.3). The main components are the stumps of the optic nerves until their convergence in the chiasma, the trigeminal nerve with the Gasserian ganglion (which, in our preparations, is always left in place unilaterally), the hypophysis, and the remnants of the meninges. The hypophysis was included in this complex to obtain comparable values for the diencephalon and for the net brain tissue of the total brain. The hypophysis was sometimes broken and thus was not available for measurement.

The complex of meninges, nerves, etc. was included in the measurements even though several of its constituent parts are not, strictly speaking, brain tissue because its parts were included when the brains were weighed. The complex was relatively small and amounted to only about 1.9% of the total brain. Its omission would not greatly distort the picture obtained of the composition of the brain. Its high variability was mainly due to minor differences in brain preparation.

The highest variability of structures consisting of true brain tissue was found in the schizocortex and the olfactory bulbs (main and accessory bulbs). The schizocortex (av. 11.5, max. 56.1) had by far the highest CVs in Tenrecinae and *Geogale* (av. 32.8). In these groups, the lamination of the schizocortex is blurred and was difficult to distinguish from the prepiriform cortex. In the olfactory bulbs, the accessory olfactory bulb had an average CV of 20.0%, with a maximum of 83.3%; the main bulb, an average of 7.7% and a maximum of 24.1%. In descending sequence were the cerebellum (av. 7.4 and max. 18.7, respectively), amygdala (7.1 and 29.3), medulla oblongata (6.9 and 18.5), hippocampus (6.7 and 25.6), neocortex (6.4 and 30.1), septum (6.3 and 17.4), striatum (6.3 and 17.4), mesencephalon (5.4 and 19.5), paleocortex (4.9 and 15.4), diencephalon (4.2 and 13.6), and telencephalon (2.5 and 11.6). In the structures measured in smaller numbers of species, the highest CVs were found in the pineal gland (av. 37.2) and subfornical body (av. 20.6). Thus, high CVs were common in very small structures. In larger structures and more complex components, the differences in variability seemed to be due mainly to true intraspecific variation and only to a small extent to difficulties in delineation. Structures that are difficult to delineate, such as diencephalon, striatum, and mesencephalon, had the same or even lower standard deviations than those that are easy to delineate (olfactory bulbs, cerebellum). These results agree with those found in Chiroptera (Stephan and Pirlot, 1970) and Primates (Stephan et al. 1988).

More detailed investigations on variability were performed on tree shrews *(Tupaia glis)* (Stephan et al. 1981a). In all, twelve brain structures were compared in 20 individuals, and the individual deviation from the mean was determined for each structure. The average CV of the total of all 240 values (twelve structures for 20 individuals) was found to be 3.9, with a range of 0.1 to 16.9. The maximum deviation of 16.9 was again found in the olfactory bulbs. Of all 240 values, 171 (71%) were below an average CV of 5, and 226 (94%), below 10. In random samples, a deviation of less than 10% from the mean for all 20 specimens was nearly always reached when two individuals were studied. In the most unfavorable sequence of theoretically possible *Tupaia* specimens, a maximum of seven brains was necessary for measurements of the olfactory bulb to arrive at an average deviation lower than 10%, five for the medulla oblongata and four for the hippocampus. However, such sequences are extremely unlikely. In the remaining nine of the twelve structures, an average deviation lower than 10% was reached even under the most unfavorable sequences when two brains were studied.

Significance tests (t-tests) were performed for cerebellum (CER, see p. 85) and neocortex (NEO, see pp. 85 and 135).

Methods of Comparing Brain Size and Volume

Requirements: In recent years there has been an increasing number of quantitative investigations on brains and brain structures in order to elucidate possible trends in brain evolution. One of the most important questions in connection with brain evolution, still open to controversy, concerns the most appropriate statistical method for calculating differences in the size of the brain and of brain parts which are biologically meaningful. This problem has been discussed recently in detail for primates (Stephan et al. 1988). The requirements postulated for the method were that it: (1) be easily applicable to large numbers of specimens and species; (2) provide similar results for species which are closely related and share similar basic morphological and ecoethological features; and (3) reflect differences in these features in an intuitively reasonable manner. Size comparisons in groups of **isoponderous** species were performed by Stephan et al. (1988). They give reliable data on relative brain size and thus can be used directly as criteria for assessing the available methods. The selected method must, of course, also be applicable to species with different body weights.

Percentages

Percentages do not meet the postulated requirements but since they provide an easy way to express size differences of brain structures, they are often used in the literature. In the text and some tables of this book, reference is made to percentages (1) of total BrW relative to BoW and (2) within the brain, of structures relative to larger complexes. Percentage data are, however, of little value for evaluating intergroup differences for the following reasons:

(1) When relating BrW to BoW, percentages are a special case in allometric methods in which the full power of the BoW is taken into account (= BoW^1). This power, however, is far too high, making it impossible to minimize the influence of BoW on relative brain size (see Section 3.2).

(2) When comparing the composition within the brain, e.g., the percentages of the five fundamental brain parts relative to the total brain, interpretation of the results is difficult because the percentage values depend not only on the size of the structure under comparison, but also to a considerable extent on the size of all the brain structures included in the reference system. This also applies, of course, to differences found between taxonomic units and to any trend which may appear. Compared with Insectivora, advanced primates have low percentages in many structures, even if the allometric indices of these structures are distinctly higher. For example, the striatum drops from 8.3 or 8.6% in Insectivora to 2.7% in man despite a more than tenfold allometric enlargement (Stephan 1979 and section 3.3.6). The percentage reduction in the striatum is due to an extreme enlargement of the neocortex in man. Therefore,

evaluating percentage values is unsatisfactory for interspecific comparisons. For intraspecific comparisons, percentages are a useful tool for a first insight into the brain composition of a species or group. Furthermore, percentages allow comparison with data taken from the literature, since the data reported are often restricted to the percentage size of brains and brain parts.

Indices

The relationship between brain and body size has been found to be best represented by the power function

$$\text{BrW} = \text{RBrS} \times \text{BoW}^{a}$$

in which BrW is the brain weight or the volume of a brain structure, RBrS the relative brain size, BoW the body weight, and "a" its power.

Best fit lines: In a log-log plot (Fig. 52), RBrS is the y-intercept, and the power "a" is the slope of the best fit line. In contrast to the investigation on primates in which the slope was determined by least-square regression methods (Stephan et al. 1988), it will here be determined by a canonical method (Rempe, 1970; Rempe and Weber, 1972). The total of 56 families or subfamilies of Insectivora, Scandentia, Chiroptera, and Primates were analysed in order to obtain a common slope. The canonical discrimination analysis gives each subfamily or family its appropriate weight when calculating the common slope so that groups for which many data are available and/or with a large body weight range have more influence on the slope than those that are poorly documented and/or have a small body weight range. In practice, both procedures give for the various families or subfamilies similar results, since the correlation is high, i.e., there is little scatter of the points about the lines. Similar results in the case of high correlation are also obtained when using other mathematical procedures to estimate best fit lines (as e.g., major axis, reduced major axis; see Stephan et al. 1988).

The most critical decision, perhaps, when applying the allometric equation is the choice of the sample or the taxonomic level. Stephan et al. (1988) showed in a review of pertinent publications that the allometric exponent or slope of the best fit line varies with the taxonomic level as well as with the systematic group. The consensus seems to be that the slopes for lower taxonomic units are usually less steep than those for higher taxonomic units (e.g., Passingham 1975; Lande 1979). It seems, however, that at least on the genus, subfamily, and family levels the slopes are similar provided the sample size and body weight range are sufficiently large. Indeed, RBrS values similar to those obtained from groups of isoponderous species resulted only when BoW powers slightly higher than 0.6 were used (Stephan et al. 1988). Such slopes have been obtained in regression analyses indiscriminately for genera, subfamilies and families, but not for species (lower slopes), suborders, or orders or across mammalian orders (generally higher values).

Stepwise increments: Just as the intraspecific lines appear to be one above the other when plotted along the best fit line for a family (see Bauchot and Stephan 1964), interspecific lines for families appear as stepwise increments along the lines within orders (see Stephan et al. 1988) and so do orders within the mammalian class. It follows that the taxonomic level at which the allometric analysis is performed affects not only the allometric exponents and the allometric coefficients but also the residuals, which reflect size differences of the brain and the brain parts.

Choice and placement of reference line: ***a) Conservative groups:*** It must be emphasized that careful and critical **determination of the slope** is the most important basic requirement when using allometric methods for size comparisons (see Stephan et al. 1988). In contrast, the **placement of the reference line,** i.e., the determination of the y-intercept, poses few problems. It depends on the purpose of the comparison and thus may be based on phylogenetic, ecoethological, or physiological considerations. The reference line can easily be displaced by parallel shifting from one reference group to another. The distances between the log-log points, and thus the ratios of RBrS of different species, are independent of the position of the reference line.

Previous studies on brain evolution in primates and other mammals have shown that meaningful results can be obtained by determining RBrS based on a conservative group (e.g., Stephan and Andy 1964; Stephan and Bauchot 1965; Bauchot and Stephan 1966, 1969; Stephan 1967a,b; Pirlot and Stephan 1970; Stephan and Pirlot 1970; Baron 1972, 1974, 1977a, b, 1978, 1979; Jolicoeur and Baron 1980; Jolicoeur et al. 1984). Since it is impossible to obtain precise information on the quantitative composition of brain in fossil mammals, thc most conservative forms of extant Insectivora have been used as a reference group ("Basal Insectivora"). In the "Basal Insectivora" Tenrecinae, Erinaceinae, and terrestrial Soricidae were combined. It is reasonable to assume that the brains of the "Basal Insectivora" are similar to those of extinct forms, which are considered to be forerunners of many mammalian orders, including Primates. Brain sizes of various species relative to those of isoponderous conservative Insectivora have been investigated since 1956 by Stephan and coauthors.

When BrW or brain volumes are plotted against BoW the lowest levels are nearly always occupied by *Geogale*. However, only extrapolated body and brain weights and one serially sectioned head with a poorly preserved brain were available from *Geogale* (Rehkämper et al. 1986). One more sectioned head is now available, but unfortunately, its brain is also poorly preserved. The next lowest size of total brain and many brain structures in Insectivora (and perhaps in all the extant eutherian mammals) was found in Tenrecinae. Therefore, as in the comparison of primate brains (Stephan et al. 1988), the Tenrecinae were chosen as the reference for comparison, and the reference lines (thick lines in Figs. 52, 54, 70, 91) were placed through the average for Tenrecinae (AvTn). All points on these AvTn lines represent a RBrS of 1 or 100%. For each species, the distance from the AvTn line indicates the extent of the difference from the Tenrecinae. Indices of 2 or 200 mean twice the relative size of brains or brain components of Tenrecinae, indices of 0.5 or 50, half. The indices can be averaged for any of the higher taxonomic units. To determine the mean index levels in Insectivora, the averages (1) of all the species under comparison (AvIS) and (2) of the families and subfamilies indicated in the tables by asterisks (AvIF) were calculated. When referring to the Insectivora level, the family/subfamily average (AvIF) was usually chosen because in the average for all species (AvIS), the Soricidae were strongly overrepresented in most cases.

The choice of a primitive group as the main reference group does not preclude using another taxonomic unit (e.g., *Homo;* Holloway and Post 1982) as a reference group for intragroup and intergroup comparisons. This is particularly appropriate when no primitive group is included in the material under comparison, or when a brain structure is not sufficiently differentiated in the "basal" group to provide precise measurements (as, e.g., the visual cortex in Insectivora).

b) Mouse-elephant line: As a reference line, several authors have used the best fit line for all mammals (often called mouse-elephant line), which predicts the brain size of average mammals for any given body size. Jerison (1973) found a slope of 0.66 and called the deviation of a particular brain size from this line the encephalization quotient (EQ). The use of larger samples has led to estimates of steeper slopes (0.72 - 0.76; e.g., Eisenberg 1981; Martin 1981, 1982; Hofman 1982; Armstrong 1982, 1983). In the Insectivora-Primates line, the slope reached as high as 0.91, thus approaching isometry. If such a steep slope is used, the relative size of the brain or a brain part appears systematically larger in species with a smaller body weight than in species with a larger body weight belonging to the same family. Furthermore, the values for relative size of the brain and brain parts obtained from the mouse-elephant line are often devoid of any biological meaning and may result in such oddities as similarity of relative size values for the total brain and many brain parts in man and in small prosimians, or higher values in small prosimians than in apes.

c) Congeneric species: The comparative brain size indices (CBS) proposed by Clutton-Brock and Harvey (1980) and Mace et al. (1981) are used to compare "congeneric groups," i.e., congeneric species which "share the same ecological category and social system" within a family. They express the distance of the mean BrW of a congeneric group from the best fit line for the family. Although all genera within a family may be similar in particular characteristics as claimed by Harvey (1988), it has been shown that adaptive radiation can be observed in small taxonomic units like genera (Hutterer 1985; Baron and Stephan in press). For example, in the genus *Sorex* we find forms that are scansorial (*alpinus*), terrestrial (*minutus*), semifossorial (*unguiculatus*), semi-aquatic (*palustris*), and even adapted to subdesert conditions (*merriami, arizonae*). Similar examples can be found in other genera (Hutterer 1985). Brain size and brain composition differ considerably within these taxonomic units. Generic values may express the general trend of brain evolution in a genus but otherwise are of little biological significance.

Interstructural analysis: Another fallacy in the study of brain evolution may result from allometric analyses relating the volume of a brain part to the whole brain (e.g., Passingham 1975) and from analyses relating measurements of two brain structures (e.g., Deacon 1988). For evolutionary interpretation, such methods are improper and/or the conclusions, unfounded.

Comparisons relating a part to the whole become more spurious as the proportion of the part to the whole increases. For instance, comparing the size of the neocortex, which constitutes one of the major parts of the mammalian brain, -particularly the primate brain-, with the whole brain, results in a best fit line with very small residuals because of the strong autocorrelative effect of the neocortex. For example, interstructural analyses of neocortex to brain yielded slopes of 1.116, 1.098 and 1.279 and correlation coefficients of 0.942, 0.996 and 0.984 in Insectivora, in Primates, and in the two groups together, respectively (Frahm et al. 1982). In the overall analysis, man has a smaller neocortex than would be predicted from the regression line, a mathematically indisputable but biologically dubious conclusion. In fact, such relations do not permit any predictions about phylogenetic trends and merely indicate that some brain parts increase faster than others.

Interstructural analyses of subdivisions of the brain must also be interpreted with caution since the effect of body size is to increase the correlation coefficient. Furthermore, many brain components do not develop independently of one another because they belong to the same functional unit by virtue of direct fiber connections. They thus may be highly correlated independently of the taxonomic position of the species (e.g., Stephan and Andy 1969; Baron 1979). Analyses may, however, reveal diverging trends for structures with no or little functional relationships, as for example Stephan and Andy (1969) showed in the case of the olfactory bulb and the hippocampus. Conclusions from interstructural analyses on typical evolutionary trends in various taxonomic units may therefore be very hazardous.

The inferential sterility with respect to species-specific biological differences, and the fallacy of conclusions such as mentioned above, are the reason we do not use higher-order taxonomic grouping nor do we base our interspecific comparisons on interstructural analyses. Instead, we use samples of closely related species despite the greater difficulty in determining a satisfactory common slope and in selecting a reference line.

Interstructural comparisons can reveal evolutionary trends if they are based on correlations of ratios of residuals or size indices. Indeed, correlations of size indices may be positive or negative or drop abruptly and reveal evolutionary compensation, regression or progression depending on the taxonomic unit (Stephan and Andy 1969; Jolicoeur and Baron 1980; Jolicoeur et al. 1984).

3 Comparative Brain Characteristics

3.1 Comparative Macromorphology

Linear Measurements

Intraspecific variability was investigated in the four external linear measures shown in Figs. 1 and 2 and listed in Table 2. The CVs were averaged in the 35 species in which more than one individual was measured. The largest variabilities, were found to be 3.6 (%) for hemisphere length (HemL) and brain height (BH). In these two measures, it is difficult to arrive at identical values in all individuals of a species. In HemL, the caudal pole is often difficult to fix with the calipers, especially in species where it is low ventrally and projects between the lobulus petrosus ('flocculus') and the cerebellar hemispheres, e.g., in *Chrysochloris* (Figs. 8 and 20). In BH the lowest and highest extremities are not always at the same level (perpendicular to the longitudinal axis of the brain), and thus may be measured at different angles to this axis. The average CV of brain length (BL) is somewhat lower (3.3) than that of HemL and BH. BL may be influenced by the cerebellar uvula, which in different individuals may project to different levels beyond the dorsally adjacent posterior lobe of the cerebellar vermis. The lowest variability in CVs was found for brain width (BW, av. CV = 2.9). BW is easy to measure in all individuals of a species. In Chiroptera, similar differences in CVs have been found in corresponding measures by Stephan and Nelson (1981).

Brain Proportions

The general brain proportions were expressed as proportion indices (= PI; see Material and Methods). These indices are the relationship between a linear measure on the brain and the cube root of brain volume, i.e., the length of the edge of a cube which has a volume equal to that of the brain. In order to facilitate the comparisons, some of which were based on large numbers of brains (Table 2), the measures were averaged for each species and then expressed relative to the cube root of the standard brain volume (Tables 3 and 4).

Brain Length (BL): The shortest brains in the Insectivora were found in the Chrysochloridae (PI 1.39), the longest in *Solenodon* (2.08). The highest PI was 1.5 times larger than the lowest. Relatively short brains, with PIs lower than 1.55, were found in *Oryzorictes* (Tenrecidae) and in *Neomys anomalus* and *Anourosorex* (Soricidae); long brains, with PIs higher than 1.94, were found in *Tenrec* and *Echinops* (Tenrecidae) in *Echinosorex*, *Atelerix algirus* and *Hemiechinus* (Erinaceidae), and in *Suncus murinus,* which, for a Soricidae, has a very long brain. In general, the Crocidurinae have longer brains than the Soricinae, with just a slight overlap.

Length of Cerebral Hemispheres (HemL): The shortest hemispheres were found in *Setifer* (0.83) and generally in Tenrecinae (0.87), the longest in *Galemys* (1.30) and generally in Desmaninae (1.27). The highest PI was 1.57 times larger than the lowest. In Soricidae, the Soricinae have longer hemispheres than the Crocidurinae, with a slight overlap. This is surprising, since the relation was reversed in brain length. Thus, when HemL was compared directly to BL, the differences became more pronounced: 0.49 - 0.61 for Crocidurinae and 0.61 - 0.71 for Soricinae.

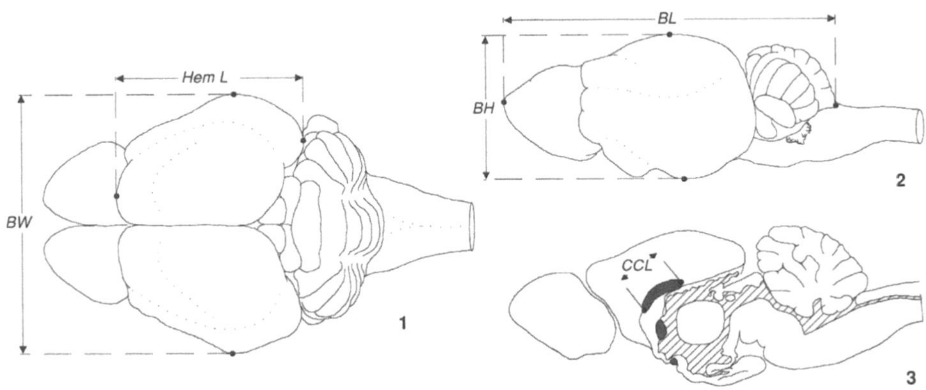

Figures 1 - 3: Measuring points on the Insectivora brain

BH	brain height	BL	brain length
BW	brain width	CCL	corpus callosum length
HemL	hemisphere length		

Brain Width (BW): In Insectivora, brain width always corresponded to the width of the cerebral hemispheres. The smallest widths were found in *Tenrec* (PI = 1.19) and *Potamogale* (1.21), the greatest in *Scalopus* (1.68) and *Chrysochloris stuhlmanni* and in *Anourosorex* (1.65). The highest PI was 1.41 times larger than the lowest.

Brain Height (BH): In general, brain height corresponded to the height of the cerebral hemispheres. In just a few cases, was the brain stem - cerebellar complex slightly higher, e.g., in *Scalopus* (Fig. 25). The lowest BHs were found in *Scalopus* (PI = 0.68), more generally in Talpinae (av. 0.72), and in both species of *Suncus* (av. 0.71), the highest in Chrysochloridae (av. PI =

1.02). The largest PI was 1.51 times greater than the smallest. In Soricidae, the Soricinae have higher brains on the average than the Crocidurinae, with a slight overlap.

Of the four measurements under investigation, the width of the brains (hemispheres) had the lowest interspecific variation in Insectivora, and the length of the hemispheres, the highest. HemL showed a positive correlation with encephalization, and the more highly encephalized brains in general had the longer hemispheres. The association between increase in brain size and increase in relative length and/or height of the hemispheres rather than increase in relative width would suggest some selective restriction on changes in width, perhaps because such changes would produce major rearrangements in jaw musculature and hence in feeding mechanisms. In Chiroptera, the restriction of brain width has been found to be still stronger than in Insectivora (Stephan and Nelson 1981).

Length of the Corpus Callosum (CCL): Corpus callosum length gives no information about brain proportion since it depends on the amount and distribution of crossing neocortical fibers. The shortest CCLs were found in *Geogale* (PI = 0.07, i.e., 7% of CR) and in Tenrecinae (average 16%), the longest in Desmaninae (66%), *Potamogale* (60%), and Talpinae (48%) (Tables 3, 4). Thus the percentage values are 4.1 times larger in Desmaninae than in Tenrecinae, 3.8 times in *Potamogale* and 3.0 times in Talpinae. In the same groups, the differences in the indices of NEO volumes were 6.3, 4.6, and 3.7, respectively (Tables 44, 45) and thus were distinctly higher. Thus, CCL gives some hints but no precise information on NEO size. Areas of the transversally sectioned corpus callosum (CCA) and the derived indices are more similar to the indices for the NEO volumes (cf. Tables 44 and 45 with 72 and 73).

The very large interspecific differences in CCL are immediately obvious when comparing the midsagittal reconstructions of *Geogale* and *Tenrec* (Figs. 41 and 40) with those of *Potamogale* and *Galemys* (Figs. 43 and 48).

Macromorphological Details

There were differences in the transition of medulla oblongata (OBL) to spinal cord. The transition could be slow and steady (Tenrecidae, Solenodontidae, Erinaceidae) or abrupt (Chrysochloridae, Talpidae, Soricidae). OBL was well developed in all Insectivora, especially in Potamogalinae, where it occupies nearly half of the brain as seen from below.

Characteristics which appear to be related to brain development include exposure of the mesencephalic tectum and development of sulci and fissures in cerebrum and cerebellum.

Exposure of the Mesencephalic Tectum: The mesencephalic tectum was completely or nearly completely exposed in *Geogale* and, among Tenrecinae, in *Echinops* and *Tenrec*. Large parts were exposed in the other Tenrecinae, in *Solenodon*, Erinaceidae, Chrysochloridae, *Suncus*, and *Sylvisorex*, and small parts in *Microgale* and most Crocidurinae. No exposure of the mesencephalic tectum was found among Oryzorictinae in *Oryzorictes* and *Limnogale*, in the Potamogalinae, Talpidae, and most Soricinae. A fully covered tectum was found when neocortex (NEO), rostrally, and/or cerebellum (CER), caudally, were large and thus completely obscured the tectum. Different participation in the cover by NEO and CER can be seen in the midsagittal reconstructions (Figs. 40 - 51). In *Galemys* (Fig. 48), the NEO covers nearly the whole tectum, whereas in *Blarina* (Fig. 50), large caudal parts are covered by CER. Large NEO and CER indicate advanced or progressive brain characteristics. In the most advanced brains, complete exposure was never found, and in the most conservative brains, cover was never complete. There was, however, no close correlation, since size of tectum and the brain-skull relationship also seemed to influence the cover of the mesencephalic tectum. The general impression was that there is less space for the brain in the skulls of Soricidae, Talpidae, and Chrysochloridae than in such larger forms as Tenrecinae, *Solenodon*, and Erinaceidae. Small species have a much higher percentage of brain size relative to body size (max. = 4.23%) than larger species (min. = 0.30%) (see Table 6 and Section 3.2). Such differences may parallel variations in the extent of the mesencephalic cover. More brain-skull relationships are dicussed in connection with cerebellar fissuration (see next page: lobulus petrosus).

Cerebral Sulci: There are no cerebral sulci in small Insectivora and thus the cerebral hemispheres are lissencephalic. The sulcus rhinalis, which, when present, macroscopically separates the piriform lobe from the neocortical pallium, was either absent or only discernible as a small impression. In larger brains, its rostral part was more prominent and, in Erinaceidae, it even extended into caudal parts of the hemisphere. When neopallial sulci were recognizable, they mainly appeared as shallow depressions in two places. (1) A rostral (orbital) sulcus may lie behind and parallel to the junction with the olfactory bulbs. Laterally, it may bend caudally then run somewhat distant from the piriform lobe. This sulcus was found in most larger brains *(Tenrec, Setifer, Solenodon,* Erinaceidae) but was particularly prominent in *Microgale dobsoni* and *M. talazaci* (Fig. 6). (2) A dorsal depression may be found in the caudal half of the hemispheres parallel to the interhemispheric fissure. It is still shallower and was identifiable only in the largest brains, i.e., in those of *Echinosorex* and *Solenodon*.

Cerebellar Fissuration: In the smaller Insectivora (*Geogale* and Soricidae), the surface of the cerebellum is smooth, with very few recognizable fissures. No fissures on the external surface were found in the smallest of the shrews, *Suncus etruscus* (Fig. 51), whose arbor vitae on its hidden (rostral and ventral) surface is subdivided by three fissures only. The longest, the fissura prima (2 in Figs. 40 - 51), separates the anterior (I-V) from the posterior lobe (VI-IX), and the shortest, the fissura posterolateralis (4), separates the posterior lobe from the nodulus (X). The fissura praeculminata (1) subdivides the anterior lobe, whereas no subdivision at all was found in the posterior lobe. Based on the description of Fons et al. (1984), the cerebellum of *Suncus etruscus* may be the least subdivided of all mammalian species. Seen from above, the vermis of this and other small Insectivora is quite broad. It is large when compared with the cerebellar hemispheres and separated from them by shallow depressions. A fissura secunda (3 in Figs. 40 - 51) is the next fissure to appear. It subdivides the posterior lobe and is the only additional fissure in *Geogale,* small Soricidae and Oryzorictinae and, surprisingly, in Chrysochloridae. Farther along is an additional fissure in the posterior lobe *(Echinops)* and the beginning of a subdivision of the anterior lobe *(Hemicentetes, Setifer, Microgale, Talpa).* In the largest Insectivora, the cerebellar hemispheres are somewhat larger and more distinctly separated from the vermis (Figs. 7, 9 - 11). The strongest fissural subdivision of the vermis was found in *Potamogale* (Fig. 43), *Solenodon* (Fig. 45), and *Echinosorex* (Fig. 46). In the arbor vitae, about 7 primary and many more secondary fissures are present. Bauchot and Stephan (1970) have described great intraspecific constancy in the larger species of tenrecoid Insectivora with regard to the primary subdivision of the cerebellum, but great individual variability with regard to the secondary subdivision of the various lobes and lobuli. This appears valid for all Insectivora and may extend to all mammalian forms.

Great complexity was also found in those parts of the cerebellum lying lateroventral to the hemispheres. In several species, e.g., in *Potamogale* (Fig. 19), there was an exposed fissure, running mainly longitudinally and separating large ventral parts from the cerebellar hemispheres. Based on the investigations of Brauer (1969), we believe that the separated lobe is the dorsal component of the paraflocculus (paraflocculus dorsalis), which in other species may be smaller or is included in the ventrally adjoining stalked part, or both. This stalked part is inserted in the fossa subarcuata of the petrosal and may be called "lobulus petrosus." Very often it is incorrectly called "flocculus." The stalked part may consist of the ventral component of the paraflocculus (paraflocculus ventralis) alone or may also include parts of the dorsal paraflocculus. This interpretation is supported by the fact that it was especially large when no dorsal part or only a small one was readily apparent. *Micropotamogale ruwenzorii,* whose dorsal paraflocculus is almost completely hidden, has a relatively large lobulus petrosus, which makes up more than one third of the cerebellar surface when viewed laterally. In contrast, *Potamogale,* whose highly visible dorsal paraflocculus is very large, has a relatively small lobulus petrosus, which makes up only one fifth of the lateral cerebellar surface (Fig. 19). Similar relationships were found in *Echinosorex.* In some species, there was no stalked lobulus petrosus (in *Setifer*), or it was very small (in *Echinops*). In these cases, all components of the paraflocculus were conspicuous in the side view. A similar variability in size, shape and position of the paraflocculus and its parts has been described by Brauer (1969) in rodents and by Stephan and Nelson (1981) in Chiroptera.

Obviously, interspecific differences in cerebellar development are strongly influenced by the size of the animal. Allometric volume comparisons have shown, however, that independent of body size differences, the complexity of the animal's locomotory interactions with its environment are highly interrelated with cerebellar development (see Sections 3.3.3 and 7.7).

Macromorphological Brain Characteristics in Subfamilies and Families

Tenrecinae brains showed substantial interspecific variation, which appeared to be larger than in any other subfamily (or family), except Oryzorictinae. Some brains were long, flat, and narrow, with short hemispheres, as in *Tenrec* (Figs. 4, 16, 28, 40), while others were relatively short, high and broad, as in *Hemicentetes.* The mesencephalic tectum may be fully exposed, as in *Tenrec* and *Echinops,* or covered in large parts, as in *Setifer* and *Hemicentetes.* The cerebellar lobulus petrosus may be sessile, as in *Setifer,* nearly so, as in *Echinops,* elongated, as in *Tenrec,* or mushroom-like, i.e., stalked with a big head, as in *Hemicentetes.* The characteristics described were combined in a variety of ways in the various species, making them all distinctly different from one another. The visual nerves were well developed in all species.

Geogalinae brains (Figs. 5, 17, 29, 41) are known only from reconstructions. Based on the figures given by Rehkämper et al. (1986), they are relatively short, broad, and flat, with an exposed tectum and an elongated lobulus petrosus.

Oryzorictinae were close to the insectivoran average in terms of linear brain measurements. As in Tenrecinae, there was large interspecific variability. The brain of *Microgale talazaci* was relatively long and narrow (Figs. 6, 18, 30, 42), that of *Oryzorictes,* short and broad. *Limnogale* had relatively small olfactory bulbs. The tectum was partly exposed in *Microgale* spp. and fully covered in *Oryzorictes* and *Limnogale.* The lobulus petrosus was uniformly mushroom-like, with a thin stalk and a big head. The visual nerves were thin but well preserved in *Microgale* and *Limnogale,* but were not recognizable in *Oryzorictes.*

Potamogalinae brains were more uniform than those of Tenrecinae and Oryzorictinae but showed variation in the lobulus petrosus, which in *Micropotamogale* was larger than in *Potamogale*, as described above. The brains were relatively long and slender; in *Potamogale* (Figs. 7, 19, 31, 43), the brain was the narrowest found in Insectivora, after that of *Tenrec*. The olfactory bulbs were relatively small and the tectum was fully covered. The visual nerves were small but easy to recognize.

Chrysochloridae brains had the most extreme shape found in all Insectivora groups. The brains were extremely short, high, and wide (Figs. 8, 20, 32, 44). Brain width was greater than brain length, which, except for *Anourosorex*, is unique in the Insectivora. The brain appeared to be strongly compressed in the anteroposterior direction. Despite this compression, the tectum was almost fully exposed and the impression was that this is mainly because of poor cerebellar development. Its vermis was found to be one of the least fissurated among Insectivora. In Chrysochloridae (and Talpidae), the lobulus petrosus was large and almost totally enclosed by the dorsally adjacent brain structures (piriform lobe and cerebellar hemispheres). Visual nerves were never recognizable in our preparations.

Solenodontidae brains (Figs. 9, 21, 33, 45) were the longest and among the narrowest found. The cerebral sulci were among the best developed of Insectivora. The tectum was only partially covered; the cerebellar hemispheres were well fissurated, relatively large, and well separated from the vermis. The lobulus petrosus was elongated but had no thin stalk. The visual nerves were small but distinct.

Erinaceidae were close to the insectivoran average in terms of linear brain measures and thus may be considered, together with those of the Oryzorictinae, to be the most "typical" in Insectivora. This was valid for both Echinosoricinae (Figs. 10, 22, 34, 46) and Erinaceinae (Figs. 11, 23, 35, 47). The tectum was always exposed and the lobulus petrosus generally elongated, but only in *Hylomys* did it have a thin stalk and a big head. The visual nerves were well developed in all species.

Talpidae had certain brain characteristics which clearly differentiated between Desmaninae and Talpinae. The shape seen from above was similar in the two subfamilies. The cerebral hemispheres were relatively small rostrally and attained relatively large widths far caudally. *Scalopus* reached the extreme width among Insectivora. Hemisphere length showed a different picture. Desmaninae (Figs. 12, 24, 36, 48) had the longest hemispheres of all Insectivora and the structures covered large parts of the cerebellum. In Talpinae (Figs. 13, 25, 37, 49), more of the cerebellum was visible. However, in all mole brains, the tectum was fully covered. In side view, the brains of Talpinae were the flattest of all Insectivora, and those of Desmaninae were clearly higher. The cerebellar lobulus petrosus was large in all moles and always mushroom-like.

In all cases, it appeared to be strongly pressed into the corner between the piriform lobe and cerebellar hemispheres. Small visual nerves could be preserved in Desmaninae but not in Talpinae.

Soricidae brains were generally similar to those of Talpinae. The two soricid subfamilies, however, shared different characteristics with talpine brains. Crocidurine brains (Figs. 15, 27, 39, 51) were next to Talpinae in flatness, but their lateral contour was quite different. In Crocidurinae, the greatest width of the cerebral hemispheres was reached far rostrally. Then the contour ran in a more or less straight line caudally, parallel to the interhemispheric fissure. Thus, unlike Talpidae, Crocidurinae had no distinct lateral pole in their hemispheres. In soricine brains (Figs. 14, 26, 38, 50), the lateral contour was similar to that of mole brains, running far laterally to at least half of the fronto-caudal extension of the hemispheres and then bending sharply medially. Thus, a distinct lateral pole was formed which was situated mostly in the beginning of the posterior half of the hemispheres. The cerebral hemispheres were broader (especially in *Anourosorex*), higher (especially in *Neomys anomalus*), and longer than in Crocidurinae (see Tables 3, 4). In general, the tectum was covered more in Soricinae than in Crocidurinae. Characteristics of the lobulus petrosus were similar to those of Talpidae. The visual nerves were small but could mostly be preserved in the preparations.

Selected references: Leche (1905, 1907), Clark (1932), Stephan and Spatz (1962), Bauchot and Stephan (1970), Stephan and Andy (1982).

Figures 4 - 51 (on the following pages): Brains of the species representing the 12 subfamilies and families included in this book, from dorsal (Figs. 4 - 15), left side (Figs. 16 - 27), ventral (Figs. 28 - 39), and midsagittal reconstructions (Figs. 40 - 51). Abbreviations for Figs. 40 - 51 are given on p. 43. Linear enlargement as indicated.

Figures 4, 16, 28, 40	*Tenrec ecaudatus*	Tenrecinae
Figures 5, 17, 29, 41	*Geogale aurita*	Geogalinae
Figures 6, 18, 30, 42	*Microgale talazaci*	Oryzorictinae
Figures 7, 19, 31, 43	*Potamogale velox*	Potamogalinae
Figures 8, 20, 32, 44	*Chrysochloris stuhlmanni*	Chrysochloridae
Figures 9, 21, 33, 45	*Solenodon paradoxus*	Solenodontidae
Figures 10, 22, 34, 46	*Echinosorex gymnurus*	Echinosoricinae
Figures 11, 23, 35, 47	*Erinaceus europaeus*	Erinaceinae
Figures 12, 24, 36, 48	*Galemys pyrenaicus*	Desmaninae
Figures 13, 25, 37, 49	*Scalopus aquaticus*	Talpinae
Figures 14, 26, 38, 50	*Blarina brevicauda*	Soricinae
Figures 15, 27, 39, 51	*Suncus etruscus*	Crocidurinae

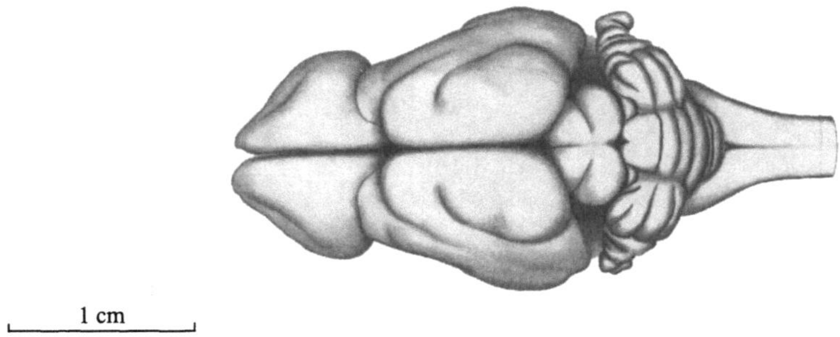

Figure 4 ***Tenrec ecaudatus***

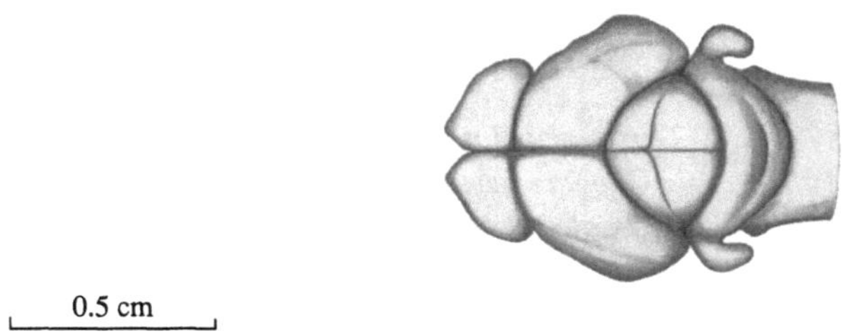

Figure 5 ***Geogale aurita***

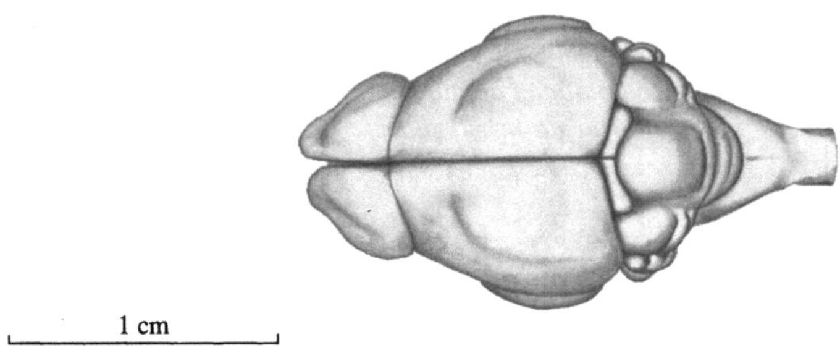

Figure 6 ***Microgale talazaci***

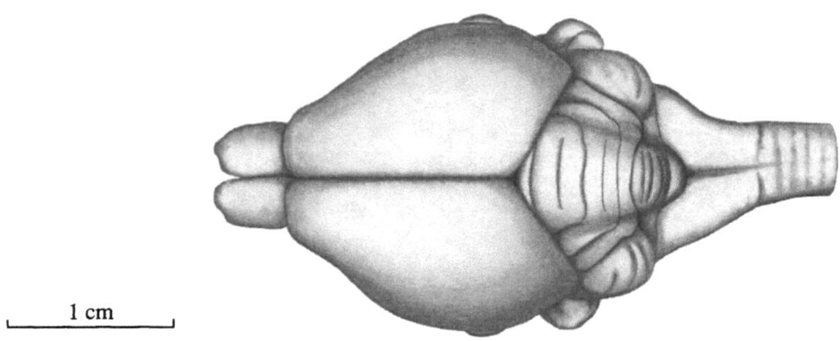

Figure 7 ***Potamogale velox***

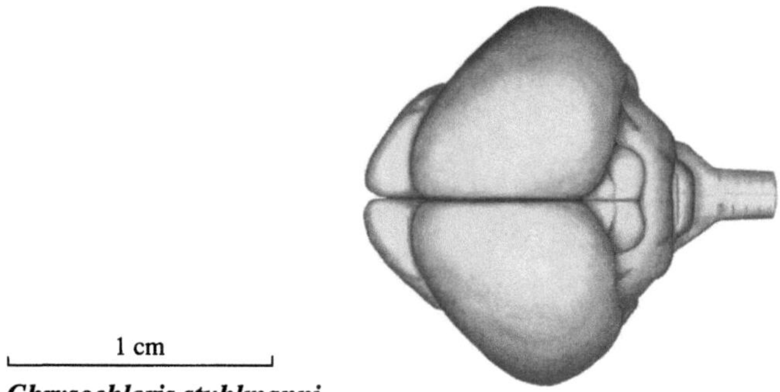

Chrysochloris stuhlmanni

Figure 8

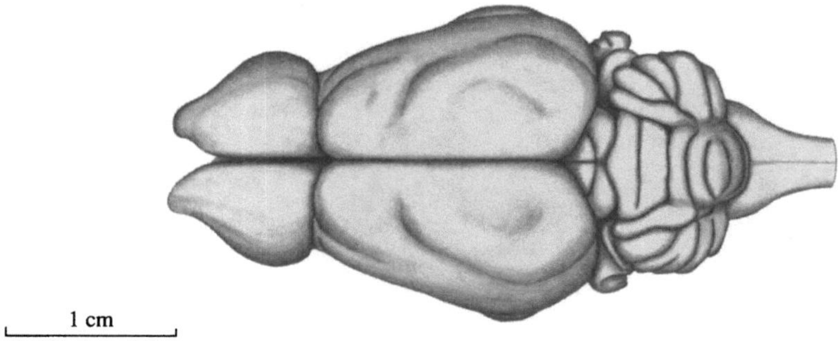

Solenodon paradoxus

Figure 9

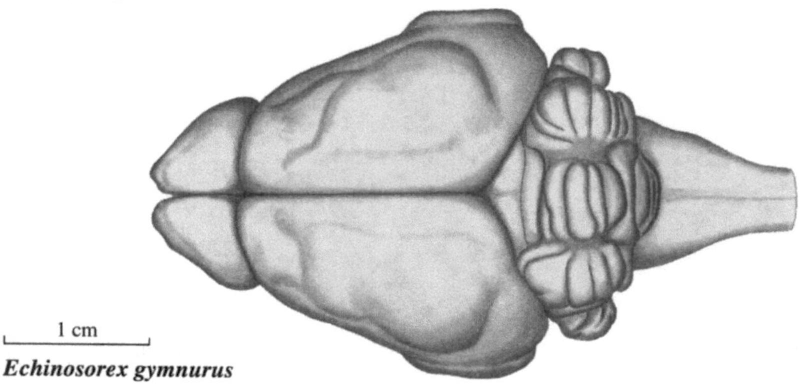

Echinosorex gymnurus

Figure 10

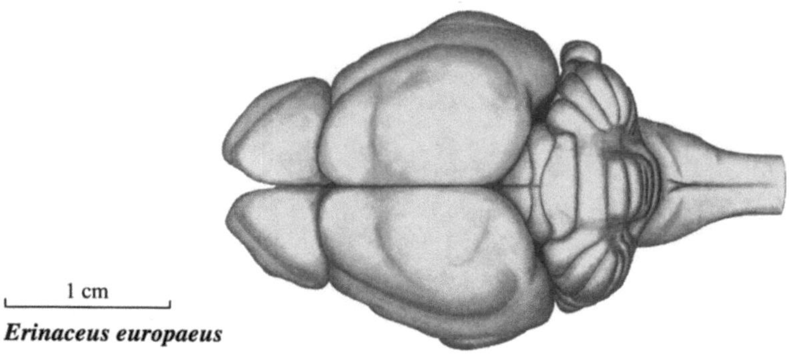

Erinaceus europaeus

Figure 11

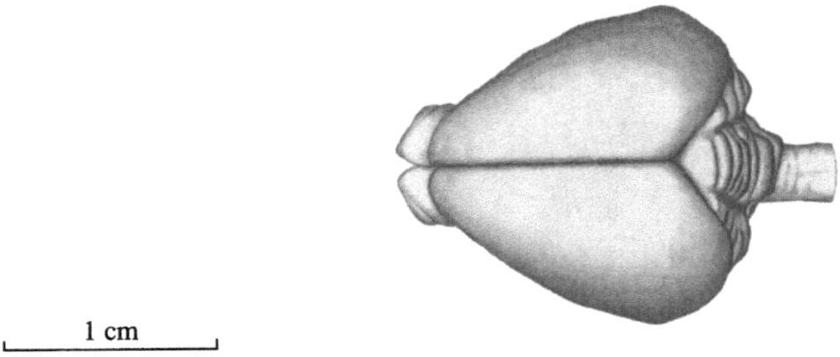

Figure 12 ***Galemys pyrenaicus***

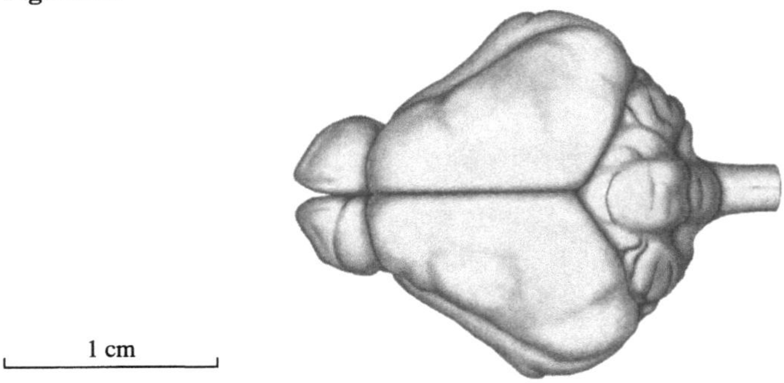

Figure 13 ***Scalopus aquaticus***

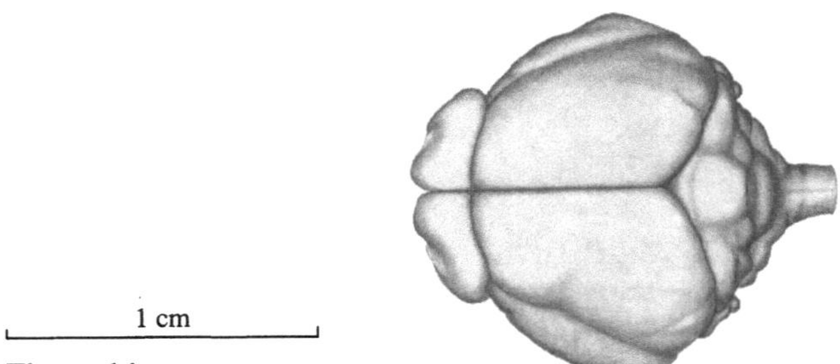

Figure 14 ***Blarina brevicauda***

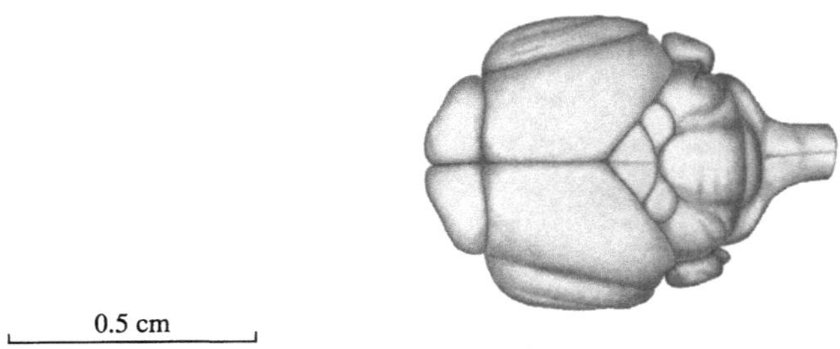

Figure 15 ***Suncus etruscus***

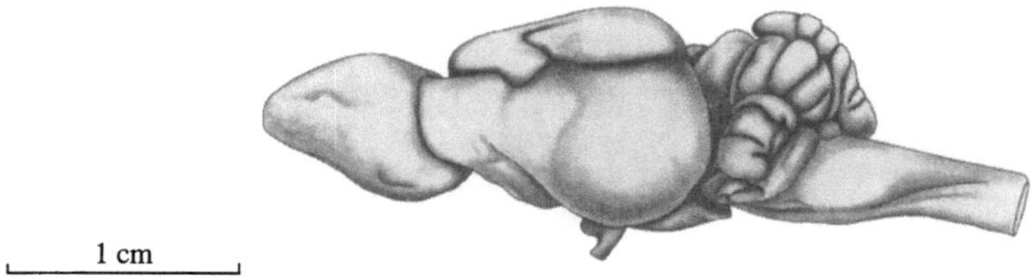

Tenrec ecaudatus **Figure 16**

Geogale aurita **Figure 17**

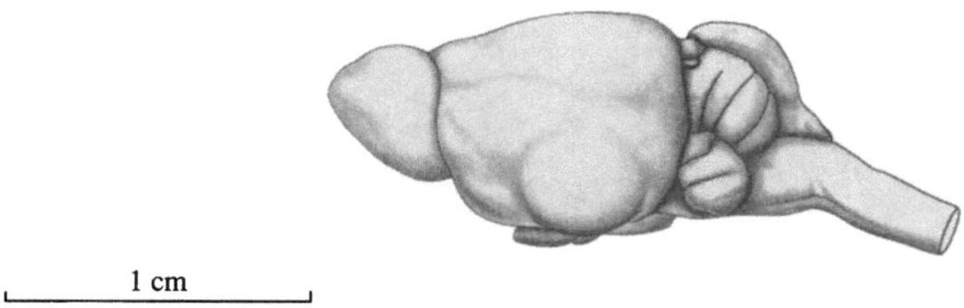

Microgale talazaci **Figure 18**

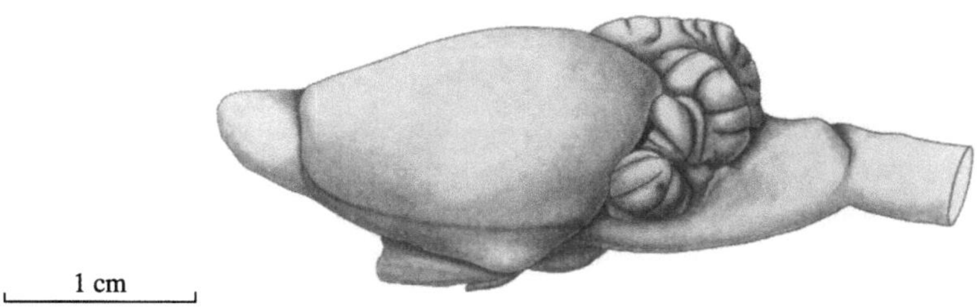

Potamogale velox **Figure 19**

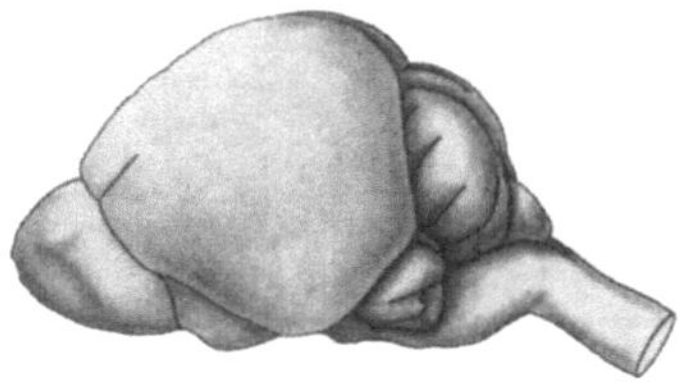

1 cm

Figure 20 *Chrysochloris stuhlmanni*

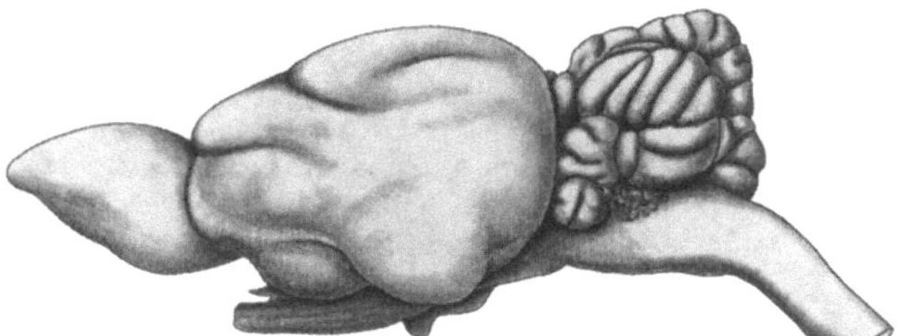

1 cm

Figure 21 *Solenodon paradoxus*

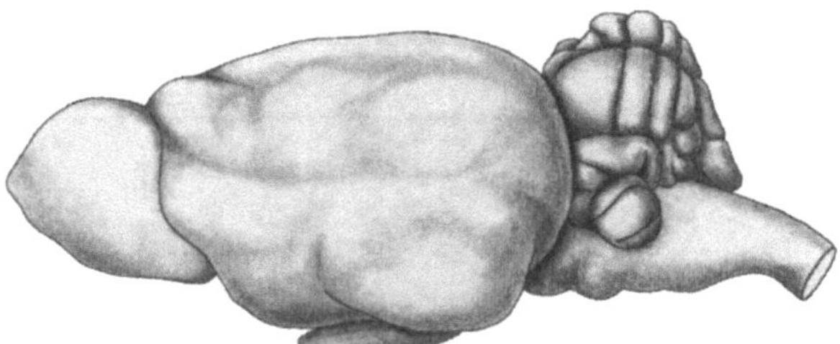

1 cm

Figure 22 *Echinosorex gymnurus*

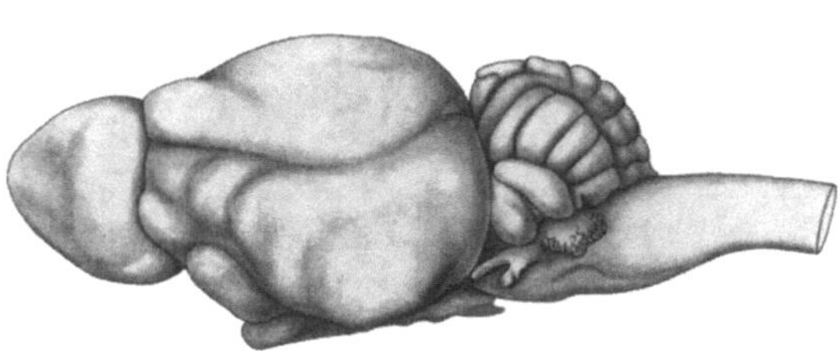

1 cm

Figure 23 *Erinaceus europaeus*

1 cm

Galemys pyrenaicus

Figure 24

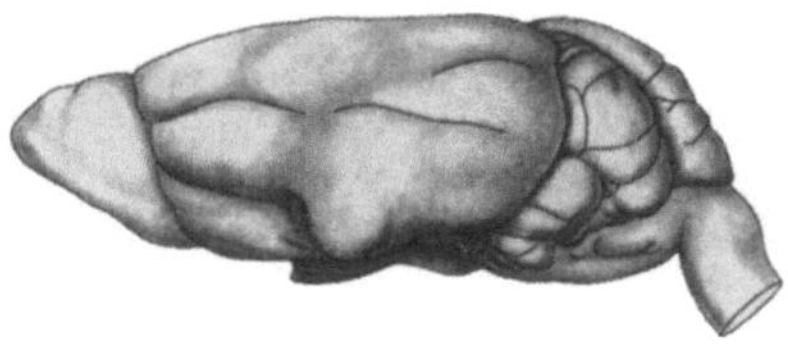

1 cm

Scalopus aquaticus

Figure 25

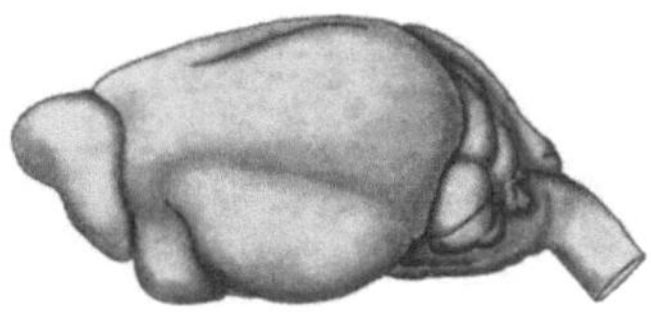

1 cm

Blarina brevicauda

Figure 26

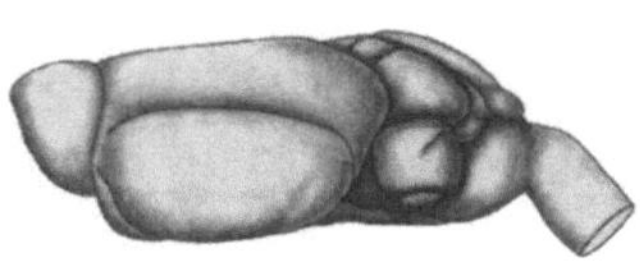

0.5 cm

Suncus etruscus

Figure 27

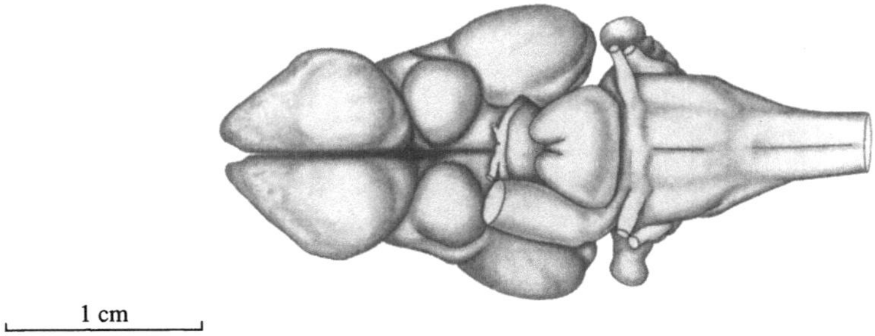

Figure 28 ***Tenrec ecaudatus***

Figure 29 ***Geogale aurita***

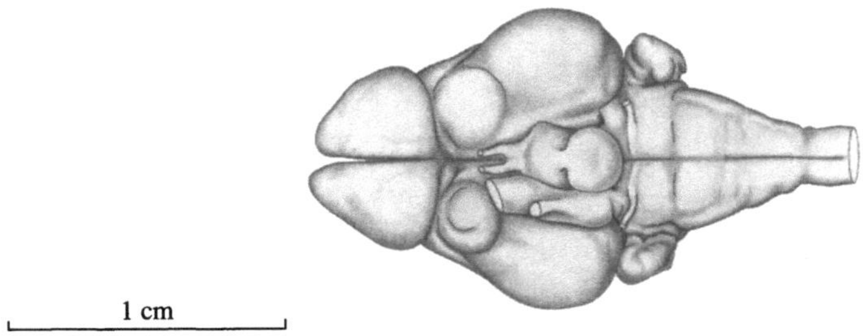

Figure 30 ***Microgale talazaci***

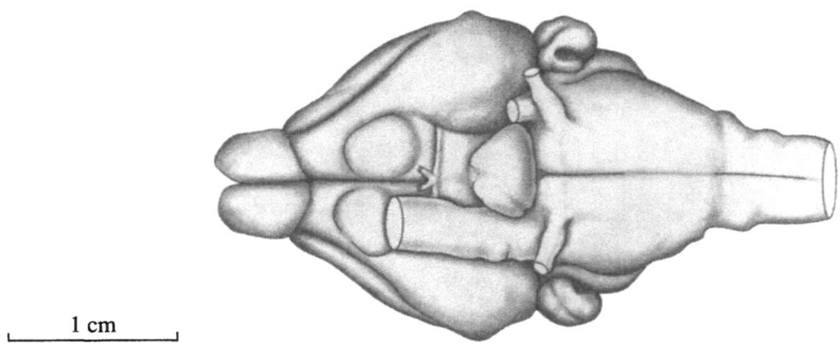

Figure 31 ***Potamogale velox***

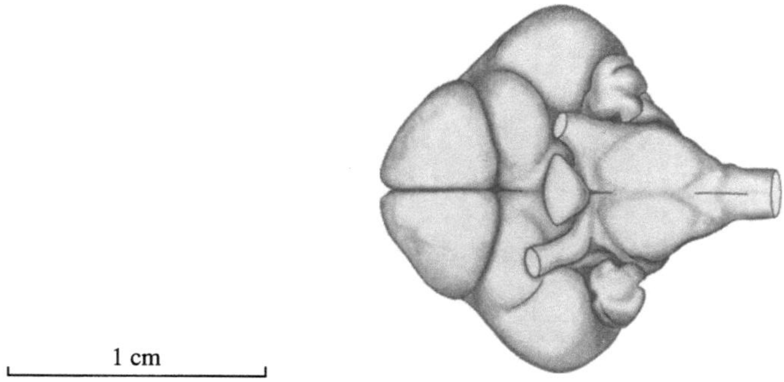

Chrysochloris stuhlmanni

Figure 32

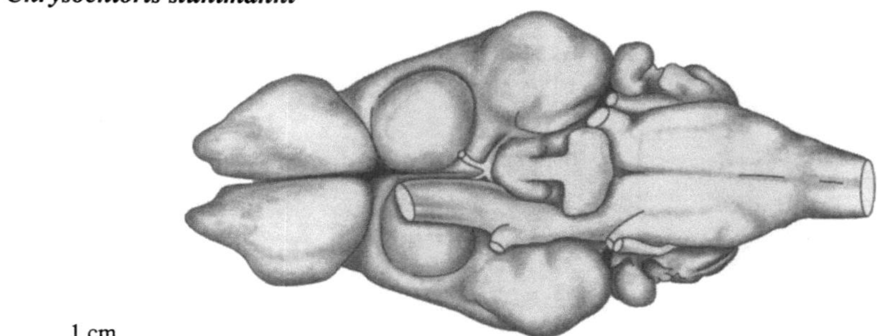

Solenodon paradoxus

Figure 33

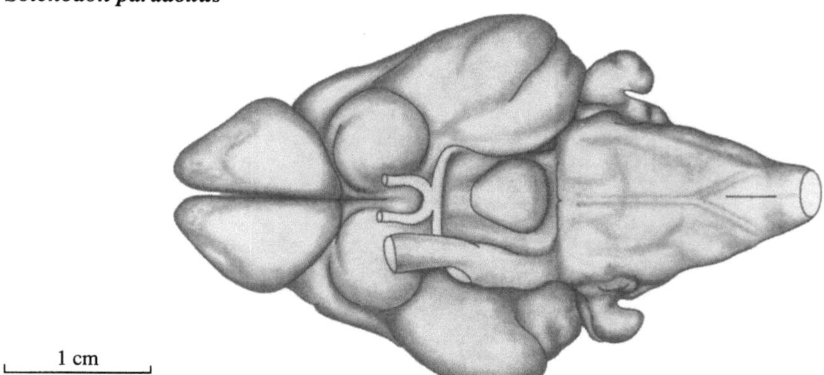

Echinosorex gymnurus

Figure 34

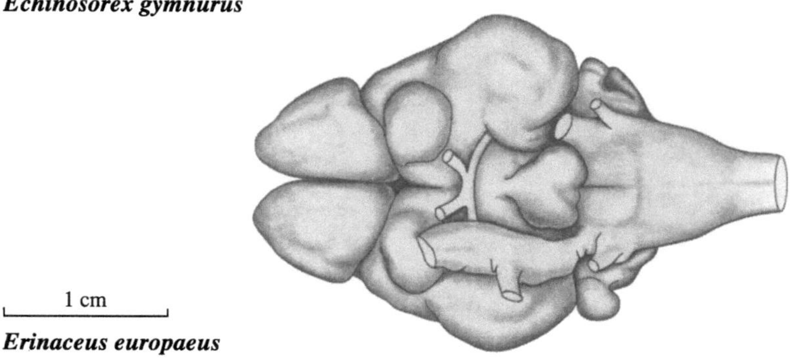

Erinaceus europaeus

Figure 35

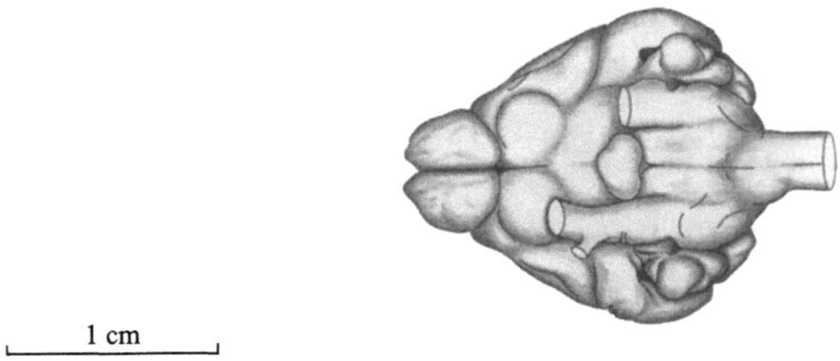

Figure 36 ***Galemys pyrenaicus***

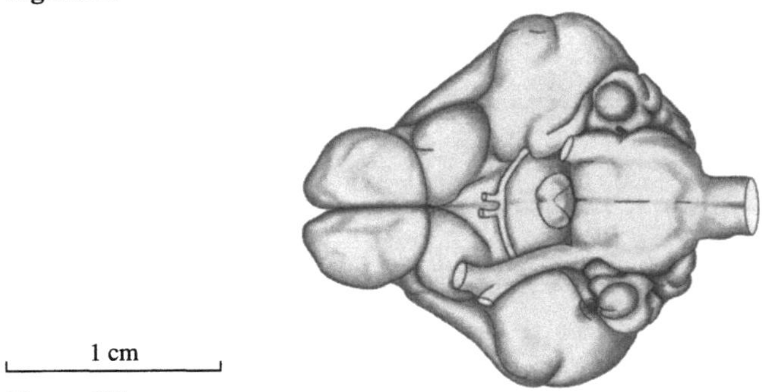

Figure 37 ***Scalopus aquaticus***

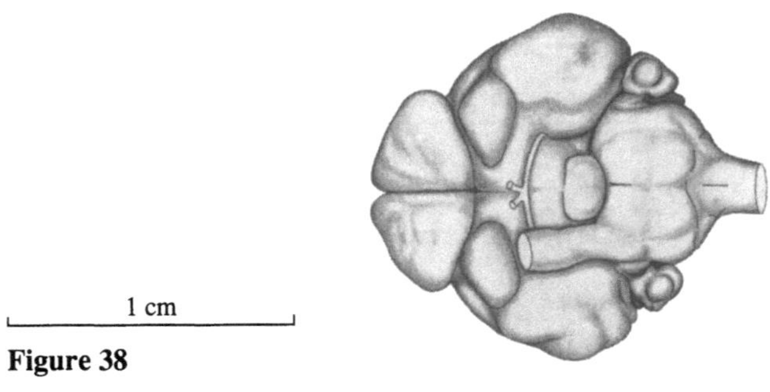

Figure 38 ***Blarina brevicauda***

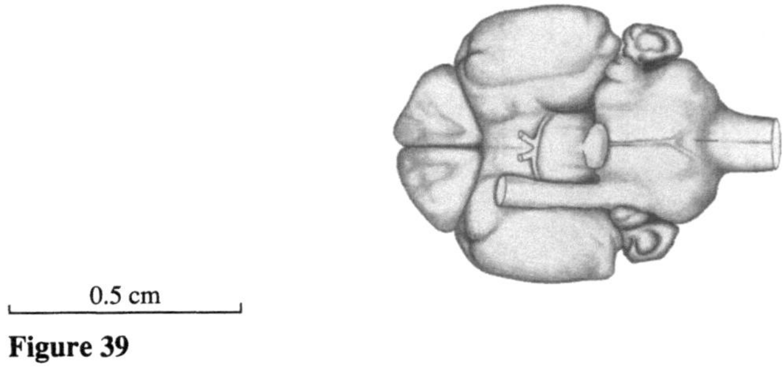

Figure 39 ***Suncus etruscus***

Figures 40 - 51 (on the following pages): Midsagittal reconstructions. The ventricular system is diagonally hatched; the surface of the cerebral hemispheres adjacent to the midsagittal plane is marked by dots; cerebral commissures and optic chiasma including the supraoptic commissures are dark.

Abbreviations:

AM	aqueductus mesencephali	LTC	lamina tectoria
CC	corpus callosum	MOB	main olfactory bulb
CHO	chiasma opticum	MTC	mesencephalic tectum
CM	corpus mamillare	NEO	neocortex
COA	commissura anterior	OBL	medulla oblongata
COC	commissura posterior	SEP	septum telencephali
COH	commissura habenulare	SFB	subfornical body
EPI	epiphysis	SOC	supraoptic commissures
HA	hippocampus anterior	THA	thalamus
HR	hipp. retrocommissuralis	VPO	ventral pons
HYP	hypophysis	V III	third ventricle
LT	lamina terminalis	V IV	fourth ventricle

Subdivision of the cerebellar vermis (arbor vitae):

1	fissura preculminata	IV	lobulus ventralis culminis
2	fissura prima	V	lobulus dorsalis culminis
3	fissura secunda	VI	declive
4	fissura postpyramidalis	VII	tuber vermis
I	lingula	VIII	pyramis
II	lobulus centralis ventralis	IX	uvula
III	lobulus centralis dorsalis	X	nodulus

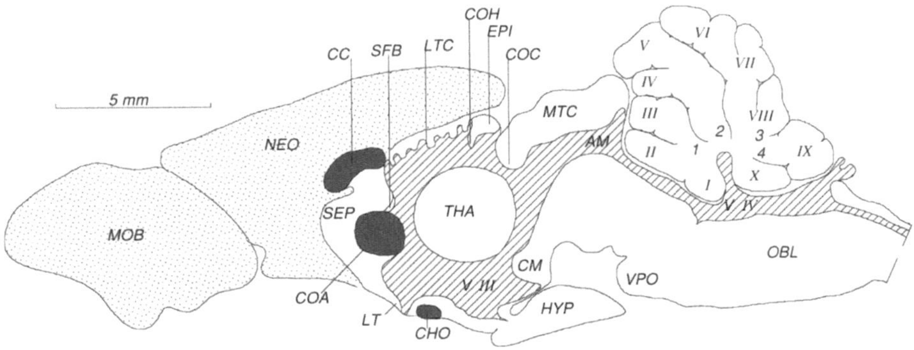

Figure 40 ***Tenrec ecaudatus***

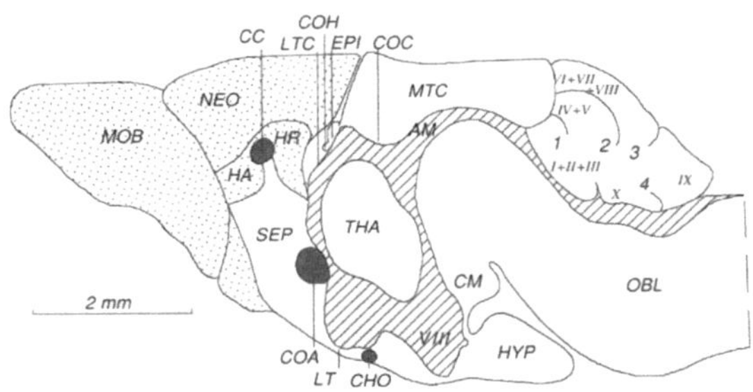

Figure 41 ***Geogale aurita***

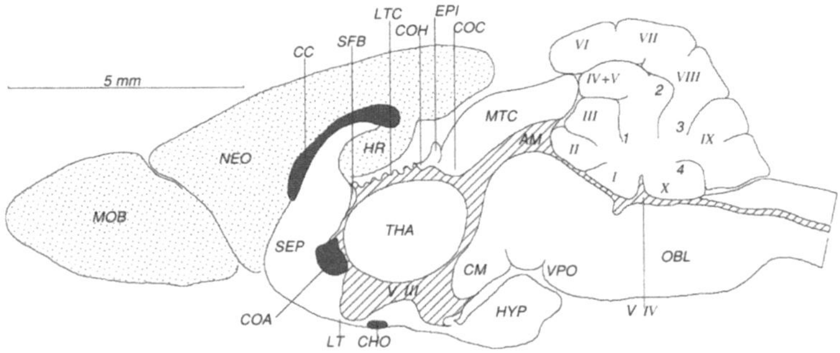

Figure 42 ***Microgale talazaci***

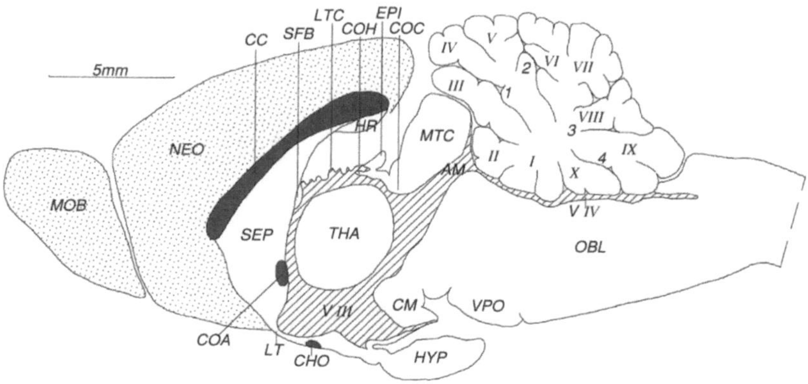

Potamogale velox **Figure 43**

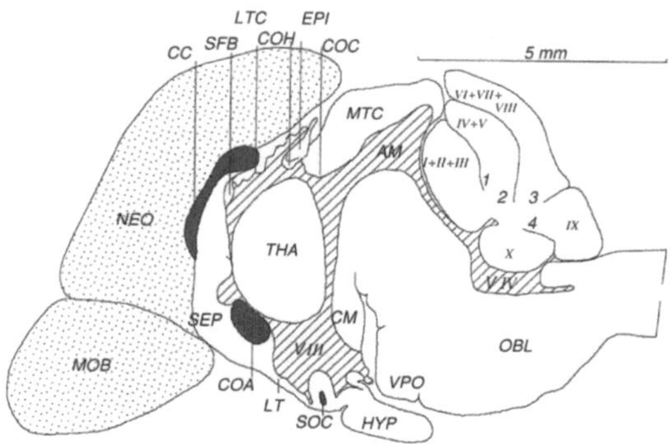

Chrysochloris stuhlmanni **Figure 44**

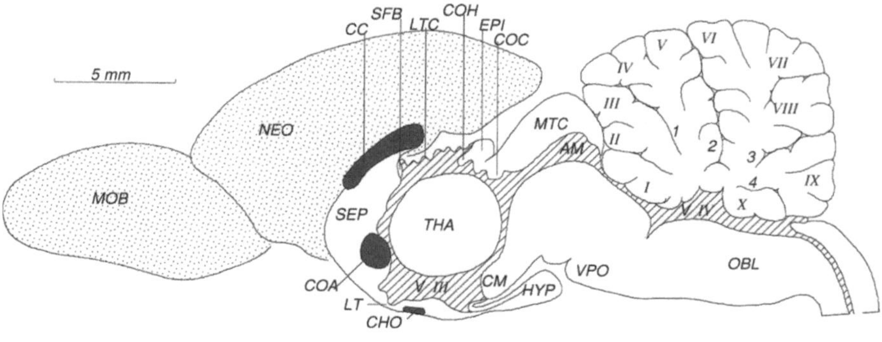

Solenodon paradoxus **Figure 45**

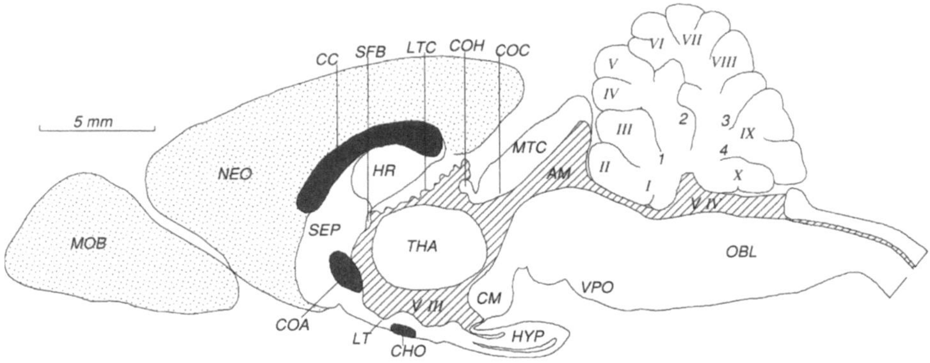

Figure 46 ***Echinosorex gymnurus***

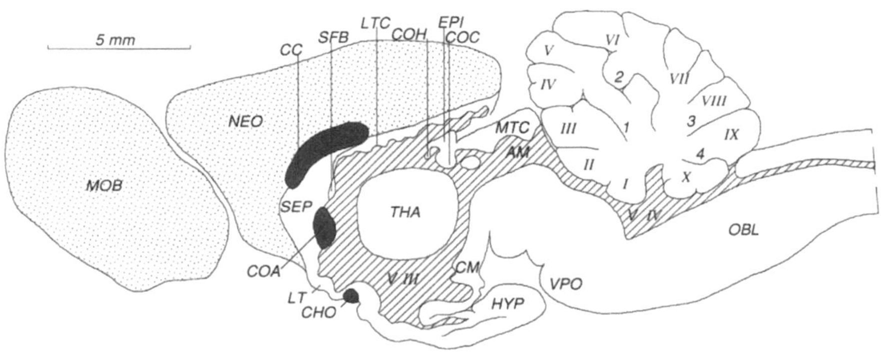

Figure 47 ***Erinaceus europaeus***

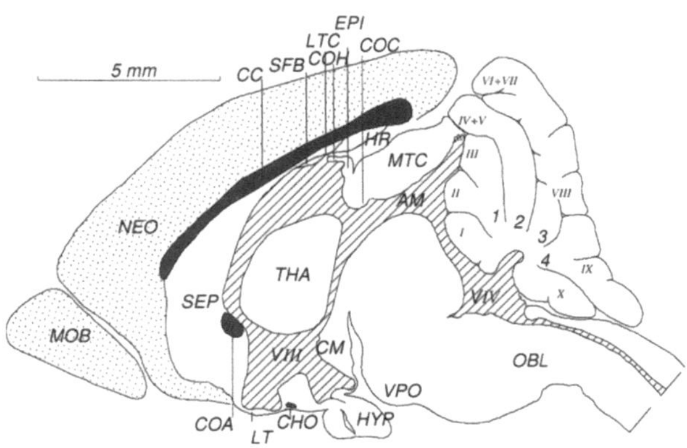

Figure 48 ***Galemys pyrenaicus***

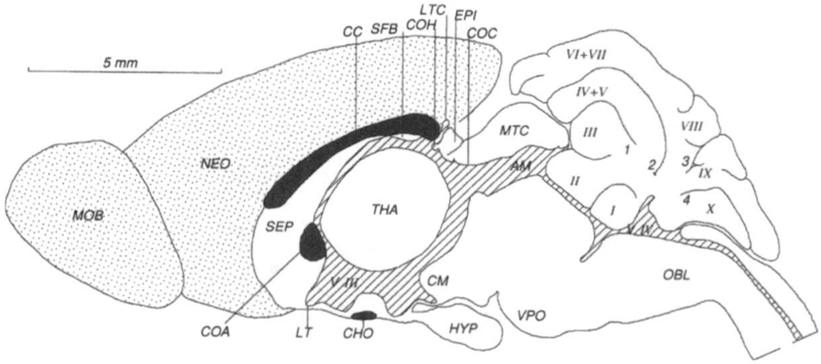

Scalopus aquaticus **Figure 49**

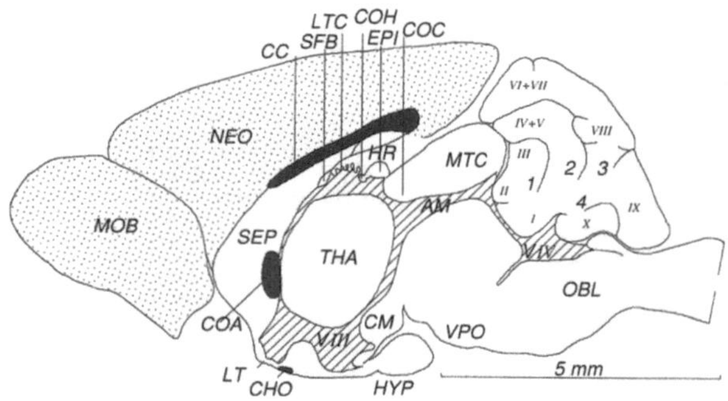

Blarina brevicauda **Figure 50**

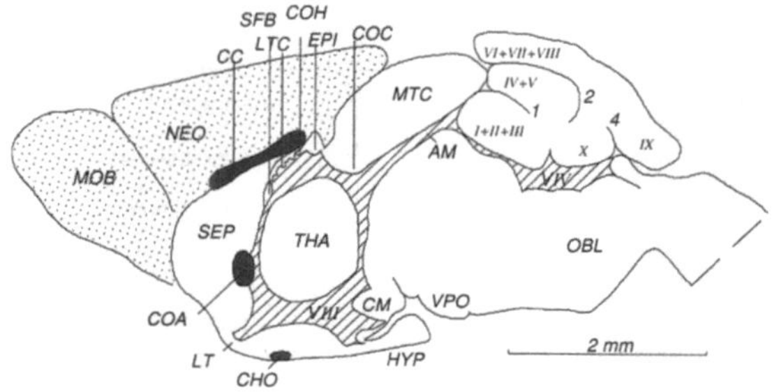

Suncus etruscus **Figure 51**

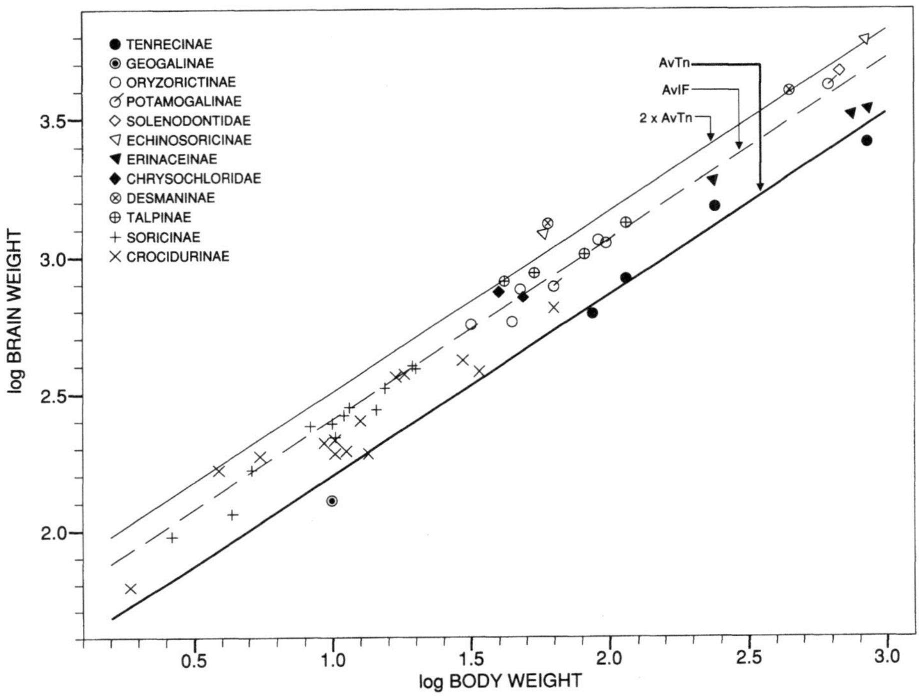

Figure 52

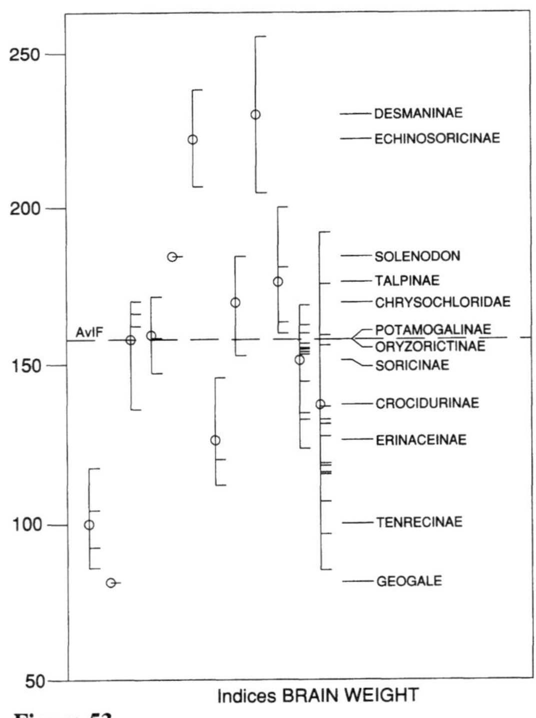

Figure 53

3.2 Total Brain Size and Comparisons

The measures of total brain size used were brain weight (Tables 5, 6) and net brain volume (Table 8). Brain weight (BrW) was preferred for broad comparisons, including brains not cut in serial sections, whereas net brain volume (NET) was used when volumes of brain components were measured and could be added together to produce total NET volume.

Brain Weight (BrW)

Brain weight in Insectivora varied from 59 to 6,460 mg. The lowest value was found in an adult female of *Suncus etruscus* having a body weight of 1.72 g, while the highest value came from an adult male of *Echinosorex gymnurus* with a body weight of 895 g.

Percentages: Relative to body weight, the percentage was 3.33 for *Suncus etruscus* and 0.74 for *Echinosorex*. However, these were not the extreme values. The maximum was found in *Sylvisorex granti*, whose BrW was on the average 4.23%, and the minimum, in *Tenrec ecaudatus*, with 0.30% (Table 6). The differences confirm the rule that larger species in general have lower percentage brain size and smaller species, higher percentages. Such differences

---

Figure 52: Brain weight (in milligrams) plotted against body weight (in grams) in a double logarithmic scale. The slope of the reference line was determined by a canonical analysis on subfamilies and families. The thick reference base line (AvTn) is placed through the average of the four Tenrecinae (see text). Its equation is

$$\log BrW = 1.544 + 0.66 \times \log BoW$$

The AvTn line represents the average Tenrecinae, and all points on it have an index of 1 or 100(%). The thinner parallel line (2 x AvTn) corresponds to an index of 200 in the scale given in Fig. 53. The broken parallel line (AvIF) is at a distance of 1.58 x AvTn and corresponds to the average found in the twelve families and subfamilies.

Figure 53: Brain weight (encephalization) indices. The indices are numerical values of the distances from the AvTn given in Fig. 52. Each of the small horizontal bars represents the encephalization index for a species. On the left, the range (vertical bars) and the mean (o) of the various taxonomic units are given, while on the right, the sequence of the means is scaled and the names of the units are given. The broken AvIF line corresponds to the average found in the twelve families and subfamilies and has an index of 158 (see Table 7).

exist even in closely related forms, as e.g., in *Suncus etruscus* (3.33%) compared with *S. murinus* (1.13%), whose BoW was nearly 20 times larger (Table 6). For comparison, man has 2.05%, which is just in the middle of the Insectivora range. Percentages are therefore less useful for evaluating interspecific or intergroup differences (see also Material and Methods). On the average, the 53 species of Insectivora (AvIS) had 1.71% brain relative to body weight. The average of the twelve subfamilies and families (AvIF) was lower (1.36%). The difference is due to the great number of shrews in the species average (27 out of 53). All the small-bodied shrews had high percentages of BrW to BoW (Tables 6 and 7).

Encephalization Indices: To minimize the differences in relative brain size within narrowly related groups, and to arrive at a well founded base to evaluate the differences between groups, detailed analyses were carried out using the canonical method as described by Rempe (1970). These analyses resulted in a BoW power of 0.66. A reference line with a slope of 0.66 placed through the average of the Tenrecinae (AvTn line in Fig. 52) has a y-intercept of 1.544. With reference to this AvTn line, i.e., to the average of the Tenrecinae (= 1 or 100), the encephalization indices (EI) varied from 81 for *Geogale* to 255 for *Galemys*. High values similar to those of *Galemys* (i.e., > 200) were found in *Desmana* and in the two Echinosoricinae; low values, i.e., in the range of Tenrecinae, were found in some Crocidurinae and in *Erinaceus europaeus* (Fig. 53). Within a family, the EIs of subfamilies may differ markedly. In Tenrecidae, the Geogalinae were the lowest, followed in ascending order by Tenrecinae, Oryzorictinae, and Potamogalinae. In Soricidae, the Soricinae had on the average somewhat higher EIs than the Crocidurinae; in Talpidae, the Desmaninae were clearly higher than the Talpinae, and, in Erinaceidae, the Echinosoricinae were much higher than the Erinaceinae (Table 7). Similar results were obtained when the net volumes of the total brains were compared.

Net Brain Volume (NET)

The net brain volume (NET) is composed of pure brain tissues, i.e., ventricles, meningeal membranes and nerves were excluded (Table 8). NET is the sum of all the measured structures and averages 96.9% of the total or gross volume for all the species (AvIS). Therefore, it is a good reflection of the total brain. On the average, only 1.2% is made up of ventricles and 1.9% of miscellaneous structures, such as meninges, nerves, and hypophysis (Table 9). In serially sectioned brains, these structures may be present to various extents or may be altogether absent.

NET indices (Tables 6 and 7) were nearly the same as the encephalization indices (EIs). They were by far the largest in Desmaninae and Echinosoricinae. In Desmaninae, the high indices are surprising, since the observed decrease in the olfactory structures should have reduced brain size. Thus, non-olfactory brain structures must be enlarged in Desmaninae. A second grouping in NET indices was made up of *Solenodon*, Talpinae, Chrysochloridae, Potamogalinae, Oryzorictinae, and Soricinae. These forms were quite close to the subfamily average, which is 160. All species of Talpinae were above this average; Chrysochloridae, Potamogalinae and Oryzorictinae averaged on or somewhat above the AvIF but contain some species below the average. In Potamogalinae, the NET indices were reduced by a strong decrease in the olfactory structures. Clearly below the subfamily average were Crocidurinae and Erinaceinae. Some species of the Crocidurinae were as low as the Tenrecinae. A still lower position than in Tenrecinae was found in *Geogale*.

The differences in the relative brain size of the various taxonomic units, as shown in Fig. 53, were caused by variations in specific brain parts as will be discussed in more detail in the following sections.

3.3 Comparison of Brain Components

Introductory Remarks

Percentage values will be given mainly to demonstrate the **intra**specific composition of a brain or brain complex. For **inter**specific comparisons, percentage values are not very informative since they do not depend on size changes in the structure under consideration alone, but to a large degree also on size changes in other brain structures included in the reference system. This will be shown in several extreme cases. Therefore, evaluation of relative size in interspecific comparisons in this book will be based on allometric size comparisons only. Subsequently, the word "**size**" will always be used in the sense of **relative allometric size** (here expressed by size indices relative to the average Tenrecinae = 100) if other attributes such as percentage or volume size are not specified.

In previous investigations (e.g., Stephan 1961), cortical size was expressed as **surface** dimension, since this measure may well characterize cortical extension. For a general comparison, however, which includes both cortical and non-cortical structures, **volumes** are the most universal measure, since surface measures of non-cortical structures tend to be meaningless. Additional surface measurements may be profitable in some cortices, such as hippocampal components, whose delimitation is difficult in the very broad molecular layer (Stephan and Manolescu 1980; Stephan et al. 1987c) and in the thin olfactory cortices, where inner borders may be blurred. Delimitation may become even more difficult when the cortices are reduced, as in the case of the olfactory cortices of water-adapted species.

Size Comparison
(Tables 10-12)

The NET brain (= 100%) of a typical Insectivora (average of family and subfamily averages; AvIF) is composed of

			range
medulla oblongata	(OBL)	13.8%	(8.9 - 27.8)
mesencephalon	(MES)	6.4%	(4.3 - 10.4)
cerebellum	(CER)	13.3%	(9.5 - 18.2)
diencephalon	(DIE)	7.4%	(6.2 - 9.5)
telencephalon	(TEL)	59.1%	(45.1 - 68.1)

The largest interspecific variation was found in OBL, whose percentage in *Limnogale* is 2.2 times (19.5 versus 8.9) and in *Geogale* even 3.1 times (27.8 versus 8.9) larger than in *Parascalops* and *Chrysochloris asiatica* (8.9%). Since, however, the volume measurements for *Geogale* are based on poor material, they may be unreliable and the derived volumes, percentages, and indices must be interpreted with caution.

By way of comparison, **in man** the percentage is much higher in TEL (85.0%) and consequently lower in all other brain parts (0.8% in OBL, 0.6% in MES, 11.0% in CER, and 2.7% in DIE). However, none of these fundamental brain parts is smaller in primates than in isoponderal Insectivora (see Section 3.3.6).

Most of the components discussed here are composed of a variety of nuclear complexes, nuclei, and fiber bundles or, in cortices, of regions, areas, and laminae. Not all of the smaller structures could be shown in the stereotaxic atlas of the *Atelerix* brain (see pp. 503 - 553) since it would have become too detailed and thus too cumbersome to be useful. Details on the position of structures not shown in the atlas are given in the text.

More than 90 complexes and structures were measured. The tables are summarized in Table 1. The allometric size indices showed clear differences in (1) their range of variation, (2) the average values reached in the 50 (or fewer)

species and twelve (or fewer) subfamilies or families, and (3) their recognizable trends and characteristics in species and/or groups of species.

3.3.1 Medulla Oblongata (OBL)

Delimitation
(Atlas plates AP 0 - P 10)

The most rostral parts of the spinal cord (SPC) were included in the medulla oblongata (OBL). The separation was made artificially such that the length of SPC remaining with the brain equaled its width, i.e., formed a square (viewed from below). This artificial border of OBL was easy to define when the transition of the medulla oblongata to the spinal cord was abrupt, as in Chrysochloridae, Talpidae, or Soricidae. It was more difficult when the transition was slow and gradual, as in Tenrecidae, Solenodontidae, and Erinaceidae. In such cases, the rostral edge of the square (spinal cord) was placed at a level followed caudally by no more conspicuous tapering.

The rostral border of the OBL complex, as understood here, was set for practical purposes and included parts of the metencephalon and mesencephalic tegmentum. It is almost impossible to separate the substantia reticularis of the met- and mesencephalon from that of OBL in an identical manner in all species. Therefore, we have included the dorsal pons and the formatio reticularis as a whole in OBL, which thus extended with its rostral projection nearly to the mesencephalic nucleus ruber (RUB). In contrast to the dorsal pons, which contains nuclear and fiber structures associated with cranial nerves V - VIII, the ventral pons (VPO) contains massive longitudinal and transverse fiber systems closely connecting the pontine grey to the cerebellum (CER). Consequently, VPO was included in CER for volume measurements.

Size Comparison
(Tables 10-12; Figs. 54, 55)

Of the five fundamental brain parts, OBL was on the average the second largest, representing 13.8% of the total NET brain (see table on p. 52). The size indices for OBL may be described as follows:

Averages: 132 in the 50 species, 139 in the twelve subfamilies and families. This was the lowest level reached of the five fundamental brain parts.

Range: 86 - 240. The highest value *(Potamogale)* was 2.8 times larger than the lowest *(Tenrec)*; this is relatively narrow variation.

Small size: Tenrecinae (100), Erinaceinae (102), Chrysochloridae (105).

Large size: *Limnogale* (216), Potamogalinae (207), *Galemys* (203).

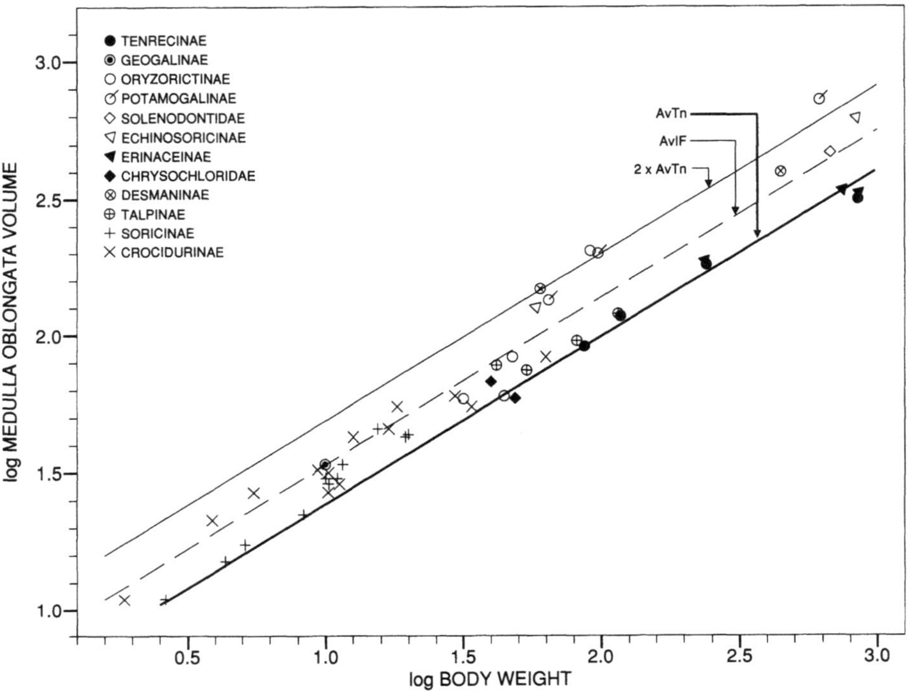

Figure 54

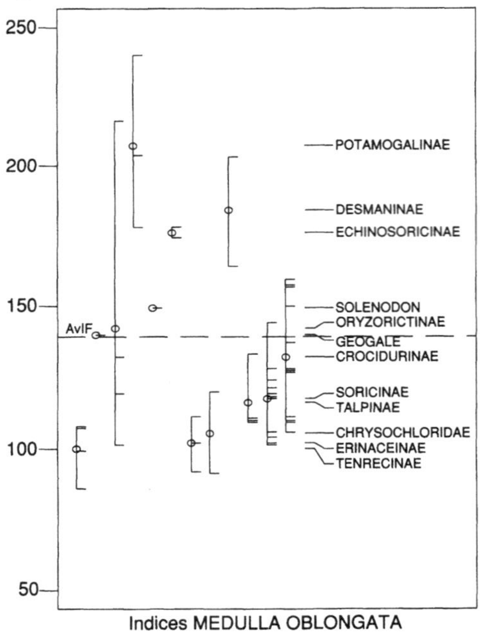

Figure 55

Figure 54: Medulla oblongata volume (in mm^3) plotted against body weight (in grams) in a double logarithmic scale.
The equation of the AvTn line is log OBL volume = 0.776 + 0.61 x log BoW. The broken line (AvIF) is in a distance of 1.39 x AvTn.
For additional explanations see Fig. 52.

Figure 55: Size indices of medulla oblongata (OBL). The indices are numerical values of the distances from the reference line given in Fig. 54. The broken line (AvIF) has an index of 139 (see Table 12). For additional explanations see Fig. 53.

Characteristics and trends: There was a clear tendency for the water-adapted species to have strongly enlarged OBL with indices reaching from 164 in *Desmana* to 240 in *Potamogale*. In terrestrial forms, only the Echinosoricinae reached into this range. The indices of the water shrews *Neomys* spp. were considerably lower but, compared with terrestrial Soricinae, they also had a clearly larger OBL (av. 136 versus 112).

Medulla Oblongata Components

Only slightly more than one third of the medulla oblongata (36%) could be delimited with sufficient precision to be measured in all species. The remaining two thirds consisted of reticular components, fibers, and nuclei which are either poorly circumscribed or too small to allow for exact measurement.

In all, 20 OBL nuclei or complexes were measured and, except for vestibular nuclei and trigeminal complex (Baron et al. 1988, 1990), are published here for the first time. The largest was the sensory trigeminal complex (TR), occupying in Insectivora more than one sixth of the total OBL on the average. The smallest measured components were the locus coeruleus (0.1%) and some of the motor nuclei (0.3 - 0.5%). The average percentages for the main components or groups of components relative to OBL were as follows: sensory trigeminal complex (TR) = 18.4%, nucleus of solitary tract (TSO) = 2.0%, funicular nuclei (FUN: FGR + FCM + FCE) = 1.9%, relay nuclei (RLN: REL + INO + PRP + COE) = 2.6%, motor nuclei (MOT: V + VII + X + XII) = 2.3%, vestibular nuclei complex (VC: VM + VI + VL + VS) = 4.5%, and auditory nuclei (AUD: OLS + DCO + VCO) = 4.4%.

Complexus sensorius trigeminalis (TR)

Topography, Delimitation, Structural Characteristics, Fiber Connections
(Atlas P 2 - P 10; Fig. 56)

The sensory nuclei of the trigeminal nerve (TR) extend from the mesencephalon down to the rostral end of the dorsal horn of the first cervical segment of the spinal cord. The complex is usually divided into three main parts: the nucleus mesencephalicus nervi trigemini (mesencephalic sensory nucleus) (TRM), the nucleus sensorius principalis nervi trigemini (principal sensory nucleus) (TRP), and the nucleus spinalis nervi trigemini (nucleus of the spinal tract) (TRS). TRM consists of a narrow band of round cells scattered at the lateral margin of the griseum centrale, making it impossible to measure the volume of this nucleus. Cell counts would provide a better index of the size, but, unfortunately none have so far been done in Insectivora. TRP and TRS were combined for measurement since it is difficult to separate them precisely in all species.

TRP begins rostrally at the caudal pole of the medial parabrachial nucleus and merges caudally with the rostral part of the subnucleus oralis of TRS. It is composed of two divisions which are different in cell type and cell arrangement. The ventral division contains loosely distributed medium-sized fusiform and multipolar neurons, the dorsal division contains more densely packed and slightly smaller cells. The supratrigeminal nucleus mentioned by Lorente de Nó (1922, ex Darian-Smith 1973) is here considered a part of TRP, and hence was included in the TR volume. It is an aggregate of loosely packed cells located at the dorsomedial angle of TRP.

TRS is continuous with TRP rostrally and with the dorsal horn of the spinal cord caudally. The terminations from the trigeminal nerve are somatotopically organized (Tracey 1985). TRS may be subdivided into three continuous regions: subnuclei oralis, interpolaris, and caudalis (Olszewski 1950). The subnucleus oralis is not structurally uniform and its medial border is poorly demarcated from the adjacent reticular nuclei. The rostral part contains some large multipolar cells, but most neurons are similar to those of the nucleus principalis. The subnucleus interpolaris, extending from the subnucleus oralis to the level of the obex, consists of at least two distinct cell groups. The marginal region contains medium to large multipolar cells. The rest of the subnucleus consists of smaller neurons with spherical perikarya such as described in the mouse by Aström (1953). The subnucleus caudalis, extending from the level of the obex to the first cervical root, is differentiated into three layers corresponding to the first four cytoarchitectonic laminae of the spinal cord. The marginal zone contains a few moderately large neurons and many smaller cells. The next zone is the gelatinous layer, corresponding to laminae II and III of the spinal cord. The predominant neurons are small spindle-shaped cells with little cytoplasm. The deepest part, corresponding to lamina IV, is characterized by medium to large neurons.

The primary afferents, in which representations of the vibrissae occupy a large part, terminate somatotopically in TRP and in the interpolar and caudal subnuclei of TRS (Belford and Killackey 1979; Arvidsson 1982; Bates and Killackey 1983). In contrast, the oral subnucleus does not appear to receive any significant projection from the vibrissae; its input seems rather to be from the nasal and oral cavities, at least in the cat (Wall and Taub 1962). The marginal zone of the caudal subnucleus receives noxious and thermal input (Tracey 1985). The supratrigeminal nucleus receives proprioceptive afferents (Jerge 1963 a, b).

Efferent projections from TRP and all three subnuclei of TRS are to the nucleus ventralis posteromedialis of the thalamus (VPM) (Torvik 1957; Carpenter and Hanna 1961; Fukushima and Kerr 1979; Erzurumlu and Killackey 1980). Some nuclei project to other regions of the thalamus (Stewart and King 1963; Lund and Webster 1967; Smith 1973; Shigenaga et al. 1979), to the superior colliculus (Huerta et al. 1983), to brain stem motor nuclei (Erzurumlu and Killackey 1979; Travers and Norgren 1983), the zona incerta (Smith 1973), and the inferior olive (Huerta et al. 1983). In the rat, about 70% of the neurons of the interpolar subnucleus of TRS project directly to the vermal region of the cerebellum (Watson and Switzer 1978).

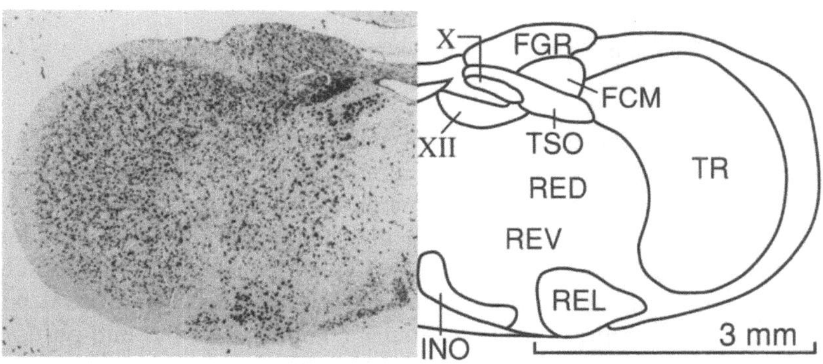

Figure 56

Transversal section through the medulla oblongata at the level of the sensory trigeminal complex in *Potamogale velox*. Cresyl violet stain.

Abbreviations:

FCM	nucleus fasc. cuneatus medialis	REV	nucleus reticularis ventralis
FGR	nucleus fascicularis gracilis	TR	compl. sensorius nervi trigemini
INO	complexus olivaris inferior	TSO	nucleus tractus solitarii
RED	nucleus reticularis dorsalis	X	ncl. dorsalis motorius nervi vagi
REL	nucleus reticularis lateralis	XII	nucleus nervi hypoglossi

Size Comparison
(Tables 13 - 15; Fig. 57)

Of the measured OBL components, TR is by far the largest, representing on the average 18.4% relative to the total OBL, ranging from 11.6% in *Atelerix algirus* to 24.6% in *Sorex cinereus*. The size indices may be described as follows:

Averages: 159 in the 30 species, 165 in the eleven subfamilies and families. This is the third highest level reached among all the 20 OBL structures.

Range: 81 - 342. The highest value *(Potamogale)* is 4.2 times larger than the lowest *(Atelerix algirus)*. This is a relatively wide variation. The maximum is the highest of all 600 values (= 20 OBL structures measured in 30 species).

Small size: Erinaceinae (87), *Talpa* (99), Tenrecinae (100).

Large size: Potamogalinae (332), Desmaninae (279), *Ruwenzorisorex* (217), *Neomys* (194).

Characteristics and trends: The five semiaquatic forms are among the six species with the largest TR. The largest intrafamily differences were found in Tenrecidae between Potamogalinae and Tenrecinae (332 : 100) and in Talpidae between Desmaninae and *Talpa* (279 : 99). Soricidae, both Soricinae and Crocidurinae, have relatively large TR size. The very large TR of *Potamogale* is shown in Fig. 56.

Nucleus tractus solitarii (TSO)

Topography, Delimitation, Structural Characteristics, Fiber Connections
(Atlas P 5 - P 8; Fig. 56)

The rostral limit of the nucleus tractus solitarii (TSO) is at a level near the rostral end of the nucleus vestibularis inferior, where it merges ventrolaterally with the dorsomedial area of the trigeminal nucleus of the spinal tract. At the caudal end, which is slightly posterior to the caudal pole of the nucleus dorsalis motorius nervi vagi, the nuclei of the two sides located dorsal to the central canal are fused, forming the commissure nucleus of Cajal. The rostral portion of the nucleus lies ventral to the vestibular complex, and the mid-portion lies beneath the ependyma of the fourth ventricle, replacing the nucleus vestibularis medialis (VM) in a caudal direction. The caudal portion bends ventromedially with the concomitant enlargement of the nucleus gracilis. Cell type and cell arrangement are not uniform throughout the structure. The nucleus is composed primarily of small and medium-sized cells.

Primary afferents from the four branchial nerves (V, VII, IX, and X) making up the solitary tract project in a topographic order. The caudal region of TSO receives viscerosensorial fibers from the internal organs; the rostral region receives gustatory input. Efferents project to various motor nuclei. The rostral part of the nucleus projects to the facial, trigeminal, and hypoglossal nuclei, the caudal part to the dorsal nucleus of the vagus and the nucleus ambiguus. Other efferents go to the contralateral ventral horn, various nuclei of the medulla implicated in cardiovascular and respiratory control, to the locus coeruleus, and others. The thalamus, hypothalamus, and amygdala also receive projections from TSO (for ref., see Bystrzycka and Nail 1985).

Size Comparison
(Tables 13 - 15)

TSO represents an average of 2.0% of the total OBL. Its size indices may be described as follows:

Averages: 151 in the 30 species, 156 in the eleven subfamilies and families. This is the fourth highest level reached among all the 20 structures of OBL.

Range: 84 - 234. The highest value *(Galemys)* is 2.8 times larger than the lowest *(Setifer)*. This is relatively low variation.

Small size: Tenrecinae (100).

Large size: Desmaninae (228), *Neomys* (215), Potamogalinae (206), *Sylvisorex* and *Solenodon* (191).

Characteristics and trends: Clear trends: (1) Progression relative to Tenrecinae. All four Tenrecinae are among the seven species with the lowest indices (besides two Crocidurinae and *Atelerix algirus*). (2) Progression in all semi-aquatic species.

Funicular nuclei (FUN)

Topography, Fiber Connections
(Atlas P 6 - P 10)

The funicular nuclei, or nuclei of the dorsal columns, are positioned at the rostral end of the dorsal columns of the spinal cord and include the gracile (FGR), medial cuneate (FCM), and external cuneate (FCE) nuclei. FGR receives primary input via the fasciculus gracilis from sacral and lumbar levels, and FCM and FCE, via the fasciculus cuneatus from thoracic and cervical levels of the spinal cord (Kahle 1976; Nieuwenhuys et al. 1980). The nuclei are somatotopically organized. FGR and FCM project mainly to the lateral part of the ventroposterior thalamic nucleus (VPL). Unlike FCM, the FCE projects primarily to the cerebellum. The FCE is the forelimb equivalent of Clarke's column, which receives primary afferent termination from the hindlimb and also projects to the cerebellum (Tracey 1985).

The funicular nuclei are small OBL components which together on the average amount to 1.9% of the OBL. Their allometric size is large in Talpidae and small in Soricidae. In most shrews they are even smaller than in Tenrecinae.

Nucleus fascicularis gracilis (FGR)

Topography, Delimitation, Structural Characteristics
(Atlas P 8 - P 10; Fig. 56)

The nucleus fascicularis gracilis (FGR) extends from the caudal pole of the nucleus vestibularis medialis rostrally to about the level of the first cervical segment. Caudally, FGR is embedded in the fasciculus gracilis. Cells of various sizes and shapes are present within the nucleus. Medium-sized cells predominate.

Size Comparison
(Tables 13 - 15)

FGR on the average represents 28% of the FUN and about 0.55% of the OBL. Its size indices may be described as follows:

Averages: 131 in the 30 species, 143 in the eleven families and subfamilies. This is a middle level (8th) of the 20 OBL structures.

Range: 67 - 207. The highest value *(Solenodon)* is 3.1 times larger than the lowest *(Tenrec)*; this is a relatively low variation.

Small size: Tenrecinae (100), Soricinae (100), Erinaceinae (103).

Large size: *Solenodon* (207), *Sylvisorex megalura* (191), Desmaninae (184), Potamogalinae (181).

Figures 57 - 60: Size indices of medulla oblongata components.

Figure 57: Complexus sensorius trigeminalis (TR)
Figure 58: Complexus olivaris inferior (INO)
Figure 59: Nucleus vestibularis superior (VS)
Figure 60: Nucleus cochlearis dorsalis (DCO)

The indices are based on AvTn lines (index = 100) the equations of which are, respectively,

$$\log \text{TR volume} = -0.001 + 0.59 \times \log \text{BoW} \quad (1.65)$$
$$\log \text{INO volume} = -1.340 + 0.61 \times \log \text{BoW} \quad (2.05)$$
$$\log \text{VS volume} = -1.149 + 0.51 \times \log \text{BoW} \quad (1.16)$$
$$\log \text{DCO volume} = -0.867 + 0.50 \times \log \text{BoW} \quad (1.13)$$

The distances of the AvIF broken lines are given in brackets. For additional explanations see Fig. 53.

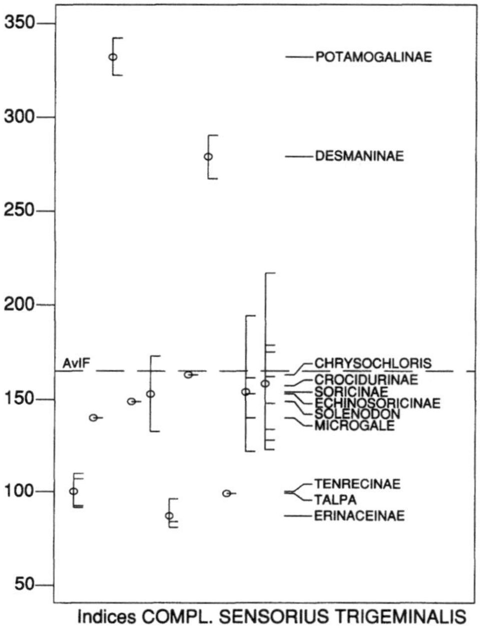

Figure 57

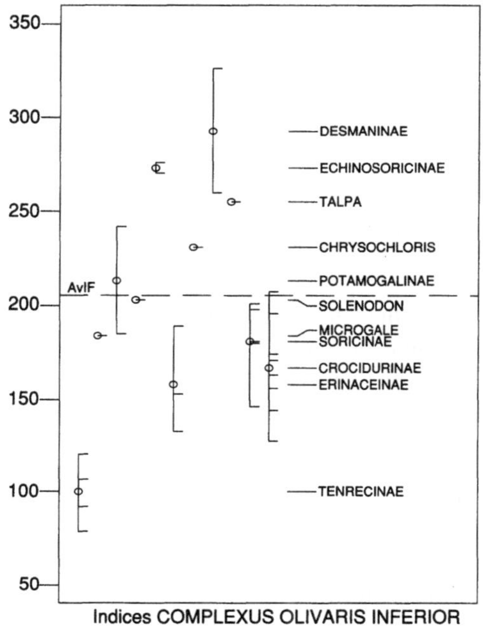

Figure 58

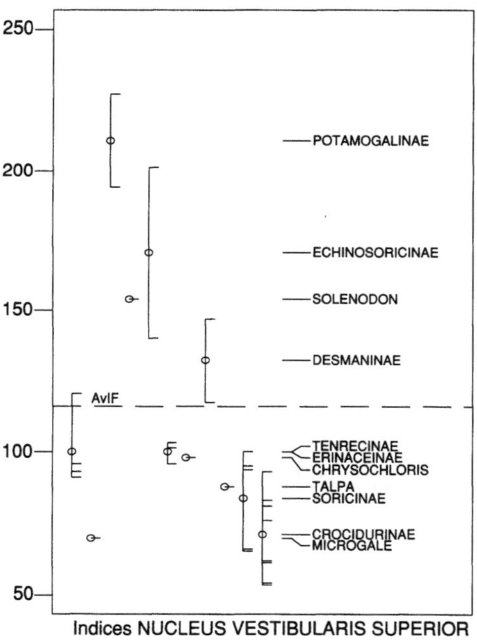

Figure 59

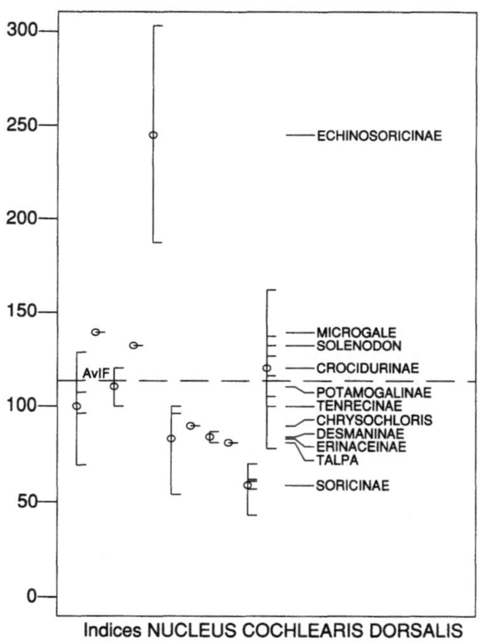

Figure 60

Characteristics and trends: Four of the five water-adapted species are among the highest. The fifth (*Neomys*) has the largest FGR in Soricinae, which, on the average, have a small FGR size, lower than in Crocidurinae and on the same level as Tenrecinae.

Nucleus fascicularis cuneatus medialis (FCM)

Topography, Delimitation, Structural Characteristics

(Atlas P 7 - P 10; Fig. 56)

The nucleus fascicularis cuneatus medialis (FCM) has its rostral end near the caudal pole of the nucleus vestibularis inferior and extends to a level slightly rostral to the caudal pole of the nucleus fascicularis gracilis (FGR). Cells are similar but slightly larger than those found in FGR. The caudal end appears as a cellular ridge between FGR and the trigeminal complex (TR) at the ventral edge of the fasciculus cuneatus. At more rostral levels, it projects dorsally into the fasciculus cuneatus and in many species merges at its rostral region with the nucleus fascicularis cuneatus externus (FCE), dorsolaterally, and FGR, dorsomedially. This region of FCM may therefore be poorly delineated, particularly from FCE.

Size Comparison

(Tables 13 - 15)

FCM is larger than FGR and on the average amounts to 43% of the funicular nuclei (FUN) and 0.84% of OBL. Its size indices may be described as follows:

Averages: 118 in the 30 species, 147 in the eleven subfamilies and families. As in FGR, this is a middle level (10th) among the 20 OBL structures measured. The large difference between the two averages is due to the small size in shrews, which comprise 13 of the 30 species.

Range: 57 - 320. The highest value *(Talpa europaea)* is 5.6 times larger than the lowest *(Crocidura russula)*. This is the third highest variation among all OBL structures measured. The maximum (320) is the third highest among all OBL structures in all species.

Small size: Shrews (88) [Soricinae somewhat lower (79) than Crocidurinae (93)], Tenrecinae (100), Erinaceidae (101).

Large size: Talpidae (258) [*Talpa* (320) still distinctly higher than Desmaninae (227)], *Solenodon* (189).

Characteristics and trends: All shrews have smaller FCM size than Tenrecinae on the average, except *Sylvisorex* and *Ruwenzorisorex*, which have slightly higher indices.

Nucleus fascicularis cuneatus externus (FCE)

Topography, Delimitation, Structural Characteristics
(Atlas P 6 - P 7)

FCE appears rostrally at a level lateral to the caudal region of the nucleus vestibularis inferior and extends caudally to a level situated caudal to the obex. Dorsally, FCE covers the trigeminal tract and also extends lateroventrally into the inferior cerebellar peduncle. The most prominent cells of the nucleus are relatively large, deeply stained multipolar neurons which tend to be grouped in small clusters.

Size Comparison
(Tables 13 - 15)

FCE is similar in size to FGR and represents about 29% of FUN and 0.53% of the total OBL. Its size indices may be described as follows:

Averages: 111 in the 30 species, 127 in the eleven families and subfamilies. This is a low level (15th) among the 20 OBL structures.

Range: 23 - 245. The highest value *(Hylomys)* is 10.7 times larger than the lowest *(Crocidura russula)*. This is the highest variation of all 20 OBL structures under comparison.

Small size: Soricidae (70), Potamogalinae (98), *Solenodon* (98), Tenrecinae (100).

Large size: *Hylomys* (245), Echinosoricinae (209), Erinaceinae (194), Talpidae (165).

Characteristics and trends: No clear trends. The large differences are not related to different taxonomic or adaptation groups. Both low and high values are found in Soricinae, Crocidurinae, Potamogalinae, and Tenrecinae. This indicates that the large variations in this nucleus may be partly due to difficulties in its delineation.

Relay Nuclei (RLN)

The four following nuclei (REL, INO, PRP, COE) are comprised under the name "relay nuclei." This grouping is more or less artificial since the various nuclei belong to different functional systems and thus their size changes in evolutionary trends and adaptational processes also may be quite different. They are all relatively small structures and together make up 2.6% of total OBL.

Nucleus reticularis lateralis (REL)

Topography, Delimitation, Structural Characteristics, Fiber Connections
(Atlas P 7 - P 8; Fig. 56)

The nucleus reticularis lateralis (REL) is a cell group lying ventrally in OBL between the trigeminal complex laterally and the inferior olivary complex medially. The rostral pole of REL, which is formed by the most rostral cells of the so-called subnucleus subtrigeminalis, lies at a level close to the caudal end of the nucleus facialis. However, cells at the rostral pole merge with the reticular formation, which sometimes makes this limit difficult to discern. The subnucleus subtrigeminalis is considered a part of REL. REL extends caudally to a level near the caudal end of the inferior olivary complex. The caudal portion is a very conspicuous ovoid group of cells. In transversal sections, the middle portion extends mediolaterally and splits rostrally into two parts. The large rostromedial part merges with the nucleus reticularis medialis dorsally. This part of REL is poorly delineated in some species. The lateral part, which extends more rostrally, fuses with the subnucleus subtrigeminalis lying ventromedial to the trigeminal complex.

The most conspicuous cells of REL are large multipolar neurons similar to those of the paramedian reticular nucleus. There are also medium and small cells predominant in the small ventrolateral region, sometimes called "parvocellular nucleus" (Taber 1961), and in the subtrigeminal subnucleus.

REL seems to receive projections from the spinal cord, red nucleus, cerebellar nuclei, and cerebral cortex (Flumerfelt and Hrycyshyn 1985). Large numbers of axons from spinal neurons originate mainly in the cervical enlargement and rostral lumbar segments (Menétrey et al. 1983). REL is an important precerebellar relay nucleus which sends most of its information by mossy fibers to the cerebellum, where it is employed in the cerebellar control of motor activity (Chan-Palay et al. 1977; Flumerfelt and Hrycyshyn 1985).

Size Comparison
(Tables 16 - 18)

REL amounts to about 0.8% of the total OBL. Its size indices may be described as follows:

Averages: 170 in the 30 species, 182 in the eleven families and subfamilies. This is the second highest level among all of the 20 OBL structures.

Range: 72 - 317. The highest value *(Hylomys)* is 4.4 times larger than the lowest *(Setifer)*. This indicates medium variation. The maximum (317) is the fourth highest reached in all OBL nuclei.

Small size: Tenrecinae (100).

Large size: Echinosoricinae (279), Potamogalinae (250), *Sylvisorex* (239), *Solenodon* (229).

Characteristics and trends: Relative to Tenrecinae, there are progressive trends which, however, differ from those found in other progressive brain structures (such as cerebellum, diencephalon, striatum, and neocortex) in that high indices are dispersed nearly equally in many groups. Desmaninae, always high in the progressive structures mentioned, are relatively low in REL, and Crocidurinae, sometimes regarded as "basal" forms, are relatively high.

Complexus olivaris inferior (INO)

Topography, Delimitation, Structural Characteristics, Fiber Connections
(Atlas P 6 - P 8; Fig. 56)

The inferior olive (INO) is a complexly folded cellular lamina in the ventral part of OBL just dorsal or dorsolateral to the very small pyramids. The complex extends from the caudal pole of the nucleus facialis, rostrally, to a level slightly rostral to the caudal pole of the nucleus reticularis lateralis, caudally. In mammals, INO is composed of three large subdivisions (Kooy 1916) and four smaller subdivisions (Brodal 1940). At least the three major nuclei can be distinguished in Insectivora on topographical grounds. The cells are relatively small and uniform in size throughout the complex.

INO receives afferents from many areas between the lumbar spinal cord and the cerebral cortex (for ref. see Flumerfelt and Hrycyshyn 1985). Spinal projections come mainly from contralateral cervical, thoracic, and lumbar segments (Swenson and Castro 1983a,b). The deep cerebellar nuclei, many mesencephalic regions, such as the superior colliculus, the ventral tegmental area, and others, and diencephalic regions, such as zona incerta, also project to INO. Finally, INO receives fibers from the cerebral motor cortex (Swenson and Castro 1983a,b). All neurons of INO project topographically to the cerebellum. The olivary complex is in fact the sole source of climbing fibers to the cerebellum (for ref., see Flumerfelt and Hrycyshyn 1985).

Size Comparison
(Tables 16 - 18; Fig. 58)

The inferior olive (INO) makes up on the average about 1.1% of the total OBL. Its size indices may be described as follows:

Averages: 185 in the 30 species, 205 in the eleven subfamilies and families. This is the highest level reached by any of the 20 OBL structures under consideration.

Range: 79 - 326. The highest value *(Galemys)* is 4.1 times larger than the lowest *(Setifer)*. This is medium variation. The maximum (326 in *Galemys)* is the second highest reached among all OBL components.

Small size: Tenrecinae (100).

Large size: Desmaninae (293), Echinosoricinae (273), *Talpa* (255), *Potamogale* (242), *Chrysochloris* (231).

Characteristics and trends: Clear trends: (1) Progression relative to Tenrecinae. All the four lowest values are from Tenrecinae. (2) Progression in all water-adapted and subterranean species. Trends are similar to those found in other progressive brain structures.

Nucleus prepositus hypoglossi (PRP)

Topography, Delimitation, Structural Characteristics, Fiber Connections
(Atlas P 4 - P 6)

The nucleus prepositus hypoglossi (PRP) is located lateral to the medial longitudinal fasciculus beneath the ependymal layer of the 4th ventricle. It extends from the genu of the VIIth nerve, rostrally, to the rostral pole of the nucleus nervi hypoglossi, caudally. Throughout its entire length, the nucleus lies medial to the nucleus vestibularis medialis. Relatively small cells predominate in the rostral part of the nucleus, while medium-sized cells are more numerous in the caudal part.

There are conflicting results as to the fiber connections of PRP. Most recent studies have shown that nuclei of the pretectal area which receive primary optic fiber connections, such as the nucleus of the optic tract and the olivary pretectal nucleus, project directly to PRP (Cazin et al. 1982). Efferent projections go to the three ocular muscle nuclei (Graybiel and Hartwieg 1974), the vestibular complex (Rubertone and Mehler 1980,) and the cerebellum (Blanks et al. 1983). The fiber connections indicate that, in addition to the direct pathway to the oculomotor nuclei, the prepositus hypoglossi also constitutes an indirect relay to these motor nuclei via connections with the vestibular complex and the cerebellum (Mehler and Rubertone 1985).

Size Comparisons
(Tables 16 - 18)

The PRP nucleus is relatively small and represents on the average about 0.7% of the total OBL. Its size indices may be described as follows:

Averages: 136 in the 30 species, 145 in the eleven subfamilies and families. This is the seventh highest level reached among the 20 structures.

Range: 83 - 219. The highest value *(Hylomys)* is 2.6 times larger than the lowest *(Setifer)*. This is relatively low variation.

Small size: Tenrecinae (100), *Talpa* (102), *Suncus* spp. (105), *Sorex* spp. (109), Erinaceinae (110).

Large size: Echinosoricinae (209), Desmaninae (188), Potamogalinae (187), *Solenodon* (174).

Characteristics and trends: Some progressive trends. Tenrecinae and Erinaceinae are lowest. However, in shrews, which in previous studies were included in the "Basal Insectivora," very high values were found in addition to low ones. Echinosoricinae and semiaquatic species have high PRP size.

Nucleus locus coeruleus (COE)

Topography, Delimitation, Structural Characteristics, Fiber Connections
(Atlas P 2.5)

The very short locus coeruleus (COE) lies at the ventrolateral border of the caudal part of the central grey. The lateral border is formed by the parabrachial and, most caudally, by the superior vestibular nucleus. Cells of the mesencephalic trigeminal nucleus lie at the lateral margin or penetrate the lateral peripheral zone of the nucleus. COE is made up of densely packed small to medium-sized, well-stained cells.

The afferent projections to COE, which are extremely diverse, come from the insular cortex, amygdala, thalamus, and spinal cord, as well as from nuclei in the hypothalamus, reticular formation, and OBL. COE projects widely throughout the central nervous system. The major projection of COE, known as the dorsal noradrenergic bundle, goes to various mesencephalic, diencephalic, and telencephalic grisea. Other efferent fibers go to the cerebellum. Descending fibers project to various nuclei of the medulla and to the entire length of the spinal cord (for ref., see Loughlin and Fallon 1985).

Size Comparison
(Tables 16 - 18)

The locus coeruleus (COE) is a small nucleus representing only 0.1% of the total OBL. Its size indices may be described as follows:

Averages: 120 for the 29 species, 128 for the eleven families and subfamilies. This is a low level, the 13th among the 20 structures.

Range: 63 - 240. The highest value *(Micropotamogale ruwenzorii)* is 3.8 times larger than the lowest *(Tenrec)*. This is medium variation.

Small size: *Suncus* spp. (73), *Sorex* spp. (96), Tenrecinae (100), *Chrysochloris* (104), Soricidae with both subfamilies (105).

Large size: Potamogalinae (200), *Hylomys* (194).

Characteristics and trends: No clear trends. Potamogalinae and Echinosoricinae are high; all other averages for families and subfamilies are quite close together, but there is extremely wide variation in Crocidurinae. This may reflect difficulties in measuring exact volumes since this nucleus is represented in a few sections only.

Motor Nuclei (MOT)

The motor nuclei (V, VII, X, XII) are small and together amount to 2.3% of the total OBL. However, they are well circumscribed and large enough to be measured. The allometric size of the whole complex is strongly influenced by nucleus VII, which makes up slightly more than half the total. Thus, MOT shows characteristics similar to VII, i.e., distinctly largest size in Echinosoricinae and Potamogalinae and large size in the other water-adapted species. Size is small (except for nucleus X) in subterranean species and many terrestrial and/or fossorial shrews. The various nuclei may, however, behave differently.

Nucleus motorius nervi trigemini (mot V)

Topography, Delimitation, Structural Characteristics, Fiber Connections
(Atlas P 2.5 - P 3)

The nucleus motorius nervi trigemini (mot V) lies at the level of the locus coeruleus medial to the principal sensory trigeminal nucleus. In mammals, it is usually divided into a dorsolateral group and a smaller ventromedial group. These groups can also be distinguished in Insectivora. The cells are large, dark-staining multipolar neurons grouped together with moderate cell density. The dorsolateral division innervates the jaw-closing muscles (the masseter, temporalis, and medial pterygoid); the ventromedial division innervates the jaw-opening muscles (the anterior digastric and mylohyoid). The tensor tympani is also innervated by trigeminal motoneurons (for ref., see Travers 1985).

Size Comparison
(Tables 19 - 21)

This small nucleus on the average amounts to 18.5% of MOT and 0.4% of the total OBL. Its size indices may be described as follows:

Averages: 121 in the 30 species, 133 in the eleven subfamilies and families. This is a middle level (12th) among the 20 OBL structures.

Range: 70 - 203. The highest value *(Hylomys)* is 2.9 times larger than the lowest *(Crocidura russula)*. This is relatively low variation.

Small size: Crocidurinae (97), Erinaceinae (98), Tenrecinae (100).

Large size: Echinosoricinae (197), *Galemys* (173), Potamogalinae (170).

Characteristics and trends: Echinosoricinae and water-adapted species (except *Desmana*) have a large motor V. In shrews, the average in Soricinae is higher than in Crocidurinae, but there is high variability in both subfamilies. This may reflect difficulties in measuring exact volumes since this nucleus is represented in some species in only a few sections.

Nucleus nervi facialis (mot VII)

Topography, Delimitation, Structural Characteristics, Fiber connections
(Atlas P 3.5 - P 4)

The nucleus nervi facialis (mot VII) extends from a level slightly anterior to the caudal tip of the superior olivary complex to the level of the rostral pole of the inferior olivary complex. In Insectivora, as in other mammals (Papez 1927; Baron 1970; Watson et al. 1982), this nucleus can be divided into several groups. There are, however, no cytoarchitectural differences between them. All cells are large multipolar neurons projecting mainly to superficial muscles of the head and neck. Most of these muscles are involved in facial expression. Other muscles, innervated by the facial nerve, include the stapedius muscle of the middle ear (for ref. see Travers 1985).

Size Comparison
(Tables 19 - 21)

This nucleus is by far the largest of the four motor nuclei under comparison. On the average it amounts to 51% of MOT and 1.2% of the total OBL. Its size indices may be described as follows:

Averages: 127 in the 30 species, 133 in the eleven subfamilies and families. This is a medium level (11th) among the 20 OBL structures.

Range: 76 - 282. The highest value (*Potamogale*) is 3.7 times larger than the lowest (*Tenrec*). This is moderate variation.

Small size: *Talpa* (93), terrestrial Soricinae (97), *Chrysochloris* (98), Tenrecinae (100).

Large size: Potamogalinae (230), Echinosoricinae (200), Desmaninae (163).

Characteristics and trends: Large or very large size in four of the five water-adapted species. The index of the fifth species, the water shrew *(Neomys)* is lower but, compared with terrestrial Soricinae, it is higher (129 versus 97). Subterranean species and several shrews have a small nucleus mot VII.

Nucleus dorsalis motorius nervi vagi (mot X)

Topography, Delimitation, Structural Characteristics, Fiber Connections
(Atlas P 6 - P 8; Fig. 56)

The nucleus dorsalis motorius nervi vagi (mot X) extends from a level at the caudal pole of the nucleus prepositus hypoglossi, rostrally, to a level near the caudal end of the nucleus tractus solitarii on both sides of the central canal, caudally. It always lies between the nucleus tractus solitarius and the nucleus nervi hypoglossi and is separated from the latter by a small area with low cell density. In the caudal portion, it occupies the ventral edge, in the rostral portion, the ventromedial aspect of the nucleus tractus solitarii. The neurons, which are the preganglionic cells of the vagal nerve, send axons to the parasympathetic ganglia of the thoracic and abdominal viscera.

Size Comparison
(Tables 19 - 21)

This is the smallest of the four motor nuclei under consideration. On the average it amounts to 11% of MOT and 0.3% of the total OBL. Its size indices may be described as follows:

Averages: 140 in the 30 species, 138 in the eleven subfamilies and families. This is an average level (12th) among the 20 OBL structures.

Range: 88 - 238. The highest value *(Blarina)* is 2.7 times larger than the lowest *(Tenrec, Setifer)*. This is relatively low variation.

Small size: *Sorex cinereus* (99), Tenrecinae (100).

Large size: *Blarina* (238), *Hylomys* (223), *Sylvisorex* (182), *Ruwenzorisorex* (181).

Characteristics and trends: This nucleus is largest in shrews. Soricinae, however, have a wide variation.

Nucleus nervi hypoglossi (mot XII)

Topography, Delimitation, Structural Characteristics, Fiber Connections
(Atlas P 7 - P 8; Fig. 56)

The nucleus nervi hypoglossi (mot XII) extends from the caudal end of the nucleus prepositus hypoglossi to a level slightly caudal to the caudal pole of the nucleus dorsalis motorius nervi vagi (mot X) ventrolateral to the central canal. The nucleus is composed predominantly of large, well-stained, multipolar neurons. Subdivisions into longitudinal cell columns have been described by several authors (e.g., Taber 1961; Travers 1985), but could not be recognized in Nissl-stained transversal sections of Insectivora brains. The nucleus innervates tongue retractor and tongue protruder muscles (Chibuzo and Cummings 1982).

Size Comparison
(Tables 19 - 21)

This small nucleus on the average amounts to 19% of MOT and 0.45% of the total OBL. Its size indices may be described as follows:

Averages: 116 in the 30 species, 116 in the eleven subfamilies and families. This is a low level (16th) among the 20 OBL structures.

Range: 72 - 183. The highest value *(Hemiechinus)* is 2.5 times larger than the lowest *(Talpa)*. This is relatively low variation.

Small size: *Talpa* (72), *Crocidura* spp. (83), *Microgale* (90), Tenrecinae (100), Potamogalinae (103).

Large size: Echinosoricinae (169), Erinaceinae (166), *Galemys* (151).

Characteristics and trends: This nucleus is by far largest in hedgehogs and moon rats and smallest in the European mole and several Crocidurinae.

Vestibular Complex (VC)

Topography, Delimitation, Structural Characteristics, Fiber Connections
(Atlas P 3 - P 6)

The vestibular complex (VC) consists of four major vestibular nuclei (medial VM, inferior VI, lateral VL, and superior vestibular nucleus VS) and a number of associated smaller cell groups, which have been included in our measurements in the adjoining major nuclei. Cytoarchitecturally, the exact borders between the four main nuclei are, despite occasional ambiguities, generally clear and precise enough in Nissl stained sections to allow for volume measurements.

All four nuclei receive varying numbers of primary vestibular fibers. The fibers forming the ascending bundle come from the cristae ampullares and project mainly to VS. The fibers forming the descending bundle come from the maculae of the sacculus and utriculus. The fibers end mainly in VI and VM. Only a few seem to terminate in VL (Walberg et al. 1958; Stein and Carpenter 1967; Gacek 1969; Mehler and Rubertone 1985). Another important afferent system arises in the cerebellum (Walberg et al. 1962; Haines 1977; Carleton and Carpenter 1983). VL receives axons mainly from the Purkinje cells of the paravermal region (Nieuwenhuys et al. 1978, 1980). Spino-vestibular fibers terminate in various regions of the vestibular complex (Pompeiano and Brodal 1957; Rubertone and Haines 1982).

The efferents are as diversified as the afferents. Vestibulospinal fibers originate mainly in VM and VL. VM axons project only as far as thoracic levels, whereas VL axons project downward to lumbar levels. Vestibulocerebellar fibers coming mainly from VM and VI terminate in different parts of the "vestibulocerebellum." All vestibular nuclei send fibers to the oculomotor nuclei. The main source of vestibular input to these nuclei, however, is VS (for ref., see Carleton and Carpenter 1983; Mehler and Rubertone 1985).

Size Comparison
(Tables 22 - 24)

The entire vestibular complex (VC) amounts 4.5% of the total OBL. Its size indices in Insectivora may be described as follows:

Averages. 127 in the 30 species, 139 in the eleven subfamilies and families. This is a medium level for the OBL complexes.

Range: 81 - 247. The highest value *(Hylomys)* is 3.0 times larger than the lowest *(Microsorex)*. This is relatively low variation.

Small size: *Talpa* (99), Tenrecinae (100), Soricidae (103).

Large size: Echinosoricinae (241), Potamogalinae (197), Desmaninae (182), *Solenodon* (173).

Characteristics and trends: Clear progressive trends in Echinosoricinae and in all water-adapted species. The allometric size of VC is almost twice as large in the semiaquatic Potamogalinae as in the terrestrial Tenrecinae (Baron et al. 1988). Similarly, the vestibular complex of semiaquatic Desmaninae is 1.8 times larger than that of the fossorial *Talpa*. In Soricinae, *Neomys* is at the top of the index scale.

Nucleus vestibularis medialis (VM)

Topography, Delimitation, Structural Characteristics
(Atlas P 3.5 - P 6)

The nucleus vestibularis medialis (VM) extends from a level near the caudal pole of the locus coeruleus (COE), rostrally, to the rostral pole of the nucleus fascicularis gracilis (FGR), caudally. In front of FGR is a small cell group, "g" (Brodal and Pompeiano 1957), which is included in our VM measurements. Throughout its rostro-caudal extent, the dorsomedial edge of the triangular VM lies beneath the ependyma of the fourth ventricle. Medially, VM is bordered by the nucleus prepositus hypoglossi except for the most rostral part, which is bounded by the nucleus nervi abducentis, and the caudal end, which is bordered by the nucleus and fibers of the solitary tract. The dorsolateral side of the most rostral part of VM is covered by VS, and the lateral border is bounded from rostral to caudal by VL and VI. Ventrally, VM borders the reticular formation, except at the most rostral part, where it overlies the supratrigeminal nucleus, and at the most caudal part, where it is bounded by the nucleus of the solitary tract. The cells are in general more compactly arranged than in the neighbouring VI and are predominantly medium-sized. In the rostral part, however, most of the cells are small. In the ventrolateral region of the middle part are a few relatively large cells similar to those found in the adjacent VL.

Size Comparison
(Tables 22 - 24)

VM, the largest of the vestibular nuclei (about 41%), amounts to 1.85% of the total OBL. Its size indices in Insectivora may be described as follows:

Averages: 141 in the 30 species, 152 in the eleven subfamilies and families. This is the fifth highest level among the 20 OBL structures.

Range: 82 - 281. The highest value *(Hylomys)* is 3.4 times larger than the lowest *(Microsorex)*. This is moderate variation.

Small size: *Microsorex* (82), Tenrecinae (100).

Large size: Echinosoricinae (276), Desmaninae (206), *Solenodon* (185), Potamogalinae (184).

Characteristics and trends: They are similar to those described for the total complex, i.e., large size in Echinosoricinae, water-adapted forms, and in *Solenodon*.

Nucleus vestibularis inferior (VI)

Topography, Delimitation, Structural Characteristics

(Atlas P 5 - P 6)

The nucleus vestibularis inferior (VI) lies lateral to VM and caudal to VL. It extends to the dorsal surface of OBL except at its most rostral region, where it is covered by the caudal portion of VL and partly by the caudal portion of the dorsal cochlear nucleus. Laterally, it is bounded by fibers of the inferior cerebellar peduncle and, more caudally, by the FCE. The ventral border is formed by the nucleus and fibers of the solitary tract. The cell group x (Brodal and Pompeiano 1957, 1958) is included with VI in our measurements. It lies ventrolateral to VI almost over its entire length and overlaps partially with the rostral part of FCE.

The loosely arranged cells are predominantly medium in size. The cells are interstitial to the descending root fibers of the vestibular nerve, a feature distinguishing it clearly from VM.

Size Comparison

(Tables 22 - 24)

VI is on the average somewhat smaller than VM and amounts to about 30% of VC and 1.33% of the total OBL. Its size indices may be described as follows:

Averages: 129 in the 30 species, 142 in the eleven subfamilies and families. This is a medium level (9th) among the 20 OBL structures.

Range: 69 - 249. The highest value *(Potamogale)* is 3.6 times larger than the lowest *(Sorex araneus)*. This is a moderate variation.

Small size: *Sorex* spp. (83), *Suncus* spp. (83), *Chrysochloris* (97), *Microgale* (97), Tenrecinae (100).

Large size: Potamogalinae (223), Echinosoricinae (220), Desmaninae (207), *Solenodon* (188).

Characteristics and trends: Similar to VM. Progression in all water-adapted species. Progressive trends relative to Erinaceinae and Tenrecinae. Large variation may occur even within genera (e.g., *Sorex, Crocidura*).

Nucleus vestibularis lateralis (VL)

Topography, Delimitation, Structural Characteristics
(Atlas P 3.5 - P 4)

The nucleus vestibularis lateralis (VL) lies in its rostro-caudal extension between VS and VI. Medially, VL is bounded throughout its length by VM and, laterally, by the inferior cerebellar peduncle. The ventral border is formed rostrally by the trigeminal complex and caudally by the anterior portion of VI. VL is composed of very characteristic multipolar giant cells. However, there are cells similar to those in VM and VI as well.

Size Comparison
(Tables 22 - 24)

This small nucleus on the average amounts to about 16% of VC and 0.71% of the total OBL. Its size indices may be described as follows:

Averages: 119 in the 30 species, 128 in the eleven subfamilies and families. This is a low level (14th) among the 20 OBL structures.

Range: 64 - 284. The highest value *(Echinosorex)* is 4.4 times larger than the lowest *(Talpa, Suncus etruscus)*. This is a medium variation.

Small size: *Talpa* (64), *Suncus* spp. (75), Soricidae (97), Tenrecinae (100).

Large size: Echinosoricinae (266), *Sylvisorex* (164), Potamogalinae (163).

Characteristics and trends: High indices are from water-adapted species, but still higher values were found in the terrestrial Echinosoricinae and *Sylvisorex*. These species also had high VM indices. VL size was very low in *Talpa*.

Nucleus vestibularis superior (VS)

Topography, Delimitation, Structural Characteristics
(Atlas P 3 - P 3.5)

The nucleus vestibularis superior (VS) is the most rostral part of the cell column lateral to VM and lies beneath the superior cerebellar peduncle. The medial border is formed by VM, except for the most rostral part; the lateral border, mainly by the inferior cerebellar peduncle, and the ventral border, rostrally, by the supratrigeminal nucleus and, caudally, by the anterior portion of VL.

VS is composed of medium to small cells often grouped in strands by fiber bundles running in a dorsomedial direction.

Size Comparison
(Tables 22 - 24; Fig. 59)

VS is the smallest of the vestibular nuclei. On the average it amounts to about 13% of VC and 0.61% of the total OBL. Its size indices may be described as follows:

Averages: 104 in the 30 species, 116 in the eleven subfamilies and families. This is the fourth lowest level among the 20 OBL structures.

Range: 54 - 227. The highest value *(Potamogale)* is 4.2 times larger than the lowest *(Crocidura russula* and *Suncus murinus)*. This is a medium variation.

Small size: *Microgale* (70), Crocidurinae (71), Soricinae (84), *Talpa* (88), *Chrysochloris* (98), Tenrecinae (100), Erinaceinae (100).

Large size: Potamogalinae (211), Echinosoricinae (171), *Solenodon* (154), Desmaninae (132).

Characteristics and trends: High indices in water-adapted species and in Echinosoricinae. There is a large variation within subfamilies (Echinosoricinae, Crocidurinae).

Auditory Nuclei (AUD)

Topography, Delimitation, Structural Characteristics, Fiber Connections
(Atlas P 2 - P 4)

The auditory components of the medulla oblongata (OBL) comprise the dorsal (DCO) and ventral (VCO) cochlear nuclei and the superior olive (OLS). The cochlear nuclei complex (CON) is located at the dorsolateral border of the rostral part of OBL near its junction with the pontine region. It is composed of a dorsal nucleus (nucleus cochlearis dorsalis, DCO) and a ventral nucleus (nucleus cochlearis ventralis, VCO), with a pars anterior, pars posterior, and pars interstitialis.

The dorsal cochlear nucleus (DCO), an oval-shaped structure in transversal sections, lies in the caudal part of the CON dorsolateral to the inferior cerebellar peduncle and to the caudal edge of the ventral cochlear nucleus (VCO). This nucleus is organized in layers characterized by particular cell types. The outer molecular layer is composed of sparsely scattered small cells. The intermediate layer contains medium-sized fusiform cells intermingled with small granular cells. The inner, or polymorphic, layer has a predominance of medium and small-sized neurons. The granular cells of the intermediate layer merge with the intercalated cell masses at the border between the DCO and the anterior part of the VCO. In our measurements, the small-celled intercalated masses were included in the dorsal nucleus.

In the ventral cochlear nucleus (VCO), the anterior part is separated from the posterior part by the cochlear nerve root: the anterior part being penetrated by ascending fibers and the posterior part, by descending fibers. The large and medium-sized cells of various forms are rather uniformly scattered throughout the pars posterior of the VCO. The cells of various sizes and shapes in the pars anterior of the VCO show a gradient of increasing density from the ventrocaudal to the rostrodorsal region. Scattered medium-sized cells within the radix nervi cochlearis form the pars interstitialis of the VCO.

All cochlear nuclei receive fibers from the auditory nerve and are tonotopically organized. The neurons of the dorsal parts of the nuclei, are tuned to high frequencies and receive fibers from the lateral end of the cochlea, whereas the ventral parts are tuned to low frequencies and receive fibers from the apical end (Powell and Cowan 1962; Sando 1965).

The cochlear nuclei complex (CON) sends three separate pathways. The trapezoid body, composed of fibers originating in the pars anterior of the VCO, projects to the medial nucleus of the olivary complex (OLS) bilaterally and to the lateral nucleus of the OLS ipsilaterally. Collaterals go to the nucleus of the trapezoid body. Other fibers of the trapezoid body ascend contralaterally as lemniscus lateralis to the nuclei lemnisci lateralis (NLL) and the inferior colliculi (INC). The intermediate acoustic stria (stria of Held), formed by axons of the posterior part of the VCO, projects to OLS. DCO sends fibers to the contralateral INC via the dorsal acoustic stria (stria of Monakow) (Beyerl 1978; Webster 1985).

The superior olive (OLS), located in the ventral OBL caudal to the pons, is composed of four nuclei: the lateral nucleus of the superior olive (nucleus olivaris superior lateralis), the medial nucleus (nucleus olivaris superior medialis), the preolivary nucleus (nucleus praeolivaris), and the nucleus of the trapezoid body (nucleus corporis trapezoidei). OLS extends from the caudal end of the ventral nucleus of the lateral lemniscus to the rostral pole of the facial nucleus. The medial nucleus was supposed by Masterton et al. (1975) to be the chief binaural time-analyzing center for sound localization.

The lateral nucleus of the superior olive, which extends throughout the length of the complex, is formed of densely-packed, medium-sized cells. The characteristic S-shape described in many mammals is not conspicuous in Insectivora. The medial nucleus does not consist of the typical column of fusiform cells found in Primates and other mammals. Its boundary is difficult to trace, since its medium-large cells are irregularly distributed in the space between the nucleus of the trapazoid body and the lateral nucleus. The preolivary nucleus is composed of loosely packed cells bordering the medial and lateral nuclei ventrally, laterally and rostrally. The nucleus of the trapezoid body is located medial to the lateral olivary nucleus in the rostral region of the complex and, after the appearance of the medial olivary nucleus, medial to this nucleus in the caudal region. It consists of a prominent cluster of loosely packed globular shaped cells lying between the fibers of the trapezoid body.

Size Comparison

The auditory nuclei (DCO, VCO, OLS) account for an average of 4.4% of the total OBL. Their allometric size obviously depends to a large extent on the influences of adaptational processes. Large auditory nuclei are found in Echinosoricinae, *Potamogale, Sylvisorex megalura, Solenodon* and, among Tenrecinae, in *Echinops,* which otherwise is generally very low in brain development. Small auditory nuclei were found in Soricinae and Talpidae. In Echinosoricinae, the auditory nuclei (AUD) are about 2.1 times larger than in Erinaceinae, and in Crocidurinae, they are nearly two times larger than in Soricinae. The differences apply to all nuclei and (nearly) all species of these subfamilies.

Nucleus cochlearis dorsalis (DCO)

Topography, Delimitation, Structural Characteristics
See above and Atlas P 4

Size Comparison
(Tables 25 - 27; Fig. 60)

DCO represents on the average 26% of AUD and 1.1% of the total OBL. Its size indices may be described as follows:

Averages: 107 in the 30 species, 113 in the eleven subfamilies and families. This is the third lowest level (18th) among the 20 OBL structures.

Range: 43 - 303. The highest value *(Echinosorex)* is 7.0 times larger than the lowest *(Microsorex)*. This is high variation.

Small size: Soricinae (59), *Talpa* (81), Erinaceinae (83), Desmaninae (84), *Chrysochloris* (90), Tenrecinae (100).

Large size: Echinosoricinae (245), *Sylvisorex* (162; and the Crocidurinae in general, 120), *Microgale* (139).

Characteristics and trends: No clear trends. Relatively large values in Tenrecinae. Extremely large differences in Erinaceidae, in which the Echinosoricinae have nearly three times larger DCOs than the Erinaceinae (245 versus 83). Large differences between Crocidurinae and Soricinae (120 versus 59) were also found.

Nucleus cochlearis ventralis (VCO)

Topography, Delimitation, Structural Characteristics
See above and Atlas P 2.5 - P 4

Size Comparison
(Tables 25 - 27)

VCO represents on the average 33% of AUD and 1.5% of the total OBL. Its size indices may be described as follows:

Averages: 90 in the 30 species, 95 in the eleven subfamilies and families. This is the lowest level among the 20 OBL structures.

Range: 38 - 177. The highest value *(Hylomys)* is 4.7 times larger than the lowest *(Microsorex)*. This is a relatively high variation.

Small size: Soricinae (46), *Talpa* (56), Desmaninae (64), *Chrysochloris* (71).

Large size: Echinosoricinae (173), *Echinops* (141), *Scutisorex* (134), *Solenodon* (128), *Microgale* (126).

Characteristics and trends: No clear trends. Relatively large values in Tenrecinae. There is, however, wide variation and the largest differences among all the 20 OBL structures were found in this subfamily. The VCO indices reach from 67 in *Tenrec* to 141 in *Echinops*. A large drop again was found between Echinosoricinae and Erinaceinae and between Crocidurinae and Soricinae. Very low values were found in Talpidae.

Complexus olivaris superior (OLS)

Topography, Delimitation, Structural Characteristics
See above and Atlas P 2 - P 3

Size Comparison
(Tables 25 - 27)

The OLS complex represents on the average 41% of AUD and 1.8% of the total OBL. Its size indices may be described as follows:

Averages: 104 in the 30 species, 112 in the eleven subfamilies and families. This is a low level (19th) among the 20 OBL structures.

Range: 40 - 223. The highest value *(Hylomys)* is 5.6 times larger than the lowest (*Microsorex*). This is one of the highest variations among the 20 OBL structures, which, however, is due to the strongly deviating maximum values. If the two maxima are excluded (from *Hylomys* and *Echinosorex*), the quotient shrinks to 4.0 (159 : 40), a value falling within the average range.

Small size: Soricinae (54), *Erinaceus* (59), *Talpa* (76), Tenrecinae (100).

Large size: Echinosoricinae (207), *Potamogale* (159), *Sylvisorex* (148), *Hemiechinus* (147).

Characteristics and trends: No clear trends. Several groups are much lower than Tenrecinae, whose indices are otherwise the lowest for many structures. Large differences were found between Echinosoricinae and Erinaceinae (207 versus 102) and Crocidurinae and Soricinae (105 versus 54).

3.3.2 Mesencephalon (MES)

Topography, Delimitation, Structural Characteristics
(Atlas A 3 - P 2.5)

The OBL/MES complex contains the centers of most cranial nerves and is essential for basic life functions. The mesencephalon (MES) or midbrain can be divided into three levels. The basal portion is formed by the cerebral peduncle (crura cerebri) and the substantia nigra. This part corresponds to the ventral pons of the metencephalon and the pyramids and inferior olivary complex of the medulla oblongata (OBL). Its development parallels that of the neocortex. The tegmentum, the middle level of MES, is the rostral continuation of the dorsal pons. The rostral parts of the formatio reticularis reaching into MES are included in OBL in our measurements (see p. 53). The highest level, i.e., the roof of MES, is called the tectum (quadrigeminal plate).

The MES tegmentum contains the motor nuclei of the third and fourth cranial nerves, which innervate the ocular muscles, and the accessory oculomotor nucleus (Edinger-Westphal), which innervates the sphincter pupillae and ciliary muscles. The nucleus ruber is, like the substantia nigra, an important relay station of the extrapyramidal motor pathway. The nuclei lemnisci lateralis (NLL) are the only tegmental part which was measured in ten species. The two components of the tectum, the superior colliculi and the inferior colliculi, were measured separately in the same ten species as NLL. The colliculi are macroscopically visible between the cerebral hemispheres and the cerebellum when the tectum is exposed (see Section 3.1). The superior colliculi serve as a visual reflex center, while the inferior colliculi are relay stations for the auditory system.

In histological sections, a small periventricular structure, the subcommissural body (SCB), is conspicuous in the dorsorostral part of the mesencephalic aqueduct, where it is attached to the posterior (tectal) commissure (COC in Figs. 40 - 51). It is a modified ependyma, which in most species is smooth and recognizable only by a slight thickening, whereas in others it may be folded and/or better differentiated. It was found to be well developed in *Talpa europaea*, with a broad band of cells lying at different distances from the lumen of the ventricle and with cilia reaching into it (Fig. 61).

Size Comparison
(Tables 10 - 12; Fig. 62)

The mesencephalon (MES) is on the average the smallest of the five fundamental brain parts, accounting on the average for 6.4% of the total brain (see table on p. 52). The size indices may be described as follows:

Averages: 136 in the 50 species, 150 in the twelve subfamilies and families. This is the second lowest level (next to OBL) of the five fundamental brain parts.

Range: 88 - 266. The highest value *(Hylomys)* is three times larger than the lowest *(Tenrec)*; this is a medium variation.

Small size: Tenrecinae (100), terrestrial Soricinae (103), *Suncus* spp. (107).

Large size: Echinosoricinae (254), Desmaninae (201), Potamogalinae (183).

Characteristics and trends: MES is by far largest in Echinosoricinae, but very large also in Desmaninae and Potamogalinae. Its size seems to be influenced by water-adaptation, since in subfamilies with both terrestrial and water-adapted forms (Oryzorictinae, Soricinae), the water-adapted are the highest or at least among the highest. In *Oryzorictes* and some terrestrial Soricinae, the MES indices are even below the tenrecine level.

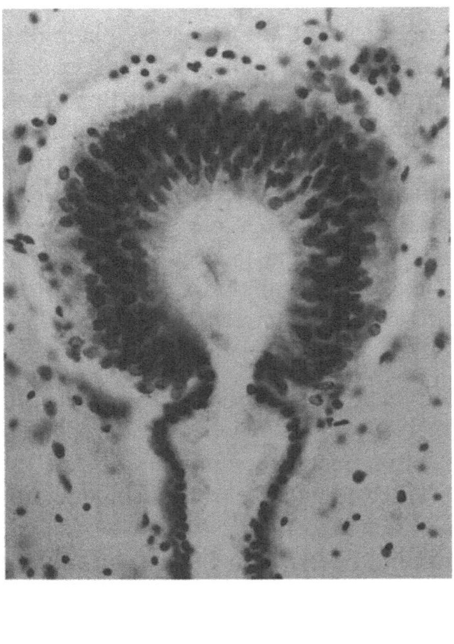

Figure 61: Transversal section through the subcommissural body in *Talpa europaea*. Cresyl violet stain. Enlargement 350 x.

Mesencephalon Components

Topography, Delimitation, Structural Characteristics, Fiber Connections
(Atlas A 3 - P 2.5)

The nuclei lemnisci lateralis (NLL) form two grey regions in the lemniscal fibers. The ventral nucleus of the lateral lemniscus, which fuses caudally with the OLS, is made up of medium-sized neurons similar to those in the lateral nucleus of the superior olive. The dorsal nucleus of the lateral lemniscus is composed of relatively large, deep-staining neurons intermingled with medium-sized cells arranged in clusters separated by fiber bundles of the lateral lemniscus. In the ventral region of this nucleus is a higher concentration of slightly smaller cells.

The colliculus inferior (INC) is not clearly laminated in Nissl preparations. The cells are small and medium-sized. Based on Nissl stain, no valid criteria have been recognized for further subdivision of INC. The INC is part of the auditory system. It receives fibers from most of the brain stem auditory nuclei, and all these fibers converge to form the lateral lemniscus. The most important fibers come from the contralateral complex of cochlear nuclei, the contralateral and ipsilateral lateral nucleus of the OLS, the ipsilateral medial nucleus of the OLS, the ipsilateral ventral nucleus of the NLL, and the ipsi- and contralateral dorsal nuclei of the NLL (Nieuwenhuys et al. 1978, 1980; Webster 1985).

The colliculus superior (SUC) shows a horizontally laminated organization in seven strata which, however, are not always clearly differentiated in Insectivora (e.g., short-tailed shrew; Huber and Crosby 1943). The predominant cells in all strata are small to medium-sized. The superficial layers, the stratum griseum superficiale and the stratum opticum, are innervated by retinal axons, and cells in these layers in turn project to deeper layers, where other sensory systems are represented. This organization seems to be found in all mammals (Stein 1981; Sefton and Dreher 1985). SUC seems to be involved in the spatial localization of visual stimuli but not in their identification (Schneider 1969). Furthermore, the SUC has been shown to be involved in directing eye movements (McHaffie and Stein 1982), turning behavior (DiChiara et al. 1982), and locomotor exploration (Dean et al. 1982).

Size Comparison
(Tables 28 - 30)

Measurements of mesencephalon components were limited to ten species only and were originally performed to find a reference base for comparison with Chiroptera (Baron 1972, 1974, 1977a). Therefore, only conservative "basal" forms were selected for these investigations, and these, of course, give no insight into the wide variation that really exists in Insectivora. This is obvious when the MES data from the selected ten species are compared with those given above for the total 50 species (Tables 10 - 12). The range of indices in the limited material is much lower (88 - 125) and, thus, the averages are also much lower (103 for the ten species and 102 the four subfamilies; see Table 74). Only species represented in the lower part of the scale in Fig. 62 are included in the following comparisons of MES components.

In the ten conservative species, MES is composed of about 60% tegmentum, including the cerebral peduncles, which are, however, very small in Insectivora, and 40% tectum. Of the components of the tegmentum (MTG), only the nuclei of the lateral lemniscus (NLL) were delimited. They represent 6.7% of MTG and 4% of MES. In the tectum (MTC), superior and inferior colliculi are on the average of equal size (50% relative to MTC, 20% relative to MES).

In the two large mesencephalic components, MTG produce higher average allometric indices (108/110) than MTC (97/94) (Table 30). The largest MTG size was found for *Crocidura* spp. (122), the smallest for Tenrecinae (100) and Soricinae (107). The only MTG component measured, i.e., the nucleus lemnisci lateralis (NLL), is small (54) in *Erinaceus*. NLL reach its largest size in *Crocidura* spp. (128). In the tectum (MTC), the colliculi show very different trends. A relatively large superior colliculus (SUC) was found in the shrews, both Soricinae (134) and Crocidurinae (142), whereas the inferior colliculus (INC) is small in the two shrew subfamilies, but only half the size in Soricinae (47) as in Crocidurinae (90). In *Erinaceus*, SUC is larger (115) than in Tenrecinae, whereas INC is smaller (67). Tenrecinae have a relatively large INC compared with other conservative Insectivora.

The very small subcommissural body was measured in 23 species sufficiently to gain fairly good insight into its variation in Insectivora, and thus the data will be presented in more detail.

Subcommissural body (SCB)

The size indices for this very small periventricular structure may be described as follows (Tables 40 - 42):

Averages: 203 in the 23 species, 260 in the ten subfamilies and families. Compared with the OBL structures, in which a comparable number of species was investigated, this is very high.

Range: 23 - 857. This is very high variation, with the highest value *(Scalopus)* more than 37 times larger than the lowest *(Potamogale)*.

Small size: *Potamogale* (23), Soricinae (48), Oryzorictinae (64).

Large size: Talpinae (630), *Solenodon* (553), *Chrysochloris asiatica* (427), Desmaninae (349).

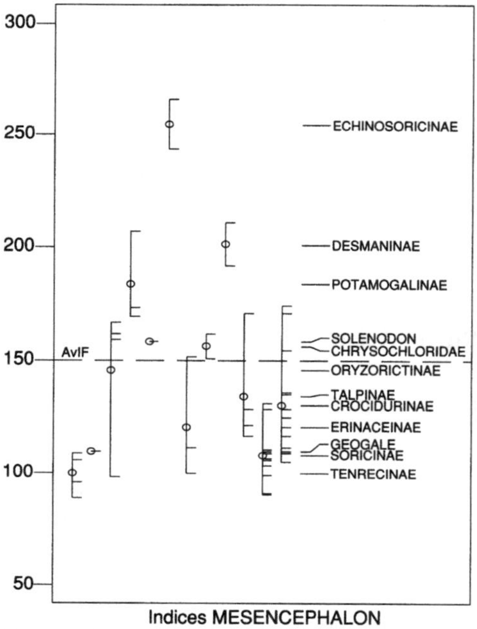

Figure 62

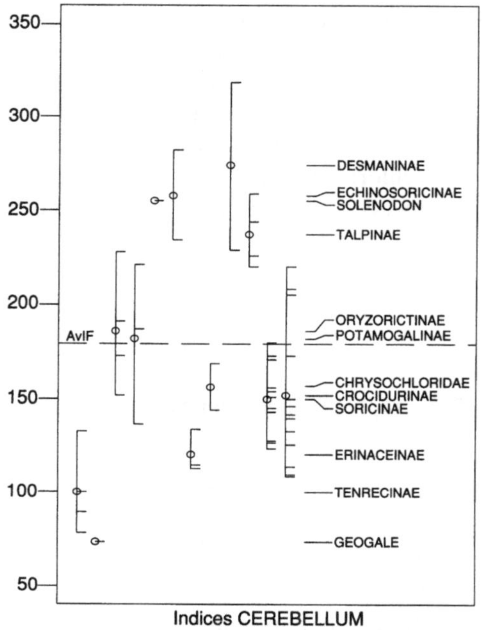

Figure 63

Figures 62 - 64: Size indices for fundamental brain parts

Figure 62: Mesencephalon (MES)
Figure 63: Cerebellum (CER)
Figure 64: Diencephalon (DIE)

The indices are based on AvTn lines (index = 100) whose equations are, respectively,

log MES vol = 0.481 + 0.58 x log BoW (1.50)

log CER vol = 0.552 + 0.68 x log BoW (1.79)

log DIE vol = 0.350 + 0.65 x log BoW (1.80)

The distances of the AvIF broken lines are given in brackets. For additional explanations, see Fig. 53.

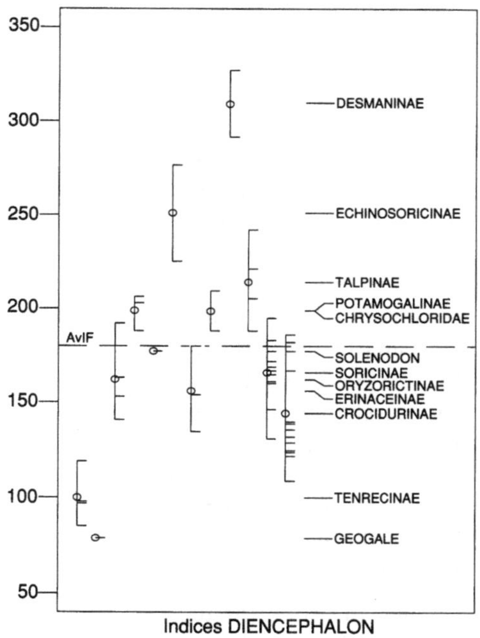

Figure 64

Characteristics and trends: No clear trends. Among Tenrecidae, SCB is largest in Tenrecinae and smallest in Potamogalinae *(Potamogale)*. The small size in *Potamogale* obviously does not, however, depend on water adaptation, since SCB is large in the semiaquatic *Desmana* and *Galemys*. The largest size and structural differentiation was found in the subterranean forms (Chrysochloridae, Talpinae; see Fig. 61, *Talpa*) and in *Solenodon*. There was, however, high variability in subfamilies, e.g., in Crocidurinae (40 - 144) and even genera, e.g., in *Crocidura* (40 and 100).

3.3.3 Cerebellum (CER)

Topography, Delimitation, Structural Characteristics, Fiber Connections
(Atlas P 1 - P 8)

Included in the cerebellar complex for volume measurement were the cerebellum with the central cerebellar nuclei as well as the brachium and nuclei pontis (ventral pons, VPO). The structures of the dorsal pons were included in the OBL volumes.

The cerebellum can be expected to serve as a kind of computer for spatial and temporal coordination of posture and movements, for muscular tonus and body balance. It receives afferent fibers from virtually all sensory systems, particularly from the somatosensory and vestibular, but also from the auditory and visual systems (Eccles et al. 1967). The pathways through which CER receives afferent fibers include the tractus spinocerebellaris, cuneocerebellaris, vestibulocerebellaris, and olivocerebellaris. The cerebral cortex projects to CER via the pontine nuclei and through pontocerebellar tracts. The various cerebellar pathways are connected with different parts, called the archi-, paleo-, and neocerebellum. The flocculonodular lobe and the nucleus medialis are considered parts of the archicerebellum. Because of its direct afferents from the vestibular organ and the vestibular nuclei, the archicerebellum is also referred to as vestibulocerebellum. The paleocerebellum is formed of the anterior lobe and parts of the vermis, including the pyramis and uvula (VIII and IX in Figs. 40 - 51). This structure receives fibers from the anterior and posterior spinocerebellar tracts and from the cuneocerebellar tract. For this reason, the paleocerebellum is also called spinocerebellum (Brodal 1981). The neocerebellum develops between the paleo- and archicerebellum, the phylogenetically older parts of the cerebellum. It receives fibers originating in the neocortex and passing through the pons. The olivocerebellar tract is the only pathway projecting to the entire cerebellar cortex.

Since the study of the cerebellar cortico-nuclear projections by Jansen and Brodal (1940), it has generally been accepted that the cerebellum can be divided into three longitidinal zones or components: the vermal zone, projecting to the nucleus medialis (MCN), the paravermal or intermediate zone, projecting to the nucleus interpositus (ICN), and the lateral zone, comprising the hemispheres and projecting to the nucleus lateralis (LCN) (Kuhlenbeck 1975; Eccles 1982). The fourth component, the flocculonodular lobe, projects directly to the vestibular nuclei. Direct cerebellovestibular fibers also come from the anterior lobe. The lateral zone or hemisphere-lateralis component is involved in the enormously complex organization and control of movement, whereas the medial zone is primarily related to the spinal cord functions of muscle tone and posture, and to automatic movements (Eccles 1982). The flocculonodular lobe (vestibulocerebellum) is concerned with regulation of the spatial orientation of the body and its parts.

Significance tests (T-tests) for differences in cerebellum and neocortex between the various subfamilies and families of Insectivora

-- **Cerebellum** (above diagonal)

	N	Ten	Geo	Ory	Pot	Sol	Ech	Eri	Chr	Des	Tal	Sor	Cro
Tenrecinae	4		ns	**	*	**	**	ns	*	**	**	**	*
Geogalinae	1	*		ns	ns		ns	ns	ns	ns	**	**	ns
Oryzorictinae	4	**	ns		ns	ns	ns	*	ns	ns	*	*	ns
Potamogalinae	3	**	ns	*		ns	ns	ns	ns	ns	ns	ns	ns
Solenodontidae	1	**		ns	ns		ns	**	ns	ns	ns	**	*
Echinosoricinae	2	**	*	*	ns	*		**	*	ns	ns	**	**
Erinaceinae	3	**	ns	ns	ns	ns	**		ns	*	**	ns	ns
Chrysochloridae	2	**	ns	ns	ns	ns	ns	ns		ns	**	ns	ns
Desmaninae	2	**	ns	**	ns	ns	ns	**	ns		ns	**	**
Talpinae	4	**	*	*	ns	ns	ns	*	ns	*		**	**
Soricinae	11	**	**	ns	ns	ns	*	ns	ns	**	*		ns
Crocidurinae	13	**	**	ns	**	ns	**	ns	**	**	**	**	

Neocortex (below diagonal) --

N number of species
ns not significant
* P < .05; significant at the 5% level
** P < .01; significant at the 1% level

Significant differences in the 65 combinations of the twelve families or subfamilies were found for the cerebellum in 19 combinations at the 1% level and in nine cases at the 5% level. No significance could be found in 37 cases. Failure to find significant differences may indicate: (1) they were really too small, or (2) the number of species was too small to reach significance, even when there were large size differences. Tenrecinae had significantly smaller CER than all other groups except Erinaceinae and *Geogale*.

We were not able to separate the three zones in the cerebellar cortex and the flocculonodular lobe adequately in all species to allow for measurement. However, some measurements on the cerebellar nuclei (Tables 31 - 33) may reflect the relative size of the three longitudinal zones. Such measurements have been made on ten conservative species for comparison with Chiroptera (Baron 1978) and Primates (Matano et al. 1985b). In these forms, CER is composed of about 1% ventral pons (N = 2) and 3% cerebellar nuclei (N = 10). The remaining 96% is cerebellar cortex (incl. the fibers of the white substance). Thus, the size relations for the total cerebellar cortex may be similar to those of CER as a whole.

Size Comparison
(Tables 10 - 12; Fig. 63)

In Insectivora, CER averages 13.3% of the total brain (see table on p. 52). The size indices may be described as follows:

Averages: 167 in the 50 species, 179 in the twelve subfamilies and families. These are the second highest averages (next to DIE) reached by any of the five fundamental brain parts.

Range: 78 - 319. The highest value (*Galemys*) is 4.1 times larger than the lowest (*Echinops*). This is the highest variation found in any of the five brain parts. It is even higher (4.4) when *Geogale* is included; the *Geogale* value (73), however, was derived from poor material.

Small size: *Geogale* (73), Tenrecinae (100), *Suncus* spp. (109), Erinaceinae (121).

Large size: Desmaninae (274), Echinosoricinae (258), *Solenodon* (255), Talpinae (237).

Characteristics and trends: Of all the fundamental brain parts compared so far, CER reflects the progressive trends in Insectivora most clearly. The most conservative states were found in *Geogale* and Tenrecinae, but low levels also in Erinaceinae and most Soricidae. The most derived or progressive states were found in Desmaninae, *Solenodon*, and Echinosoricinae. Since averages and range of variation are also high, CER must be regarded as one of the most progressive structures in Insectivora.

Cerebellar Nuclei

Topography, Delimitation, Structural Characteristics, Fiber Connections
(Atlas P 3.5 - P 5)

The cerebellar nuclei (TCN) are embedded in the ventral white matter of CER. The nucleus medialis cerebelli (MCN) is situated near the median plane. Ventrally, it lines the roof of the fourth ventricle, and medially, it merges with the nucleus interpositus (ICN) over most of its length. The predominant cells are somewhat loosely arranged medium-sized neurons. The nucleus interpositus (ICN) is composed of cells similar to those in the medial nucleus. An anterior and posterior nucleus, often described in mammals, could not be distinguished with certainty. The ventral side of the rostral part of the ICN merges with superior and lateral vestibular nuclei. The cells of the nucleus lateralis cerebelli (LCN) are also loosely arranged, but seem to be a little larger than the cells of the other nuclei. Again, the nucleus is not sharply delineated from the adjoining ICN. The three cerebellar nuclei were delineated mainly on topographical grounds.

The nucleus medialis (MCN) sends fibers via the fasciculus uncinatus to the vestibular complex, the reticular formation, and the ventral nuclei of the thalamus. Other fibers go to the spinal cord. Fibers from the ICN end mainly in the magnocellular part, and fibers from the LCN, mainly in the parvocellular part of the nucleus ruber. Both also send fibers to the ventral nuclei of the thalamus, which in turn sends fibers to the cerebral cortex.

Size Comparison
(Tables 31 - 33)

Measurements on cerebellar components were limited to ten conservative species. Thus, only species represented in the lower part of Fig. 63 are included in the following comparisons. In the limited material, CER indices range from 78 - 146 (instead of 73 - 319) and the averages are much lower than those described above for the more complete material: 113 for the ten species and 115 for the four subfamilies (instead of 167 and 179). Thus, they are not representative of the Insectivora on the average, and similar differences may account for the measured nuclei as well.

In the percentage composition of the cerebellar nuclei (TCN), the interposed (ICN) is on the average the largest (about 40% relative to TCN and 1.2% relative to CER), followed by the lateral (LCN; 35% and 1.0%, respectively) and medial nuclei (MCN; 25% and 0.7%, respectively).

The allometric indices for the cerebellar nuclei (TCN) are slightly larger in Erinaceinae (124) and Soricinae (119) than in Crocidurinae (103) and Tenrecinae (100). This is mainly because of the medial nucleus, whose indices are 155 in Soricinae and 147 in Erinaceinae. The interposed nucleus is small in all shrews, both Soricinae (87) and Crocidurinae (83), whereas the lateral nucleus is relatively large (140 in Soricinae, 131 in Crocidurinae).

In comparison with primates (see Section 3.3.6), clear trends towards enlargement are present in the cerebellar nuclei from Insectivora to primates and from prosimians to man. The strongest progression, reaching almost CER values, and in non-human simians even surpassing them, was found for LCN, the lowest for ICN. Similar trends are already present to a small extent in Insectivora, and it can be deduced that in Insectivora, too, the lateral longitudinal zone (cerebellar hemispheres) may be the most progressive.

Deductions about trends in the flocculonodular lobe (vestibulocerebellum) may be obtained from the size of the vestibular nuclei, which form part of OBL (Tables 22 - 24).

3.3.4 Diencephalon (DIE)

Topography, Delimitation, Structural Characteristics
(Atlas A 8 - A 2)

The diencephalon (DIE) is a highly complex brain part which is usually divided into four horizontal zones. From dorsal to ventral, they are: epithalamus (ETH), thalamus (THA), subthalamus (STH), and hypothalamus (HTH). These zones, too, are more or less complex. Their characteristics will be discussed under their respective headings.

Size Comparison
(Tables 10 - 12; Fig. 64)

In Insectivora, the diencephalon (DIE) on the average accounts for 7.4% of the total brain (see table on p. 52). Next to the mesencephalon, this is the second smallest of the five fundamental brain parts. The size indices may be described as follows:

Averages: 169 for the 50 species, 180 for the twelve subfamilies and families. This is the highest level reached in the five fundamental brain parts.

Range: 85 - 327. The highest value (*Galemys*) is 3.8 times larger than the lowest (*Tenrec*). When *Geogale* (79) is included, even though the value was based on poor material, this range (4.1) becomes one of the largest for the five brain parts.

Small size: *Geogale* (79), Tenrecinae (100), *Suncus* spp. (120), *Crocidura* spp. (132).

Large size: Desmaninae (310), Echinosoricinae (251).

Characteristics and trends: Clear general trends similar to those found in CER. DIE size seems to be influenced by water adaptation.

Diencephalon Components

Topography, Delimitation, Structural Characteristics
(Atlas A 8 - A 2)

The diencephalon has been subdivided in two quantitative investigations, one by the authors and the other by Bauchot (1963, 1979b), as follows:

After the four main components (ETH, THA, STH, HTH), the authors subdivided two fiber components, the optic tract (TRO) and the internal capsule (CAI) to the site at which the CAI passes through the diencephalon. In STH, pallidum (PALL = globus pallidus) and corpus luysi (LUY) were separated. For technical reasons (i.e., difficulties in accurate delimitation), the zona incerta and the two fields of Forel, both parts of STH, were incorporated with HTH.

Bauchot (1963, 1979b) gave a comparison of diencephalic components in 20 species. His subdivision is based on somewhat different criteria than those described in the preceding paragraph. In addition to the four main components, he separated two smaller parts, the nucleus reticularis (RTH) and the pretectal region (PTE), both of which were included in THA in our measurements. Bauchot's THA subdivision is considerably more detailed than ours (seven components). In STH, he included the zona incerta, which in our data is measured with the HTH. Finally, Bauchot did not measure the fiber structures separately.

Size Comparison
(Tables 34 - 36, 65 - 67)

Percentages: The diencephalon of an Insectivora (= 100%) is composed on the average of

	ETH	THA	STH	HTH	others
Authors	2.0	47.9	6.8	35.8	7.5
Bauchot	1.8	40.7	28.2	24.2	5.1

The authors investigated ten selected conservative species only, while Bauchot (1979b) also included progressive species. This may explain some of the differences, but they may be due mainly to differences in the criteria used in delineating the components. Thus, the parts included by Bauchot in STH are much more extensive, and those included in HTH, more limited.

Indices: Based on our selected material, ranges and averages of the allometric indices may be representative for conservative species only. This becomes obvious when the DIE indices of the limited material (N = 10) are compared with those of the larger material given above (N = 50). In the limited material, the range is from 85 - 146 (instead of 79 - 327, as given above) and the averages are 117 for the ten species and 124 for the four subfamilies (instead of 169 and 180 for the larger number of species and groups). Similar restrictions are certainly valid for the DIE components.

Of the six main components listed in Tables 34 - 36, the highest indices were found for the subthalamus (STH) in *Sorex* spp. (196) and the optic tract (TRO) in the hedgehog (172). The lowest values for the main components were also found in TRO, i.e., 80 in shrews, with Crocidurinae still lower than Soricinae (69 versus 98). The high position of the hedgehog *(Erinaceus)* and the low position of the shrews is confirmed by area measurements of the sagittally sectioned chiasma opticum (Tables 71 - 73). In those tables markedly higher values than in Erinaceinae (av. index 188) were found in Echinosoricinae (341), markedly lower indices than in shrews (av. index 53), in the subterranean Talpinae (15) and Chrysochloridae (19) and in some mainly fossorial forms, e.g., *Anourosorex* (17) and *Oryzorictes* (27).

Values on the level of Tenrecinae were found for the hypothalamus of Crocidurinae (101 in Table 36). Relatively high values in shrews were found for Soricinae in the subthalamus (196) and thalamus (151) and, in Crocidurinae, for the thalamus (138) and capsula interna fibers (127). In hedgehogs, the highest values after TRO were found for the thalamus (156).

In the average indices for the six main components of DIE, the highest progression is reached by the thalamus (127 for the ten species, 136 for the four subfamilies). In descending order follow: subthalamus (127/132), capsula interna fibers (113/115), epithalamus (111/111), hypothalamus (106/110), and optic tract (97/110).

In the indices based on Bauchot's data (Tables 65 - 67), the highest progression is reached by the subthalamus (218 for the 20 species, 244 for the ten subfamilies and families). In descending sequence follow: thalamus (160/177), nucleus reticularis (156/172), hypothalamus (136/144), epithalamus (130/140), and pretectal region (131/138). In both studies (Bauchot's and here), thalamus and subthalamus are more progressive than epithalamus and hypothalamus. This was similarly described for the progression in primates (see Section 3.3.6).

Epithalamus (ETH)

Topography, Delimitation, Structural Characteristics, Fiber Connections
(Atlas A 4 - A 2)

The epithalamus (ETH) is composed of structures located on and around the roof of the third ventricle. These structures include the anterior choroid plexus, the habenular nuclei, the stria medullaris (as far as it runs in epithalamic levels), and the pineal body (epiphysis, EPI). The choroid plexus is not included in ETH volumes; instead, it was added to the meninges, etc. in the complex of "rests" (Tables 8 and 9). The habenulae of the two sides are interconnected by the habenular commissure (COH in Figs. 40 - 51). The nucleus medialis habenulae (HAM) has an ovoid shape in transversal section. At its dorsal surface is a thin lining of ependymal cells, that separate it from the ventricular cavity. The dark staining small neurons making up this nucleus are densely packed. The moderately staining cells of the nucleus lateralis habenulae are less densely packed.

The habenula seems to receive afferents via the stria medullaris from the septum and lateral hypothalamus (Herkenham and Nauta 1977). Efferents from the habenula have been traced chiefly in the caudal direction and end largely in the interpeduncular nucleus (Brodal 1981). The functional role of the stria medullaris-habenula system is not clear. Judging from its fiber connections, however, it must serve to connect structures in the anterior brain stem with the hypothalamus and limbic structures (Brodal 1981).

Size Comparison
(Tables 34 - 36, 65 - 67)

Both in Bauchot's and our own measurements, ETH was found to have a low progression. The indices for the 20 species converted from Bauchot's data vary from 82 in *Tenrec* to 210 in *Galemys*. Low values were found in *Potamogale* (96), Tenrecinae (100), and Crocidurinae (105), high values in *Chrysochloris* (205), Desmaninae (185), and *Solenodon* (168).

Of the ETH components, EPI and HAM were measured separately in 38 species. These species cover eleven of the twelve subfamilies or families (except Geogalinae) and thus may provide insight into the variation in Insectivora.

Nucleus habenularis medialis (HAM)

Its size indices may be described as follows (Tables 40 - 42):

Averages: 153 for the 38 species, 162 for the eleven subfamilies and families.

Range: 77 - 245. The highest value *(Hylomys)* is 3.2 times larger than the lowest *(Setifer)*. Compared with the oblongata structures, for which a comparable number of species was investigated, this is low variation.

Small size: *Atelerix* (89), Tenrecinae (100).

Large size: Echinosoricinae (229), *Solenodon* (209), Chrysochloridae (198).

Characteristics and trends: No clear trends.

Epiphysis (EPI)

Its size indices may be described as follows (Tables 40 - 42):

Averages: 50 in the 38 species, 62 in the eleven subfamilies and families.

Range: 0 - 203. The variation is very high since in several species (*Echinosorex*, *Potamogale*) there is little or no pineal gland. High variation is also found in subfamilies (Echinosoricinae 0 and 130; Crocidurinae 4 - 57; Tenrecinae 66 - 129) and even genera (*Chrysochloris*, 18 and 77).

Small size: *Echinosorex* (0), *Potamogale* (0.5), *Myosorex* (4), *Micropotamogale ruwenzorii* (6).

Large size: *Solenodon* (203), *Hylomys* (130), Tenrecinae (100).

Characteristics and trends: Compared with Tenrecinae, the pineal gland is more or less reduced in the other groups (except *Solenodon* and *Hylomys*). There is, however, very large intragroup variability.

Thalamus (THA)

Topography, Delimitation, Structural Characteristics
(Atlas A 6 - A 2)

The thalamus (THA) lies between the third ventricle medially and the posterior limb of the internal capsule laterally. In the interthalamic adhesion (massa intermedia), the thalami of the two sides are connected. THA is subdivided by a more or less conspicuous lamina of myelinated fibers, the internal medullary lamina, into anterior, medial, and lateral nuclear groups. The intralaminar nuclear group is intercalated in the medullary lamina. In the midline, this group of nuclei is continuous with the midline nuclear group. The lateral border of THA is formed by the reticular nucleus. Finally, the caudolateral region is formed by the so-called metathalamus, made up of the medial and lateral geniculate bodies. Of the many thalamic nuclei, only those which were measured will be described.

THA and the cerebral cortex are interconnected by reciprocal fiber connections. All sensory information to the neocortex, except olfactory, passes through THA. Thus, quantitative analysis of thalamic nuclei or nuclear groups may indicate the relative size of these areas (just as the relative size of the various longitudinal zones of the cerebellar cortex may be deduced from the size of the cerebellar and vestibular nuclei; see pp. 85, 86). This is important, since the relative size of neocortical areas cannot be determined by our methods for all Insectivora and mammalian lines because of difficulties in homologization and, thus, in accurate delimitation for reliable measurement. Therefore, Bauchot's dorsal nuclear group (DOT), which corresponds mainly to the lateroposterior (pulvinar) and laterodorsal thalamic nuclei (see Stephan et al. 1980), is related to the association cortex of the parietal, temporal, and occipital lobes and may reflect the degree of progression in these neocortical areas. The ventral complex (VET), which corresponds to the ventral nuclei of the lateral group (see Stephan et al. 1980), contains (1) the specific motor relay nucleus, which is interposed between the primary motor cortex and other motor centers, such as cerebellum, pallidum, and the substantia nigra (Mehler 1971; Rinvik 1975; Carpenter et al. 1976; Strick 1976; Künzle 1976; Catsman-Berrevoets and Kuypers 1978; Faull and Mehler 1985) and (2) the specific somatosensory relay nuclei, which receive input from the entire body, including the mysticial vibrissae, and project in a topographical manner to the somatosensory cortices (Faull and Mehler 1985; Tracey 1985). The medial nuclei (MLT), with the medio-dorsal nucleus as its main component (Stephan et al. 1980), receive visceral input from the preoptic and tuberal regions of HTH and project to the neocortical frontal lobe (Kahle 1976; Faull and Mehler 1985). Through this circuit, the affective component of behavior, which is mainly determined by unconscious visceral and somatic stimuli, may become conscious (Kahle 1976), a function which seems to be of increased importance in humans. The anterior nuclei (ANT) have fiber connections which are related to HTH and limbic structures and also project to the prefrontal cortex (Nieuwenhuys et al. 1980). The corpora geniculata mediale (CGM) and laterale (CGL) receive auditory and visual input, respectively, and project to corresponding cortical areas (Sefton and Dreher 1985; Webster 1985). The medial geniculate body, which is more or less round in transverse section, is located at the lateral border of the junction of the mesencephalon and the diencephalon. Its cells are of medium size and ovoid to polygonal in shape. CGM receives afferents mainly from the ipsilateral inferior colliculus and only a few from the contralateral side.

Size Comparison
(Tables 34 - 39, 65 - 70)

Both in our own and Bauchot's measurements, the thalamus was found to have a high progression. In the thalamus of the ten conservative species, medial (CGM) and lateral (CGL) geniculate bodies were measured and, in the lateral geniculate body, dorsal and ventral parts (GLD and GLV) as well. All the species belong to the conservative forms and thus the material investigated provides no insight into the full variation in Insectivora.

Compared with Tenrecinae, the lateral geniculate body (CGL) is larger in hedgehogs and smaller in shrews. The differences are larger in the dorsal part (GLD), reaching from 62 in *Crocidura russula* to 245 in *Erinaceus* (Table 38). This is a relatively broad variation.

Bauchot delineated and measured seven components (nuclear groups) of THA. Their average percentages relative to THA (100%) are 6.2% for anterior (ANT), 12.4% for medial (MLT), 13.2% for median (MNT), 11.6% for dorsal (DOT), 29.0% for ventral (VET) nuclear groups, 23.0% for medial geniculate bodies (CGM), and 4.6% for lateral geniculate bodies (CGL). Apparently, Bauchot included more structures in CGM than we did in our subdivision. According to our measurements, CGM is on the average only 11.4% of THA.

Bauchot's measurements (1979b) revealed clear differences in the size of the various nuclear groups of THA (Tables 68 - 70).

Nuclei anterior thalami (ANT)

Averages: 168 for the 20 species, 187 for the ten subfamilies and families. This is a relatively low level (fifth) among Bauchot's seven THA nuclear groups.

Range: 66 - 330. The highest value *(Galemys)* is 5.0 times larger than the lowest *(Tenrec)*. This is medium variation for THA components.

Small size: *Suncus* (97), Tenrecinae (100), Crocidurinae (118).

Large size: *Galemys* (330), *Microgale* (324).

Characteristics and trends: Slight trends toward enlargement, with a very high index in *Microgale*.

Nuclei mediales thalami (MLT)

Averages: 195 for the 20 species, 220 for the ten subfamilies and families. This is a medium level (third) among Bauchot's seven THA nuclear groups.

Range: 84 - 466. The highest value *(Galemys)* is 5.5 times larger than the lowest *(Tenrec)*. This is medium variation for THA components.

Small size: Tenrecinae (100), Crocidurinae (131), Soricinae (150).

Large size: Desmaninae (420), *Potamogale* (304).

Characteristics and trends: Clear tendency toward enlargement from conservative to advanced forms.

Nuclei mediani thalami (MNT)

Averages: 191 for the 20 species, 219 for the ten subfamilies and families. This is a medium level (fourth) among Bauchot's seven THA nuclear groups.

Range: 77 - 451. The highest value *(Galemys)* is 5.9 times larger than the lowest *(Tenrec)*. This is a broad variation for THA components.

Small size: *Suncus* (95), Tenrecinae (100), Crocidurinae (118).

Large size: Desmaninae (392), *Chrysochloris* (318), *Solenodon* (289), *Talpa* (270).

Characteristics and trends: Tendency toward enlargement, with relatively high indices in *Solenodon* and subterranean forms.

Nuclei dorsales thalami (DOT; Fig. 65)

Averages: 288 for the 20 species, 337 for the ten subfamilies and families. This is the highest level (first) among Bauchot's seven THA nuclear groups.

Range: 85 - 899. The highest value *(Galemys)* is 10.6 times larger than the lowest *(Tenrec)*. This is large variation for THA components.

Small size: Tenrecinae (100).

Large size: Desmaninae (749), *Potamogale* (524), *Chrysochloris* (421).

Characteristics and trends: Clear tendency toward enlargement. Large size in species with semiaquatic and subterranean adaptations, in most species even exceeding NEO enlargement.

Nuclei ventrales thalami (VET; Fig. 66)

Averages: 105 for the 20 species, 110 for the ten subfamilies and families. This is a low level (sixth) among Bauchot's seven THA nuclear groups.

Range: 86 - 146. The highest value *(Solenodon)* is only 1.7 times larger than the lowest *(Neomys)*. This is the lowest variation in THA components.

Small size: Crocidurinae (92), Soricinae (94), Tenrecinae (100).

Large size: *Solenodon* (146), *Potamogale* (122), *Atelerix* (121).

Characteristics and trends: Nearly no variation, i.e., progression. Relatively small in all Soricidae. No clear tendency towards enlargement or reduction.

Corpus geniculatum mediale (CGM; Fig. 67)

Averages: 257 for the 20 species, 285 for the ten subfamilies and families. This is a high level (second) among Bauchot's seven THA nuclear groups.

Range: 77 - 564. The highest value *(Galemys)* is 7.3 times larger than the lowest *(Tenrec)*. This is high variation for THA components.

Small size: Tenrecinae (100), Erinaceinae (103).

Large size: Desmaninae (515), *Potamogale* (497), *Limnogale* (476) *Neomys* (473).

Characteristics and trends: Clear relations to water adaptation: largest in all five semiaquatic species.

Corpus geniculatum laterale (CGL; Fig. 68)

Averages: 98 for the 20 species, 96 for the ten subfamilies and families. This is the lowest level (seventh) among Bauchot's seven THA nuclear groups.

Range: 57 - 148. The highest value *(Erinaceus)* is 2.6 times larger than the lowest *(Talpa)*. This is low variation for THA components.

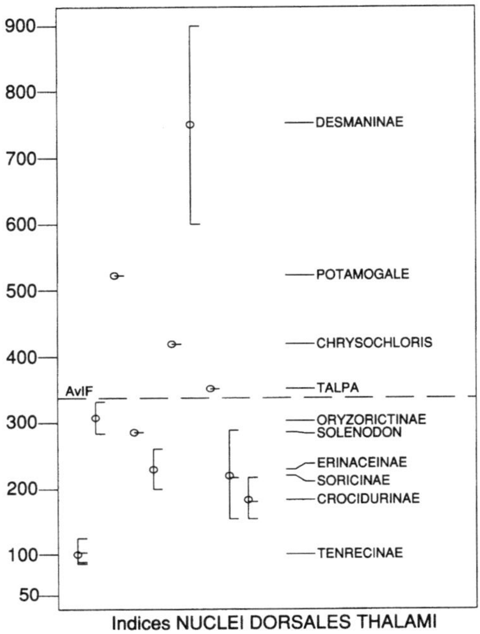

Figure 65

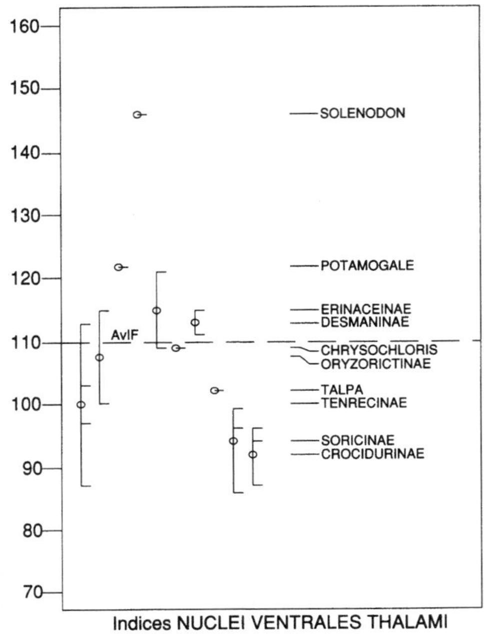

Figure 66

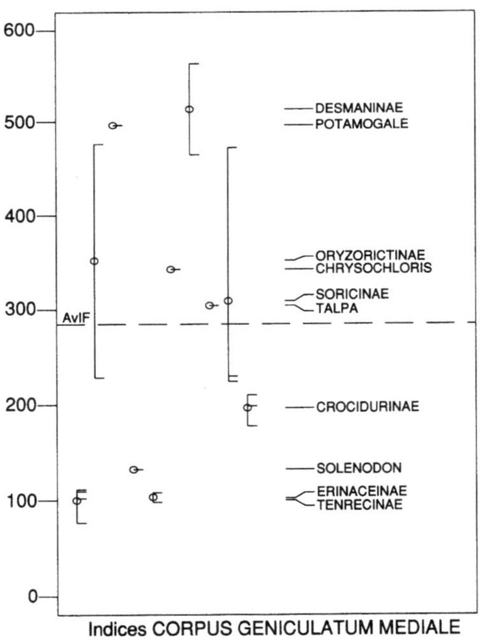

Figure 67

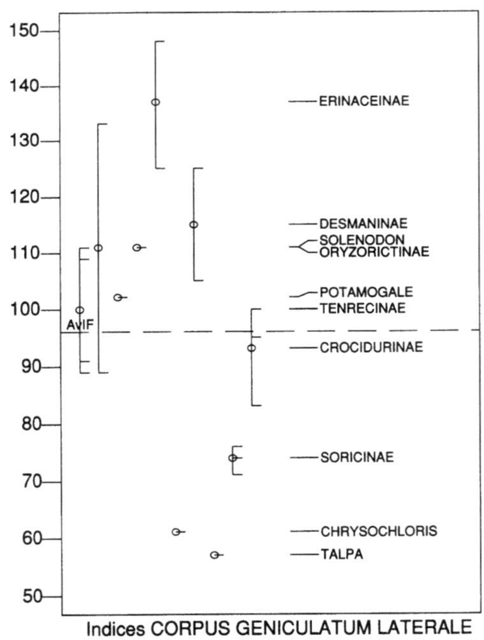

Figure 68

Small size: *Talpa* (57), *Chrysochloris* (61), Soricinae (74).
Large size: Erinaceinae (137), *Microgale talazaci* (133), *Desmana* (125).
Characteristics and trends: Smallest size in subterranean species. Smaller in Soricinae than in Crocidurinae and in Insectivora on the average than in Tenrecinae. Largest size in Erinaceinae, for which also the largest eyes were found. Strong interrelationships with other visual structures.

Subthalamus (STH)

Topography, Delimitation, Structural Characteristics, Fiber Connections
(Atlas A 7 - A 3)

The subthalamus (STH) is composed, in our measurements, of the nucleus subthalamicus Luysi (LUY) and the globus pallidus (PALL), including the nucleus entopeduncularis, which is considered to be homologous to the pars interna of the pallidum in Primates. PALL lies medial to the putamen. In contrast to the striatum, the majority of the loosely grouped neurons are relatively large. The nucleus entopeduncularis is composed of loosely grouped large and medium-sized cells interspersed among fibers of the pedunculus cerebri. LUY lies between the zona incerta, dorsally, and the pedunculus cerebri, ventrally. It contains a homogeneous collection of densely-packed, medium-sized, dark-staining neurons.

Figures 65 - 68: Size indices for diencephalic components based on volume data of Bauchot (1979b).

Figure 65: Nuclei dorsales thalami (DOT)
Figure 66: Nuclei ventrales thalami (VET)
Figure 67: Corpus geniculatum mediale (CGM)
Figure 68: Corpus geniculatum laterale (CGL)

The indices are based on AvTn lines (index = 100) whose equations are, respectively,
$\log \text{DOT volume} = -1.281 + 0.68 \times \log \text{BoW}$ (3.37)
$\log \text{VET volume} = -0.377 + 0.64 \times \log \text{BoW}$ (1.10)
$\log \text{CGM volume} = -0.996 + 0.72 \times \log \text{BoW}$ (2.85)
$\log \text{CGL volume} = -1.046 + 0.60 \times \log \text{BoW}$ (0.96)

The distances of the AvIF broken lines are given in brackets. For additional explanations, see Fig. 53.

The main afferents to PALL are from the striatum. The information received by the striatum from the cerebral cortex, thalamus, substantia nigra and amygdaloid complex is funneled through PALL. In fact, PALL and the substantia nigra are the only well documented targets of striatal efferent projections (Brodal 1981). The quantitatively most important efferent system is the pallidothalamic projection. PALL is therefore an important link in the neocortex - striatum - globus pallidus - thalamus - neocortex circuit. In addition, PALL participates in secondary circuits involving the striatum and the thalamus, on the one hand, and LUY, on the other. Finally, through the pallidohabenular projection, PALL is linked with the limbic system (Nieuwenhuys et al. 1978, 1980; Brodal 1981; Heimer et al. 1985). LUY receives, in addition to pallidal input, afferents originating in the motor and premotor areas of the cerebral cortex, in the thalamus, and in various monoaminergic cell groups such as the locus coeruleus (for ref. see Heimer et al. 1985). The two major projection targets are PALL and the substantia nigra (Nauta and Cole 1978; Ricardo 1980).

Size Comparison
(Tables 34 - 39, 65 - 67)

The subthalamus was found to have a high progression both in our own and Bauchot's measurements. In the STH of ten conservative species, PALL and LUY were measured. Because of problems in accurate delimitation, the zona incerta and the two fields of Forel, which are part of STH, were incorporated with HTH.

Compared with Tenrecinae, both parts of STH are largest in Soricinae, and smaller, but still distinctly progressive in Erinaceinae and Crocidurinae. In all the groups, LUY is more progressive than PALL.

The indices for STH of the 20 species converted from Bauchot's data vary from 87 in *Tenrec* to 528 in *Galemys*. Relatively low values were found in *Suncus* (124; and generally in Crocidurinae, 151), high values in the Desmaninae (494), *Chrysochloris* (336), and *Talpa* (334). Thus, next to Desmaninae the subterranean forms have a relatively large STH.

Hypothalamus (HTH)

Topography, Delimitation, Structural Characteristics, Fiber Connections
(Atlas A 8 - A 3)

The hypothalamus (HTH) forms the base of the diencephalon, bordering rostrally on the lamina terminalis and merging caudally with the periventricular and tegmental grey of the mesencephalon at the level of the caudal pole of the mamillary bodies. The preoptic area lies lateral to the rostral part of the third ventricle. Although, according to Nieuwenhuys et al. (1978, 1980) the preoptic area is of telencephalic origin, it is closely related to the hypothalamus structurally and, thus, was included here for measurement. At the level of the tuber cinereum, the floor of the third ventricle evaginates ventrally, caudal to the optic chiasma, to form the infundibular portion of the neurohypophysis. The mamillary bodies are the most caudal parts of HTH.

Cytoarchitecturally, HTH is characterized by heterogeneity of cellular size, shape, and staining properties in most nuclei and regions. This suggests that the parts participate in more than one functional system (Bleier and Byne 1985). Over its whole rostrocaudal extent, HTH may be divided into three zones (Crosby and Woodburne 1940): periventricular, medial and lateral. The periventricular zone is composed of small cells often arranged in layers parallel to the ventricle border. The medial cell-dense zone contains several distinct nuclei. Through the less cell-dense lateral zone pass fiber bundles between telencephalon and mesencephalon.

HTH possesses elaborate reciprocal connections with the midbrain and limbic system, as well as major efferent projections to the neurohypophysis. It functions as a link in neural systems coordinating a variety of autonomic, endocrine, behavioral and circadian functions (Bleier and Byne 1985).

Size Comparison
(Tables 34 - 36, 65 - 67)

Both in our own and Bauchot's measurements, HTH was found to have a low progression. HTH was not subdivided. The hypophysis (HYP) was not included in HTH volumes, since it is not preserved in all brains because of difficulties in preparation, especially in higher primates. HYP is included in a complex called "rest," together with meninges, nerves, and other remnants.

The indices of the 20 species converted from Bauchot's data vary from 88 in *Tenrec* to 225 in *Galemys*. Values close to those of Tenrecinae were found in the Crocidurinae (104). The highest values were found in the Desmaninae (219), and in several forms with indices between 160 and 150 (*Talpa, Potamogale, Limnogale,* Erinaceinae).

3.3.5 Telencephalon (TEL)

Topography, Delimitation, Structural Characteristics
(Atlas A 12.5 - AP 0)

The telencephalon (TEL) is a highly heteromorphic brain part with purely cortical (neocortex, schizocortex, hippocampus) and purely subcortical subdivisions (striatum), as well as a mixture of cortical and subcortical elements (paleocortex, septum, amygdala). The olfactory bulbs may be considered a special type of allocortex. The characteristics of the various main parts will be described under their respective headings.

Size comparison

Percentages: The telencephalon is by far the largest of the five fundamental brain parts, representing in the species an average of 60.0% and in the families and subfamilies an average of 59.1% of the total brain (see table on p. 52). The percentages (AvIF) of the 8 main components (TEL = 100%) are as follows:

olfactory bulbs	(BOL)	14.0%	(4.3 - 23.5)
paleocortex	(PAL)	19.4%	(6.5 - 29.0)
striatum	(STR)	8.6%	(5.8 - 11.4)
septum	(SEP)	2.8%	(2.0 - 3.8)
amygdala	(AMY)	5.4%	(3.2 - 7.5)
hippocampus	(HIP)	15.7%	(10.4 - 20.9)
schizocortex	(SCH)	6.0%	(3.7 - 8.6)
neocortex	(NEO)	28.1%	(13.1 - 53.3)

The largest interspecific variation is found in BOL, whose percentage is 5.5 times larger in *Setifer* than in *Potamogale*. Other structures with a relatively broad range of variation (> 4 times) are PAL and NEO, and with a relatively narrow range of variation (≤ 2) are SEP, STR, and HIP. Percentages of BOL lower than 9.2 were found only in semiaquatic Insectivora but, surprisingly, a value of 9.2% relative to TEL was found in *Ruwenzorisorex*.

For comparison, **in man**, the percentage is much higher in NEO (94.6%) and consequently lower in all other brain parts (highest in striatum = 2.7%; all other parts are lower than 1%; BOL is only 0.01%). Thus, as already mentioned several times, interpretation of percentage values in interspecific comparisons is questionable, since percentages do not depend solely on size changes in the structure under consideration, but to a large extent also on size changes in other brain structures included in the reference system. Thus, the low percentage size of STR in man (2.7%) compared with that in Insectivora (8.6%) does not mean that this structure is smaller in man. In fact, its allometric size in man (1393, see Section 3.3.6) is 6.7 times larger than in the average Insectivora (207, Table 45) (see also p. 150). Therefore, since evaluation of percentage size for interspecific comparisons gives no reliable results, the interspecific comparisons here are based on allometric size indices only.

Indices: (Tables 10 - 12; Fig. 69): The size indices for TEL may be described as follows:

Averages: 159 in the 50 species, 161 in the twelve subfamilies and families. This is the median level reached in the five fundamental brain parts.

Range: 83 - 260. The highest value *(Galemys)* is 3.1 times larger than the lowest *(Tenrec);* this is a medium variation. When the very low value of *Geogale* (66, which, however, is based on poor material) is included, the range of variation becomes distinctly larger (3.9).

Small size: *Geogale* (66), Tenrecinae (100), Erinaceinae (126).

Large size: Desmaninae (227), Echinosoricinae (225).

Characteristics and trends: The trends and details of groups are similar to those for the total brain, in which the TEL represents three fifths (cf. Figs. 53 and 69). Since the components of TEL may have quite different tendencies in their relative size, a discussion of the individual components will be more informative than consideration of the whole complex.

Figure 69: Size indices for the telencephalon (TEL). The indices are based on the AvTn line (index = 100) whose equation is log TEL volume = 1.242 + 0.68 x log BoW.
The distance of the AvIF broken line is 1.61 x AvTn and, thus, the index is 161. For additional explanations, see Fig. 53.

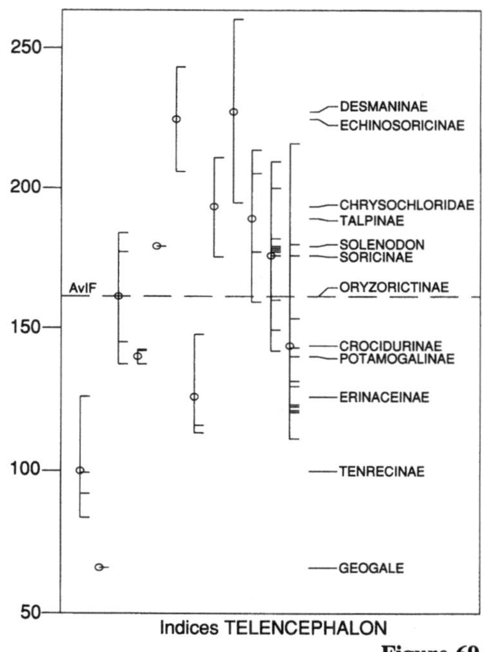

Figure 69

Telencephalon Components

In most of the main components of the telencephalon, subdivisions were delimited and measured. Such subdivisions may be nuclei (PAL, SEP, AMY), cortical regions (PAL, HIP), or layers (MOB, AOB, NEO).

Olfactory bulbs (BOL)

The olfactory bulbs (BOL = main + accessory bulb) in Insectivora on the average account for 8.3% of the NET brain and 14.0% of TEL. The BOL are the fourth largest TEL structure, having on the average nearly the same size as the hippocampus. Only 0.5% of the BOL is accessory olfactory bulb (AOB), all the rest (99.5%) is main olfactory bulb (MOB). Thus, BOL and MOB sizes are nearly identical.

Main olfactory bulb (MOB)

Topography, Delimitation, Structural Characteristics, Fiber Connections
(Atlas A 12.5; Figs. 73 - 75)

The main olfactory bulb (MOB) is the primary brain center in the main olfactory system and serves as a relay station for all olfactory impulses between the olfactory mucosa and higher olfactory centers, which belong primarily to the paleocortex (PAL). Consequently, MOB size may be regarded as an important indicator of the significance of the olfactory system in a given species. In terrestrial Insectivora, olfaction may be the most important sensory system in searching for food, but, even when this dominant importance is lost, as in

---

Figure 70: Main olfactory bulb volume (in mm^3) plotted against body weight (in grams) in a double logarithmic scale. The equation of the AvTn line is

$$\log \text{MOB volume} = 0.681 + 0.64 \times \log \text{BoW}.$$

The broken AvIF line is at a distance of 0.91 x AvTn. For additional explanations, see Fig. 52.

Figure 71: Size indices for main olfactory bulb (MOB). The indices are numerical values of the distances from the AvTn line given in Fig. 70. The broken AvIF line has an index of 91. For additional explanations, see Fig. 53.

Figure 72: Size indices for accessory olfactory bulb (AOB). The indices are based on the AvTn line (index = 100), whose equation is

$$\log \text{AOB volume} = -1.903 + 0.64 \times \log \text{BoW}.$$

The distance of the AvIF broken line is 1.58 x AvTn and, thus, the AvIF index is 158. For additional explanations, see Fig. 53.

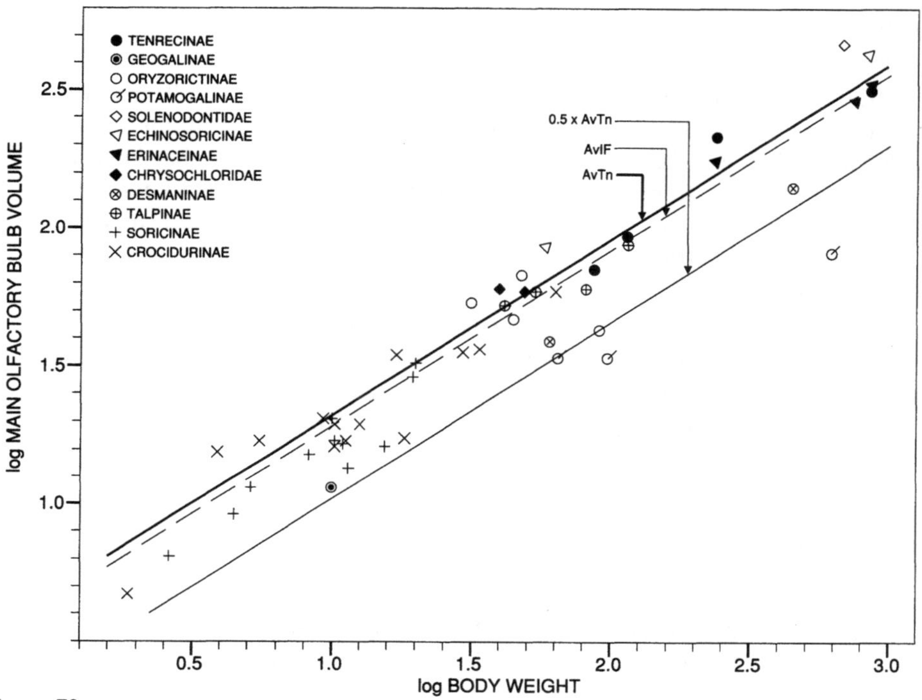

Figure 70

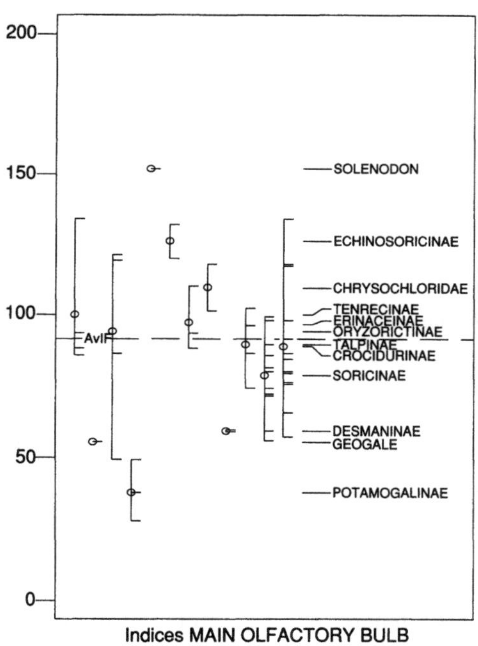

Figure 71

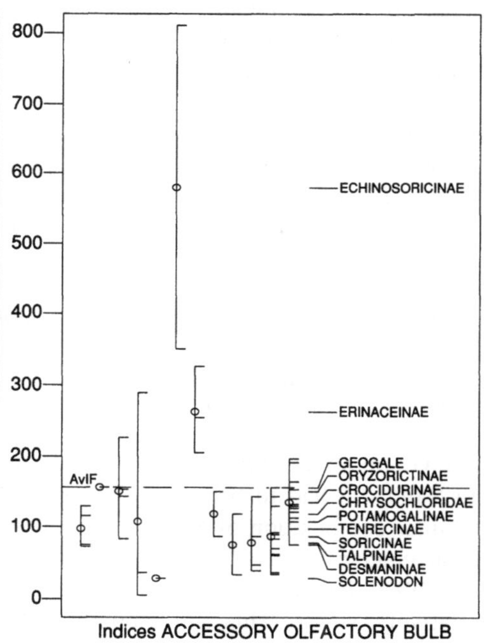

Figure 72

semiaquatic Insectivora or in simians, olfaction may maintain its importance in social behavior.

MOB is the most rostral part of the brain, projecting in all Insectivora beyond the cerebral hemispheres, with which it is connected by the olfactory peduncle. MOB is a layered structure. The number of layers that have been differentiated varies from two (Walter 1861) to nine (Winkler and Potter 1914). The most widely used subdivision is that of Cajal (1911), who discriminates six layers and a periventricular fiber zone (Figs. 73 - 75). For a detailed description of the microscopic anatomy of MOB, see Stephan (1975, pp. 215 - 270).

The efferent targets of the MOB include the retrobulbar region, prepiriform cortex, olfactory tubercle, amygdala, and entorhinal cortex. Afferent axons originate both in ipsilateral and contralateral olfactory structures. Contralateral fibers originate mainly in the retrobulbar region (RB) (ref. see Switzer et al. 1985).

Size comparison
(Tables 46 - 48; Figs. 70, 71)

Averages: 87 in the 50 species, 91 in the twelve subfamilies and families. This is the lowest level in all the measured telencephalic components.

Range: 28 - 152. The highest value *(Solenodon)* is 5.4 times larger than the lowest *(Potamogale)*. This is average variation.

Small size: Potamogalinae (38), *Limnogale* (49), *Geogale* (55), *Neomys* spp. (58), Desmaninae (60).

Large size: *Solenodon* (152), *Setifer* (134), Echinosoricinae (126), *Sylvisorex* spp. (126).

Characteristics and trends: All water-adapted forms have small MOBs. In terrestrial groups, MOB is small in *Geogale*, in shrews except *Sylvisorex* and *Myosorex*, and in Talpinae. Large MOBs are found in certain groups, such as *Solenodon* and among Tenrecinae, in *Setifer*.

Main olfactory bulb components

Topography, Delimitation, Structural Characteristics
(Atlas A 12.5; Figs. 73 - 75)

For volume measurements in a great variety of species (incl. primates), layers 1+2 and 4-6 had to be combined, since their separation is often difficult. The laminar composition of MOB is relatively stable, i.e., the components show no clear change in size from well developed to strongly reduced MOB. Baron et al. (1983) have, however, pointed out that in diurnal simians, layers 4-6 are relatively small. This corresponds with the general observation (from light-microscopy) that the granular layer (layer 6) is reduced and decomposed in higher primates, especially in man.

Size comparison
(Tables 46 - 48)

The largest components of MOB are the external fiber and glomerular layers (L1+2) and the combined inner cell layers (mitral and granular layers including the internal plexiform layer, L4-6). The two complexes represent on the average 35% - 39% of the total MOB. The smallest is the periventricular zone (PvZ), which on the average accounts for 3%, but in small species (shrews) has lower percentages (0.8% - 2.2%) than in larger species (3.4% - 7.7%). The size indices may be described as follows:

Averages: The averages are similar for all the four MOB components. They are lowest for PvZ (59 for the 43 species, 80 for the twelve subfamilies and families) and highest for L3 (97 and 94).

Range: The ranges differ more than the averages. They are highest for PvZ (6 - 226; quotient = 37.7), high also for L1+2 (25 - 197; quotient = 7.9) and lower for the other components (31 - 170 for L3, quotient = 5.5; 29 - 147 for L4-6, quotient = 5.1).

Small size: *Potamogale* for all four components (24 - 31).

Large size: *Solenodon* for all four components (118 - 223); Echinosoricinae for all four components (119 - 169); for L1-6 also in *Setifer* (125 - 143); for PvZ next to *Echinosorex* (226) and *Solenodon* (223) in Erinaceinae (183) and *Tenrec* (182).

Characteristics and trends: Trends for L1+2, L3, and L4-6 are similar to those found for the total MOB. In contrast, for PvZ the size of the animal (or brain, or MOB) seems to have a large influence. Thus, PvZ is very small in shrews (Fig. 75) and clearly larger in larger species as, e.g., *Echinosorex* (Fig. 73). In Tenrecinae, whose species differ widely in body size, the PvZ indices increase from the smallest (*Echinops*, 38) to the largest species (*Tenrec*, 182).

Accessory olfactory bulb (AOB)

Topography, Delimitation, Structural Characteristics, Fiber Connections
(Atlas A 12.5, Figs. 73, 74)

Sensory input to the accessory olfactory bulb (AOB) is from the vomeronasal or Jacobson's organ through the vomeronasal nerve. Jacobson's organ is a special differentiation of the olfactory mucosa. AOB thus is the primary brain center in the accessory olfactory system and may serve as a good indicator of the significance of the vomeronasal system in a given species. The efferent projections are by way of the lateral olfactory tract to parts of the medial and cortical amgydaloid regions and of the bed nucleus of the stria terminalis

(Scalia and Winans 1975; Skeen and Hall 1977). Comparative data support the suggestion that the vomeronasal system may be involved in sexual and/or social interaction (Stephan et al. 1982).

AOB is situated dorsomedially in the transitional zone between MOB and the olfactory peduncle and is in general much smaller than MOB. Its lamination is similar to that of MOB and, as in MOB, six layers were differentiated: (1) vomeronasal fibers, (2) glomerular, (3) external plexiform, (4) mitral, (5) internal plexiform, and (6) internal granular (Figs. 73, 74). Detailed descriptions of the microscopic anatomy and architectonics of the AOB have been given by Stephan (1965, 1975). MOB fibers, which contribute to the lateral olfactory tract, run diffusely or in clear bundles through the lower parts of AOB. These fibers may penetrate the AOB between layers 5 and 6 or under layer 6. The differences between species were interpreted by Stephan (1965) as the consequence of mechanical factors. Switzer et al. (1980) used them as a phylogenetic indicator.

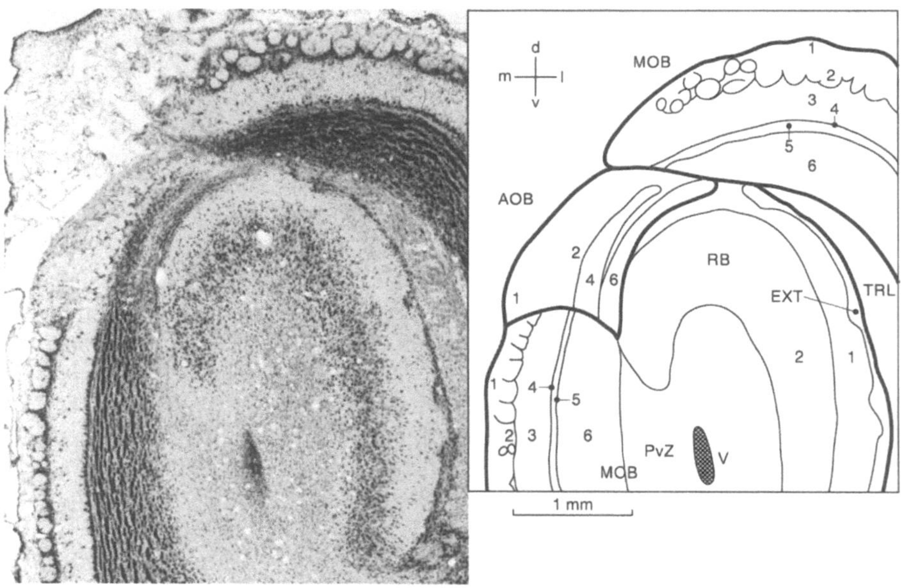

Figure 73

Figures 73 - 75: Transversal sections through the olfactory bulbs of *Echinosorex gymnurus* (Fig. 73), *Hylomys suillus* (Fig. 74), and *Suncus etruscus* (Fig. 75). *Suncus etruscus* has no cell-poor periventricular zone (PvZ) as is found in *Echinosorex* and *Hylomys*. In none of the three species is there a large lumen of the olfactory ventricle as is found in tenrecs, solenodons, and hedgehogs (e.g. in *Atelerix*, as shown in Atlas A 12.5).

Abbreviations:

AOB	accessory olf. bulb
EXT	pars externa of RB
MOB	main olfactory bulb
PRPI	regio prepiriformis
PvZ	periventricular zone of MOB
RB	regio retrobulbaris with molecular layer and cell layer
TOL	tuberculum olfactorium
TRL	tractus olfactorius lateralis
V	olfactory ventricle
1	layer of olfactory (MOB) or vomero nasal (AOB) fibers
2	glomerular layer
3	external plexiform layer
4	mitral cell layer
5	internal plexiform layer
6	internal granular layer

Direction in sections:

d	dorsal
l	lateral
m	medial
v	ventral

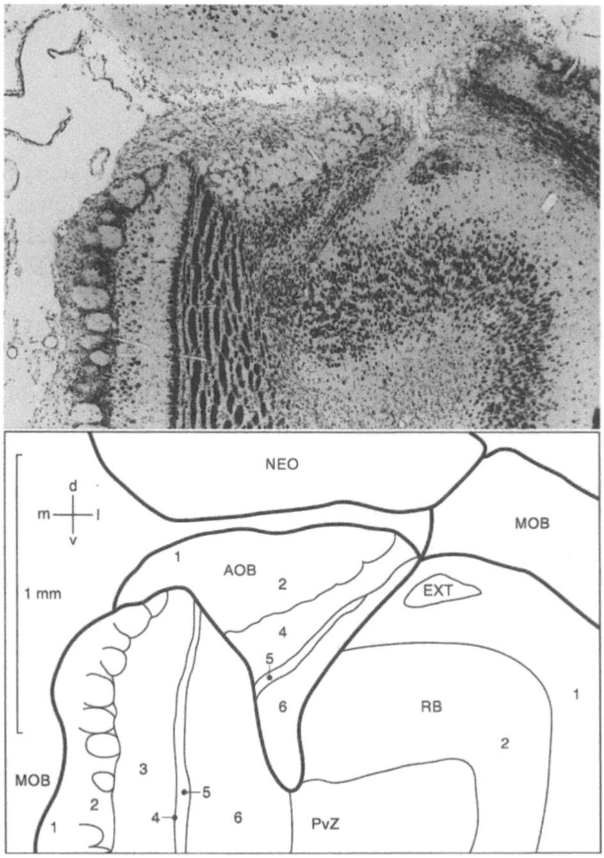

Figure 74

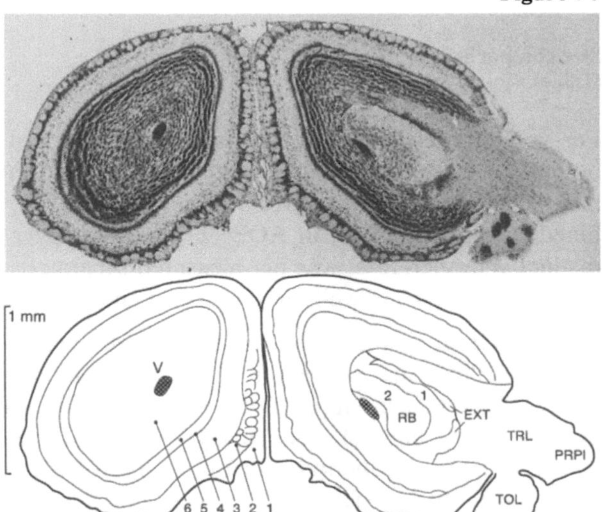

Figure 75

Size comparison
(Tables 49 - 51; Fig. 72)

The small accessory bulb on the average represents only 0.49% of BOL. Its size and size differences may be described as follows:

Averages: 138 in the 50 species, 158 in the twelve subfamilies and families. This is a much higher level than reached by MOB and a medium level for TEL components.

Range: 6 - 810. The highest value *(Hylomys)* is 135 times larger than the lowest (*Micropotamogale ruwenzorii*). Next to the pineal gland (Table 41), this is by far the highest range found in any of the brain structures. Such broad ranges in variation occur when small structures are strongly reduced in certain species.

Small size: *Micropotamogale* spp. (22), *Solenodon* (28), terrestrial Soricinae (77; and especially *Cryptotis* and *Anourosorex*, 35, 36), Talpidae (79; and especially *Desmana*, *Scalopus* and *Parascalops*, 35, 41, and 47).

Large size: *Hylomys* (810), *Echinosorex* (349), *Potamogale* (288), Erinaceinae (261).

Characteristics and trends: AOB clearly differs in its index pattern from that of the MOB. There is wide variability even in subfamilies and no clear trends could be found with respect to ecological adaptations or taxonomic affiliations. In fact, size variations in the AOB in the species investigated seem to be independent of terrestrial, subterranean, and semiaquatic life habits. For example, the AOB index is high in *Potamogale* (288), but very low in *Micropotamogale* spp. (6 and 37).

Accessory olfactory bulb components

Size comparison
(Tables 49 - 51)

For volume measurements, layers 1+2 and 3-5 had to be combined, since the borders of the individual layers are less clear and their separation is difficult in many species. Unlike the MOB, where layers 1+2, 3 and 4-6 had to be combined for measurement, in AOB (1) the third layer (external plexiform) is not distinct and, therefore, was not measured separately, whereas (2) the sixth layer (internal granular) is well circumscribed and could be measured separately.

The average percentages for the various components relative to the total AOB (= 100%) are 34.2% for L1+2, 49.7% for L3-5, and 16.1% for L6. The indices show that in large AOBs, as e.g., in Echinosoricinae (Figs. 73, 74), the increase in the internal granular layer (L6) is distinctly larger than in the other layers. In *Hylomys*, the index of L6 is 15.9 times larger than that of the average Tenrecinae, that of L3-5 7.7 times and that of L1+2 6.4 times. Similar relations were found in primates in the prosimian Cheirogaleinae, whose AOBs and especially internal granular layers were found to be extremely large.

Paleocortex (PAL)

Topography, Delimitation, Structural Characteristics
(Figs. 76, 77; Atlas A 12.5 - A 3)

The surface structures of the piriform lobe are combined under the term "paleocortex." These are the retrobulbar region (RB), the olfactory tubercle (TOL), the prepiriform region (PRPI), and the periamygdaloid region (PAM). Superficial parts of the septum and the diagonal band of Broca may also be included. The problem whether these structures are in fact cortical, corticoid, or non-cortical has been discussed in detail by Stephan (1975) and will not be dealt with here. Volume and surface measurements were made of most of these structures, except for PAM, which is not separable from the deeper parts of the amygdala. Thus, PAM size is known only from surface measurements (Stephan 1961).

The most caudal parts of the piriform lobe are not included in PAL, since they have a fundamentally different laminar pattern in the more advanced forms (especially in higher Primates). These parts were combined under the term "schizocortex" (SCH) and were, in the cortical subdivision of Filimonoff (1947) as "periarchicortex," more closely related to the archicortex (=hippocampus, HIP).

On the other hand, two fiber bundles and a clearly subcortical structure were included in the volume measurements of PAL for practical reasons, i.e., to avoid an overly detailed split into tiny components. These structures are (1) the superficial lateral olfactory tract (TRL), (2) the deep internal tract contributing to the anterior commissure (COA), and (3) the substantia innominata (SIN). Thus, PAL is subdivided for measurement into six parts, three of which (RB, PRPI, TOL) are olfactory cortices, two are fiber bundles (TRL, COA), and one, a subcortical component (SIN).

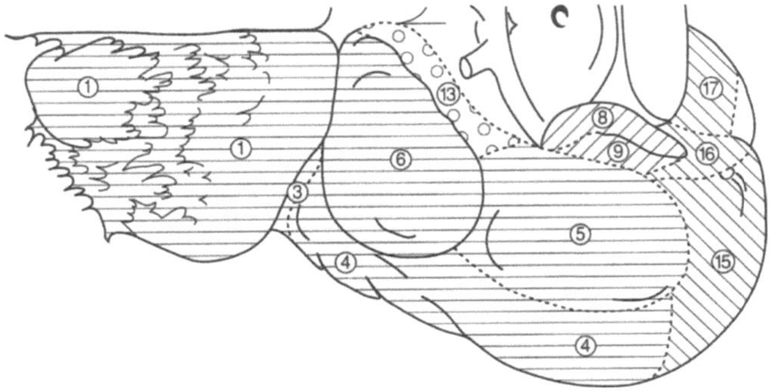

Figure 76

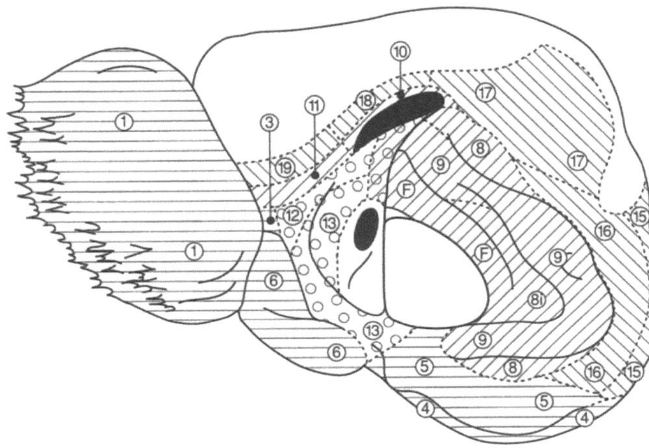

Figure 77

Figures 76 - 77: Maps of allocortical regions of the hedgehog cerebral hemisphere in ventral (Fig. 76) and medial (Fig. 77) views. Commissures are in black; neocortex and diencephalon are in white. Horizontally hatched are bulbi olfactorii and paleocortex; diagonally ascending hatched is archicortex; diagonally descending hatched is periarchicortex; structures with uncertain affiliation (septum, diagonal band of Broca) are marked by open circles.

Abbreviations:

1	bulbus olfactorius	10	hippocampus supracommissuralis
2	bulbus olfactorius accessorius	11	hippocampus praecommissuralis
3	retrobulbar region (ncl. olf. ant.)	12	septum
4	prepiriform region	13	diagonal band of Broca
5	periamygdaloid region	14	insular cortex (not visible)
6	olfactory tubercle	15	entorhinal + perirhinal cortex
7	subiculum (not visible)	16	presubicular + parasubicular cortex
8	cornu ammonis	17	retrosplenial cortex
8i	cornu ammonis inversum	18	periarchicortical cingular cortex
9	fascia dentata	19	subgenual cortex
F	fimbria-fornix complex		

Size comparison
(Tables 43 - 45; Fig. 78)

Percentages: Next to the neocortex, the paleocortex is the second largest component of TEL in the average Insectivora. It constitutes on the average 11.5% of the NET brain and 19.4% of TEL.

The average volume percentages of the components of PAL (PAL = 100%) are as follows (AvIF):

retrobulbar region	(RB)	13.9%
prepiriform region	(PRPI)	57.3%
olfactory tubercle	(TOL)	19.0%
lateral olfactory tract	(TRL)	3.4%
anterior commissure	(COA)	3.1%
substantia innominata	(SIN)	3.3%

For size determination of the three main olfactory cortices (RB, PRPI, TOL), surface measurements may be more reliable than volume measurements, since their inner border may be blurred. Therefore, the area measurements (performed in eleven species) were also taken into account. The percentage compositions (olfactory cortices = 100%) in volume and area measurements are as follows:

		Volumes	Areas	Areas
retrobulbar region	(RB)	15.4	18.7	14.9
prepiriform region	(PRPI)	63.5	53.5	43.6
olfactory tubercle	(TOL)	21.1	27.8	22.0
periamygdaloid cortex	(PAM)	-.-	-.-	19.5

The larger percentage size of TOL in area measurements may be explained by the fact that the strongly twisted stratum densocellulare was measured. The larger percentage size of PRPI in volume measurements is certainly due to its thickness. The similar composition in the two types of measurement confirms the utility of both.

The superficial parts of the amygdala (AMY) may be considered cortical or corticoid and may be termed periamygdaloid cortices (PAM). They are not separable from the deeper parts of AMY for volume measurement and, thus, their dimension is known only from area measurements. When PAM is included in the olfactory cortices, it amounts to about one fifth of the structure (19.5%, see above table).

Indices: In the average Insectivora, the indices of PAL are slightly higher than in the reference group, the Tenrecinae. The averages are 103 in the 50 species and 104 in the twelve subfamilies and families. Next to BOL or MOB, this is the lowest level of all brain structures. All PAL components have similarly low averages (Tables 53, 54). The largest component by far is the prepiriform cortex (PRPI) and, thus, its size characterization is similar to that of PAL.

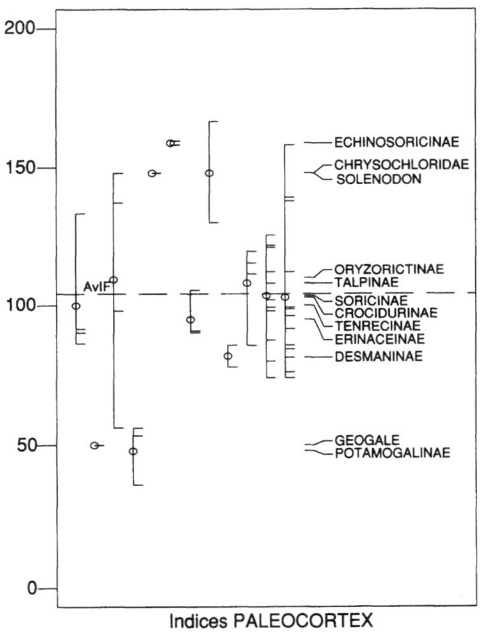

Figure 78

Figure 78: Size indices for the paleocortex (PAL). The indices are based on the AvTn line (index = 100), whose equation is log PAL volume = 0.799 + 0.63 x log BoW.

The broken AvIF line is at a distance of 1.04 x AvTn. For additional explanations, see Fig. 53.

Paleocortical Components

Prepiriform region (PRPI)

Topography, Delimitation, Structural Characteristics, Fiber Connections
(Atlas A 12.5 - A 3; Figs. 82 - 84)

The prepiriform cortex (PRPI) lies in the piriform lobe caudal to the retrobulbar region (RB) and lateral to the olfactory tubercle (TOL) and the periamygdaloid region (PAM). PRPI may be superficially separated from PAM by a shallow sulcus semiannularis (SAN in Fig. 83). It covers most of the piriform lobe except a caudal part, which is covered by schizocortex (SCH = entorhinal + perirhinal + presubicular + parasubicular cortices). Not included in PRPI are the superficial parts of the amygdaloid complex (anterior amygdaloid area, cortical and medial amygdaloid nuclei), which belong to AMY and may be called periamygdaloid cortex (PAM; 5 in Figs. 76, 77). PRPI is distinguishable from PAM by its distinctive arrangement in three layers; dorsal parts are underlain by the claustrum (Atlas A 6, A 8). Its laminar differentiation is well developed in the macrosmatic Insectivora. For a detailed description of the microscopic anatomy of PRPI, see Stephan (1975, pp. 427 - 464).

PRPI is the most extensive recipient of direct olfactory bulb projections (ref. see Stephan 1975; Heimer 1978; Switzer et al. 1985). Significant input also comes from the olfactory areas surrounding the prepiriform cortex (Haberly and Price 1978a,b). Other afferent fibers come from a variety of subcortical structures, including the amygdala, midline thalamic nuclei, hypothalamus, and various brain stem nuclei. Efferent fibers from PRPI reach other olfactory areas such as the olfactory bulb, the retrobulbar region (anterior olfactory nucleus), and various paleocortical areas. Subcortical structures receiving efferent PRPI fibers include the striatum, amygdala, thalamus, and hypothalamus (ref. see Switzer et al. 1985).

Size comparison
(Tables 52 - 54)

The volumes of PRPI represent on the average 57.3% of PAL, 11.1% of TEL and 6.6% of NET brain. The size indices may be described as follows:

Averages: 94 in the 50 species, 98 in the twelve subfamilies and families.

Range: 35 - 159. The highest value *(Echinosorex)* is 4.5 times larger than the lowest *(Potamogale)*. This is an average variation.

Small size: *Geogale* (44), Potamogalinae (45), *Limnogale* (51), *Neomys* spp. (61), *Suncus* spp. (70), Desmaninae (76).

Large size: Echinosoricinae (154), *Solenodon* (148), *Microgale* spp. (142), *Sylvisorex granti* (138), Chrysochloridae (131), *Setifer* (127).

Characteristics and trends: No clear trends relative to Tenrecinae; strong parallels with MOB. Small size in all water-adapted forms; large size in a variety of species. The largest variation is found in Oryzorictinae, with very low values in the water-adapted *Limnogale* and very high values in *Microgale* spp. Wide variation is also found in the purely terrestrial Crocidurinae, in which *Suncus* spp. have low and *Sylvisorex* spp. nearly two times larger PRPI size.

Retrobulbar region (RB) and **Olfactory tubercle** (TOL)

Topography, Delimitation, Structural Characteristics, Fiber Connections
(Atlas A 12.5 - A 8; Figs. 76, 77)

The retrobulbar region (RB) corresponds to the "anterior olfactory nucleus" of most Anglo-American authors. Its place is in the peduncle connecting the olfactory bulb with the hemisphere. In the macrosmatic Insectivora, this peduncle is thick, and the RB lies like a tube around the olfactory ventricle or its vestiges, separating the internal from the lateral olfactory tract. Its rostral part penetrates the olfactory bulb; more caudally, it lies free on the surface and borders on PRPI, laterally, and on TOL, medially.

In the macrosmatic Insectivora, RB has attributes of cortical structures, such as the laminar organization of its afferents. RB has strong similarities with PRPI, whose cortical structure is not questioned. Gurdjian (1925) actually considered RB a less differentiated rostral projection of the piriform cortex (PRPI). Three layers are recognizable, with the third (polymorphic or multiform) layer not identifiable throughout. This, together with differences in cell distribution, makes it possible to differentiate certain areas. In the periphery of RB a highly characteristic "pars externa" is found (EXT in Figs. 73 - 75). A poorly separated third layer or its lack is one of the main features distinguishing RB from PRPI. For a detailed description of the microscopic anatomy of RB, see Stephan (1975, pp. 289 - 309).

From RB originates the olfactory component of the anterior commissure, which projects with its main fibers to the contralateral main olfactory bulb (MOB) and with collaterals to the ipsilateral MOB. Coordination of the bulbi on the two sides is by way of RB. Apart from the main bulb, RB receives major afferents from a variety of cortical areas including PRPI and areas of the hippocampus (subiculum and CA1), and from subcortical structures including nuclei of the septum, amygdala, and hypothalamus. Efferent RB fibers reach all olfactory structures. Like other olfactory structures, it projects to hypothalamic, septal, and amygdaloid nuclei (Switzer et al. 1985).

The olfactory tubercle (TOL) borders rostrally on RB and laterally on PRPI. In its position on the ventral surface of the hemisphere, TOL is closely (spatially and functionally) related to the ventral parts of the striatum (nucleus accumbens) and to the substantia innominata. Its position is stable, even when, in Primates, neighboring structures strongly shift. Because it receives direct afferents from the main olfactory bulb, TOL is considered to belong to the olfactory cortices (for ref., see Stephan 1975). Apart from MOB, it receives afferents from widespread areas of other olfactory cortices and from various structures of the limbic system, including hippocampus, schizocortex and amygdala. In addition, it receives fibers from the intralaminar thalamus and from dopamine-containing areas of the mesencephalon (Newman and Winans 1980). Although our knowledge about efferents from TOL is still scanty, it does seem that TOL sends fibers to the ventral globus pallidus, substantia nigra, and thalamus (Heimer et al. 1985).

Despite its close functional relationships with the striatum and the limbic system, the size of TOL depends largely on that of the olfactory structures. It is prominent in the macrosmatic Insectivora, whereas in microsmatic species *(Potamogale)*, it may be suppressed. In macrosmatic Insectivora, TOL is well differentiated and laminated; the dense-celled layer is twisted. Highly characteristic are irregularly placed islands of densely packed cells (islands of Calleja). In species with reduced olfaction (based on the relative size of MOB), the molecular layer becomes thinner and the cellular layer smoother and/or dissolves. For a detailed description of the microscopic anatomy of TOL, see Stephan (1975, pp. 309 - 339).

Size comparison
(Tables 52 - 54)

RB and TOL are smaller components of PAL (13.9 and 19.0%) which in general show similar trends to PRPI and MOB. Therefore, only some of the stronger index deviations of RB and TOL from PRPI will be mentioned:

RB has stronger enlargement than PRPI in all Soricidae.

TOL is larger than PRPI in semiaquatic species, in *Solenodon*, Chrysochloridae, and Talpinae, and in shrews mostly larger in Soricinae, but smaller in Crocidurinae (except *Myosorex*).

Lateral olfactory tract (TRL) and **Commissura anterior** (COA)

Topography, Delimitation, Structural Characteristics
(Atlas A 12.5 - A 8)

The lateral olfactory tract (TRL) is the purely ipsilateral efferent bundle of the olfactory bulbs projecting mainly to the olfactory cortices (RB, PRPI, TOL, PAM). In addition, the bundle contains a small contingent of centrifugal fibers coming from the diagonal band and other structures and running to MOB. TRL runs on the surface of RB and PRPI. Its core can be followed from MOB to the nucleus of the lateral olfactory tract (NTO) lying in the rostral part of the periamygdaloid cortex (PAM); its delineation posed no problems and was based on fiber-stained sections.

The anterior commissure (COA) is concentrated with all its fibers in the midsagittal plane (Figs. 40 - 51). Three components can be differentiated: (1) a small stria terminalis component, (2) an olfactory component (pars anterior), which is relatively large in Insectivora and small in simian Primates, and (3) a non-olfactory component, which is relatively small in Insectivora and large in Primates. In Primates, the non-olfactory fibers increase markedly, ultimately leading to a much greater relative size of COA than in Insectivora. The olfactory component is the internal olfactory tract, running in the center of the olfactory peduncle, originating in RB and projecting with its main fibers to the contralateral MOB and with collaterals to the ipsilateral MOB. Strong non-olfactory components do not seem to exist in Insectivora and Prosimians, since a reduction in the olfactory centers (MOB, RB), as in semiaquatic Insectivora and prosimian Indriidae, results in an equally strong reduction in COA.

Size comparison
(Tables 52 - 54)

TRL and COA are small fiber components (3.4% and 3.1% relative to PAL) having large size indices in Echinosoricinae, *Solenodon,* and Chrysochloridae, and small ones in Potamogalinae, Desmaninae, *Geogale* and many shrews. Their trends are similar to those of MOB and the olfactory cortices (PRPI, RB, TOL). Some stronger deviations are as follows:

TRL is smaller in Soricidae and Talpidae.

COA is smaller in Oryzorictinae and Desmaninae, and larger in Echinosoricinae.

The **area dimensions** of the midsagittally sectioned COA (Tables 71 - 73) were measured and found to be largest in *Myosorex* (1.5x AvTn), *Chrysochloris, Echinosorex, Solenodon, Setifer,* and *Microgale* (1.2 - 1.3 times). The lowest values were found in Potamogalinae, *Neomys*, Desmaninae, and *Limnogale*, i.e., in semiaquatic species, and in *Sorex minutus* and *Erinaceus europaeus*. Very small size is recognizable also in the midsagittal plane, as e.g., in *Potamogale* (Fig. 43) and *Galemys* (Fig. 48). In Insectivora, there are clear interrelations between size of COA and size of MOB.

The smaller non-olfactory and stria terminalis components of COA could not be recognized in the serial sections of all species. Thus, they could not be used for comparison throughout.

Substantia innominata (SIN)

Topography, Delimitation, Structural Characteristics
(Atlas A 8 - A 7)

The substantia innominata (SIN) is a loosely-packed, large-celled area between the diagonal band and olfactory tubercle, superficially, and the striatum (nucleus accumbens) centrally. Its delineation from the deep multiform layer of TOL is difficult in some places. A close relationship to the olfactory system is unlikely, since, in Insectivora, SIN does not follow the decrease from terrestrial to semiaquatic species found in all olfactory structures.

Size comparison
(Tables 52 - 54)

SIN is a relatively small component of PAL (3.3%). Its size indices may be described as follows:

Averages: 96 in the 50 species, 99 in the twelve subfamilies and families. This is low for TEL components.

Range: 45 - 154. The highest value *(Galemys)* is 3.4 times larger than the lowest *(Limnogale)*. This is relatively low variation.

Particularities and trends: There are no strong differences between the groups, but wide variations in the groups. This indicates difficulties in determination and delineation of this structure. Unlike the olfactory components of PAL, SIN appears to be relatively large in Desmaninae.

Striatum (STR)

Topography, Delimitation, Structural Characteristics, Fiber Connections
(Atlas A 10 - A 6)

The striatum (STR) is the largest subcortical motor center of the numerous interconnected motor circuits between the neocortex and the spinal cord, which formerly were known as the "extrapyramidal motor system" as distinguished from the pyramidal system. However, recent investigations have shown that the concept of two largely independent motor systems should be abandoned. The pyramidal tract has been shown to be the most important output system of "extrapyramidal" circuits too (Nieuwenhuys et al. 1978, 1980).

The main components of STR are the nucleus caudatus and the putamen. Included in the measured STR are those parts of the internal capsule which are fully enclosed by these structures, as well as the nucleus accumbens. The designation of the caudate-putamen complex as STR is especially relevant in Insectivora, where a large part of the complex consists of cell stripes dispersed between the fiber bundles of the capsula interna. Thus, the distinction between the two nuclei is largely topographic in Insectivora in contrast to Primates, where the caudatum is clearly separated from the putamen by the internal capsule. The cytoarchitectural features of the two parts are identical. Caudatum and putamen constitute dense aggregations of medium-sized, moderately stained cells.

Caudatum and putamen receive afferents from practically all neocortical areas. Other major projection systems come from the intralaminar thalamic nuclei and from the substantia nigra. The nucleus accumbens receives fibers from the basolateral amygdaloid body. Efferent fibers from the entire STR reach the globus pallidus and the substantia nigra (for ref., see Nieuwenhuys et al. 1978, 1980; Heimer et al. 1985).

Size comparison
(Tables 43 - 45; Fig. 79)

STR on the average represents 5.1% of the NET brain and 8.6% of TEL. Its size and size differences may be described as follows:

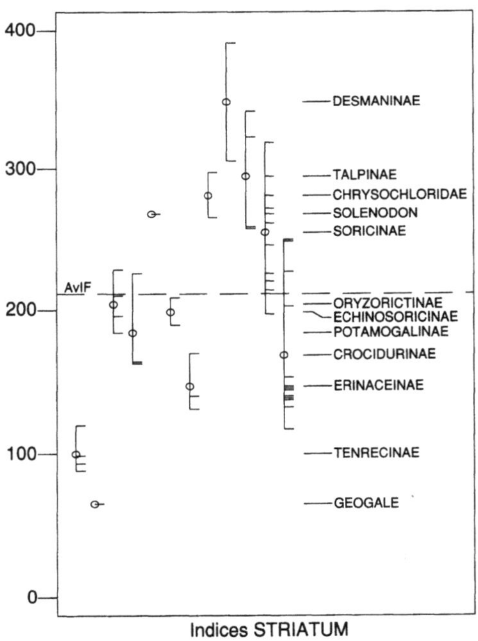

Figure 79

Figure 79: Size indices of the striatum (STR). The indices are based on the AvTn line (index = 100) whose equation is log STR volume = 0.089 + 0.67 x log BoW. The broken AvIF line is in a distance of 2.10 x AvTn and, thus, the AvIF index is 210. For additional explanations see Fig. 53.

Averages: 207 in the 50 species, 210 in the twelve subfamilies and families. This is the second highest level reached by TEL components (next to NEO).

Range: 88 - 391. The highest value *(Galemys)* is 4.4 times larger than the lowest *(Tenrec)*. This is medium variation. When *Geogale* is included (66, but poor material), the range becomes clearly larger (quotient 5.9).

Small size: *Geogale* (66), Tenrecinae (100), *Suncus* spp. (131), *Crocidura* spp. (143), Erinaceinae (147).

Large size: Desmaninae (348), Talpinae (295), Chrysochloridae (281), *Neomys* (279), *Solenodon* (268).

Characteristics and trends: In Soricidae, the STR of Soricinae is on the average larger than that of *Crocidura* spp., *Suncus* spp., and *Scutisorex* (similar relations were found in the neocortex). Clear trends: (1) general evolutionary trend with smallest size in the most conservative forms, such as *Geogale*, Tenrecinae, and certain Crocidurinae, and large size in progressive forms (Desmaninae, *Solenodon*); (2) large size in subterranean forms (Talpinae, Chrysochloridae).

Septum (SEP)

Topography, Delimitation, Structural Characteristics, Fiber Connections
(Atlas A 10 - A 7; Fig. 80)

The septum (SEP) is an important relay station in the limbic system. This central position implies that it may have an important role in visceral and behavioral mechanisms. SEP arises embryologically from the commissural plate and extends anteriorly a short distance into the medial walls of the cerebral hemispheres. It is bounded laterally by the lateral ventricles, dorsally by the corpus callosum, rostrally by the precommissural hippocampus (HP), caudally by the anterior commissure and the third ventricle, and ventrally by TOL and the nucleus accumbens (part of STR). These boundaries are not all sharp and rigid. Some structures extend beyond them, e.g., the diagonal band ventro-caudo-laterally behind TOL. SEP of Insectivora primarily consists of cellular elements and does not possess the thin dorsal glial and fiber component ("septum pellucidum") found in the higher Primates and man as a result of considerable expansion of the corpus callosum.

Included in SEP is the diagonal band of Broca (DIAG) in its whole extent and the very small subfornical body (SFB). SFB is always placed in the median sagittal plane where ventricular and extraventricular liquor spaces meet and, thus, may be related to liquor control. In general, SEP is subdivided into lateral and medial nuclear groups. Andy and Stephan (1959) proposed a more detailed subdivision into four groups (dorsal, ventral, medial, and caudal). The caudal group includes the triangular nucleus (SET; Fig. 80) and the bed nuclei of the anterior commissure and of the stria terminalis. The stria terminalis nuclei extend into the preoptic area and anterior hypothalamic region. The nucleus triangularis septi (SET) seems to be closely related to the fibers of the anterior hippocampal commissure (psalterium ventrale). Parts of the medial septum (including the diagonal band) may be considered corticoid (periseptal region). The microscopic anatomy of these parts has been described in more detail by Stephan (1975, pp. 401 - 427).

The septum receives its major input from the hippocampus, amygdala and various parts of the hypothalamus and midbrain. It projects back to the same regions. Other efferent fibers reach the habenular nuclei and the thalamus (for ref., see Stephan 1975; Bleier and Byne 1985).

Size comparison
(Tables 43 - 45; Fig. 85)

The telencephalic septum is a small component representing in the average Insectivora 1.7% of NET and 2.8% of TEL. Its size may be described as follows:

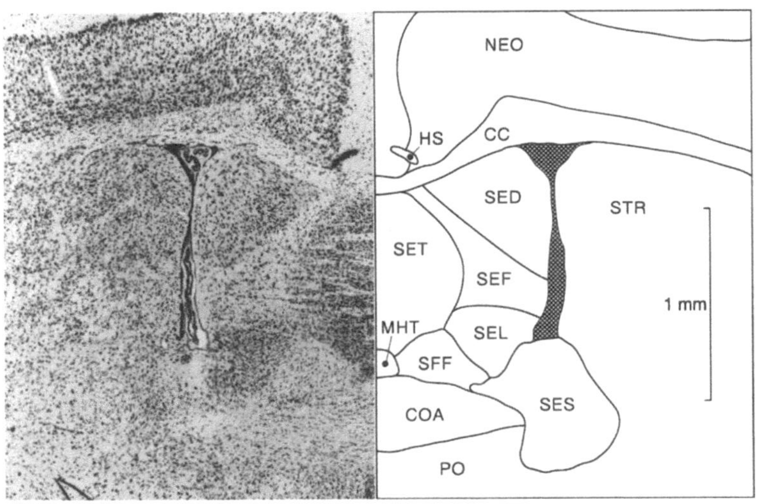

Figure 80

Transversal section through the septum of *Crocidura flavescens* at the level of the anterior commissure (COA). Abbreviations:

CC	corpus callosum	SED	nucleus dorsalis septi
COA	commissura anterior	SEF	nucleus fimbrialis septi
HS	hipp. supracommissuralis	SEL	nucleus lateralis septi
MHT	nucleus medianus hypothalami	SES	nucleus of stria terminalis
NEO	neocortex	SET	nucleus triangularis septi
PO	area preoptica	SFF	fimbria-fornix complex
		STR	striatum

Averages: 132 for the 50 species, 139 for the twelve subfamilies and families. This is a low level, the fifth of the eight main TEL components.

Range: 81 - 223. The highest value *(Galemys)* is 2.8 times larger than the lowest *(Tenrec)*. This is low variation which however becomes medium (3.3) when *Geogale* (67) is included.

Small size: *Geogale* (67), *Suncus* spp. (95), Tenrecinae (100), *Crocidura* spp. (101).

Large size: *Galemys* (223), Echinosoricinae (186), *Talpa micrura* (185), *Solenodon* (168).

Characteristics and trends: Clear but not very strong tendency toward enlargement, with small size in the most conservative forms and large size in the most progressive.

Septum components

The nine subdivisions proposed by Andy and Stephan (1959) were measured in three species of Insectivora. The largest components are the diagonal band of Broca (DIAG) with 29.2%, the nucleus dorsalis septi (SED = 20.4%), and the bed nucleus of the stria terminalis (SES = 13.2%). Size comparisons are worthwhile only with three species of Primates measured in the same way (see Section 3.3.6).

Two very small SEP structures were measured in 38 species. One of these, the triangular nucleus (SET; Fig. 80), is related to the fibers of the hippocampal commissure and the other, the subfornical body (SFB; Atlas A 7), is placed superficially in the median sagittal plane, where the lamina tectoria attaches to the septum and thus ventricular and extraventricular liquor spaces meet.

Nucleus triangularis septi (SET)

Size comparison
(Tables 40 - 42)

The size of SET may be described as follows:

Averages: 162 in the 38 species, 166 in the eleven subfamilies and families. This is a relatively high level, higher than that of the total septum.

Range: 75 - 258. The highest value *(Parascalops)* is 3.4 times larger than the lowest *(Setifer)*. This is narrow variation.

Small size: *Hemiechinus* (96), Tenrecinae (100), *Erinaceus* (105), *Solenodon* (119).

Large size: *Parascalops* (258), *Neomys* (249), Chrysochloridae (235), *Limnogale* (223), Echinosoricinae (214).

Characteristics and trends: No clear trends.

Subfornical body (SFB)

Size comparison
(Tables 40 - 42)

The size of SFB may be described as follows:

Averages: 75 in the 38 species, 83 in the eleven subfamilies and families. This is the lowest level found in TEL.

Range: 25 - 189. The highest value *(Micropotamogale ruwenzorii)* is 7.6 times larger than the lowest *(Talpa europaea)*. This is high variation.

Small size: *Talpa europaea* (25), *Sorex* spp. (28), *Microgale dobsoni* (33).

Large size: *Micropotamogale ruwenzorii* (189), *Hemiechinus* (146), *Tenrec* (128), *Galemys* (126).

Characteristics and trends: Relatively large size in Tenrecinae and clearly smaller size in most other Insectivora. Small size, however, was not always found in advanced species such as *Galemys*, whose SFB value is high. Thus, no clear trend.

Amygdala (AMY)

Topography, Delimitation, Structural Characteristics, Fiber Connections
(Atlas A 7 - A 2; Figs. 81 - 84)

The amygdala (AMY) plays an important role not only in the modulation of endocrine functions, visceral effector mechanisms, and emotional components of behavior through its connections with the hypothalamus, but also because of its influence on cortical cognitive functions (Mishkin and Aggleton 1981). Its role as a link between the hypothalamus and cortex requires that AMY be a major site of convergence for vegetative and sensory information.

AMY lies at the base of the piriform lobe medial to the caudal part of PRPI (Figs. 82 - 84). Many complexes, nuclei and areas were subdivided. Several of them could be accurately delineated for measurement in different species. In all species (from Insectivora through Primates), it was possible to distinguish between two amygdaloid divisions whose borders are characterized by a cell-poor lamina intermedia, intercalated cell masses, and different cell sizes and densities in the structures on either side (Stephan and Andy 1977).

The centromedial amygdaloid complex, or simply medial amygdala (MAM), is composed of the central and medial amygdaloid nuclei and of the anterior amygdaloid area. The intercalated cell masses, located at and in part marking the border with the cortico-basolateral amygdaloid complex, or simply lateral amygdala (LAM), were arbitrarily included with MAM. LAM is composed of lateral, cortical, and basal amygdaloid nuclei; the basal nucleus may be further

Figure 81: Map of periamygdaloid surfaces (PAM) in the long-eared desert hedgehog *(Hemiechinus auritus)*. The levels of Figs. 82 - 84 are marked.

Directions:
c caudal
l lateral
m medial
r rostral

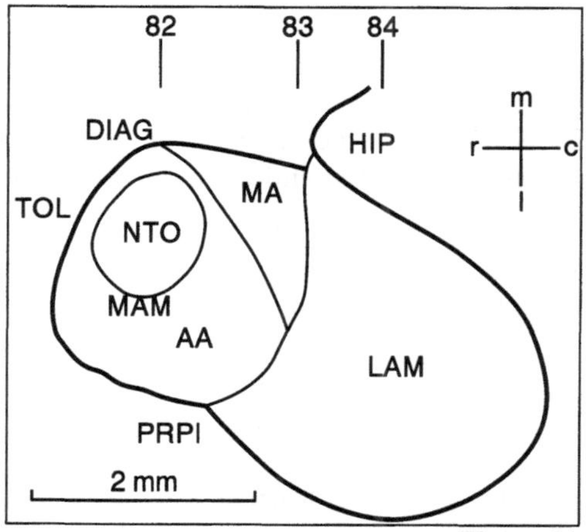

Figure 81

Figures 82 - 84: Transversal sections through the amygdala of *Hemiechinus auritus* at the levels marked in Fig. 81. Enlargement 14.4 x.

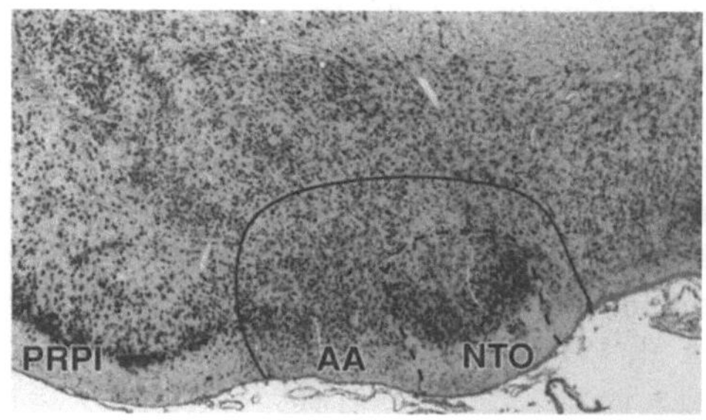

Figure 82

Abbreviations for Figs. 81 - 84:

AA	anterior amygdaloid area
BA	basal amygdaloid nucleus
CAM	cortical amygdaloid nucleus
CEA	central amygdaloid nucleus
DIAG	diagonal band of Broca
HIP	hippocampus
LA	lateral amygdaloid nucleus
LAM	lateral amygdaloid division
MA	medial amygdaloid nucleus
MAM	medial amygdaloid division
NTO	bed nucleus of lateral olfactory tract
PRPI	regio prepiriformis
SAN	sulcus semiannularis
TOL	olfactory tubercle

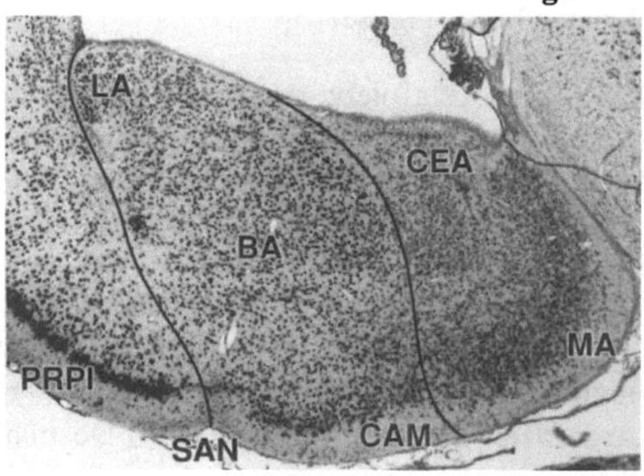

Figure 83

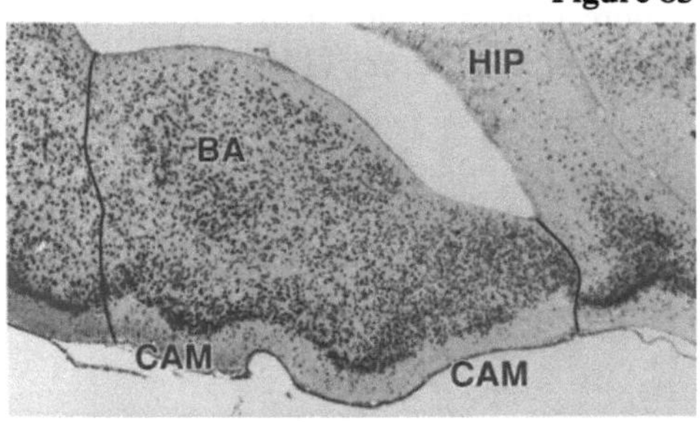

Figure 84

subdivided into accessory, magnocellular and parvocellular parts. Of the finer parts, only the bed nucleus of the lateral olfactory tract (NTO), which belongs to MAM, and the magnocellular part of the basal nucleus (MCB), which belongs to LAM, could be delineated with sufficient accuracy for measurement.

Superficial parts of AMY have a distinct molecular layer and, therefore, were considered cortical. In MAM, such surfaces exist in the anterior amygdaloid area and the medial nucleus, and in LAM, in the cortical nucleus. They were grouped under the term "periamygdaloid" allocortex (PAM). The PAM surface was measured and compared in a few species. Its percentage size in the complex of olfactory cortices (RB, PRPI, TOL, PAM) is 19.5% (see table on p. 111), i.e., nearly one fifth. In the Night Monkey (*Aotus trivirgatus*, the only simian so far investigated), the percentage of PAM is much higher (24.1%). This percentage increase, however, is due to a strong reduction in RB (14.9% in Insectivora, 1.9% in *Aotus*). For a detailed description of the microscopic anatomy of the superficial parts of AMY, see Stephan (1975, pp. 339 - 401).

The amygdala receives afferent fibers from (1) the main olfactory system (olfactory bulb, olfactory cortices), (2) accessory olfactory system, (3) hypothalamus, and (4) neocortical regions. Efferent fibers reach (1) the septum, (2) hypothalamus, (3) thalamus, (4) formatio reticularis and certain brain stem nuclei, and (5) various cortical areas (for ref. see Stephan 1975; Nieuwenhuys et al. 1978, 1980; Stephan et al. 1987a).

Size comparison
(Tables 43 - 45; Fig. 86)

AMY is a relatively small brain component, representing 3.2% of the NET brain and 5.4% of TEL. Its size may be described as follows:

Averages: 114 in the 50 species, 122 in the twelve subfamilies and families. Next to the olfactory structures, this is the lowest level.

Range: 70 - 194. The highest value *(Hylomys)* is 2.8 times larger than the lowest *(Hemicentetes* and *Suncus etruscus)*. When *Geogale* (61) is included, the quotient is 3.2. This variation is relatively low.

Small size: *Geogale* (61), *Suncus* spp. (74), Potamogalinae (84).

Large size: Echinosoricinae (186), Desmaninae (161), *Solenodon* (160).

Characteristics and trends: General tendency toward enlargement from the more conservative to the more advanced forms. Strong deviations, such as large size in *Setifer* (149), may be due to olfactory influences.

Medial amygdaloid division (MAM)

Size comparison
(Tables 55 - 57)

This group of centromedial nuclei and anterior amygdaloid area represents on the average about 45% of AMY in Insectivora. Its size may be described as follows:

Averages: 114 in the 50 species, 117 in the twelve subfamilies and families. This is a low level, slightly lower than that of AMY.

Range: 69 - 191. The highest value *(Hylomys)* is 2.8 times larger than the lowest *(Hemicentetes)*. When *Geogale* (64) is included, the quotient is 3.0. This variation is relatively low.

Small size: *Geogale* (64), *Hemicentetes* (69), Potamogalinae (80), *Suncus* spp. (81), *Limnogale* (92), Tenrecinae (100).

Large size: Echinosoricinae (173), *Chrysochloris stuhlmanni* (166), *Setifer* (149), Desmaninae (147).

Characteristics and trends: Tendency toward general enlargement (e.g., Desmaninae) seems to be intensified with strong development of olfaction (e.g., Echinosoricinae, *Setifer*).

A small component of this group (about 1/15 of MAM) is the bed nucleus of the lateral olfactory tract (NTO), which varies widely from 4 in *Desmana* to 204 in *Sylvisorex granti*. This nucleus is very small in all semiaquatic forms and in *Ruwenzorisorex* (45), and largest in *Sylvisorex* spp. (172), *Myosorex* (154), *Setifer* (153), *Hemiechinus* (151), and *Chrysochloris stuhlmanni* (150). Echinosoricinae are not measured so far. When olfactory structures become reduced, this nucleus is even more affected than the MOB.

Lateral amygdaloid division (LAM)

Size comparison
(Tables 55 - 57)

Size and size differences of this cortico-basolateral group of amygdaloid nuclei may be described as follows:

Averages: 115 in the 50 species, 127 in the twelve subfamilies and families. This is a slightly higher level than that of the total amygdala.

Range: 62 - 204. The highest value *(Galemys)* is 3.3 times larger than the lowest *(Suncus etruscus)*. When *Geogale* (58) is included, the quotient becomes 3.5. The variation is low to medium.

Small size: *Geogale* (58), *Suncus* spp. (69), Potamogalinae (88), and *Crocidura* spp. (91).

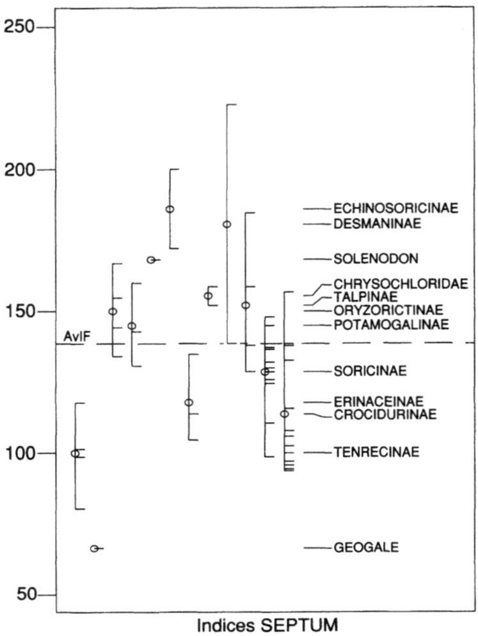

Figure 85

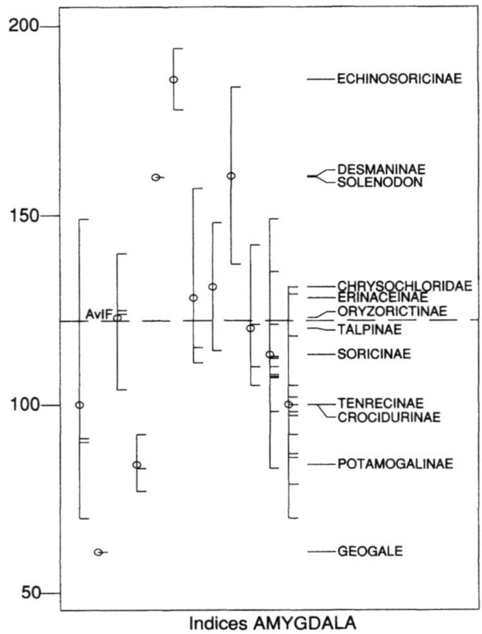

Figure 86

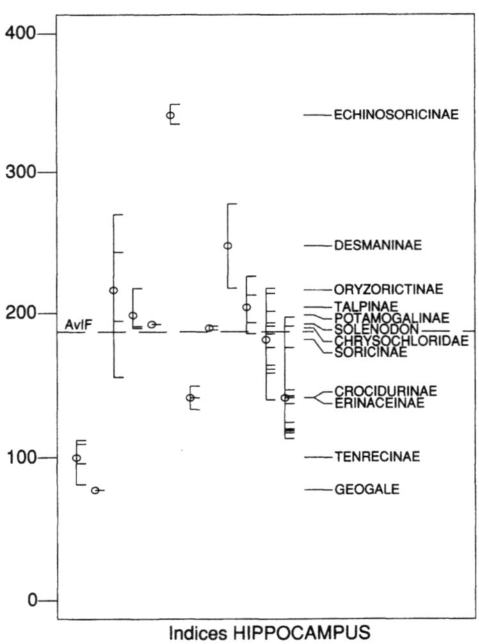

Figure 87

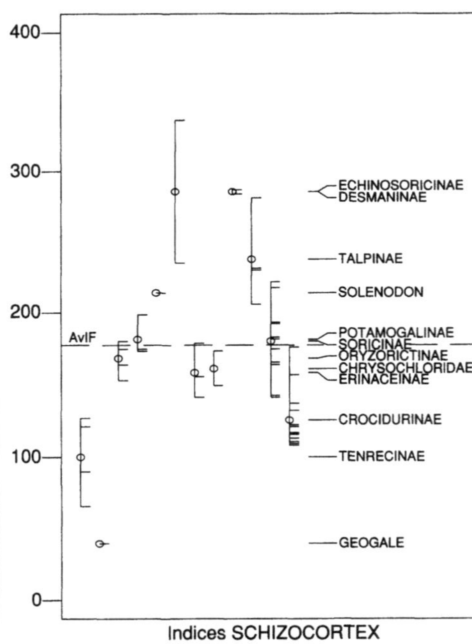

Figure 88

Large size: Echinosoricinae (198), *Solenodon* (187), Desmaninae (173), *Hemiechinus* (173).

Characteristics and trends: Tendency toward general enlargement is more distinct than in MAM; olfactory influences appear lesser.

A conspicuous component of LAM (about one fifth of the structure) is the magnocellular part of the basal nucleus (MCB). Its minimum size was found in *Suncus etruscus* (40), the maximum (242) in *Galemys*. The largest value is 6.1 times larger than the smallest, a relatively wide variation. The indices of MCB compared with those of LAM are lower in most forms, but larger in *Sorex minutus*, Talpidae, *Echinops*, and *Neomys*.

Hippocampus (HIP)

Topography, Delimitation, Structural Characteristics, Fiber Connections

(Atlas A 10 - A 1; Figs. 89, 90)

The hippocampus (HIP) is an important part of the limbic system which is concerned with very complex functions such as emotional reactions and general endogenous stimulation, as well as attentiveness, vigilance and memory (e.g., Hassler 1964).

HIP is a crescent-shaped structure occupying most of the caudomedial walls of the cerebral hemispheres and may be subdivided into three parts: HP (precommissural hippocampus), lying rostral and/or ventral to the rostrum of the corpus callosum; HS (supracommissural hippocampus; indusium griseum), lying directly on the surface of the corpus callosum; and HR (retro- or postcommissural hippocampus), behind the septum and corpus callosum, extending ventral to the amygdala. The first two parts may be combined under the term HA (hippocampus anterior), the third is the main part, and, if sectioned trans-

---

Figures 85 - 88: Size indices for telencephalic components.

Figure 85: Septum (SEP)
Figure 86: Amygdala (AMY)
Figure 87: Hippocampus (HIP)
Figure 88: Schizocortex (SCH)

The indices are based on AvTn lines (index = 100), whose equations are, respectively,

$$\log \text{SEP volume} = -0.129 + 0.61 \times \log \text{BoW} \quad (1.39)$$

$$\log \text{AMY volume} = 0.232 + 0.60 \times \log \text{BoW} \quad (1.22)$$

$$\log \text{HIP volume} = 0.581 + 0.57 \times \log \text{BoW} \quad (1.86)$$

$$\log \text{SCH volume} = 0.062 + 0.64 \times \log \text{BoW} \quad (1.78)$$

The distances of the AvIF broken lines are given in brackets. For additional explanations, see Fig. 53.

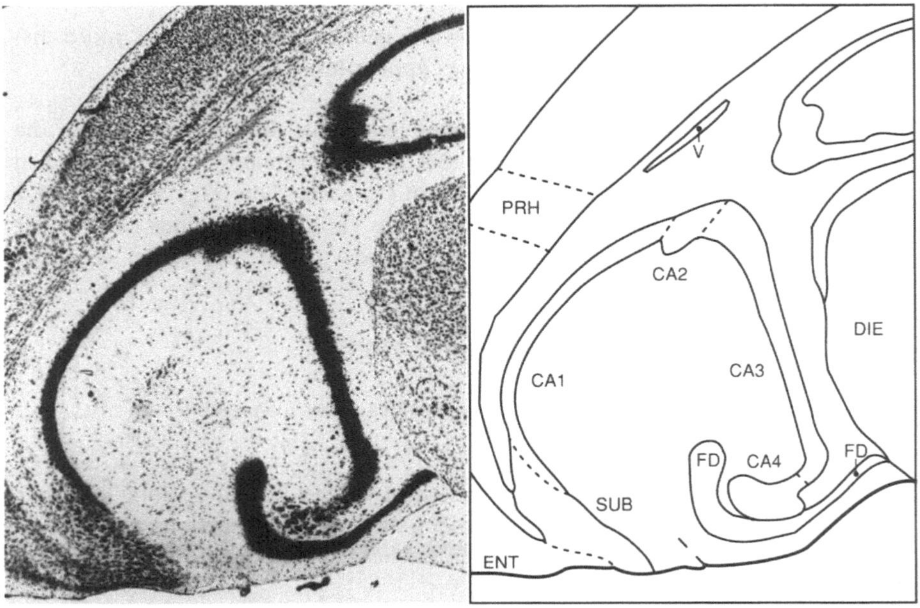

Figure 89

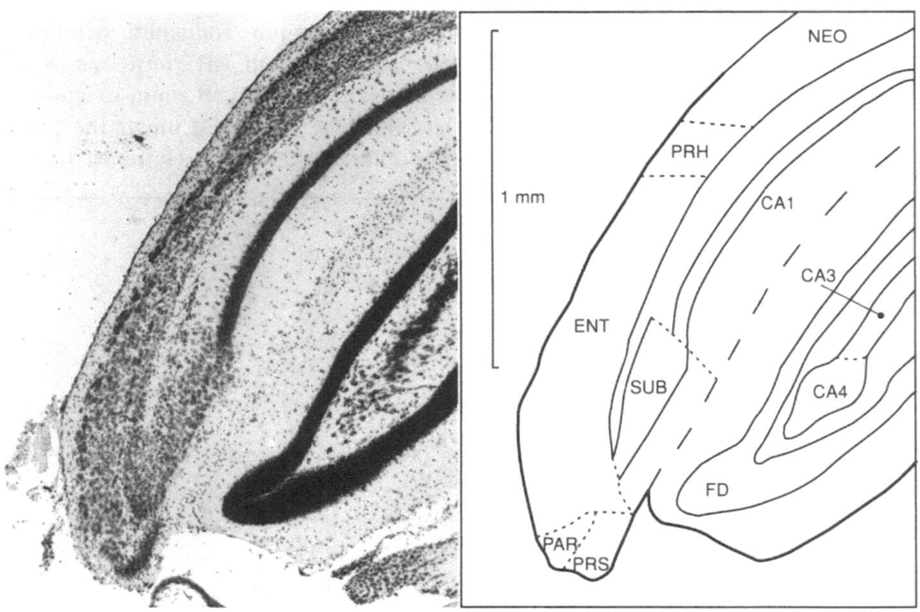

Figure 90

verse to its longitudinal axis, three components can be readily differentiated: fascia dentata (FD), Ammon's horn or cornu ammonis (CA), and subiculum (SUB). CA can be subdivided into an upper and a lower region, and into a third or end blade, which is surrounded by FD (Cajal 1893; Koelliker 1896). Of the possible cornu ammonis subdivisions that of Lorente de Nó (1934) into CA1 - CA4 is the most widely accepted.

CA4 is in the end blade; CA3 corresponds to the lower region; CA1 to the upper region, and CA2 forms a kind of transitional zone in which the upper and lower regions meet. Stephan (1975) added another transitional zone (CA0) between SUB and CA1, which in several species (as e.g. the prosimian *Lepilemur*) is distinct and highly characteristic. For a detailed description of the microscopic anatomy of HIP, see Stephan (1975; pp. 494 - 642).

Although all parts consist of three characteristic layers, i.e., the stratum moleculare, the stratum pyramidale or granulare, and the stratum multiforme, they contain marked cytoarchitectural differences. The FD, which forms a cap over the free edge of CA, contains a densely packed layer of granular cells. Pyramidal cells predominate in CA and SUB.

HIP has afferent connections with the septum, hypothalamic nuclei, parahippocampal gyrus, and the dopaminergic and serotoninergic centers of the brain stem. Efferent fibers can be divided into several pathways. The precommissural fibers of the fornix extend to the septum, gyrus rectus, frontal cortex, and nucleus accumbens. The postcommissural fibers project mainly into the hypothalamus, where the majority of fibers terminate in the mamillary bodies. A smaller number of fibers reach the anterior nucleus of the thalamus and the nuclei of the stria terminalis (Swanson and Cowan 1975, 1977; Meibach and Siegel 1975; Rosene and VanHoesen 1977). The postcommissural fornix is part of a closed system known as Papez circuit (hippocampus —> mamillary body —> anterior nucleus of the thalamus —> cingulate gyrus and cingulum —> hippocampus). SUB also projects to the amygdala and to cortical regions, including the area entorhinalis, regions of the temporal, frontal, and cingulate cortex (for ref. see Stephan 1975; Nieuwenhuys et al. 1978, 1980; Bayer 1985).

Figures 89 - 90: Transversal sections through hippocampus retrocommissuralis (Fig. 89; enlargement 28.5x) and schizocortex (Fig. 90; enlargement 47.3x) in *Sorex minutus*.

Abbreviations:

CA1	ammons horn area CA1	NEO	neocortex
CA2	ammons horn area CA2	PAR	parasubicular region
CA3	ammons horn area CA3	PRH	perirhinal region
CA4	ammons horn area CA4	PRS	presubicular region
DIE	diencephalon	SUB	subiculum
ENT	entorhinal region	V	ventricle
FD	fascia dentata		

Bilateral lesions of HIP cause a dramatic loss of recent memory. The lack of memory disorders following transsection of the fornix may indicate the importance of the direct subiculo-cortical efferents (Rosene and VanHoesen 1977).

Size comparison
(Tables 43 - 45; Fig. 87)

HIP is the third largest component of TEL next to NEO and PAL. It represents on the average 9.3% of NET and 15.7% of TEL. Its size may be described as follows:

Averages: 175 in the 50 species, 186 in the twelve subfamilies and families. Next to NEO and STR, this is the third highest level in the eight main parts of TEL.

Range: 82 - 348. The highest value *(Echinosorex)* is 4.2 times larger than the lowest *(Tenrec)*. When *Geogale* (77) is included, the quotient is 4.5 times higher. This variation is medium.

Small size: *Geogale* (77), Tenrecinae (100), *Suncus* spp. (116).

Large size: Echinosoricinae (340), *Microgale* spp. (257), Desmaninae (247).

Characteristics and trends: General tendency toward enlargement from the more conservative to the more advanced forms is obvious, but the extremely large size in Echinosoricinae and the relatively small size in *Solenodon* indicate an influence of specialization.

Hippocampus Components

Size comparison
(Tables 58 - 60)

By far the largest component is the **retrocommissural hippocampus** (HR; Atlas A 6 - A 1), representing on the average 93.6% of HIP, whereas **supracommissural** (HS; Atlas A 8 - A 6) and **precommissural hippocampus** (HP; Atlas A 10 - A 9) are small components (1.1 and 5.3% of HIP). The size of these parts was investigated in 10 conservative species only and thus, the range and averages for the allometric HIP indices are not representative of the whole order of Insectivora, as was also the case in components of MES, CER, and DIE. In the limited number of species, the HIP range from 82 to 159 (instead of 77 - 348, as given above), and the averages are 122 for the ten species and 128 for the four subfamilies and families (instead of 175 and 186 for the larger numbers of species and groups). The three parts (HR, HP, HS) have similar average indices.

Of the "basal" forms, Tenrecinae always have the smallest average size of HIP components, and Soricidae have slightly higher indices than Erinaceidae. The highest values are reached by HS of the shrews. Since HS covers the dorsal surface of the corpus callosum, its size may be related to the size of the corpus callosum which is distinctly larger (longer) in Soricidae than in Tenrecinae and Erinaceinae (see Section 3.1 and Table 4). The lowest indices were found in HP. This rostralmost part, lying between the genu of the corpus callosum and the olfactory peduncle, is functionally related to olfaction.

Area measurements were performed in a limited number of species for some components of HR, such as subiculum (SUB), some CA areas, and fascia dentata (FD). On the average (N = 15), FD area is slightly more than one third (37.7%) of HR area, and SUB+CA slightly less than two thirds (62.3%). SUB (N = 10) is slightly more than one fifth (20.7%) of SUB+CA (CA = 79.3%), and, within CA areas, CA1 and CA2/3 are of about the same size (47% - 48%), whereas the transitional zone (CA0) between SUB and CA is a small component, making up slightly less than 5%. Size indices based on these area measurements reveal the least progression for CA2/3 (102) and the highest for CA0 (133) and CA1 (128). FD and SUB+CA have about equal average indices in Insectivora. The highest values were found in *Galemys*. Echinosoricinae, whose HIP indices are by far the highest, have not yet been investigated.

Schizocortex (SCH)

Topography, Delimitation, Structural Characteristics, Fiber Connections
(Atlas A 5 - AP 0; Figs. 89, 90)

The schizocortex (SCH) is one of the main sources of fibers to HIP and thus is closely related to it functionally. SCH comprises entorhinal (ENT), perirhinal (PRH), and presubicular (PRS, including parasubicular) cortices and the underlying white matter. These cortices are characterized by the presence of one or more, almost cell-free layers or sublayers (= "schizocortex"). ENT covers the most caudal part of the piriform lobe. Its laminar differentiation is poor in Insectivora. The whole intermediate zone of the band of cells is cell-poor and appears pale in Nissl-stained material. PRH is a transitional zone between ENT and NEO. PRS covers the medial hemispheric wall in a stripe between HIP and NEO. For detailed descriptions of the microscopic anatomy of these fields, see Stephan (1975; pp. 642 - 758).

Schizocortical structures, such as those of the gyrus parahippocampalis, contain mainly the entorhinal cortex. They receive afferent fibers from large areas of the neocortex and send efferents to the hippocampus. The fiber connections show that SCH is not only topographically and cytoarchitecturally, but also functionally, a transitional zone between the neocortex and the hippocampus.

Size comparison
(Tables 43 - 45; Fig. 88)

SCH is one of the smallest TEL components representing, 6.0% of TEL and 3.5% of NET brain. Its size may be described as follows:

Averages: 168 for the 50 species, 178 for the twelve subfamilies and families. The level is medium (fourth) for the eight main TEL components.

Range: 65 - 336. The highest value *(Hylomys)* is 5.2 times larger than the lowest *(Echinops)*. This variation is relatively broad. When the very low index of *Geogale* (40) is included, the variation becomes much larger (8.4 times).

Small size: *Geogale* (40), Tenrecinae (100), *Suncus* spp. (113).

Large size: Echinosoricinae (285), Desmaninae (285), Talpinae (237), *Solenodon* (213).

Characteristics and trends: Clear tendency toward enlargement, from the generally conservative to the generally advanced forms.

According to **area measurements** (Stephan 1961), the largest component of SCH is the entorhinal cortex (ENT), which on the average (N = 11), accounts for between one half and two thirds of SCH (57%). The presubicular cortex (PRS, including a very small parasubicular component) makes up one third (33%), and the perirhinal cortex (PRH), which is transitional between ENT and NEO, about one tenth.

Size indices based on these area measurements reveal a small SCH in *Setifer* and *Chrysochloris* and a large one in *Galemys, Talpa* and *Neomys*. In *Setifer* (the only Tenrecinae measured), small size is typical of all three SCH components, while in *Chrysochloris* PRS is of average size. Large size was found for all components in *Neomys fodiens* and *Talpa europaea*, but not for PRH in *Galemys*. In general, the results of area measurements show similar trends as found in volume measurements.

Neocortex (NEO)

Topography, Delimitation, Structural Characteristics
(Atlas A 10 - AP 0; Figs. 80, 89, 90)

The neocortex (NEO) contains the highest sites of integration in the brain. It covers the convexity of the cerebral hemispheres and is not uniform, differing both in laminar and areal differentiation. It comprises various projection and integration areas.

Rehkämper (1981) has shown that, in Insectivora, the laminar differentiation differs among species and also among areas within the same species. The most conspicuous changes occur in a laminar zone, which he called "lamina III," and concern differentiation of a granular cell layer corresponding to Brodmann's (1909) "lamina granularis interna." In all species, lamina III contains a layer of pyramidal cells. Based on the differentiation of a second layer, the granular layer below the pyramidal cells in lamina III, Rehkämper distinguished three groups of Insectivora. One group, comprising *Crocidura, Echinops,* and *Solenodon,* has no granular layer in lamina III. In the second group, including *Chrysochloris, Talpa,* and *Erinaceus,* the granular layer is differentiated in some neocortical areas but not in others. In the third group, to which *Neomys, Desmana,* and *Potamogale* belong, the granular layer is present in all neocortical areas. Based on cytoarchitectural and topographical criteria, Rehkämper distinguished four areas in NEO: two frontal, a caudomedial, and a caudolateral. The subdivision of the two frontal areas is based simply on the existence of cell groups near the deep layer in area 1. These cell groups may be considered part of the claustrum.

Comparative quantitative investigations of areal components (fields, regions) are difficult to make because it is difficult to establish homologies and clearly differentiate the areas. Unambiguous borders around clearly homologous areas are preconditions for comparative quantitative studies. Therefore, in our measurements, NEO is regarded as an areal unit, despite the fact that, in Insectivora too, different fields or regions have been described (see Section 4.4.2). To our knowledge, only one study of area measurements which included Insectivora has been carried out. It was performed by Brodmann (1913) who measured the precentral region and area striata of the hedgehog. We were not able to delimit NEO areas in the great variety of Insectivora species accurately enough to allow for exact measurement. Area striata size was inferred from Primates using size of the lateral geniculate bodies in both Primates and Insectivora (Frahm et al. 1984).

In our volume data for the total NEO, the underlying white matter and the corpus callosum are included, as are the insular (claustrocortical), subgenual, cingular and retrosplenial regions of the transitional cortices. Detailed descriptions of these transitional cortices have been given by Stephan (1975; pp. 464 - 494, 758 - 838).

Size comparison
(Tables 43 - 45; Figs. 91, 92)

NEO is the largest of the eight main TEL components, representing 28.1% of TEL. Relative to NET brain, it accounts for 16.6% and thus is also larger than any of the non-telencephalic brain components. Its size and size differences may be described as follows:

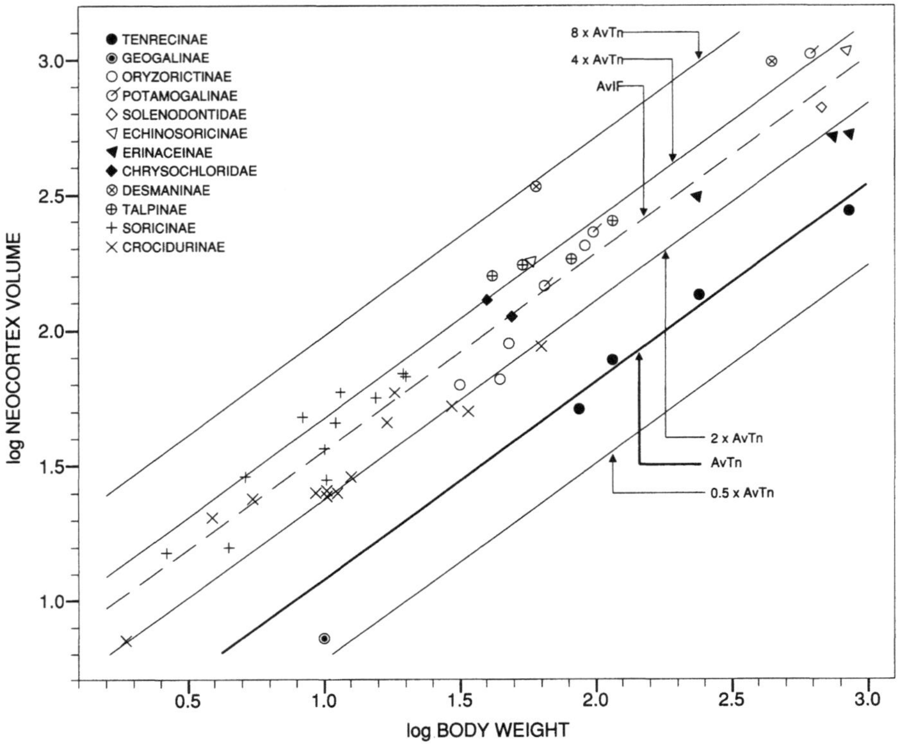

Figure 91

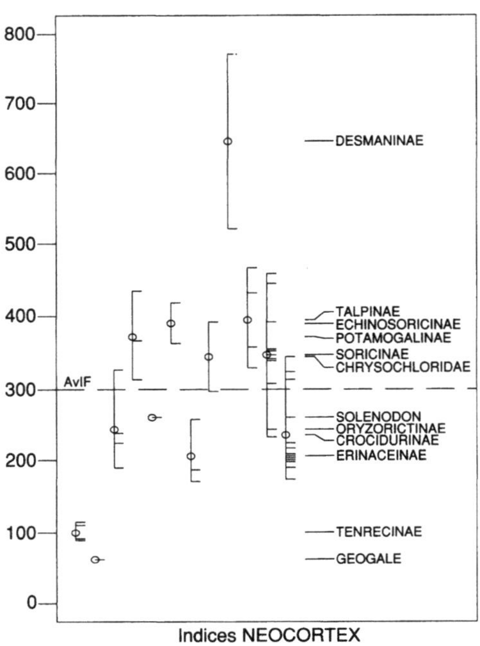

Figure 92

Figure 91: Neocortex volume (in mm^3) plotted against body weight (in grams) in a double logarithmic scale. The equation of the AvTn line is log NEO volume = 0.346 + 0.73 x log BoW.

The broken AvIF line is at a distance of 3.00 x AvTn. For additional explanations, see Fig. 52.

Figure 92: Size indices for neocortex (NEO). The indices are numerical values of the distances from the AvTn line given in Fig. 91.

The broken AvIF line has an index of 300. For additional explanations, see Fig. 53.

Averages: 292 for the 50 species, 300 for the twelve subfamilies and families. This is the highest average level reached in any of the brain components, telencephalic or other.

Range: 89 - 772. The highest value *(Galemys)* is 8.7 times larger than the lowest *(Tenrec, Echinops)*. This variation is very high. It becomes even larger when *Geogale* (61) is included (12.7 times). This is the highest variation found in any of the larger brain components.

Small size: *Geogale* (61), Tenrecinae (100). No overlap with other forms.

Large size: Desmaninae (645). No overlap with next groups: Talpinae (396), Echinosoricinae (391), Potamogalinae (370).

Characteristics and trends: NEO reflects the general evolutionary trends most clearly and thus is the best indicator of general brain evolution. Large differences exist between subfamilies of the same family: in Tenrecidae there are wide steps from Geogalinae (61) to Tenrecinae (100) to Oryzorictinae (244) and finally to Potamogalinae (370); in Erinaceidae, from Erinaceinae to Echinosoricinae (205 to 391); in Talpidae, from Talpinae to Desmaninae (396 to 645); and in Soricidae, from Crocidurinae to Soricinae (234 to 347). Most of these steps are significant (see next paragraph). In Crocidurinae, there are large differences between two groups: one includes *Crocidura* spp., *Suncus* spp., and *Scutisorex* (av. index is 201) and the other, *Sylvisorex* spp., *Ruwenzorisorex,* and *Myosorex* (av. index is 309). Similar differences have already been mentioned for other structures, e.g., STR.

Significance tests (T-tests) for differences between the twelve subfamilies and families (see table on p. 85) revealed, in 65 combinations for neocortex, significant differences in 22 combinations at the 1% level and in eleven combinations at the 5% level. No significance could be found in 32 cases. Tenrecinae have significantly smaller NEO than all other subfamilies and families (1% level) except for *Geogale*, which has a significantly smaller NEO at the 5% level. The frequent significance of Crocidurinae (eight in eleven comparisons) may be due to the relatively large number of species in this subfamily. Significant differences between Crocidurinae and Soricinae could be demonstrated since Soricinae, too, are represented by a fairly large number of species. Thus, as is well known, both the degree of differences and the number of species substantially affect significance tests.

Neocortex laminar components

NEO was subdivided into white (NW) and grey matter (NG) and within the grey matter the molecular layer (NG1) was separated from the remaining, more cell-dense layers (NG2-6). NEO is composed on the average of 9.6% NW and 90.4% NG, and NG is composed of 25.6% NG1 and 74.4% NG2-6.

Size comparison
(Tables 61 - 63)

Averages: The highest averages were found for NG2-6 (335 for the 50 species, 343 for the twelve subfamilies and families), and the lowest for NG1 (209 and 214). In the increasing NEO, the molecular layer (NG1) increases much less than the cell-dense layers. NG (296 and 301) and NW (266 and 284) have intermediate averages.

Characteristics: NW increases more than NG with increasing size (BoW, or BrW, or NEO size). When this occurs in narrowly related groups, it is reflected by a higher slope, i.e., a larger BoW power (0.83 for NW versus 0.72 for NG; see Tables 62, 63). Increasing NW size independent of BoW may also occur in more advanced species. Species with the largest NEO indices (Desmaninae, Echinosoricinae) have higher NW than NG indices.

Neocortex areal components, area striata

Size comparison
(Table 64)

We were not able to delimit NEO areas and regions precisely enough over the whole spectrum of Insectivora to measure their volumes. However, we estimated the volumes of the area striata grey (ASG) from the volumes of the dorsal part of the corpus geniculatum laterale (GLD) based on the close relationship between GLD and ASG in primates (Frahm et al. 1984). This relationship follows the equation log GLD = -0.8801 + 0.8476 x log ASG. Using this equation, the GLD volumes given in Table 37 result in the ASG volumes in Table 64. In the paper mentioned, the size of GLD in Insectivora was determined on one side only, but was erroneously inserted in the primate equation, in which GLD and ASG were from the two sides of the brain. Thus, the ASG volumes for Insectivora, obtained by Frahm et al. (1984) were too small.

Percentages: Relative to total neocortex grey (NG = 100%) ASG is 12.6% on the average in Insectivora, but the variation was broad, from 4.5% in *Sorex araneus* to 27.6% in *Erinaceus europaeus*. This latter value, estimated indirectly from GLD size, may be too high and differs enormously from the 12.4% found by Brodmann (1913) for the European hedgehog. The size of the visual fields given in cortical maps resulting from electrophysiological (Lende and Sadler 1967; Lende 1969) and cytoarchitectural investigations (Rehkämper 1981) (see Section 4.4.2 and Figs. 98 and 99 on p. 216) seem to support the figures obtained by Brodmann. Thus, the relations between GLD and ASG found in Primates may be of limited value for determining ASG size in Insectivora. The rather large volume shown for ASG in Table 64 of course inflates the indices given in this Table, whereas the relative differences between the various taxonomic units are less affected.

Indices: The size indices (tenrecine level = 100) may be described as follows:
Averages: 99 in the ten species, 136 in the four subfamilies.
Range: 35 - 350. The higest value *(Erinaceus europaeus)* is 10 times larger than the lowest *(Crocidura russula).* Compared with other structures, this is wide variation.
Small size: Soricinae (39), Crocidurinae (55).
Large size: *Erinaceus* (350).
Characteristics and trends: The results confirm the general impression and the data obtained from other visual structures (Tables 35, 38, 69, 72) indicating that, in Insectivora, hedgehogs have relatively good, and shrews, relatively poor visual capacities (see also Herter 1933; Polyak 1957, p. 961). So far, there are no volume measurements on the visual cortex of subterranean forms.

3.3.6 Comparison of Size of the Structures in Insectivora and Primates

Some size characteristics of the various brain structures in Insectivora are summarized in Table 74, which presents the maximum and minimum indices, relations between maximum and minimum as an indication of size variation, averages of the 50 (or fewer) species and of the twelve (or fewer) families and/or subfamilies, and their coefficients of variation (CV) as an additional measure of size variation. To facilitate comparison of some main complexes, the averages obtained for the species were scaled (Fig. 93). Species averages (= AvIS) instead of family and subfamily averages (= AvIF) were chosen to allow for comparison with corresponding data on primates (Figs. 94 - 97). The numbers of species of Insectivora on which the averages are based are given in Tables 10 - 70. The comparison with primates may provide an insight into the direction and intensities of trends beyond the variation in Insectivora. The size of the brain parts in primates is also referred to the tenrecine base (AvTn) and thus the indices are directly comparable. In primates they are, in general, averaged for 18 species of prosimians and 26 species of non-human simians, and are also given for man. Deviating numbers of species are given in brackets after the average index.

The Five Fundamental Brain Parts

In **Insectivora** (Fig. 93), the highest averages and thus progressive enlargements in the five fundamental brain parts are found in diencephalon (DIE) and cerebellum (CER). These structures are about 1.7 times larger in the average Insectivora than in the average Tenrecinae. In descending sequence are telencephalon (TEL), mesencephalon (MES), and medulla oblongata (OBL). The average enlargement of the OBL is 1.32x AvTn.

	Tenre-cinae	Insecti-vora	Pro-simians	NH-simians	*Homo*
OBL	100	132	153	176	187
MES	100	136	261	308	432
CER	100	167	459	633	2057
DIE	100	169	523	706	1107
TEL	100	159	457	921	3250
N	4	50	18	26	1

In the three groups of **Primates** (Fig. 95), the five fundamental brain parts have different sequences and distinctly stronger differences in progression compared with Insectivora. All the fundamental brain parts are clearly enlarged. As in Insectivora, the smallest increase is in OBL. In man, OBL is only 1.87x AvTn and 1.42x AvIS (187/132). Similarly, in all primate groups, OBL is in the lowest position of the fundamental brain parts (Fig. 95). The second lowest place is uniformly occupied by the mesencephalon (MES) in Insectivora and all primate groups. Its increase is, however, stronger than that of OBL. In man, MES is 4.32x AvTn and 3.18x AvIS (432/136). The other fundamental brain parts (TEL, DIE, CER) occupy different places in the various taxonomic groups. In Insectivora, the telencephalon (TEL) occupies the middle (third) place, in prosimians, the second highest, and in non-human simians and man, the highest places. In man, TEL is about 32x AvTn and 20x AvIS (3250/159). In Insectivora and prosimians, the diencephalon (DIE) occupies the highest position, in non-human simians the second highest, and in man the third highest place. In man, DIE is about 11x AvTn and 6.5x AvIS (1107/169). In Insectivora, prosimians and man, the second highest place is occupied by the cerebellum (CER). In non-human simians, CER has only the middle (third) position. In man, CER is nearly 21x AvTn and 12x AvIS (2057/167).

All fundamental brain parts have their largest indices in man. There are, however, very large differences in the extent of increase. Compared with the average Insectivora, the increase is lowest in OBL (1.42x AvIS). In ascending order follow MES (3.18x), DIE (6.55x), CER (12.3x), and TEL (20.4x). Thus, in primates, TEL is by far the most progressive brain part (Fig. 95). This is in contrast to Insectivora, where, on the average, DIE and CER are highest.

Medulla oblongata

In **Insectivora** (Fig. 93, and table on p. 142), most parts of OBL show a moderate increase. The highest progression is found for the inferior olive (INO), which is on the average 1.85x AvTn. Averages >1.4 are also found for the nucleus reticularis lateralis (REL, 1.7x AvTn), sensory trigeminal complex (TR, 1.6x), nucleus tractus solitarii (TSO, 1.5x), nucleus vestibularis medialis (VM) and nucleus dorsalis motorius nervi vagi (mot X) 1.4x. An average slightly smaller than in the Tenrecinae is found for the nucleus cochlearis ventralis (VCO, 0.90x AvTn).

Abbreviations for Figures 93 - 97 (on the following pages):

AMY	amygdala	OLS	superior olive
ANT	nuclei anteriores thalami	PAL	paleocortex
AOB	accessory olfactory bulb	PRP	nucleus prepositus hypoglossi
CER	cerebellum	REL	nucleus reticularis lateralis
CGL	corpus geniculatum laterale	SCH	schizocortex
CGM	corpus geniculatum mediale	SEP	septum telencephali
COE	locus coeruleus	STR	striatum
DCO	dorsal cochlear nucleus	TEL	telencephalon
DIE	diencephalon	TR	complexus sensorius nervi trigemini
DOT	nuclei dorsales thalami	TSO	nucleus tractus solitarii
FCE	nucleus fascicularis cuneatus externus	VCO	nucleus cochlearis ventralis
FCM	nucleus fascicularis cuneatus medialis	VET	nuclei ventrales thalami
FGR	nucleus fascicularis gracilis	VI	nucleus vestibularis inferior
HIP	hippocampus	VL	nucleus vestibularis lateralis
INO	inferior olive	VM	nucleus vestibularis medialis
MES	mesencephalon	VS	nucleus vestibularis superior
MLT	nuclei mediales thalami	V	nucleus motorius nervi trigemini
MNT	nuclei mediani thalami	VII	nucleus nervi facialis
MOB	main olfactory bulb	X	nucleus dorsalis motorius nervi vagi
NEO	neocortex		
OBL	medulla oblongata	XII	nucleus nervi hypoglossi

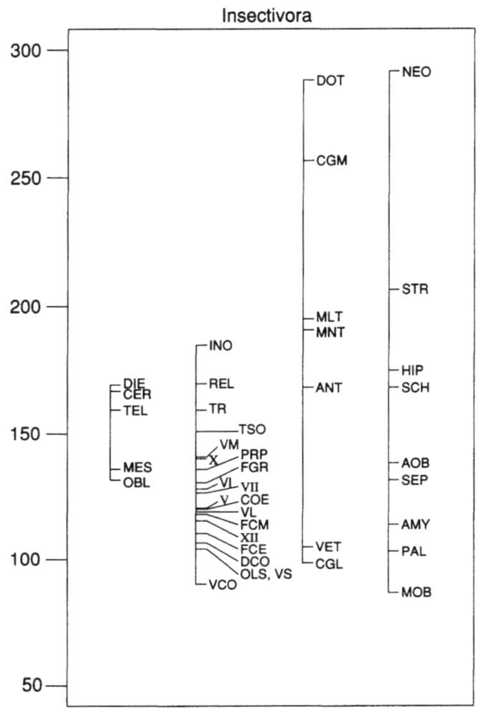

Figure 93

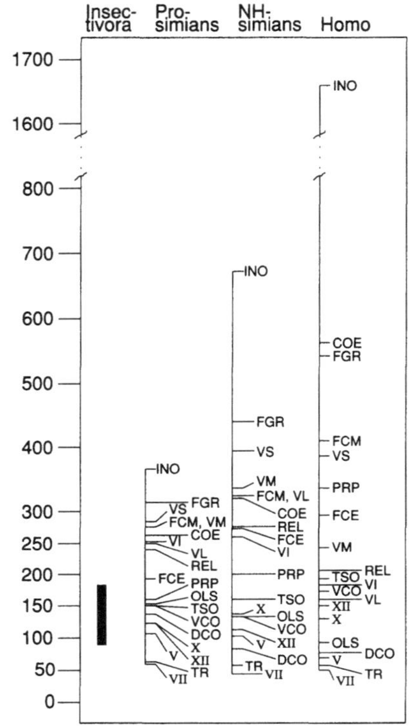

Figure 94

Figure 93: Average size indices for the brain structures in Insectivora. From left to right: the five fundamental brain parts (50 species), the twenty medulla oblongata components (30 species), the seven thalamus components based on volume measurements by Bauchot (1979b) (20 species), and the nine main telencephalic structures (50 species). The average SI for each of the structures in the four Tenrecinae is uniformly given the value 100. Abbreviations on p. 139.

Figures 94 - 97: Average size indices for the brain structures in Insectivora, prosimians, non-human simians, and man. The average SI for each of the structures in the four Tenrecinae is uniformly given the value 100. Abbreviations on p. 139.

Figure 94: The twenty medulla oblongata components, from left to right: Insectivora (30 species), prosimians (18 species), non-human simians (26 species), and man.

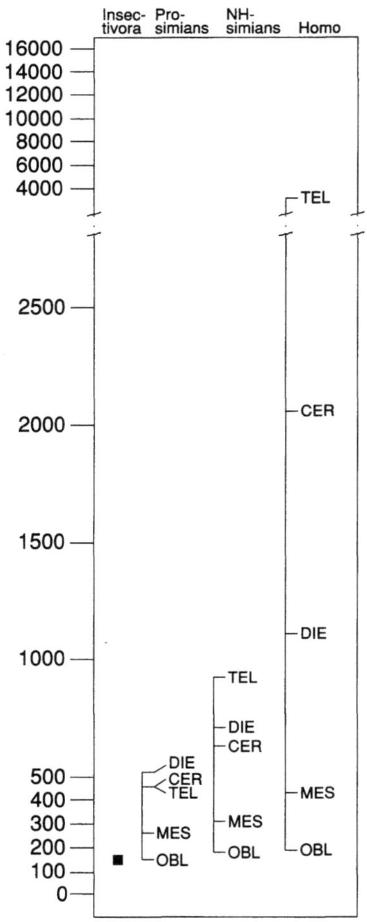

Figure 95

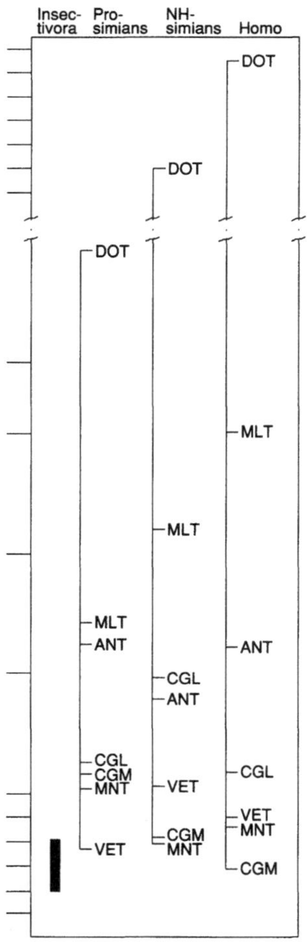

Figure 96

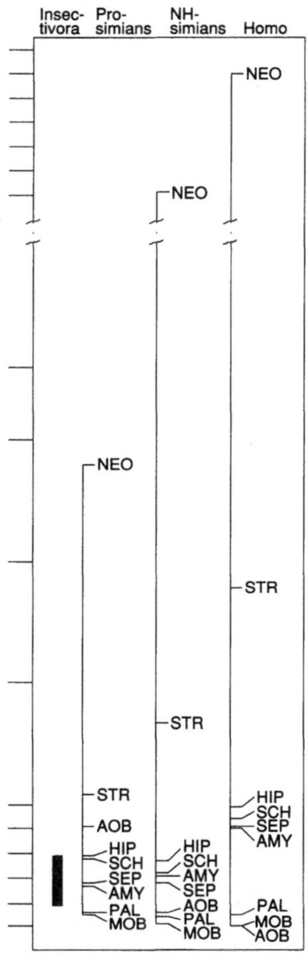

Figure 97

Figure 95: The five fundamental brain parts, from left to right: Insectivora (50), prosimians (18 species), non-human simians (26 species), and man.

Figure 96: The seven thalamus components based on volume measurements by Bauchot (1979b), from left to right: Insectivora (20 species), prosimians (18 species), non-human simians (20 species), and man.

Figure 97: The nine main telencephalic structures, from left to right: Insectivora (50 species), prosimians (18 species), non-human simians (26 species), and man.

	Tenre-cinae	Insecti-vora	Pro-simians	NH-simians	*Homo*
OBL	100	132 (50)	153	176	187
TR	100	159	62	56	58
TSO	100	151	151	160	193
FGR	100	131	312	440	543
FCM	100	118	277	324	409
FCE	100	111	192	274	295
REL	100	170	240	277	206
INO	100	185	367	672	1660
PRP	100	136	160	200	338
COE	100	120 (29)	262	319	563
mot V	100	121	106	102 (25)	70
mot VII	100	127	59	43	50
mot X	100	140	122	136	130
mot XII	100	116	125	114	151
VM	100	141	277	336	243
VI	100	129	253	261	183
VL	100	119	250	323	161
VS	100	104	284	395	388
OLS	100	104	153	135	94
DCO	100	107	138	82	78
VCO	100	90	151	135	174
N	4	30	18	26	1

(N = number of species; differences are given in brackets)

In **Primates** (Fig. 94), many parts of OBL show a moderate increase. There are, however, several structures that are increasingly progressive from Insectivora through prosimians and non-human simians to man. These are, first, the inferior olive (INO), which increases in man to nearly 17x AvTn. Other strongly progressive structures are the nucleus locus coeruleus (COE, 5.6x), nucleus fascicularis gracilis (FGR, 5.4x), and nucleus fascicularis cuneatus medialis (FCM, 4.1x).

Regressive in all primate groups are the nucleus nervi facialis (mot VII, 0.43 - 0.59x) and the sensory trigeminal complex (TR, 0.56 - 0.62x AvTn); in simians and man we also find a regressive nucleus cochlearis dorsalis (DCO, 0.82 and 0.78x), and in man, regressive complexus olivaris superior (OLS, 0.94x) and nucleus motorius nervi trigemini (mot V, 0.70x).

Conformity to the progressive trends is found only in the inferior olive, which has the highest indices both in Insectivora and all groups of primates. The strongest differences are in the sensory trigeminal complex (TR), which is very progressive in Insectivora, but highly regressive in all groups of primates.

TR: The sensory trigeminal complex is distinctly smaller in primates than in Insectivora, with no trend from prosimians to man. Thus, the functional importance of the trigeminal system should be lesser in primates than in Insectivora. This may be mainly because of the decreasing importance of somatosensory input from the orofacial region in feeding and orientation in primates. In Insectivora, the marked enlargement of OBL in water-adapted forms is largely due to the strong enlargement of TR.

TSO: There is a slight progression of the nucleus tractus solitarii from Tenrecinae to primates which, however, does not exceed the range already existing in Insectivora. The small range of variation of TSO may be explained by its relation to visceral functions.

FGR: Clear progressive trends of the nucleus fascicularis gracilis from Insectivora to primates and from prosimians to man.

FCM: Clear progressive trends in the nucleus fascicularis cuneatus medialis from Insectivora to primates and from prosimians to man. In man, FCM is slightly more than 4x AvTn. In all groups, FCM is somewhat less progressive than FGR.

FCE: The nucleus fascicularis cuneatus externus has similar trends than FCM but a lower progression.

REL: The nucleus reticularis lateralis is relatively large in Insectivora. In primates, slight trends toward enlargement relative to Tenrecinae are found. However, all primate values are within the range of Insectivora species (Tables 17, 18) with man's value only slightly above the Insectivora average (206 versus 170).

INO: Of all the OBL components, the complex of the inferior olive has the most pronounced tendency toward enlargement in all taxonomic groups: Insectivora, prosimians, non-human simians, and man. In man, it is nearly 17x AvTn.

PRP: The nucleus prepositus hypoglossi shows a slight tendency toward enlargement in non-human primates and a stronger one in man. In terms of mean values, man has the highest PRP index, which clearly surpass the maximum reached in Insectivora *(Hylomys)* (219 versus 338).

COE: In Insectivora, the locus coeruleus is relatively small and only slightly above the tenrecine level. In primates, COE shows a very clear tendency toward enlargement, especially in man, whose COE index (5.6x AvTn) is the second highest (after INO) of all OBL components.

mot V: The nucleus motorius nervi trigemini is larger in Insectivora than in primates. In non-human primates (prosimians and simians), the average indices are at the tenrecine level, whereas in man, they show a clear reduction (0.70x AvTn).

mot VII: In Insectivora, the nucleus nervi facialis is twice as large as in primates. This is surprising, considering its involvement in facial expression, which in primates and man is highly developed. In Insectivora, however, especially in semiaquatic species, the larger size may be related to finely tuned motor control of the vibrissae.

mot X: In Insectivora and all primates, the nucleus dorsalis motorius nervi vagi has about the same size, slightly above the level of Tenrecinae. No obvious trends in size alteration are present.

mot XII: The nucleus nervi hypoglossi is similar to the mot X.

VC: The vestibular complex tends to enlarge from Insectivora to primates, but in man, it is smaller than in non-human simians. This applies to all the nuclei except for the superior vestibular nucleus (VS), which, in man, is at the same high level than in non-human simians.

OLS: A slight increase relative to the tenrecine level is found in the indices for the complex of the superior olive in prosimians and non-human simians. Man's index is slightly below the tenrecine level.

DCO: Trends similar to those in OLS are found in the nucleus cochlearis dorsalis, but apart from man, there is also a reduction in non-human simians.

VCO: Trends similar to those in OLS are found in the nucleus cochlearis ventralis, but there is no reduction in man. Instead, man's index is twice as large as in Insectivora.

Mesencephalon

The steady moderate increase in the MES indices from Tenrecinae through Insectivora, prosimians and non-human simians to man was mentioned earlier in conjunction with the five fundamental brain parts (see table on p. 138 and Figs. 93 and 95). In man, MES is slightly more than 4.3x AvTn. This may be mainly because of an increase in the cerebral peduncles, which, however, were not measured separately. Measurements of mesencephalic tectum and tegmentum were performed for Insectivora, but not for Primates. The subcommissural body (SCB) was also measured.

	Tenre-cinae	Insecti-vora	Pro-simians	NH-simians	*Homo*
MES	100	136	261	308	432
SCB	100	203 (23)	1973	1452 (20)	21
N	4	50	18	26	1

(N = number of species; differences are given in brackets)

SCB: In the average Insectivora, the index for the subcommissural body is distinctly larger than in Tenrecinae. SCB is also well developed in prosimians and non-human simians (in New World forms better than in Old World forms), but strongly reduced in man. There is no clear trend.

Cerebellum

The strong development of CER in the average Insectivora compared with Tenrecinae and the strong increase in CER in Primates was mentioned earlier in conjunction with the five fundamental brain parts (see table on p. 138 and Figs. 93 and 95). The increase in primates is steady up to non-human simians, but is then very steep from non-human simians to man. In humans, the ratio of afferent to efferent fibers in the cerebellum was found to be 40 : 1 (Heidary and Tomasch 1969). This input/output ratio explains the substantial coordinative role assumed by CER in all motor functions (Kretschmann and Weinrich 1986).

	Tenre-cinae	Insecti-vora	Pro-simians	NH-simians	*Homo*
CER	100	167 (50)	459	633 (26)	2057
TCN	100	107	337	505	770
MCN	100	117	480	546	363
ICN	100	94	206	312	209
LCN	100	119	429	757	1863
N	4	10	18	25	1

(N = number of species; differences are given in brackets)

In Insectivora (only "basal" forms measured), the cerebellar nuclei show nearly no increase compared with Tenrecinae, but there are slight differences between the interposed nucleus (lowest) and the lateral nucleus (highest). The same sequence is found in Primates, but all the nuclei are now distinctly more progressive with the lateral nucleus by far highest. In man LCN is nearly 19x AvTn.

Diencephalon

The structural units of the diencephalon have the following average indices in Insectivora and Primates:

Authors' Measurements

	Tenre-cinae	Insecti-vora	Pro-simians	NH-simians	*Homo*
DIE	100	169 (50)	523	706 (26)	1107
ETH	100	111	189	216	222
HAM	100	153 (38)	213 (18)	178 (20)	119
EPI	100	50 (38)	217 (18)	311 (20)	1585
THA	100	127	670	942	1412
STH	100	127	636	1168	1678
PALL	100	125	631	1160	1624
LUY	100	160	706	1273	2487
HTH	100	106	218	266	346
CAI	100	113	598	758	1426
TRO	100	97	1562	3391	2372
N	4	10	10	14	1

(N = number of species; differences are given in brackets)

Indices based on Bauchot's measurements (recalculated)

Main components					
ETH	100	130	212	193	303
THA	100	160	692	1027	1785
STH	100	218	601	800	1064
HTH	100	136	236	276	202
RTH	100	156	371	412	391
PTE	100	131	859	1101	801
Thalamic components					
ANT	100	168	1125	892	1108
MLT	100	195	1206	1603	2014
MNT	100	191	519	292	360
DOT	100	288	2756	5920	15077
VET	100	105	275	527	396
CGM	100	257	578	324	188
CGL	100	98	630	982	589
N	4	20	18	19	1

The clear enlargement of DIE in the average Insectivora compared with Tenrecinae and the strong increase in DIE in primates was mentioned earlier in conjunction with the five fundamental brain parts (see table on p. 138 and Figs. 93 and 95). The DIE size indices are highly progressive in all primates and increase continuously up to man. Its position among the five fundamental brain parts, however, may be different: in Insectivora and prosimians, DIE has the highest progression of the five fundamental brain parts; in non-human simians, the second highest, after TEL; and in man, only a moderate progression, since TEL and CER are more progressive (see table on p. 138).

In our own investigations on DIE components, most of the measurements were restricted to ten conservative species of **Insectivora** (e.g., Baron and Pirlot 1978; Baron 1979) and were not representative of the variation in Insectivora. This is obvious when the averages of the total DIE for all the material are compared with the averages for the conservative species and with Bauchot's data, which are based on 20 species including progressive species.

In **Primates**, we measured only a few diencephalic structures in a limited number of species. Whenever the same complexes were measured by Bauchot as well, trends in size changes should be similar despite different subdivisions (see section 3.3.4, pp. 89, 90), so long as the structures were homologous or equivalent in the two investigations. Indeed, in both studies, all the main parts of DIE are progressive in primates and clearly ascending through prosimians and non-human simians to man. Despite different delimitations, THA and STH are much more progressive than HTH and ETH in Bauchot's and the present investigation. In both studies, man's indices are between 10.6 and 18x AvTn for THA and STH and between 2.0 and 3.5x for HTH and ETH. THA subdivision was performed by Bauchot only.

EPI: In the average Insectivora, EPI has only half the size as in Tenrecinae (0.50x AvTn). All average values for primates are above the maximum found in Insectivora. There is a clear progression: prosimians and non-human simians are at about the same level (2 - 3x AvTn); man's index is higher than the average for all primates and is nearly 16x AvTn.

HAM: Relatively large in Insectivora. Slight progression in non-human primates, but only the tenrecine level in man. Thus, no clear trends.

PALL, LUY: The subthalamus is very progressive in primates, and the corpus luysi is distinctly more progressive than the globus pallidus. There is a clear tendency toward enlargement from Insectivora to primates and from prosimians to man. In man, PALL is about 16x AvTn and LUY, 25 times.

Thalamus measurements of Bauchot

ANT: The anterior nuclei of the thalamus are in a medium position in Insectivora (Fig. 93). In primates, a distinct enlargement is found, with the highest average (about 11x AvTn) already reached in prosimians (Fig. 96).

MLT: The nuclei mediales thalami have the second highest enlargement of the seven THA structures in all groups of primates and, in Insectivora, are in the third highest position. The size in the average Insectivora is nearly two times that of the Tenrecinae (1.95x AvTn), that of man, about 20 times.

MNT: The nuclei mediani thalami are of medium size in Insectivora, and show a clear progression in primates, which is, however, the lowest (non-human simians) or second lowest (prosimians, man) of the seven THA parts (Fig. 96).

DOT: The nuclei dorsales thalami are the most progressive of all the brain structures so far measured. This is true of all three groups of primates. In all groups, the progression is even slightly greater than that of the neocortex (cf. Figs. 93, 96 and 97).

VET: The average indices for the nuclei ventrales thalami are very low in Insectivora and primates. In Insectivora, the average is close to the tenrecine level and is the lowest next to the corpus geniculatum laterale; in prosimians, it is the lowest of the THA structures. The enlargement in non-human simians is up to 5.3x AvTn and in man, up to 4x, but in both groups, the average indices are still among the lowest compared with other thalamic structures.

CGM: In Insectivora, the corpus geniculatum mediale has the highest position after DOT. In primates, the highest progression is reached in prosimians (5.8x AvTn), whereas simians have clearly lower indices. In man, CGM has the lowest index of all the thalamic components: only 1.9 times larger than in the average Insectivora (AvIS).

CGL: In Insectivora, the corpus geniculatum laterale has the lowest average index, slightly below the tenrecine average. In primates, there is a clear tendency toward enlargement, with the largest size in non-human simians, but in prosimians, too, it is slightly larger than in man. Man's index is 5.9x AvTn.

Quantitative data on thalamic nuclei or nuclear groups may be used to deduce the relative size of the neocortical areas, similar to the deductions based on the relative size of the cerebellar and vestibular nuclei for the various longitudinal zones of the cerebellar cortex (see Section 3.3.3, p. 86). This is important, since the relative size of neocortical areas cannot be determined with our methods throughout the Insectivora and mammalian lines because of difficulties in homologization and thus accurate delimitation for reliable measurement.

Bauchot (1963, 1966, 1979b) has shown that the thalamic nuclei of the dorsal (or lateral) group (DOT) attain the highest progression of all diencephalic components, which seems to be attributable primarily to hypertrophy of the pulvinar in primates and, to a greater extent, in man. The high progression of this nuclear group, which is related to the association cortex of the parietal, temporal, and occipital lobes, may reflect the degree of progression of these neocortical areas. Indeed, within the neocortex, the association areas have increased at a much faster rate than the primary sensory and motor areas from Insectivora to primates. They reach the greatest extension by far in man (see also Brodmann 1913).

The second highest progression in the thalamic components is found for the medial nuclei. MLT, which project to the neocortical frontal lobe, seem to be of increasing importance in humans.

The anterior nuclei (ANT) have fiber connections which relate them to hypothalamic and limbic structures and also project to the prefrontal cortex. They exhibit a moderate size increase, reaching maximum levels already in prosimians. Their increase is larger than in the telencephalic limbic structures (SEP, AMY, HIP, SCH) (cf. Figs. 96 and 97).

Connected with primary neocortical areas are specific thalamic nuclei and parts containing specific nuclei (CGM, CGL, VET). They belong to the most conservative regions of THA. None of these nuclei is strongly developed in primates and even less in man. This may indicate that the primary cortical areas are likewise moderately developed. With regard to their size development, the median nuclei (MNT), too, belong to the latter group of thalamic components (Fig. 96).

Telencephalon

The moderate increase in TEL in the average Insectivora compared with that of the Tenrecinae, its strong increase in non-human primates, and its extremely large increase in man was mentioned earlier in conjunction with the five fundamental brain parts (see table on p. 138 and Figs. 93 and 95). The increase in primates is steady up to non-human simians, but is then very pronounced from non-human simians to man. In man, TEL is about 32x AvTn, whereas CER, in second place, enlarges "only" by 20 times. This very strong increase raises the question of whether all the parts of TEL develop at the same rate. Any trend found in the various structures in primates would help to evaluate the size differences in the various groups and types of Insectivora.

In the average **Insectivora** (Fig. 93) the highest increase by far is found in the neocortex (NEO), which, on the average, is nearly 3 times larger than in the average Tenrecinae. In descending sequence, a second group of structures is made up of striatum (STR), hippocampus (HIP), and schizocortex (SCH) (2.1 to 1.7x AvTn), and a third group, by accessory olfactory bulb (AOB), septum (SEP), and amygdala (AMY) (1.4 to 1.1x AvTn). No change or reductions relative to Tenrecinae are found in the olfactory structures, i.e., paleocortex (PAL) and main olfactory bulb (MOB) (1.0 to 0.9x AvTn).

	Tenre-cinae	Insecti-vora	Pro-simians	NH-simians	*Homo*
TEL	100	159	457	921	3250
MOB	100	87	49	7	2
AOB	100	138	414	61 *)	0
PAL	100	103	63	38	55
STR	100	207	542	830	1393
SEP	100	132	180	183	407
AMY	100	114	170	210	401
HIP	100	175	291	268	487
SCH	100	168	280	225	443
NEO	100	292	1902	4220	13912
N	4	50	18	26	1

*) only present and averaged in New World simians (N = 13)

In **Primates** (Fig. 97), the sequence is exactly the same as in Insectivora, except for the accessory olfactory bulb (AOB), and for SEP and AMY in non-human simians.

MOB, PAL: The main olfactory bulb (MOB) occupies the lowest, and the paleocortex (PAL), the second lowest position in Insectivora, as well as in all groups of primates. In man, MOB is about 1/50 of that in the Tenrecinae (0.02x AvTn) and 1/40 of that in the average Insectivora. This is the only very strong regression found in the main TEL components. In man (and in non-human simians), PAL has about half the size as found in Insectivora. This is a clear deviation from the size reduction found in MOB, and may be because of increasing difficulties in the homologization and delimitation of the olfactory paleocortices. In water-adapted Insectivora and prosimians, the reduction in PAL parallels that of MOB (Baron et al. 1987). This may be true also for simians, but the strong reduction blurs the borders of PAL and therefore increases the possibility of including transitional tissues in the PAL structures.

AOB: The size of the accessory olfactory bulb (AOB) is independent of MOB. In most prosimians, AOB size is distinctly larger than in Insectivora, whereas in simians, there is a clear reduction, which may finally result in a complete loss. In Old World simians and man, no AOB was found (Stephan 1965; Stephan et al. 1982).

SEP, AMY, HIP, SCH: In primates, all non-olfactory TEL parts are distinctly larger than in Insectivora. Moderate enlargements are found for the limbic structures, which, in Insectivora (Fig. 93) and in each of the three primate groups (Fig. 97), have similar indices. In man, the limbic structures are 4 - 5x AvTn and about 3x AvIS. HIP and SCH are slightly larger in prosimians than in non-human simians.

STR: The second highest progression is found in Insectivora and all groups of primates for STR. In man, STR is about 14x AvTn and 6.7x AvIS. This enlargement is similar to that of CER, perhaps because of the close relationship of both complexes with the motor systems.

NEO: The highest progression by far is found for NEO. In man, NEO is nearly 140 times larger than in Tenrecinae and 50 times larger than in the average Insectivora.

Several of the components of TEL are further subdivided and will be discussed in more detail.

Paleocortex

In **Insectivora**, all parts of the paleocortex (PAL) are similar to those of Tenrecinae or slightly smaller.

	Tenre-cinae	Insecti-vora	Pro-simians	NH-simians	*Homo*
PAL	100	103	63	38	55
RB	100	107	47	9	12
PRPI	100	94	63	30	53
TOL	100	104	81	64	112
TRL	100	84	39	42	45
COA	100	95	68	126	156
SIN	100	96	130	127	117
N	4	50	18	26	1

In **Primates**, the retrobulbar region (RB), prepiriform region (PRPI), lateral olfactory tract (TRL), and olfactory tubercle (TOL) are distinctly reduced, except TOL in man, whose index is slightly higher than in Insectivora. The main reason may be the fixation of the olfactory tubercle to the ventral surface of the very strongly enlarging striatum. In man, in particular, there are considerable structural changes in the region of the anterior perforated substance and it is uncertain whether this ventral surface is entirely formed by a "true" TOL or whether there is a "naked" STR in places (Baron et al. 1987). The strongest reduction is found for RB. An increase is observed in the anterior commissure

(COA) and the substantia innominata (SIN), except for COA in prosimians, in which the olfactory fibers may still predominate. In simians, the non-olfactory COA components not only compensate for the reduction of olfactory fibers but even enlarge the COA as a whole.

RB: In microsmatic primates, the olfactory peduncle is barely more than a thin band of fibers. In it, RB is concentrated in cell-clusters on both ends, but dispersed with fewer cells over the whole extent.

PRPI: In microsmatic simians, PRPI is restricted to a relatively small part around the fossa lateralis cerebri. Its structural differentiation is poor, especially in man, and in higher primates causes increasing difficulty in its homologization and exact delineation.

TOL: In high-grade microsmatic primates and man, the dissolution of the TOL layers has progressed and it is difficult to decide whether the structures found on the surface ventral to striatum and substantia innominata are fully homologous over the whole extent with the highly characteristic TOL structures found in macrosmatic Insectivora or whether the striatum has a "naked" surface.

COA, SIN: All the fibers of COA are concentrated in the midsagittal plane (Figs. 40 - 51). Three components can be differentiated: (1) a small stria terminalis component, (2) an olfactory component (pars anterior), which is relatively large in Insectivora and small in simian primates, and (3) a non-olfactory component, which is relatively small in Insectivora and large in primates. The olfactory component is the internal olfactory tract, running in the center of the olfactory peduncle. Strong non-olfactory components do not seem to exist in Insectivora and prosimians, since a reduction of the olfactory centers (MOB, RB), as in semiaquatic Insectivora and the prosimian Indriidae, results in an equally strong reduction in COA.

Septum

The moderate increase in the septum (SEP) in primates was mentioned earlier in conjunction with the other TEL components. In man, SEP is about 4.1x AvTn and 3.1x AvIS (407/132). This enlargement is very similar to that of other limbic structures, such as amygdala (AMY), hippocampus (HIP), and schizocortex (SCH).

In three species of Insectivora and three species of primates, nine subdivisions of SEP proposed by Andy and Stephan (1959) were measured. The basis for comparison was the average of the 3 Insectivora. Of the nine components, the most progressive were the fimbria/fornix complex (SFF, 320 in the average for the three primates), the lateral (ventral) nucleus (SEL, 304), the bed nucleus of the stria terminalis (SES, 258), and the diagonal band of Broca (DIAG, 240). The only component showing a decrease was the triangular nucleus (SET). This nucleus, as well as the subfornical body (SFB), was measured in a larger number of species, including the 4 Tenrecinae.

	Tenre-cinae	Insecti-vora	Pro-simians	NH-simians	*Homo*
SEP	100	132 (50)	180	183 (26)	407
SET	100	162	148	96	7
SFB	100	75	234	193	78
N	4	38	18	20	1

(N = number of species; differences are given in brackets)

SET: The investigation confirmed and even strengthened the conclusion of a decrease in SET. In man, SET is only 1/14 AvTn and 1/22 AvIS. SET seems to be related to the fibers of the anterior hippocampal commissure (psalterium ventrale), which, in contrast to the psalterium dorsale, is much reduced in higher primates and man (Andy and Stephan 1966a, b, 1968).

SFB: The subfornical body seems to be related to the control of the cerebrospinal fluid since it is always placed in the median sagittal plane, where ventricular and extraventricular liquor spaces meet. Its highest value is in prosimians, but SFB size shows no clear trend.

Amygdala

In the average Insectivora, only a very slight progression of the amygdala (AMY) relative to Tenrecinae was found. The moderate increase of AMY in primates was mentioned earlier in conjunction with the other TEL components. In man, AMY is about 4x AvTn and 3.5x AvIS (401/114). This enlargement is very similar to that of other limbic structures, such as septum (SEP), hippocampus (HIP), and schizocortex (SCH).

	Tenre-cinae	Insecti-vora	Pro-simians	NH-simians	*Homo*
AMY	100	114	170	210	401
MAM	100	114	112	118	212
NTO	100	92 (38)	29	5 (4)	0
LAM	100	115	221	292	569
MCB	100	92 (38)	89	158	225
N	4	50	18	26	1

(deviating N in brackets)

Of the AMY components, the average indices for medial (MAM) and lateral (LAM) divisions are nearly the same in Insectivora: about 1.1 times in MAM and LAM and 0.9 times in the nucleus of the lateral olfactory tract (NTO) and the magnocellular part of the basal nucleus (MCB). In primates, they clearly diverge. In man, MAM is about 2.1x AvTn and 1.9x AvIS (212/114), and LAM is 5.7 and 4.9 times larger, respectively. Thus, LAM is more than twice as progressive as MAM. Less progressive than LAM as a whole is the magnocellular part of its basal nucleus (MCB). In man, it is 2.2x AvTn and 2.4x AvIS. In MAM of Insectivora, the bed nucleus of the lateral olfactory tract (NTO) is conspicuous (Fig. 82), becoming less and less pronounced in Primates. In many simians and in man, NTO can no longer be identified with certainty (Stephan et al. 1987a).

Hippocampus

In the average Insectivora, a slight progression of the hippocampus (HIP) relative to Tenrecinae is found. The moderate increase in HIP in primates was mentioned earlier in conjunction with the other TEL components. HIP is somewhat larger in prosimians than in non-human simians. In man, it is about 4.9x AvTn and 2.8x AvIS (487/175). This enlargement is very similar to that of the other limbic structures, such as septum (SEP), amygdala (AMY), and schizocortex (SCH).

	Tenre-cinae	Insecti-vora	Pro-simians	NH-simians	*Homo*
Volume measurements					
HIP	100	175 (50)	291 (18)	268 (26)	487
HR	100	122	272	269	502
HS	100	137	342	768	1469
HP	100	117	78	58	52
Area measurements					
SUB	100	118	169	111	386
CA0	100	133	240	214	365
CA1	100	128	283	229	822
CA2/3	100	102	188	148	240
CA+SUB	100	142 (15)	241 (5)	167 (4)	473
FD	100	149 (15)	206 (5)	178 (4)	320
N	4	10	3	3	1
(deviating N in brackets)					

Of the HIP components, the supracommissural hippocampus (HS) is slightly more progressive than retrocommissural hippocampus (HR) and precommissural hippocampus (HP) in Insectivora, but is more strongly progressive in primates. HS covers the dorsal surface of the corpus callosum at the bottom of the interhemispheric fissure and, thus, its size is certainly influenced by the extension of the corpus callosum, which is strongly progressive in primates. In contrast, HP diminishes increasingly in primates, reaching about one half in man. This is very compatible with its olfactory relationships. Because of its large relative size, HR indices are similar to those for the total HIP. Its enlargement is very similar to that of other limbic structures, such as septum (SEP), amygdala (AMY), and schizocortex (SCH).

Area measurements of HR components in a limited number of primate species reveal that CA1 is by far the most progressive hippocampal component, and CA2/3 the least. The high progression in CA1 is accompanied by major structural changes. In primates, the cells of CA1 begin to diffuse into the deep layer (stratum oriens), and in man, they are dispersed over the entire stratum oriens and reach the alveus (Stephan and Manolescu 1980; Stephan 1983).

Schizocortex

In the average Insectivora, a slight progression of the schizocortex (SCH) relative to Tenrecinae is found. The moderate increase in primates was mentioned earlier in conjunction with the other TEL components. Like the hippocampus, the schizocortex is larger in prosimians than in simians. In man, SCH is 4.4x AvTn and 2.6x AvIS. This enlargement is very similar to that of other limbic structures, such as septum (SEP), amygdala (AMY), and hippocampus (HIP).

Area measurements of SCH components in a limited number of species reveal that the presubicular region (PRS) is slightly more progressive than the entorhinal region (ENT) and the perirhinal field (PRH). In prosimians (N = 2), the average PRS index is 176, and in the Night Monkey (*Aotus trivirgatus*, the only simian so far investigated), it is 267. In ENT, the values are 172 and 174, and in PRH, 140 and 179, respectively. The enlargement is accompanied by structural differentiations which are reflected in both laminar and regional complexity. In ENT of man, Rose (1927b) differentiated 23 fields with cytoarchitectural methods, and Braak (1972), 16 fields with pigmentarchitectural methods.

Neocortex

In Insectivora, the neocortex (NEO) is by far the most progressive structure of the brain. It increases on the average more than 2.9x AvTn. In primates, there is a further and very strong and steady increase through prosimians, non-human simians and, finally, to man. In man NEO is nearly 140 times larger than in Tenrecinae. This enlargement was described earlier in conjunction with the main TEL components.

From Bauchot's measurements of various parts of the thalamus different parts of the NEO should enlarge differently (see p. 93). Based on fiber connections, one would expect the association cortices (connected with the dorsal thalamic nuclei) to be the most progressive, the frontal cortices (connected with the medial thalamic nuclei), also to be strongly progressive, and the primary motor and sensory regions (connected with the ventral thalamic nuclei), to be the least progressive. To this last group belongs the area striata (AS). Its volumes were estimated in Insectivora species in which the corpus geniculatum laterale was measured, but this method of determination is questionable. Based on the data derived from this method, the area striata grey (ASG) in man is about 38 times larger than in the average Insectivora. Thus, the progression in AS is distinctly lower than the progression in the NEO as a whole.

	Tenre-cinae	Insecti-vora	Pro-simians	NH-simians	*Homo*
NEO	100	292	1902 (18)	4220 (26)	13912
NEO	100	292	2050	4444	13912
ASG	100	99 (10)	1740 (18)	3813 (22)	3774
NW	100	266	2766	8642	30020
NG	100	296	1928	3664	9630
NG1	100	209	932	1601	4798
NG2-6	100	335	2403	4617	11739
N	4	50	10	12	1

(deviating N in brackets)

The relative reduction in the neocortical molecular layer (NG1) and the high BoW power (slope) of the neocortical white matter (NW) found in progressive Insectivora are paralleled by the relations found in Primates (Frahm et al. 1982). In Primates, and especially in man, NW is the most progressive, and NG1 by far the least progressive component. A high progression for the cell-dense layers (NG2-6), and the very lowest for NG1 was already found in those species of Insectivora which have a large NEO. Thus, the size of NG2-6 may be the best indicator of general brain development in Insectivora.

In conclusion:

(1) The most progressive structures by far **in man** are the neocortex, with the highest centers of coordination and, based on the data of Bauchot (1979b), the dorsal nuclear group of the thalamus. These two components are about 140 - 150 times larger than in the average Tenrecinae.

(2) The corpus luysi, the cerebellum, and the medial nuclear group of the thalamus, are more than 20 times larger; and the lateral cerebellar nucleus has enlarged almost 19 times, the inferior olive 17 times, the pallidum 16 times, and the striatum 14 times. Most of these structures are closely related to the motor systems.

(3) The thalamus as a whole has enlarged about 14 times, and the diencephalon 11 times.

(4) The olfactory structures are regressive, as are some of the nuclei of the medulla oblongata, such as the sensory trigeminal complex, motor nuclei V and VII (special viscero-efferent nuclei) and the auditory nuclei. Relatively poor audition is confirmed by the index for the corpus geniculatum mediale, which is one of the smallest in man's diencephalon.

It is of interest that the highest variation by far and, thus, progression in **Insectivora** was also found for the dorsal nuclear group of the thalamus (DOT) and for the neocortex (NEO), although on a much lower level. Some interesting differences in the sequences of the structures on the index scales for Insectivora and primates are the higher positions of the corpus geniculatum mediale (CGM), and of the sensory trigeminal complex (TR) in Insectivora.

On the basis of neocortical size, the **Tenrecinae** may be expected to be the least evolved Insectivora. The conservatism of the brains of tenrecine insectivores is also reflected in the fact that the relative size of nearly all structures or complexes is lowest. This is true even for brain stem structures, which, in simians and man, show only a slight increase. Lowest relative size in Tenrecinae, however, does not extend to the olfactory structures (MOB, PAL). These structures are smaller in many groups of Insectivora. The large relative size of the olfactory structures may be a primary conservative characteristic of Tenrecinae. In other mammalian groups, e.g. in Megachiroptera, large olfactory structures may, however, be derived; this may similarly account for the very large MOB of *Solenodon* also.

4 Brain Characteristics in Taxonomic Units

General Characteristics of Insectivora

Insectivora are a heteromorphous order, some members of which seem to have departed less from the form of the generalized, primitive mammalian type, and therefore, are believed to resemble the basic stock of certain mammalian lines of descent (Walker 1975).

Insectivora are generally small. Their pentadactyl limbs (except in some species) with clawed digits are usually unspecialized and are never adapted to saltation. Most species are plantigrade, moving about on the sole of the foot with the heel touching the ground. The radius and ulna are separate, but the tibia and fibula are often fused near the ankle. The clavicles are present in all forms, except *Potamogale*, although they are highly modified in some species. With few exceptions, the skull is characterized by an elongated rostrum. The teeth are generally primitive, the lower molars usually having five pointed cusps and the uppers, three or four tubercles. The dental formula, however, is very variable.

Olfaction is one of the predominent sensory channels in Insectivora, whereas the visual capacities are generally poorly developed compared to most other mammals. Insectivora feed mainly on invertebrates and small vertebrates.

Selected references: Dobson (1882, 1883), Cabrera (1925), Weber (1928), Grassé (1955a), Findley (1967), Kingdon (1974), Walker (1975), Yates (1984), Vaughan (1985), Fons (1988), Baron and Stephan (in press).

The taxonomy of the Insectivora is still in dispute. The tree shrews (Tupaiidae) and elephant shrews (Macroscelididae), formerly placed in this order, are now considered to form separate orders (Scandentia and Macroscelidea) and are therefore excluded from Insectivora and are not considered here. The remaining forms may be classified into six families: Tenrecidae, Solenodontidae, Erinaceidae, Chrysochloridae, Talpidae, and Soricidae.

Brain Characteristics of Insectivora

The brains differ widely in shape, size, and composition, and their characteristics will be discussed in the sections on subfamilies and families. For a representative example of the topography and delineation of structures, see the stereotaxic atlas of the brain of the hedgehog *Atelerix algirus* on pp. 503 - 553.

In the earlier sections on the size of brain structures (sections 3.3.1 to 3.3.5) and on their comparison with Primates (section 3.3.6) and in the following sections (4.1 to 4.6) on general brain characteristics in taxonomic units (except the summaries, see next paragraph and section 4.7), reference is made to the average Tenrecinae (AvTn). In brain size and composition Tenrecinae were found to be the most conservative of the well represented insectivoran subfamilies. In such comparisons, the size indices depend, of course, on the special characters of this single subfamily, e.g., very small NEO and large olfactory structures. Relative to Tenrecinae, the other families and subfamilies in general had large NEO indices, but the indices for olfactory structures were quite small, even though these structures were very well developed.

Therefore, in the summaries on brain characteristics in families and subfamilies, the average tenrecine level (AvTn) is replaced by the average of the Insectivora as a whole. The measure of a brain structure in the typical Insectivora is the average value (AvIF) of the family and subfamily averages. AvIF values are given for each structure in the tables and are summarized in Table 74. They are shown by dotted lines in the diagrams and scales. In order to obtain the deviations of the families and subfamilies from the typical Insectivora, their averages were divided by the AvIF values, or, in the diagrams, each of the reference lines was shifted into the AvIF position. The averages for the Insectivora family and subfamily averages (AvIF) were chosen instead of the averages for Insectivora species (AvIS) because the shrews are strongly overrepresented in AvIS. The deviations of families and subfamilies (in Soricidae, of groups of species and of genera also) from AvIF, which are called index profiles, are illustrated in Figs. 101 - 108 (pp. 294 - 297).

The differences in brain characteristics between subfamilies belonging to a complex family may be so large that it is difficult to describe common brain characteristics valid for all its groups. Attempts will be made to bring out common brain characteristics of the complex families too. Stronger variation in brain composition, however, is not only found in complex families, but also in certain subfamilies, for example in Crocidurinae. The focus here will be on the differences, which will likely lead to a reconsideration of other morphological characters and ultimately may give rise to a reclassification of the respective taxonomic units.

To characterize the various species within a subfamily or even a smaller unit, reference is made to the average in each of these units, such as Oryzorictinae (AvOr), Potamogalinae (AvPo), Erinaceinae (AvEr), Talpinae (AvTa), Soricinae (AvSn), and in Crocidurinae the *Crocidura* group (AvCg) and the *Sylvisorex* group (AvSg).

4.1 Tenrecidae

General Characteristics

Tenrecidae underwent radial adaptation and developed a number of forms that vary widely in structure and habits. The snout is frequently long and slender, and the zygomatic arch is incomplete because of the loss of the jugal bone. The urogenital and anal openings terminate in a cloaca.

Some tenrecs bear a general resemblance to mammals as diverse as opossums, shrews, hedgehogs, muskrats, mice, and otters, and are adapted to terrestrial, fossorial, and semiaquatic life. They vary from roughly the size of a shrew to the size of a hedgehog. Members of this family inhabit forests, scrubby areas, grassland, and marshes. They shelter in burrows and, hollow logs or under rocks, and feed mainly on animal matter, although some forms may feed to a larger extent on plant matter as well.

Selected References: Grandidier and Petit (1930, 1932), Grassé (1955a), Cockrum (1962), Herter (1962a, b, 1963, 1964, 1979), Gould and Eisenberg (1966), Kingdon (1974), Walker (1975), Yates (1984), Vaughan (1985), Fons (1988).

Brain characteristics, their variation and intergroup differences are described in connection with the subfamilies. The brain characteristics strongly favor subdivision into four subfamilies, three of which (Tenrecinae, Geogalinae, Oryzorictinae) are found on Madagascar and the Comoro Islands, and one (Potamogalinae) is confined to central Africa.

4.1.1 Tenrecinae

General Characteristics

The Tenrecinae (spiny or bristled tenrecs) comprise five mostly monospecific genera *(Tenrec, Setifer, Hemicentetes, Echinops, Dasogale)*. *Dasogale* is by some authors considered synonymous with *Setifer* (see Fons 1988). They have more or less elongated snouts; the tail is very short or lacking. The eyes are small; the external ears relatively large. The pelage may be coarse and contain spines interspersed with soft hairs, as in *Tenrec*, or be spiny, as in *Setifer, Echinops*, and *Hemicentetes*. *Setifer* and *Echinops* resemble hedgehogs and have a panniculus carnosus muscle which enables them to erect the spines and roll into a protective ball.

Tenrecinae are generally nocturnal, omnivorous animals occupying a variety of habitats. Most genera live in burrows of their own construction. Only *Tenrec* is reported to hibernate during the dry season, but most species exhibit some degree of torpor at times of stress.

Brain Characteristics

Brain material is available from all genera (except *Dasogale*).

Tenrec ecaudatus, Setifer setosus, Hemicentetes semispinosus, Echinops telfairi

Brain Weight, Encephalization

Bauchot and Diagne (1973)
Bauchot and Stephan (1964, 1966)
Brodmann (1913)
Warncke (1908)

Paired data on brain and body weights on *Tenrec* are given by Warncke (1908) and by Brodmann (1913), on *Tenrec, Setifer* and *Hemicentetes* by Bauchot and Stephan (1964, 1966), and in 1966 on *Echinops* as well. Bauchot and Diagne (1973) present the ontogenetic development of brain weight in *Hemicentetes*. The largest brain, found in *Setifer*, has about 1.4 times larger indices than the smallest, found in *Tenrec* (117 versus 86; see Table 6).

Macromorphology

Bauchot and Stephan (1959, *Setifer*, 1967, 1970)
Beddard (1901, *Tenrec*)
Clark (1932, *Tenrec*)
Dräseke (1901, *Tenrec*)
Friant (1947, 1955, *Tenrec*)
Krabbe (1942, morphogenesis, *Setifer*)
Leche (1905, 1907, *Tenrec, Hemicentetes*)
Smith (1902b, *Tenrec*)
Stephan (1967b)
Stephan and Andy (1982)
Stephan and Spatz (1962, *Setifer*)

Smith (1902b) describes an endocranial cast, and Bauchot and Stephan (1967) compare endocast and brain of *Tenrec*. Brains of all four genera are described by Stephan (1967b), Bauchot and Stephan (1970), and Stephan and Andy (1982).

Most of the studies compare tenrecine brains with those of other Insectivora and mammals and report (1) features in which *Tenrec* comes close to some of the early eutherian mammals, (2) no temporal pole projecting sharply downwards, (3) a fully exposed tectum mesencephali, and (4) faint depressions in the cerebral pallium of *Tenrec* and a faint sylvian fissure (Beddard). Dräseke, however, denied its existence.

The first illustration of the midsagittal plane was given by Dräseke; subsequently many authors provided diagrams. It has been noted that the corpus callosum (CC) and the anterior commissure (COA) are similar in size. The three components of COA of *Setifer* are shown in a photoprint in Stephan and Spatz (1962).

Area Measurements

Bauchot and Stephan (1961, 1970)
Brodmann (1913)
Stephan (1961)
Stephan and Manolescu (1980)
Stephan and Spatz (1962)
Stephan et al. (1984b)

The areas of sectioned commissures or nerves are described in Bauchot and Stephan (1961, 1970) and Stephan et al. (1984b). The corpus callosum is generally small in Tenrecinae and smallest in *Echinops*, the olfactory component of COA is large, and the visual connections (nervus and chiasma) are relatively well developed in comparison to other Insectivora.

Area size of cortical surfaces are given for *Tenrec* in Brodmann (1913) and for *Setifer* in Stephan (1961) and Stephan and Spatz (1962). The very small neocortex (NEO) and non-olfactory allocortex combined with the large olfactory allocortex is considered to constitute a conservative composition of cerebral cortex. Measurements of the hippocampus (HIP) and several of its parts in the four species of Tenrecinae are presented in Stephan and Manolescu (1980). HIP and its parts are conservative as well.

Volume Measurements

Andy and Stephan (1966a)
Baron (1972, 1974, 1977a, b, 1978, 1979)
Baron et al. (1983, 1987, 1988)
Bauchot (1961, 1963, 1966, 1970, 1979a, b)
Bauchot and Diagne (1973)
Bauchot and Legait (1978)
Bauchot and Stephan (1968, 1970)
Etienne (1972)
Frahm et al. (1982, 1984)
Legait et al. (1976)
Matano et al. (1985a, b)
Stephan (1966, 1967a, b, 1969)
Stephan and Andy (1962, 1964, 1969, 1977, 1982)
Stephan and Manolescu (1980)
Stephan et al. (1970, 1981a, 1982, 1984b, 1987a, 1988)

Volumes of many brain parts have been measured by a number of authors and teams since 1961 (Andy, Baron, Bauchot, Diagne, Etienne, Frahm, Legait, Manolescu, Matano, Stephan).

Ventricle size has been investigated by Bauchot and Diagne (1973) in *Hemicentetes*. In an embryo of 0.2 g BoW, the brain ventricles made up 16.7% of the brain volume, whereas in a 10 g embryo they almost reached adult values (1.93%) (2.1% in Table 8).

Data on the five **fundamental brain parts** were given for the first time by Stephan and Andy (1964) for *Tenrec* and *Setifer* and then several times for the four tenrecine species in comparisons with other Insectivora and with Primates (Stephan 1966, 1967a, b; Bauchot and Stephan 1968; Stephan and Andy 1969, 1982; Stephan et al. 1970, 1981a, 1988; Bauchot 1979a). The largest differences in the four species of Tenrecinae are found in CER, which is about 1.7 times larger in *Hemicentetes* than in *Echinops* (133 versus 78, see Table 11). Fairly wide variation is also found in TEL, which is 1.5 times larger in *Setifer* than in *Tenrec*. Low variations are reported in MES and OBL.

In the **medulla oblongata** (OBL), data on vestibular and auditory structures are given in Baron (1972, 1974, 1977a, b), Stephan et al. (1981a), and Baron et al. (1988). The measurements on most components are published here for the first time. The vestibular structures are large in *Hemicentetes*, except for the superior nucleus (VS), which is large in *Setifer* (see Table 23). The distinctly largest auditory structures in OBL are found in *Echinops* (see Table 26), and the mesencephalic inferior colliculus (INC) (see Table 29) is also very large.

In the **mesencephalon** (MES), the tectum with both colliculi is largest in *Echinops*, whereas the tegmentum is largest in *Setifer* (Table 29).

For the **cerebellum**, data on the cerebellar nuclei are given in Baron (1978) and Matano et al. (1985a, b). The smallest are always found in *Echinops*, the largest (except for MCN) in *Hemicentetes*. MCN is clearly largest in *Setifer*.

For the **diencephalon**, data on its main parts are given in Bauchot (1963, 1966, 1979b), Bauchot and Stephan (1968), Baron (1979), and Stephan et al. (1981a). DIE is distinctly largest in *Setifer*, and this is true also of many of its parts, except, according to Baron, for TRO and ETH, which are small. *Setifer* also has a small nucleus medialis habenulae (HAM, Table 41) (Stephan 1969; Stephan et al. 1981a). CAI is especially large in *Setifer*; TRO is largest in *Tenrec*, ETH in *Hemicentetes* (Table 35), and LUY in *Echinops* (Table 38). In contrast, ETH in Bauchot's data is largest in *Setifer* (Table 66). Legait et al. (1976) provide data on the epiphysis (EPI), and Bauchot (1961, 1970) and Bauchot and Legait (1978) on the hypophysis (HYP). Small HYPs and anterior lobes are found in *Tenrec* and *Echinops*, relatively large ones in *Setifer* and *Hemicentetes*. EPI is small in *Tenrec* and large in *Setifer* and *Echinops* (see Table 41).

Data on **visual structures** are given in Frahm et al. (1984) and Stephan et al. (1984b). The large size of TRO in *Tenrec* is not mirrored by other visual structures such as CGL (Tables 38, 69) and SUC (Table 29). CGL and the derived indices for the area striata (ASG) are relatively large in *Setifer*, while SUC is large in *Echinops*.

Various parts of the **telencephalon** (TEL) have been measured and compared. Data are given in Stephan and Andy (1964, 1969, 1982), Stephan (1966, 1967a, b), Bauchot and Stephan (1968), and Stephan et al. (1970, 1981a, 1988). The parts of TEL may roughly be classified into (1) olfactory structures (MOB, AOB, PAL), (2) limbic structures (SEP, AMY, HIP, SCH), (3) striatum (STR), and (4) neocortex (NEO).

Additional data on **olfactory structures** are given for the main olfactory bulb (MOB) in Etienne (1972) and Baron et al. (1983). MOB is by far largest in *Setifer* and is smallest in *Echinops* (Table 47). Additional data on the accessory olfactory bulb (AOB) are presented in Stephan et al. (1982). Like MOB, AOB is largest in *Setifer*, where it is nearly 1.8 times larger than in *Hemicentetes* (131 versus 74; see Table 50). Additional data on the paleocortex (PAL) and its parts are given in Baron et al. (1987). Again, *Setifer* has by far the largest indices: on the average 1.5 times larger than in *Tenrec* (133 versus 86; see Table 44). *Setifer* is in a high position for all parts of PAL, whereas the lowest positions are mostly found for *Echinops* or *Tenrec*.

Additional data on the **septum** (SEP) are given in Stephan and Andy (1962, 1982) and Andy and Stephan (1966a), and data on the subfornical body (SFB) and triangular septal nucleus (SET), in Stephan (1969) and Stephan et al. (1981a). SEP is largest in *Setifer* and smallest in *Tenrec* (Table 44). However, SET is smallest in *Setifer* and largest in *Echinops*, and SFB largest in *Tenrec* and smallest in *Hemicentetes* (Table 41).

Additional data on the **amygdala** (AMY) are presented in Stephan and Andy (1977) and Stephan et al. (1981a, 1987a). AMY is 2.1 times larger in *Setifer* than in *Hemicentetes* (149 versus 70; see Table 44). This is true of both the medial and lateral divisions and of the magnocellular basal nucleus (MCB), whereas the nucleus of the lateral olfactory tract (NTO) parallels the olfactory structures and is smallest in *Echinops* (Table 56).

Additional data on the **hippocampus** (HIP) are given in Stephan and Manolescu (1980). The variation in HIP is relatively low in Tenrecinae. HIP is largest in *Setifer*, but only 1.4 times larger than in *Tenrec* (112 versus 82; see Table 44). The largest variation is found in the supracommissural hippocampus (HS), which is 2.5 times larger in *Setifer* than in *Echinops*. The very small HS in *Echinops* may be related to its very small corpus callosum.

Additional data on the **neocortex** (NEO) are provided in Bauchot and Stephan (1970) and Frahm et al. (1982). The variation in NEO is relatively low in Tenrecinae. NEO is only 1.3 times larger in *Setifer* than in *Echinops* and *Tenrec* (114 versus 89; see Table 62). Such minor differences are also found in all the laminar parts of NEO that have been measured.

More quantitative comparisons will be presented on p. 177 in the summary of brain characteristics of Tenrecinae and its species compared with those of the other Tenrecidae.

Histology, Fiber Connections

Many of the papers on volume measurements provide details on histology. Such details are important to allow the determination of borders between structural subdivisions of the brain. Additional details are given in:

Bartelmez (1960, ontogeny, *Setifer, Hemicentetes*)
Clark (1932, DIE, *Tenrec*)
Johnson et al. (1982, brain traits, *Tenrec, Setifer, Hemicentetes*)
Künzle (1988, retinal projections, *Setifer, Echinops*)
Müller and O'Rahilly (1980, ontogeny, *Setifer, Hemicentetes*)
Rehkämper (1981, NEO, *Echinops*)
Stephan (1965, AOB, *Tenrec, Setifer*)
Stephan and Spatz (1962, ependymal folds in MES aqueduct, *Setifer*)
Switzer et. al. (1980, AOB, *Tenrec, Setifer, Hemicentetes*)
Zilles et al. (1979, visual structures, *Echinops*)

Bartelmez (1960) and Müller and O'Rahilly (1980) report on the early development of the nervous system in *Hemicentetes* and *Setifer*. Some, mostly histological, brain features of Tenrecinae are presented and compared with those of other mammals by Johnson et al. (1982).

According to Clark (1932), the thalamic nuclei anterior (ANT) in *Tenrec* are poorly defined. The CGL is small and confined almost exclusively to the dorsal surface. In some sections, the dorsal nucleus of CGL appears to be merely a lateral extension of the main ventral nucleus of the thalamus. Some components of the hypothalamus, e.g., nucleus tangentialis, are highly conspicuous.

Stephan and Spatz (1962) describe ependymal folds in *Setifer* on the bottom of the mesencephalic aqueduct in its transitional zone to the fourth ventricle. The folds have cilia and are vascularized. They may be concerned with resorption and/or movement of the liquor. Stephan (1965) describes layers and cells in the accessory olfactory bulb of *Tenrec* and *Setifer,* as well as in a new-born *Setifer*. AOB is well differentiated but quite small, especially compared to the main bulb. This is shown in schematic drawings. Switzer et al. (1980) use the position where the MOB fibers penetrate into the AOB as a phylogenetic indicator. Zilles et al. (1979) report that in *Echinops* the most occipital field of the neocortex, which they assume to be visual cortex, has no granular cells and stripe of Gennari. Most fibers of the optic tract cross in the chiasma.

Rehkämper (1981) gives a cortical map of the neocortex of *Echinops* containing four fields. NEO has a thick lamina I that makes up about 25% of the whole thickness. Lamina II is thick and cell-dense. The caudal area A 3 is assumed to be the visual field. Künzle (1988) gives a detailed description of the retinal projections of *Setifer* and *Echinops*.

Sense Organs

Bielschowsky (1907, snout, *Tenrec*)
Broom (1915b, vomeronasal organ, *Tenrec*)
Gould (1965, echolocation, tenrecs)
Siemen (1976, eye, *Echinops*)

Bielschowsky (1907) investigates the cuticle of the snout of *Tenrec* and finds extraordinary nerve end formations. Large cells are regularly wrapped in dense fibrillary nets. These formations lie in the septum of the nose and project into the nostrils reaching about 1 cm into the nose. While they have some similarities with tactile formations, it is unlikely that they have tactile functions, but, rather are thermoreceptors. They belong to the area of innervation of the trigeminal nerve. Broom (1915b) reports on the vomeronasal organ of *Tenrec* and finds a number of unusual features, for example, the lower part of the septal cartilage was completely surrounded by the vomer. Gould (1965) finds a primitive type of echolocation in tenrecs. Siemen (1976) says that the retina of *Echinops* is made up of rods. The outer part of the retina is folded, a fact interpreted by the author as a feature of eye reduction. Color vision has been questioned by Siemen.

4.1.2 Geogalinae

General Characteristics

The single mouse-like species *(Geogale aurita)* is one of the least known mammals of Madagascar and may be rare. It is distinguished by its very prominent ear pinnae; its eyes, too, are large. *Geogale* seems to undergo seasonal torpor.

There is considerable dispute over the classification of *Geogale*. Initially, it was classified with the subfamily Tenrecinae (Milne-Edwards and Grandidier 1872). Dobson (1882) regarded *Geogale* as a separate subfamily (Geogalinae) in the family Potamogalidae. He revised this view (1883) after obtaining more information about *Microgale*, and then classified *Geogale* with the Tenrecidae, emphasizing the lack of sufficient material.

Leche (1907), Simpson (1945), Heim de Balsac and Bourlière (1955), and Heim de Balsac (1972) classified *Geogale* with the Oryzorictinae, whereas Starck (1974), Thenius (1980), and, based on brain investigations, Rehkämper et al. (1986), again, proposed a distinct subfamily.

Brain Characteristics

The only reference is Rehkämper et al. (1986) on macromorphology, cranial capacity and volume measurements.

Geogale aurita

Cranial Capacity, Encephalization

Brain (BrWs) and body weights (BoWs) are not available for *Geogale* and are merely inferred in Rehkämper et al. (1986). Determination of brain size is (1) by conversion of brain volumes obtained from serially sectioned heads and (2) by cranial capacities. Cranial capacities, taken from two skulls of the Musée d'Histoire Naturelle in Paris, indicate brain weights between 120 and 130 mg. *Geogale* has a large foramen magnum, and the very high percentage of medulla oblongata, which to a large extent seems to be located outside the cranium, point to a brain weight close to the upper limit, that is about 130 mg.

BoWs of *Geogale* have been inferred from head and body lengths, which average about 71 mm. Mammals of similar shape (tenrecs, mice, and shrews) of this size have an expected body weight of about 10 g.

Based on the estimated values (10 g body weight, 130 mg brain weight), *Geogale* has about two thirds of the relative brain size (encephalization) of an average Tenrecinae (0.81x AvTn). So far, this index is the lowest found in any extant mammalian species, much lower than in any other group of Insectivora. Until now, the Tenrecinae were at the bottom of the encephalization scale (Stephan et al. 1986a, 1988). *Geogale* has only about half the brain size as the average Insectivora (0.51x AvIF, = 81/158, see Table 7).

Macromorphology

The macromorphology of the geogaline brain is known only from a reconstruction of a serially-sectioned head (Rehkämper et al. 1986). The brain differs widely from those of the Oryzorictinae and is more similar to those of the Tenrecinae. Its tectum is totally uncovered; its interhemispheric fissure is short; its cerebral hemispheres are small, with a highly placed border between neocortex and piriform lobe (rhinal sulcus); its corpus callosum is minute, and its brain, though compact, is not as round as in Oryzorictinae. Most features must be regarded as conservative.

Volume Measurements

Rehkämper et al. (1986) performed volume measurements and compared the percentage composition of the brain with that of other Insectivora and Mammalia. Indices are given here for the first time. All the data given must be viewed with caution since the material is weak. With reference to the tenrecine level, the lowest size is for the schizocortex (SCH, 0.40x AvTn) and for the olfactory structures (PAL, 0.50x; MOB, 0.55x). The largest size is for AOB (1.57x), OBL (1.40x), and MES (1.10x). These structures are the only ones so far found to be above the tenrecine average. After a fairly wide gap, DIE and HIP (0.79 and 0.77x AvTn) are the next largest, followed by CER (0.73x). Compared with the family/subfamily average (AvIF), NEO is in the lowest place. Its AvIF is 300 and its index in *Geogale* is 61, i.e., about one fifth; that of STR is one third (210 versus 66).

In terms of all the macromorphological and quantitative characteristics of the brain mentioned, *Geogale aurita* is one of the most conservative extant mammalian species. In addition to its low encephalization, its small neocortex and striatum and large brain stem components (medulla oblongata and mesencephalon) support this interpretation. In most characteristics, the brain of *Geogale* must even be placed much lower than those of the average Tenrecinae, which have so far formed the bottom of the scales in relative brain and neocortex size.

A summary of brain characteristics of *Geogale* compared with those of the other Tenrecidae is given on p. 180.

4.1.3 Oryzorictinae

General Characteristics

Oryzorictinae show considerable variation in habits and habitats. In general, three genera are distinguished *(Oryzorictes, Microgale, Limnogale)*. *Oryzorictes* is highly fossorial, has a talpoid shape, its forefeet are stout, wider than its hindfeet, and have strong digging claws. Its diet is assumed to consist mainly of insects and other invertebrates. The shrew-like *Microgale* occurs in forested regions and in very moist meadows covered with dense grass and reeds. The tip of the tail in the long-tailed forms is usually modified for prehension, suggesting that at least some species are partially arboreal. The forelimbs are not adapted for digging, and the ear pinnae are prominent. *Limnogale* is semiaquatic with webbed hindfeet. It hunts its prey, aquatic invertebrates and amphibians, in fast-flowing streams. According to Walker, it feeds mainly on plant material, but, judging from its brain characteristics, this is highly unlikely.

Brain Characteristics

Brain material is available from all three genera.

Oryzorictes talpoides, Microgale* sp.?, *M. cowani, M. dobsoni, M. talazaci, Limnogale mergulus

Brain Weight, Encephalization

Bauchot and Stephan (1964, 1966)

The only known data on brain and body weights for *Oryzorictes* and *Microgale* are from Bauchot and Stephan (1964); Bauchot and Stephan (1966) provide data on *Limnogale* also. Only one brain of *M. cowani* is available, but it seems to be from a juvenile specimen and is therefore not included in the size comparisons presented here. Thus, only four species are considered: *Oryzorictes talpoides, Microgale dobsoni, M. talazaci,* and *Limnogale mergulus.*

The average encephalization of Oryzorictinae is 1.58 times greater than that of Tenrecinae. It is lowest in *Oryzorictes* (1.36x AvTn) and highest in *Microgale talazaci* (1.70x) (Table 6).

Macromorphology

Bauchot and Stephan (1970)
Clark (1932)
Leche (1905, 1907)
Stephan and Andy (1982)

Details on macromorphology are given in Leche (1905, 1907) and Clark (1932) for *Microgale*, and in Bauchot and Stephan (1970) and Stephan and Andy (1982) for all three genera. Leche described the rostral (orbital) sulcus, which, according to our investigations, is highly characteristic in *Microgale*, whereas Clark is doubtful as to its existence. Based on the unusually simple cerebellar arbor vitae, whose complexity depends strongly on size, Clark seems to have described a small species, comparable to our *M. cowani*.

The mesencephalic tectum is nearly or fully covered in Oryzorictinae. There is no marked border between piriform lobe and neopallium. The visual nerve is small; the trigeminal nerve is very big. *Limnogale* has smaller olfactory bulbs whereas the lobulus petrosus of the cerebellum appears to be large. Trigeminal nerve and medulla oblongata appear to be still larger than in the terrestrial forms.

Area Measurements

Bauchot and Stephan (1970)

The areas of sectioned commissures and nerves have been determined by Bauchot and Stephan (1970). The indices for the corpus callosum were distinctly higher in Oryzorictinae than in Tenrecinae, highest in *Limnogale*, and lowest in *M. cowani* and *Oryzorictes*.

Volume Measurements

Andy and Stephan (1966a)
Baron et al. (1983, 1987, 1988)
Bauchot (1970, 1979a, b)
Bauchot and Legait (1978)
Bauchot and Stephan (1968, 1970)
Frahm et al. (1982)
Legait et al. (1976)
Stephan (1966, 1967b, 1969, 1972)
Stephan and Andy (1969, 1982)
Stephan et al. (1970, 1981a, 1982, 1987a)

Since 1966, volumes have been measured by several authors and teams (Andy, Baron, Bauchot, Frahm, Legait, Stephan). Data on the five main parts of the brain are given in Stephan (1967b, 1972), Bauchot and Stephan (1968), Stephan and Andy (1969, 1982), Stephan et al. (1970, 1981a), and Bauchot (1979a).

The widest differences in the four species of Oryzorictinae are found for the medulla oblongata. OBL is 2.14 times larger in *Limnogale* than in *Oryzorictes* (216 versus 101; Table 11). Of the OBL parts, data are provided by Baron et al. (1988) on vestibular structures in *M. talazaci*: VM was found to be large, VS small.

Fairly large variation is also found in MES, which is largest in *Microgale dobsoni* and smallest in *Oryzorictes*. The subcommissural body is generally small in Oryzorictinae (Stephan 1969) and varies from 0.39x AvTn in *Oryzorictes* to 0.78x in *M. talazaci*. CER and DIE are largest in *Limnogale*.

Structures of DIE are compared in Bauchot (1979b) for *M. talazaci* and *Limnogale*. Of the thalamic components, *Limnogale* has a very large CGM (4.76x AvTn) and a small CGL (0.89x), and *M. talazaci* has large anterior nuclei (3.24x). Data on the hypophysis (HYP) in *M. talazaci* and *Limnogale* are given in Bauchot (1970) and Bauchot and Legait (1978). A large HYP with a large anterior lobe is found in *Microgale*, a small HYP with a relatively large intermediate lobe characterize *Limnogale*. Legait et al. (1976) provide data on epiphysis of five species. EPI is, on the average, relatively small (see also Table 41).

Various parts of the telencephalon have been measured and compared, e.g., olfactory structures (Stephan 1966; Stephan et al. 1982; Baron et al. 1983, 1987), septum (Andy and Stephan 1966a; Stephan and Andy 1982), amygdala (Stephan et al. 1987a), and neocortex (Bauchot and Stephan 1970; Frahm et al. 1982).

The species differ widely in structures of the main olfactory system. These structures are very small in *Limnogale* and fairly small also in *Oryzorictes*. MOB is 2.45 times larger in *Microgale* than in *Limnogale*. AOB is 2.66 times larger in *M. talazaci* than in *Oryzorictes* (226 versus 85; Table 50). SEP is largest in *Limnogale* and smallest in *Oryzorictes*, whereas AMY is smallest in *Limnogale*. NEO is largest in *Limnogale*: 1.7 times larger than in *Oryzorictes* (325 versus 189).

The representatives of Oryzorictinae have a larger relative brain size than Tenrecinae and Geogalinae. The average NET index is 1.60 times larger than in Tenrecinae. The largest indices are found for NEO (2.44x AvTn) and HIP (2.16x). The only structure smaller than in the average Tenrecinae is MOB (0.94x). However, this finding is due to *Limnogale* only, which has a very small MOB (0.49x), whereas, in the terrestrial Oryzorictinae, MOB is above the tenrecine level (1.09x). *Limnogale*, which unlike the other forms is adapted to semiaquatic life, also differs from *Oryzorictes* and *Microgale* in certain other structures and structural complexes.

More quantitative comparisons will be presented on p. 180 in the summary of brain characteristics of Oryzorictinae and its species compared with those of the other Tenrecidae.

Histology

Stephan (1965)

Stephan (1965) describes the accessory olfactory bulb in *Microgale dobsoni* and *M. talazaci* and mentions a distinct external granular layer and a very broad layer of mitral cells.

4.1.4 Potamogalinae

General Characteristics

The African otter shrews are so closely related to the tenrecs of Madagascar that they should be considered one of their subfamilies. Our investigations on the brain are compatible with this view (see p. 302). The Potamogalinae comprise two genera with three species: *Potamogale velox, Micropotamogale lamottei*, and *M. ruwenzorii*. Adaptation to water is strongest in the largest form *(Potamogale)* and weakest in the smallest, i.e., *M. lamottei*.

The giant form, *Potamogale velox*, has an otter-like appearance with a thick tail strongly compressed laterally. The flattened muzzle has stiff, white whiskers; the ear pinnae are small; the eyes are minute; and the nostrils are covered by flaps that act as valves when the animal is submerged. The short limbs have five non-webbed digits. Preferred foods appear to be aquatic vertebrates and invertebrates. The tail of *M. lamottei* is not laterally compressed as in *Potamogale velox* nor does the species have the webbed hands and feet found in *M. ruwenzorii*.

Brain Characteristics

Brain material is available from all three species.

Potamogale velox, Micropotamogale lamottei, M. ruwenzorii

Brain Weight, Cranial Capacity, Encephalization

Stephan (1959)
Stephan and Kuhn (1982)
Stephan et al. (1986b)

The first five pairs of brain and body weights for *Potamogale* are reported in Stephan (1959). Stephan and Kuhn (1982) added twelve cranial capacities (CrCs) for *Potamogale*, four for *M. lamottei*, and three for *M. ruwenzorii*. No body weights for the same individuals were available. BrW/BoW pairs (n = 4) were later given by Stephan et al. (1986b) for *M. ruwenzorii*, and eight other pairs could be added to the data on *Potamogale*. The largest brain (1.71x AvTn) was found in *Potamogale*, the smallest (1.47x AvTn) in *M. lamottei*.

Macromorphology

Bauchot and Stephan (1967, 1968)
Stephan (1961)
Stephan and Andy (1982)
Stephan and Kuhn (1982)
Stephan and Spatz (1962)
Stephan et al. (1986b)

Stephan (1961) reconstructed allocortical surface regions for the *Potamogale* brain, but the first macromorphological description with figures and a reconstruction of the median level was given by Stephan and Spatz (1962). The authors described a small olfactory bulb, low-running lateral rhinal sulcus, far downward-extending neocortex, large cerebellum, completely hidden mesencephalic tectum, small visual nerves, and enormous trigeminal nerves, which, with their projection sites, enlarge the OBL. Bauchot and Stephan (1967) compared endocranial cast and brain of *Potamogale*. A comparison with other water-adapted mammals has been given by Bauchot and Stephan (1968). Stephan and Andy (1982) have provided a detailed comparison with other Insectivora.

The first descriptions of the brains of *Micropotamogale lamottei* were given by Stephan and Kuhn (1982) and of *M. ruwenzorii* by Stephan et al. (1986b). Characteristics similar to those in *Potamogale* were described for both forms, but, for the most part, were less impressive.

Area Measurements

Bauchot and Stephan (1961, 1970)
Stephan (1961)
Stephan and Spatz (1962)

Bauchot and Stephan (1961) determined the areas of sectioned commissures and nerves in *Potamogale*. The corpus callosum is relatively large, and largest compared with other Tenrecidae and with *Solenodon* (Bauchot and Stephan 1970). The anterior commissure, especially its rostral component, is smallest, and the chiasma opticum is of about average size in the 13 measured species of Insectivora.

Stephan (1961) investigated area size of cortical surfaces for *Potamogale*. A major reduction in the size of all olfactory structures, including the precommissural hippocampus, was found. The neocortical surface is relatively large. Small paleocortex, average archicortex, and large neocortex are also mentioned by Stephan and Spatz (1962).

Volume Measurements

Andy and Stephan (1966a, b)
Baron et al. (1983, 1987, 1988)
Bauchot (1961, 1963, 1966, 1970, 1979a, b)
Bauchot and Legait (1978)
Bauchot and Stephan (1968)
Etienne (1972)
Frahm et al. (1982)
Legait et al. (1976)
Stephan (1966, 1967b, 1969, 1972)
Stephan and Andy (1962, 1964, 1969, 1982)
Stephan and Kuhn (1982)
Stephan et al. (1970, 1981a, 1982, 1986b, 1987a)

The volumes of many brain parts in *Potamogale* have been measured by several authors and teams since 1961 (Andy, Baron, Bauchot, Etienne, Frahm, Legait, Stephan). Investigations on *Micropotamogale* spp. only began in 1982 (Baron, Bhatnagar, Frahm, Kabongo, Kuhn, Stephan).

Data on the five **fundamental brain parts** are given in Stephan and Andy (1964, 1969, 1982), Stephan (1967b, 1972), Bauchot and Stephan (1968), Bauchot (1979a), Stephan et al. (1970, 1981a, 1986b), and Stephan and Kuhn (1982). The largest differences among the three species of Potamogalinae are in CER, which is 1.61 times larger in *Potamogale* than in *M. lamottei* (221 versus 137, Table 11). Relatively large variation is also found in OBL (240 versus 178).

In terms of quantitative characteristics, OBL has the highest indices in *Potamogale* and *M. ruwenzorii*, whereas, in *M. lamottei*, the index for DIE is slightly higher than that of OBL (188 versus 178).

In the **medulla oblongata** (OBL) of *Potamogale* and *M. ruwenzorii*, the sensory trigeminal complex (TR) and the nucleus reticularis lateralis (REL) have enlarged most strongly. Strong enlargements are also found in the nucleus nervi facialis (VII), inferior olive (INO), and the vestibular structures. The nucleus vestibularis inferior (VI) is especially large in *Potamogale* (Baron et al. 1988) (Table 23). No progression relative to the tenrecine level is found in the nucleus fascicularis cuneatus externus, the nucleus nervi hypoglossi, and the cochlear nuclei.

In the **mesencephalon**, the subcommissural body (SCB) is very small in *Potamogale* (0.23x AvTn) and is hard to identify in *M. ruwenzorii*.

For the **diencephalon**, data on the main parts of *Potamogale* are provided by Bauchot (1963, 1966, 1979b). DIE is larger in *Potamogale* than in other species of Tenrecidae and, within DIE, the nucleus reticularis (RTH) and the thalamus (THA) are especially large. In THA, Bauchot found very large dorsal nuclei (DOT) and medial geniculate bodies (CGM), whereas the lateral geniculate bodies (CGL) are only at the tenrecine level (Table 69). The CGMs are among the largest found in Insectivora so far measured.

Data on smaller components of the diencephalon are given for the epiphysis (EPI) by Stephan (1969), Legait et al. (1976) and Stephan et al. (1981a), on the nucleus medialis habenulae (HAM) by Stephan (1969) and Stephan et al. (1981a) and on the hypophysis (HYP) by Bauchot (1961, 1970) and Bauchot and Legait (1978). EPI is rudimentary in otter shrews (Table 41), HAM is fairly well developed (1.53 to 1.81x AvTn) and HYP has a relatively large inter-mediate lobe in *Potamogale*, but the other parts are small.

In the **telencephalon** (TEL), seven or eight main parts have been measured and compared by Stephan and Andy (1964, 1969, 1982), Stephan (1967b, 1972), Bauchot and Stephan (1968), Stephan et al. (1970, 1981a, 1986b), and Stephan and Kuhn (1982). In addition, the olfactory and limbic structures have been measured by Stephan (1966), the main olfactory bulb (MOB) by Etienne (1972) and Baron et al. (1983), the accessory olfactory bulb (AOB) by Stephan et al. (1982), the paleocortex (PAL) by Baron et al. (1987), the amygdala (AMY) by Stephan et al. (1987a), the septum (SEP) by Stephan and Andy (1962) and Andy and Stephan (1966a), and the neocortex (NEO) by Frahm et al. (1982).

A clear reduction relative to Tenrecinae is found in all structures in the main olfactory system (BOL, MOB, PAL, RB, PRPI, TOL) (Tables 44, 47, 53). The reduction is strongest in *Potamogale* and least in *M. lamottei.* For the two species of *Micropotamogale*, a clear reduction is also found in the accessory olfactory bulb (AOB), whereas AOB is well developed in *Potamogale* (2.88x AvTn). The relative size of MOB and AOB laminar components, however, remain more or less stable (Tables 47, 50).

Besides the olfactory structures, the amygdala (AMY) is the only main TEL structure whose size is smaller than in Tenrecinae. According to Table 56, the same is true of all the parts of AMY so far measured. Apart from AMY, the other limbic structures (SEP, HIP, SCH) are distinctly progressive in otter shrews, with HIP 2.17x AvTn in *Potamogale*. The most progressive TEL structure by far in all otter shrews is the neocortex (NEO), which is 4.34x AvTn in *Potamogale*, 3.65x in *M. ruwenzorii*, and 3.12x in *M. lamottei* (Table 44). In NEO, the cell-dense layers (NG2-6) are much more progressive (3.59 to 5.29x AvTn) than the molecular layer (NG1) (2.13 to 2.19x). This, however, is generally the case in species with progressive NEO development (Table 62).

Compared with the tenrecine level, all three species of otter shrews show similar brain characteristics, in particular, large medulla oblongata (OBL) and neocortex and small main olfactory structures. The large OBL can be explained mainly by strong development of the sensory trigeminal complex. The very large trigeminal nerve mainly innervates the strongly developed vibrissae of the muzzle, giving the snout a very broad appearance. Similar characteristics are found in *Limnogale* and must be related to adaptation to semiaquatic life. In most of the brain characteristics, *Micropotamogale lamottei* is closest to the average of the Tenrecinae, which are all purely terrestrial; *Potamogale velox* is farthest, and *Micropotamogale ruwenzorii* is in an intermediate position. As in *Limnogale*, the strong reduction in the olfactory structures results in relatively small TEL, whose indices are relatively close to the tenrecine average (Table 11) despite a strong enlargement of NEO. In spite of the fairly low TEL indices, the brains of the Potamogalinae have obviously more advanced than conservative characteristics.

More quantitative comparisons will be presented on p. 183 in the summary of brain characteristics of Potamogalinae and its species compared with those of the other Tenrecidae.

Histology

Rehkämper (1981)
Stephan (1965)
Stephan and Andy (1982)
Stephan and Spatz (1962)
Stolzenburg et al. (1989)

Some general remarks on histology have been given in many of the papers on volume measurements. Photoprints of brain sections and more details on histology of *Potamogale* are found in the studies cited.

Stephan and Spatz (1962) present some transverse sections at the level of the anterior commissure (COA), at the knee of the facial nerve and at the nucleus hypoglossi. They mention the very small visual nerve and very large trigeminal nerve and sensory complex and, in sagittal section, the very large corpus callosum and small anterior commissure. Stephan (1965) gives some details on the accessory olfactory bulb (AOB) and calls attention to the big vomeronasal nerve, the very clear mitral cell layer and the internal granular layer, which contains two types of granular cells. The inner group of these cells reach the remains of the olfactory ventricle. This is unique in Insectivora. Rehkämper (1981) describes the neocortex (NEO) of *Potamogale* and gives a cortical map containing four areas. With respect to the laminar components, he mentions a relatively thin molecular layer (I) and difficulty in separating layers II and III. In the caudal field (A 3), however, the lamina II cells lie in groups. In the deeper part of lamina III, granular-like cells are found, which also exist in *Desmana*. The lateral field A 4 of *Potamogale* is very large. Its rostral parts may be related to the trigeminal system. Histological sections of the large sensory trigeminal complex, the accessory olfactory bulb and the well differentiated septum have been presented by Stephan and Andy (1982). Stolzenburg et al. (1989) report on the size and density of glial and neuronal cells in the neocortex of *Potamogale* in comparison to other Insectivora.

4.1.5 Summary of Brain Characteristics in Subfamilies and Species of Tenrecidae

Tenrecidae, as defined here, i.e., including the Tenrecinae, Geogalinae, Oryzorictinae, and Potamogalinae, are an extremely complex family with regard to brain characters. Characteristics common to all species, genera or subfamilies are hard to find, even in the macromorphological features which in other complex families often show common features. Little emphasis will be placed on average indices for the entire tenrec family since they would be more confusing than helpful. Nevertheless, the values are included in all the tables dealing with average size indices (outlined in Table 1).

Index profiles for the four tenrec subfamilies are given in Figs. 101, 103, 105, and 107 (pp. 294 - 297). When the average indices for each of the subfamilies are compared with those for the whole family, the smallest differences are frequently found for Oryzorictinae, but this subfamily sometimes shows the largest deviation from the average for the tenrec family. The final brain composition of the various subfamilies seems to be less dependent on phylogenetic inertia but, to a large degree, on ecoethological adaptations.

Tenrecinae

In comparison to other families and/or subfamilies of Insectivora, Tenrecinae have the smallest brains (next to *Geogale*). NET is on the average 0.63x AvIF. All five fundamental brain parts are at similar levels: between 0.56x for CER and DIE and 0.72x for OBL (Fig. 101). The variation is larger in the finer subdivisions, but nearly all structures have smaller indices than found for AvIF. Only VCO in the OBL components (Fig. 103), CGL in the DIE components (Fig. 105), and MOB in the TEL components (Fig. 107) are above the AvIF level; PAL nearly reach this level. In Fig. 103, DCO and OLS are below AvIF but, of all the OBL components, they occupy the highest positions next to VCO. In Fig. 105, VET is only slightly below AvIF, whereas all the other components are clearly smaller. The lowest indices relative to AvIF are found in OBL for INO, in DIE for DOT, and in TEL for NEO, and the second lowest for STR.

In conclusion, Tenrecinae have very small brains for Insectivora. Their olfactory (MOB, PAL), visual (CGL), and auditory (VCO, DCO, OLS) structures are quite well represented, whereas motor (CER, INO, STR) and higher neencephalic centers (NEO, DOT) are poorly developed.

Species characteristics of tenrecine brains are based on the deviation from the average tenrecine level (AvTn).

Tenrec ecaudatus: The total brain, i.e. brain weight or NET volume, in *Tenrec* is below the average group level (0.86x AvTn for BrW, 0.85x for NET) and is the lowest in Tenrecinae. All main brain parts are also below AvTn. They vary in the five fundamental brain parts from 0.83x for TEL to 0.89x for CER and in the eight TEL parts from 0.81x for SEP to 0.91x for AMY. The OBL structures vary from 0.63x AvTn in COE to 1.14x in VL. Low size is found in VCO (0.67x), in the auditory structures as a whole (0.84x), and in the funicular nuclei (0.72x); relatively large size is found in INO (1.07x) and mot V (1.04x). In MES, NLL is very small (0.66x); SCB is large (1.23x). In DIE, all parts are below the tenrecine average. Especially low are ANT and EPI (0.66x), relatively high are HTH (0.88x) and STH (0.87x), and of the thalamic structures,

CGL (0.89x) and VET (0.87x). In TEL, the lowest indices are found for some of the limbic structures (SEP and HIP). A large SFB (1.28x AvTn) is found in SEP. Compared with other Tenrecinae, the brain of *Tenrec* is the longest, narrowest and flattest; the corpus callosum is the shortest.

Particularities: The brain of *Tenrec ecaudatus* is small compared with other Tenrecinae. Nearly all brain parts are uniformly small. The auditory structures, funicular nuclei and pineal gland are very small.

Setifer setosus: Setifer has the relative largest brain of all the Tenrecinae (1.17x AvTn for BrW, 1.19x for NET). The indices for all the main brain parts are at or above the tenrecine level (= 1.00), reaching as high as 1.26x AvTn in TEL, and varying in the TEL parts from 1.12x in HIP to 1.49x in AMY. Except for CER, all the main parts are larger in *Setifer* than in other Tenrecinae. In contrast, most of the OBL structures are below the tenrecine level; they vary from 0.66x in FCE to 1.20x in VS. The funicular nuclei as a whole are small (0.86x). In *Setifer,* the lowest values in the tenrecine subfamily are found for FCE (0.66x), REL (0.72x), INO (0.79x), PRP (0.83x), and TSO (0.84x); the highest values are in VS (1.20x), TR (1.10x), and mot V (1.08x). The auditory structures are also relatively large (1.09x). In MES, the tegmentum is large (1.15x), whereas SUC (0.79x) and SCB (0.69x) are small. In CER, there is a very large MCN (1.31x AvTn). In DIE, nearly all the parts are largest; they vary from 1.05x in RTH to 1.33x in MLT. Only HAM is small (0.77x). The visual structure CGL is large, as is the indirectly determined area striata. In TEL, all the olfactory structures (MOB, AOB, PAL) and AMY are large. The triangular nucleus of SEP is small (0.75x). The cerebral hemispheres are short and very wide.

Particularities: The brain of *Setifer setosus* is the largest found in Tenrecinae and has large olfactory, trigeminal, and auditory structures. The funicular nuclei, the lateral reticular nucleus (REL), and the inferior olive (INO) are all small.

Hemicentetes semispinosus: Brain size is very close to the average of the four Tenrecinae (1.04x AvTn for BrW and NET). There is wide variation in the indices of the fundamental brain parts, from 0.96x AvTn for MES to 1.33x for CER. The OBL as a whole is relatively large (1.08x) and many of the OBL parts are largest in *Hemicentetes* compared with the other Tenrecinae. This is true of PRP (1.24x), VI (1.23x, and the vestibular nuclei as a whole, 1.13x), INO (1.21x), REL (1.21x), and TSO (1.18x). Also large, but not the largest in the tenrecine group, are COE (1.23x), FCE (1.12x), the total of funicular nuclei (1.06x), and TR (1.07x). The DCO (0.69x), the total of auditory structures (0.78x), mot V (0.90x), and the motor nuclei as a whole (0.96x) are small. In MES, the inferior colliculus is very small (0.76x) and SCB is large (1.27x). The enlargement of CER (see above) is mirrored in large cerebellar and vestibular nuclei. In DIE, HAM (1.19x) and the thalamic anterior nuclei (ANT; 1.10x), dorsal nuclei (DOT; 1.04x) and CGM (1.02x) are large; all other parts are below the tenrecine level. EPI is small (0.80x). In TEL, the main parts vary from 0.70x in AMY to 1.21x in SCH. The smallest in *Hemicentetes* compared with the other Tenrecinae are AMY (0.70x), AOB (0.74x), and SFB (0.81x). The brain of *Hemicentetes* is the shortest and highest; the corpus callosum is the longest.

Particularities: In size, the brain of *Hemicentetes semispinosus* is the most typical of tenrecine brains. It has a very large cerebellar complex (incl. inferior olive and vestibular nuclei) and small auditory structures (incl. inferior colliculus) and accessory olfactory bulbs.

Echinops telfairi: *Echinops* has the second smallest brain (0.93x AvTn for BrW and NET), next to *Tenrec*. Compared with *Tenrec*, however, there is wider variation in the brain parts. Of the five fundamental parts, only MES has an index above the tenrecine level (1.09x AvTn). Many of the OBL parts are larger in *Echinops* than in other Tenrecinae. This is true of the funicular nuclei (1.36x), COE (1.31x), all the auditory structures (1.29x), and motor nuclei (1.17 - 1.26x, except mot V, which is 0.98x). The smallest in *Echinops* is TR (0.91x). Along with a very small cerebellum (0.78x AvTn), the cerebellar nuclei (0.78x), INO (0.92x) and the whole vestibular complex (0.89x) are also small. In MES, tectum (1.31x) and NLL (1.22x) are largest in *Echinops*, whereas tegmentum (0.90x) is small. SCB also is small (0.81x). In DIE, EPI (1.25x), HAM (1.18x), and CGL (1.11x) are markedly above the tenrecine level and CGL is even the highest in Tenrecinae. The lowest are DOT and MLT (0.86x), and HTH (0.91x). In TEL, all parts except SET (1.28x), AOB (1.16x), and SEP (1.01x) are below the tenrecine level. The low parts vary from 0.65x for SCH to 0.99x for STR. The smallest in *Echinops* are SCH (0.65x), MOB (0.85x), and NEO (0.89x). *Echinops* has the longest hemispheres of all Tenrecinae.

Particularities: The brain of *Echinops telfairi* is the second smallest found in Tenrecinae. It has very small motor components (cerebellum, cerebellar nuclei, inferior olive, vestibular complex) and small main olfactory structures (MOB, PAL), but the largest auditory and funicular components.

In conclusion, the smallest tenrecine brain, with TEL and NEO also smallest, is found in *Tenrec*. In ascending sequence follow *Echinops, Hemicentetes* and, in distinctly highest place, *Setifer*. The four tenrecine species are characterized by species differences in brain macromorphology (see Section 3.1) and brain composition. The centers of the functional systems may differ largely in size: the auditory system and the funicular and motor nuclei are distinctly largest in *Echinops* and, except the auditory parts, smallest in *Tenrec*; the vestibular system is distinctly largest in *Hemicentetes* and smallest in *Echinops*; and the main olfactory system is distinctly largest in *Setifer* and smallest in *Echinops*.

Geogalinae

The single species, *Geogale aurita,* has the smallest brain of all Insectivora. Since the differences from Tenrecinae, the second lowest group, are large, such a conclusion appears warranted in spite of the poor material. The total brain is on the average only half as large as in the average Insectivora (0.51x AvIF). The differences between the five fundamental brain parts are by far larger than in Tenrecinae and vary from 0.41x AvIF for TEL and CER to 1.01 for OBL (Fig. 101). Thus, OBL size is close to that of the typical Insectivora. No further subdivisions were made in OBL and DIE. In TEL (Fig. 107), the main structures vary from 0.20 for NEO to 0.61 for MOB. Only AOB is close to AvIF (0.99x).

Compared with Tenrecinae, the relative brain size of *Geogale* is only 4/5 (0.81x AvTn). All brain parts except AOB (1.57x), OBL (1.40x), and MES (1.10x) are lower than in the tenrecine average, with HIP and DIE closest to it (0.77x and 0.79x), and the schizocortex (SCH), and olfactory structures farthest from it (0.40 - 0.56x AvTn).

In conclusion, according to the indirectly inferred data of Rehkämper et al., *Geogale* shows the greatest concentration of conservative brain characteristics so far found in extant mammalian species. The less developed brain of *Geogale aurita* seems to have well represented brain stem structures and accessory olfactory centers. The smallest structures are SCH and NEO, and the motor structures (STR and CER) also are poorly represented.

Oryzorictinae

Oryzorictinae are most representative of the brain of the typical Insectivora. This is true for total brain (1.0x AvIF) and for most brain parts. The five fundamental brain parts vary in the narrow range of 0.90 for DIE to 1.04 for CER (Fig. 101); the DIE parts vary from 0.86 for STH and 0.90 for DOT to 1.28 for ANT (Fig. 105); the TEL parts vary from 0.81 for NEO to 1.16 for HIP (Fig. 107). There is greater variation in the OBL parts: from 0.61 for VS to 1.33 for VCO (Fig. 103). The highest values next to VCO were found for DCO and OLS, which are all auditory centers, and for some motor nuclei (mot V and mot VII). In contrast, two vestibular nuclei (VI, VS) are lowest and a third (VM) is somewhat higher, but still below AvIF (0.85). As in the Tenrecinae and Geogalinae, NEO is the smallest of the TEL structures in the Oryzorictinae. However, it is not the olfactory structures but the limbic HIP that is the largest of the TEL components in Oryzorictinae. The median position of the olfactory centers (1.03 for MOB, 1.05 for PAL) is caused solely by *Limnogale,* which in MOB is 0.54x AvIF, whereas the terrestrial Oryzorictinae have large olfactory structures. MOB is on the average 1.20x in the terrestrial species, and in *Microgale* spp. even 1.32x AvIF.

In conclusion, Oryzorictinae have the most typical insectivoran brain, i.e., deviating the least from AvIF. Auditory and hippocampal structures are well represented; vestibular and higher neencephalic structures (NEO, DOT) are poorly developed.

Species characteristics of oryzorictine brains are based on the deviation from the average oryzorictine level (AvOr), which is uniformly set at 1 for each of the structures.

Oryzorictes talpoides: The smallest brain in the Oryzorictinae is found in *Oryzorictes* (0.86x AvOr for BrW and NET). All the five fundamental brain parts are also the smallest, except CER, which is somewhat smaller in *Microgale dobsoni*. The five brain parts vary from 0.67x AvOr in MES to 0.93x in CER. OBL components were not measured. In MES, SCB is very small (0.61x AvOr); in DIE, EPI (0.80x) and HAM (0.86x) are also small. The TEL parts vary from 0.56x AvOr in AOB and 0.77x in NEO to 1.07x in SCH. AOB, NEO, and SEP (0.89x) are smaller in *Oryzorictes* than in any other species of Oryzorictinae. Below AvOr are also the main olfactory structures (0.91x in MOB; 0.89x in PAL), HIP (0.90x), and SET (0.93x). Above the oryzorictine level are SFB (1.21x), SCH (1.07x), AMY (1.02x) and STR (1.02x). SFB and SCH are largest in *Oryzorictes*. *Oryzorictes* has the shortest brain with the shortest and widest cerebral hemispheres; the corpus callosum also is shortest.

Particularities: *Oryzorictes talpoides* has the smallest brain in Oryzorictinae, with an especially small brain stem (MES, OBL). In TEL, NEO is smallest by far and the olfactory structures are relatively small.

Microgale dobsoni and ***M. talazaci:*** In size of the main parts of the brains, the two species have similar characteristics and, therefore, will be discussed together. The brain of *Microgale dobsoni* is very close to the average of the four Oryzorictinae (1.02x AvOr in BrW; 1.01x in NET), that of *M. talazaci* is somewhat higher (1.08x in BrW; 1.09x in NET). The five fundamental brain parts are close to the oryzorictine level. They vary in *M. dobsoni* from 0.82x AvOr in CER to 1.14x in MES, and in *M. talazaci* from 0.93x in OBL to 1.14x in TEL. Common features are small OBL and relatively large MES and TEL. In *M. dobsoni*, the TEL parts vary from 0.59x in SFB and 0.90x in STR to 1.29x in MOB, and in *M. talazaci*, from 0.96x in STR and SEP to 1.35x in PAL and 1.49x in AOB. Common features are relatively large olfactory structures (MOB, PAL, AOB), HIP, and AMY, and small STR, NEO, and SCH. In DIE, EPI is clearly larger in *M. talazaci* (1.29x AvOr) than in *M. dobsoni* (0.67x); similar but less pronounced differences are found for HAM (1.17x and

0.97x). The size of the main DIE parts is given in Bauchot (1979b) for *M. talazaci* and *Limnogale mergulus*. Compared with *Limnogale, M. talazaci* has large ANT and CGL, and a very small CGM. Both species of *Microgale* have the longest and highest brains with the longest and narrowest cerebral hemispheres. Both have a distinct rostral (orbital) sulcus. The corpus callosum is longer in *M. talazaci* than in any other oryzorictine species so far measured.

Particularities: *Microgale* spp. are most typical in oryzorictine brain size. They have large olfactory structures and HIP, and relatively small motor components (CER, STR). Based on Bauchot's measurements in *M. talazaci*, vision may be well developed (CGL large) and audition poorly developed (CGM small).

Limnogale mergulus: The brain size of *Limnogale* is similar to that of *Microgale* ssp. (1.05x AvOr in BrW; 1.03x in NET) but there is a strong reduction in the olfactory centers. Of the five fundamental brain parts, three are largest in *Limnogale*, MES is second largest, but TEL is smallest. By far the largest is OBL (1.52x AvOr), followed by CER (1.23x). In TEL the main parts differ strongly and vary from 0.51x in PAL to 1.33x in NEO. NEO is by far the largest in *Limnogale* compared with the other Oryzorictinae so far measured. STR (1.12x), SEP (1.11x) and SET (1.16x) are also large. The smallest in *Limnogale* are PAL (0.51x), MOB (0.52x), HIP (0.72x), and AMY (0.85x). Based on the DIE measurements of Bauchot (1979b), *Limnogale* has smaller ANT nuclei and CGL, and distinctly larger CGM than *M. talazaci*,. In shape, *Limnogale* has the flattest of all oryzorictine brains so far investigated.

Particularities: Of all the Oryzorictinae, *Limnogale* has the most deviating brain: very large OBL, NEO, and CGM (audition); large CER, DIE, STR, and SEP; very small main olfactory structures (MOB, PAL); small HIP, AMY, and CGL (vision). The small TEL caused by reduced olfactory structures is mainly responsible for the relatively small total brain in *Limnogale*, which is at the same level as *Microgale talazaci*.

In conclusion, the clearly smallest oryzorictine brain, with TEL and NEO also smallest, is found in *Oryzorictes*. In ascending sequence follow *Microgale* spp. and *Limnogale*, but this is true only for NEO, since TEL as a whole is small in *Limnogale* and this reduces the total brain size, which is not higher than in *M. talazaci*. The small TEL in *Limnogale* is caused mainly by a strong reduction in its main olfactory centers. This makes its TEL smaller than in the terrestrial forms, despite its clearly larger NEO. OBL is very large in *Limnogale*. Small olfactory structures and large OBL are also found in other water-adapted (semiaquatic) Insectivora, e.g., in otter shrews (Potamogalinae).

Potamogalinae

In contrast to all other subfamilies of Tenrecidae, Potamogalinae brains (1) are slightly above AvIF (1.01x) and (2) show much wider variations in the brain parts. Except for TEL, the fundamental parts are above AvIF and vary from 0.87 for TEL to 1.49 for OBL (Fig. 101). In OBL, the values vary as much as from 0.77 for FCE to 2.01 for TR (Fig. 103). In DIE, they vary from 0.69 for ETH to 1.74 for CGM and 1.55 for DOT (Fig. 105) and, in TEL, they vary from 0.42 for MOB and 0.46 for PAL to 1.23 for NEO (Fig. 107). In the very large OBL, the best represented components after TR are mot VII, vestibular nuclei and COE. FCE, FCM, mot XII, and auditory nuclei are relatively poorly developed. In contrast to the small auditory centers, CGM is very large, based on Bauchot's data.

In conclusion, the orofacial part of the somatosensory system (TR; see also Fig. 56) is very well represented in Potamogalinae and the higher neencephalic centers (NEO, DOT) are well developed. The olfactory structures, however, are poorly represented.

Species characteristics of potamogaline brains are based on the deviation from the average potamogaline level (AvPo), which is uniformly set at 1 for each of the structures.

Potamogale velox: *Potamogale* has the largest brain of the three species of Potamogalinae (1.08x AvPo in BrW; 1.07x in NET). All five fundamental brain parts are also largest, varying from 1.02x AvPo in TEL to 1.21x in CER. In TEL, the main parts differ strongly, varying from 0.74x AvPo in MOB to 1.17x in NEO and to an extremely large AOB (2.59x AvPo). The lowest in *Potamogale* compared with the other Potamogalinae are, after MOB, PAL (0.75x AvPo) and STR (0.89x); the highest, after NEO, is HIP (1.09x). OBL structures and some small periventricular structures were measured in *Potamogle velox* and *Micropotamogale ruwenzorii*. Compared with *M. ruwenzorii, Potamogale* has a very small epiphysis (1/12) and a small subfornical body (1/2). Relatively large OBL structures are mot VII, FCE, INO, and REL; FCM and COE are relatively small. *Potamogale* has a longer brain than *M. ruwenzorii,* with longer but slender hemispheres and a very long corpus callosum.

Particularities: *Potamogale velox* has the largest brain of otter shrews, with all five fundamental brain parts largest. NEO, motor components (CER, INO) and OBL are large; the accessory olfactory bulb (AOB) is very large; the main olfactory structures (MOB, PAL) are small and the pineal gland is rudimentary.

Micropotamogale lamottei: *M. lamottei* has the smallest brain of the three species of otter shrews (0.92x AvPo in BrW and NET), with four of the five fundamental brain parts also smallest. Only MES is still slightly smaller in *M. ruwenzorii*. The five parts vary from 0.75x AvPo in CER to 0.98x in TEL. In TEL, the main parts vary from 0.84x in NEO to 1.29x in MOB. Many TEL parts are above the AvPo level. These are, next to MOB, PAL (1.17x), AMY (1.10x), SEP (1.10x), and SCH (1.09x). All these structures are larger in *Micropotamogale lamottei* than in the other two species of otter shrews. Below the AvPo level are, after NEO, STR (0.89x AvPo) and HIP (0.95x). These structures are smallest in Potamogalinae. AOB is extremely small (0.33x AvPo).

Particularities: *Micropotamogale lamottei* has the smallest brain of the otter shrews, with a small accessory olfactory bulb (AOB), motor components (CER, STR), and neocortex (NEO), and relatively large main olfactory structures (MOB, PAL) and limbic structures (SCH, AMY, SEP).

Micropotamogale ruwenzorii: Of the three species of otter shrews, *M. ruwenzorii* brain size is closest to the potamogaline average (0.99x AvPo in BrW; 1.00x in NET). The five fundamental brain parts are intermediate between *Potamogale* and *Micropotamogale lamottei* except for MES, which is lowest in *M. ruwenzorii*. They vary from 0.92x AvPo in MES to 1.03x in CER. In TEL, AOB is very small (0.05x AvPo). The other parts vary from 0.95x in SCH and HIP to 1.23x in STR. AOB, SCH, and HIP are lowest under potamogaline level; STR is the largest. OBL parts and some periventricular structures were measured in *M. ruwenzorii* and *Potamogale velox* only. Thus, the comparative data are the opposite of those given above for *Potamogale*. *M. ruwenzorii* has a shorter brain than *Potamogale*, with shorter and wider hemispheres and a short corpus callosum.

Particularities: *Micropotamogale ruwenzorii* has an average brain size for a Potamogalinae with most brain parts intermediate between *Potamogale* and *M. lamottei*. AOB is extremely small. Some of the motor components (STR, CER) are relatively large.

A comparison of the three species of Potamogalinae with one another reveals similar brain characteristics, in particular, large neocortex and sensory trigeminal complex and small main olfactory structures. The reduction in the main olfactory structures, however, is stronger in *Potamogale velox* than in *M. lamottei*, where MOB is 1.75 times larger. In CER and NEO, the relation is reversed. Very pronounced differences exist in the accessory olfactory bulb, which is large in *Potamogale* but very small in both species of *Micropotamogale*: it is 48 times larger in *Potamogale* than in *M. ruwenzorii*.

4.2 Chrysochloridae

General Characteristics

The African golden moles inhabit sandy deserts, savannas, forests and cultivated regions. They are highly adapted to searching for food underground (fossorial life) and somewhat resemble the true moles. The bodies are elongated and cylindrical; the necks and tails are very short. The pelage has a metallic luster. The eyes are vestigial and covered with hairy skin; the ear pinnae are concealed in the pelage; the pointed snout has a leathery pad at its tip, and the nostrils are located under a fold of skin. The skull is conical rather than flattened and elongate, as in many Insectivora, especially in true moles. The short, powerful forelimbs have four fingers. The third finger is the longest and has a powerful claw; the other three vary in length with the species. Unlike the lateral stroke burrowing of talpids, golden moles burrow by means of the leathery snout and powerful thrusts of the forepaws. Deep and shallow burrows are constructed, with the depth of the burrows perhaps dependent on soil moisture. The diet of golden moles consists mostly of invertebrates.

Selected References: Grassé (1955a), Walker (1975), Herter (1979), Anderson and Jones (1984), Vaughan (1985), Fons (1988).

Brain Characteristics

The golden moles are represented in our serially sectioned brain material by three species *(Chrysochloris asiatica, C. stuhlmanni* and *Chlorotalpa leucorhina)* with one specimen only in two of them *(Chrysochloris asiatica* and *Chlorotalpa leucorhina*). All the species are similar in macromorphology, and since they correspond well with data from other species and genera reported in the literature, we believe that they are representative of the whole family. We have no body and brain weights from *Chlorotalpa leucorhina* and thus no indices can be given. However, in percentage brain composition, this species is consistent with the other two.

Chrysochloris asiatica, C. stuhlmanni

Brain Weight, Encephalization

Bauchot and Stephan (1966)
Stephan (1959)
Stephan and Bauchot (1960)

Data on nine pairs of brain and body weights of *Chrysochloris stuhlmanni* were given by Stephan (1959). In addition five more BoWs are available, four of males and one of a female.

Since the data indicate clear sex differences in that the males are distinctly heavier and have larger brains than the females, the median values between males and females were used as the standards for the species (Tables 5, 6). These standards are well founded. In contrast, there is only one pair of data from *Chrysochloris asiatica,* which was first given by Stephan and Bauchot (1960). All the data were summarized by Bauchot and Stephan (1966). No BoW/BrW data are available from *Chlorotalpa leucorhina*. Köppen (1915) gave 840 mg for the brain of an undetermined species. This is slightly above the upper limit of the range in *Chrysochloris stuhlmanni* (820 mg) and may have come from a somewhat larger species.

In encephalization, the golden moles are 1.7 times above the tenrecine level (Table 7), with *Chrysochloris stuhlmanni* somewhat higher than *C. asiatica* (1.8 and 1.5x AvTn) (Table 6).

Macromorphology

Burkitt (1938)
Clark (1932)
Dräseke (1930)
Hochstetter (1942)
Köppen (1915)
Leche (1905, 1907)
Spatz and Stephan (1961)
Starck (1961)
Stephan and Andy (1982)
Stephan and Bauchot (1960)
Stephan and Spatz (1962)

Leche (1905, 1907) was the first to describe the brain of a golden mole *(Amblysomus hottentotus)* and stressed its differences from those of all other Insectivora and eutherian mammals. He gave good illustrations of the brain in the dorsal, right, and median planes, and of the midsagittally sectioned skull. Dräseke (1930, undetermined species) mentioned large differences from the brain of *Talpa*. Later authors (Clark 1932, *Amblysomus hottentotus;* Hochstetter 1942, *C. aurea;* Stephan and Bauchot 1960; Spatz and Stephan 1961; Stephan and Spatz 1962; Stephan and Andy 1982, *Chrysochloris asiatica* and *C. stuhlmanni*) confirmed and added new information about the characteristics of the golden mole's brain, which may be summarized as follows: (1) compression in the anteroposterior direction, (2) extremely wide brain, with much greater width than length, (3) high brain, with the dorsal surface arched upwards and pronounced convexity, (4) corpus callosum not parallel to base of rhombencephalon, but nearly perpendicular to it, (5) no sulcus rhinalis lateralis except at the extreme rostral end, (6) no trace of visual structures, e.g., optic nerves and chiasma and, (7) very simple cerebellum, leaving large parts of the mesencephalic tectum uncovered.

Leche (1905, 1907) attributed the particularities of the brain to those of the skull, which obviously depend on the use of the head in digging. He pointed out similarities with *Notoryctes typhlops,* the Australian marsupial mole, "a remarkable example of the phenomenon of functional convergence" (Burkitt 1938). Spatz and Stephan (1961) also stressed the similarities in skulls and brains of species digging with the head ("Kopfwühler") and attributed them to adaptive convergence, a suggestion questioned by Starck (1961), who proposed other possible explanations.

Cell and Area Measurements

Bauchot (1963)
Bauchot and Stephan (1961)
Stephan (1960a, 1961)
Stephan and Spatz (1962)

Bauchot (1963) studied cell size, number and density of the diencephalic components in *Chrysochloris stuhlmanni* and *C. asiatica.*

Areas of sectioned commissures and crossing visual nerves were determined by Bauchot and Stephan (1961) in the same two species. The chiasma is very small and may contain only supraoptic commissures, since no nerves could be found (also Köppen 1915; Clark 1932). Köppen mentioned that the lack of the optic chiasma makes the supraoptic commissures of Gudden and Meynert well recognizable. The corpus callosum of golden moles is of medium size for Insectivora (family/subfamily average, Table 73) and was found by Bauchot and Stephan to be larger in *Chrysochloris stuhlmanni* than in *C. asiatica.* The anterior commissure is very large in *C. stuhlmanni* (Table 73), but distinctly smaller in *C. asiatica* according to the data of Bauchot and Stephan.

Data on area dimensions of cortical surfaces in *Chrysochloris stuhlmanni* have been given by Stephan (1960a, 1961) and Stephan and Spatz (1962). Compared with ten other species of Insectivora, relatively large olfactory structures were found (later confirmed by volume measurements; see Table 45). Slightly larger than the insectivoran average were the neocortex and non-olfactory allocortex, e.g., hippocampus, whereas the schizocortex was smaller.

Volume Measurements

Andy and Stephan (1966a)
Baron et al. (1983, 1987, 1988)
Bauchot (1961, 1963, 1966, 1970, 1979a, b)
Bauchot and Legait (1978)
Frahm et al. (1982)
Legait et al. (1976)
Stephan (1966, 1967b, 1969)
Stephan and Andy (1962, 1964, 1969, 1982)
Stephan et al. (1970, 1981a, 1982, 1987a)

Many parts of the brains of golden moles have been measured since 1961 by numerous authors and teams (Andy, Baron, Bauchot, Frahm, Legait, Stephan). Sometimes the measurements were restricted to the well represented *Chrysochloris stuhlmanni*; usually *C. asiatica* was included; *Chlorotalpa leucorhina* was included for the pineal gland (Legait et al. 1976) and some small periventricular organs (Stephan 1969) only.

Data on the five **fundamental brain parts** and the main parts of the telencephalon have been given for *Chrysochloris stuhlmanni* by Stephan and Andy (1964, 1969), Stephan (1966, 1967b), Stephan el al. (1970) and Bauchot (1979a); and for *C. stuhlmanni* and *C. asiatica* by Stephan et al. (1981a) and Stephan and Andy (1982). The largest average indices are found for DIE and TEL, which are 2.0 and 1.9 times larger than the tenrecine level; MES and CER are 1.6x AvTn. The lowest average size is found in OBL, which is slightly above the tenrecine level (1.05x; see Table 12). The fundamental brain parts are all larger in *C. stuhlmanni* than in *C. asiatica*. The largest difference was found for OBL, which is 1.3 times larger in *C. stuhlmanni* (120 versus 91; see Table 11). Compared with other taxonomic groups of Insectivora, the golden moles are in medium positions in the index scales (Figs. 62 - 64, 69), except for OBL, which is the third lowest being only above Tenrecinae and Erinaceinae (Fig. 55).

Components of the **medulla oblongata** have been measured only in *Chrysochloris stuhlmanni*. Baron et al. (1988) found that the vestibular nuclei of *C.stuhlmanni* are slightly larger than those of Tenrecinae, but smaller than in the average Insectivora (see also Table 24). The inferior olive is much larger than all other OBL components (2.3x AvTn) (Table 17, Fig. 58); all other components had indices lower than the NET index for *C. stuhlmanni*, which is 188 or 1.88x AvTn. Indices even lower than the tenrecine level were found for auditory nuclei (VCO 0.71x AvTn, DCO 0.90x, AUD 0.87x), for VI (0.97x), VS (0.98x), and for mot VII (0.98x). With regard to the auditory structures in *Amblysomus hottentotus*, Clark (1932) stated that the auditory nerve is of moderate size and that there is nothing in the auditory centers to parallel the enormous development of the tympanic cavity.

In the **mesencephalon**, the subcommissural body has been scaled by Stephan (1969) and found to be small or even missing in *Chrysochloris stuhlmanni* and *Chlorotalpa leucorhina* but it is larger in *Chrysochloris asiatica* than in other golden moles (see also Table 41).

In the **diencephalon**, volumes of the main parts have been given by Bauchot (1963, 1979a); of thalamus components by Bauchot (1963, 1966, 1979b); of medial habenular nuclei and epiphysis by Stephan (1969) and Stephan et al. (1981a); of epiphysis by Legait et al. (1976); and of hypophysis by Bauchot (1961, 1970) and Bauchot and Legait (1978).

Based on Bauchot's (1979b) data, *Chrysochloris stuhlmanni* has a large subthalamus (3.36x AvTn) and, in the thalamus (THA), large dorsal (4.21x), and median nuclear groups (3.18x) and corpus geniculatum mediale (CGM) (3.43x) (Tables 67, 70). Medial (2.42x AvTn) and anterior nuclear groups (2.12x) are also relatively large. Low indices in THA were found for the corpus geniculatum laterale (CGL) (0.61x AvTn). Large CGM and small CGL were mentioned by Koeppen (1915), who described CGL as rudimentary, while Clark (1932) said it was not definite.

Bauchot and Legait (1978) described a hypophysis in *C. stuhlmanni* two times larger than in *C. asiatica,* with a very small anterior lobe and a very large posterior lobe. The pineal gland was found to be three times larger in *Chrysochloris stuhlmanni* and *Chlorotalpa leucorhina* than in *Chrysochloris asiatica* (Legait et al. 1976; see also Table 41). Clark (1932) mentioned very large habenular nuclei. Based on our measurements, the medial nucleus (HAM) is two times larger than in Tenrecinae (Table 41), which is really a high level.

In the **telencephalon**, seven to nine main parts have been measured and compared (references: see above under five fundamental brain parts). Additional data have been given on olfactory structures by Stephan (1966), Stephan et al. (1982), and Baron et al. (1983, 1987); on septum by Stephan and Andy (1962, 1982), Andy and Stephan (1966a), and Stephan (1969); on amygdala by Stephan et al. (1987a); and on neocortex (NEO) by Frahm et al. (1982). Large size in golden moles was found for NEO (3.44x AvTn), striatum (STR) (2.81x), and the septal triangular nucleus (SET) (2.35x). NEO is distinctly larger in *Chrysochloris stuhlmanni* than in *C. asiatica*, STR slightly larger, and SET slightly smaller. The olfactory bulbs (both MOB and AOB) are larger in *Chrysochloris stuhlmanni* than in *C. asiatica* and are on the average slightly above the tenrecine level. Since Tenrecinae have a large MOB, that of golden moles also must be considered large. In the index scale (Fig. 71), it has the third highest position. Small size (0.64x AvTn) was found in both species for the subfornical body (SFB), which is considered part of SEP.

More quantitative comparisons will be presented on p. 191 in the summary of brain characteristics of Chrysochloridae and its species. However, the question remains whether the data on *Chrysochloris stuhlmanni* and *C. asiatica* are valid for the size of brains and brain components in the whole family of golden moles.

Histology

Clark (1932)
Köppen (1915)
Rehkämper (1981)
Stephan (1965)
Stephan and Andy (1982)
Stephan and Spatz (1962)

Details on the histology of the golden mole brain are given in many of the papers on volume measurements. Further information appears in the papers of Köppen (1915), Clark (1932), Stephan and Spatz (1962), Stephan (1965), Rehkämper (1981), and Stephan and Andy (1982).

Köppen (1915) presented drawings of several transverse sections of a brain of an undetermined species and described fiber bundles and cell groups. In the cerebral cortex, however, his "stripe of Baillarger," which he unexpectedly found in a blind animal in the caudal part of the median surface, was, in our serial sections, in a position taken by the hippocampus or transitional cortices (presubicular, retrosplenial) and clearly belongs to them. Koeppen's figures and descriptions are not clear enough to allow for a final determination.

Clark (1932) described *Amblysomus hottentotus* as primitive in many respects. The neocortex (NEO) is poorly differentiated and exhibits a pronounced development of the external granular layer (layer II). The transition of NEO to piriform lobe cortex is gradual. In the diencephalon, the habenular nuclei are very large. The thalamus is simple in structure, and the ventral nucleus is its main component. Definite lateral geniculate bodies and optic tracts were not found. Medial geniculate bodies and auditory nerves are of moderate size. Cells of the mesencephalic nucleus of the trigeminal nerve are especially conspicuous and numerous. Mediodorsal groups of such cells close to the surface of the tectum and immediately dorsal to the aqueduct were not found in other Insectivora. Stephan and Bauchot (1960) and Stephan and Spatz (1962) gave photoprints of transverse sections, and Stephan and Bauchot gave a map of the allocortical regions of *Chrysochloris stuhlmanni*. Stephan (1965) described a broad mitral cell layer and a small number of inner granular cells in the accessory olfactory bulb (AOB). The AOB position and its relation to the main olfactory bulb (MOB) were quite typical of Insectivora. Stephan and Andy (1982) showed characteristics of the various parts of the septum (SEP) in photoprints of transversal and sagittal sections. Rehkämper (1981) investigated and mapped the NEO of *C. stuhlmanni* and pointed out that the cytology of layer III is not uniform, especially in his area A3. Besides the medium-sized pyramidal cells typical of this layer, there are smaller, rounded granular cells which are concentrated in places in the lower half of layer III and give the impression of a sublaminar differentiation.

Sense Organs

Broom (1915b, vomeronasal organ)
Franz (1934, eye)
Sweet (1909, eye)

Broom (1915b) studied the vomeronasal organs (VNO) of a young *Chrysochloris asiatica* and a newborn *Amblysomus hottentotus*. The organs differed entirely in their structure from those of *Tenrec, Echinosorex*, and *Talpa*. The golden moles have mucous glands that open into the VNO all along its upper border, whereas the other forms have a large gland or numerous glands opening into it posteriorly. Broom tended to separate the golden moles from the Insectivora and to place them in a distinct order (Chrysochloridea). Sweet (1909) and Franz (1934) gave some details of the strongly reduced eyes of golden moles. Eye muscles are quite absent (Sweet).

Summary of Brain Characteristics in Chrysochloridae

The golden mole brain has strange macromorphological features (see Section 3.1), while its size is close to that of the Insectivora on the average (1.07x AvIF). Most structures also are close to the family/subfamily average (AvIF). In the deviations from AvIF, the golden moles represent in brain composition an average type of Insectivora. Index profiles on the brain composition are given in Figs. 101, 104, 105, and 107. Fig. 101 shows that in the five fundamental brain parts, too, the Chrysochloridae are close to the typical Insectivora. The five fundamental brain parts vary from 0.76x AvIF for OBL to 1.20x for TEL. OBL is relatively small and is the third lowest being above Tenrecinae and Erinaceinae (Fig. 55). In *Chrysochloris asiatica* it is the second lowest of all Insectivora species, next to *Tenrec*. OBL components vary from 0.67x AvIF for the lateral reticular nucleus (REL) to 1.13x for the inferior olive (INO) (Fig. 104); DIE structures, from 0.64x AvIF for the lateral geniculate bodies to 1.45x for the median group of thalamic nuclei (Fig. 105); and TEL components, from 0.76x AvIF for the accessory olfactory bulb (AOB) and 0.91x for the schizocortex to 1.42x for the paleocortex (PAL) (Fig. 107). In OBL, the nuclei of the fascicularis gracilis and mot V are quite large (besides INO) whereas, besides REL, the vestibular and cochlear structures are small. In TEL, the main olfactory structures are well represented (PAL 1.42x AvIF, MOB 1.20x). MOB is the third largest in Insectivora, whereas AOB is small (0.76x). Next to PAL, the striatum is best represented (1.34x). Thus, the size of the motor structures is not uniform in the golden moles. In contrast to CER, which is clearly below AvIF (0.88x), INO and STR are clearly above.

In conclusion, golden mole brain size is medium in Insectivora. The olfactory centers and some of the motor structures are well represented. The vestibular and cochlear centers are less well developed, and the visual structures are small. Visual nerves are missing.

Species characteristics of golden mole brains are based on direct comparisons between the two species so far investigated. It is not worthwhile to introduce an average golden mole level, since the larger size of a brain structure in one species is of course always associated with a smaller size in the other and vice versa. Therefore, the brain characteristics will be described together. Since details from other species reported in the literature also correspond with them, the data given may be representative of the golden mole family.

Chrysochloris stuhlmanni and ***Chrysochloris asiatica:*** The total brain and most brain parts are slightly larger in *Chrysochloris stuhlmanni* than in *C. asiatica*. The greatest differences are in the pineal gland, which is 4.28 times larger, in the accessory olfactory bulb (1.75x), in the subfornical body (1.42x), medulla oblongata (1.32x), neocortex (1.32x), and amygdala (1.30x). The triangular nucleus of the septum (1.11x) and the medial habenular nucleus (1.06x) are slightly larger in *Chrysochloris asiatica* than in *C. stuhlmanni*; distinctly larger is the subcommissural body, which is one of the largest in *Chrysochloris asiatica* but was not found in *C. stuhlmanni*.

4.3 Solenodontidae

General Characteristics

Solenodons (a single genus) are restricted to Cuba and Haiti. They look like large, stout shrews. The nostrils on the elongate bare-tipped snout open to the side. The eyes are relatively small, and the partly naked ear pinnae extend beyond the pelage. The moderately long tail and five-toed feet are nearly hairless. Claws of the forefoot are larger and more curved than those of the hindfoot. Secretions from glands in the armpit and the groin have a goat-like odor. The submaxillary glands produce a toxic saliva.

Solenodons inhabit forests and brushy areas, are mainly nocturnal and live (sometimes together with other individuals) in burrows, caves, hollow trees, or similar shelters. They are not completely plantigrade since, when moving, only their toes come in contact with the ground. Solenodons are generalized omnivorous feeders which hunt various invertebrates and vertebrates in leaf litter and other vegetation debris, rooting and probing the ground with their snout or using their claws. They also eat various fruits and vegetables.

Selected references: Grassé (1955a), Cockrum (1962), Walker (1975), Herter (1979), Anderson and Jones (1984), Vaughan (1985), Fons (1988).

Brain Characteristics

Solenodon paradoxus

Brain Weight, Encephalization

Bauchot and Stephan (1966)
Boller (1969)

Data on three pairs of brain and body weights were given by Bauchot and Stephan (1966) and one by Boller (1969). Additional BoWs (1000 - 1300 g) were given by Mohr (1936b, c, d, 1938). However, they are from animals in long-term captivity which may have been overweight. The encephalization of *Solenodon* is high (1.84 times that of Tenrecinae). This is the third highest in Insectivora, after Desmaninae and Echinosoricinae.

Macromorphology

Allen (1910)
Bauchot and Stephan (1967, 1970)
Boller (1969)
Clark (1932)
Stephan and Andy (1982)

In the *Solenodon* brain, the mesencephalic tectum is uncovered; the trigeminal nerves are large; the visual nerves are small. The great length of the brain (Table 4) is already apparent in the drawing by Allen (1910) and he mentioned that the cerebellum was long as well. Clark (1932), in his detailed description of macromorphology, pointed out the large size of the olfactory bulb, the shallow depression in the frontal neocortex (sulcus orbitalis), and the distinct rostral and caudal pieces of the sulcus rhinalis lateralis. Compared with *Tenrec*, *Solenodon* has a more extensive neocortex, a greater length and other clear differences in the corpus callosum, and a smaller anterior commissure. Boller (1969) found the corpus callosum to be about 28% of the hemisphere length, which is more than in the hedgehog but less than in *Sorex araneus* and *Talpa*.

Bauchot and Stephan (1967) compared endocast and brain of *Solenodon* and (1970) described the large fissuration of the cerebellum. Reconstructions of the midsagittal plane were given by Boller (1969) and Bauchot and Stephan (1970). Stephan and Andy (1982) gave a comparative outline of many brain characteristics and, like Stephan (1966) and Boller (1969), denied the existence of an accessory olfactory bulb (AOB).

Area Measurements

Bauchot and Stephan (1970)
Boller (1969)

Area size of some sectioned commissures were given by Boller (1969) and Bauchot and Stephan (1970) and of cortical surfaces by Boller. Boller found a three times larger corpus callosum than anterior commissure and average percentage size of the large cortical regions compared with eight other species of Insectivora. Bauchot and Stephan (1970) found an average corpus callosum index compared with ten other species of Insectivora.

Volume Measurements

Andy and Stephan (1966a)
Baron et al. (1983, 1987, 1988)
Bauchot (1970, 1979a, b)
Bauchot and Legait (1978)
Frahm et al. (1982)
Legait et al. (1976)
Stephan (1966, 1967b, 1969)
Stephan and Andy (1982)
Stephan et al. (1970, 1981a, 1982, 1987a)

Volumes of many brain parts have been measured since 1966 by several authors and teams (Andy, Baron, Bauchot, Frahm, Legait, Stephan). Data on the five fundamental brain parts and/or the main parts of the telencephalon of *Solenodon* were given by Stephan (1967b), Stephan et al. (1970, 1981a), Bauchot (1979a), and Stephan and Andy (1982).

Additional data on the medulla oblongata (OBL) were given for vestibular components (VC) by Baron et al. (1988), on the mesencephalon (MES) for the subcommissural body (SCB) by Stephan (1969) and Stephan et al. (1981a), and on the diencephalon (DIE) by Bauchot (1979b). Measurements on smaller diencephalic components were performed by Stephan (1969) on medial habenular nucleus (HAM) and epiphysis (EPI), by Legait et al. (1976) on EPI, and by Bauchot (1970) and Bauchot and Legait (1978) on hypophysis (HYP).

From the very first, attention was called to the large size of the cerebellum (CER). CER has the third largest index, after *Galemys* and *Hylomys*: 2.55x AvTn (Table 11). In the medulla oblongata (OBL), whose index is of average size, the nucleus fascicularis gracilis (FGR) is the largest in any Insectivora. The nucleus reticularis lateralis (REL) is large, and the nucleus fascicularis cuneatus externus (FCE) is small (Tables 14, 17).

In the mesencephalon, the subcommissural body (SCB) is the second largest found in Insectivora (next to *Scalopus*): 5.5x AvTn. Based on Bauchot's measurements, the diencephalic epithalamus (ETH) is relatively large and the third largest after *Galemys* and *Chrysochloris* (Table 66). Large size is also typical of the medial habenular nucleus (HAM) and the pineal gland (EPI), which is the largest by far among all Insectivora. In the thalamus (THA), the highest indices are found for the median and dorsal nuclei (MNT and DOT): 2.9x AvTn. For DOT, however, this is only a medium position in Insectivora since this nuclear group is very progressive in several families or subfamilies (Fig. 65). In contrast, the ventral nuclei (VET), which are only 1.5x, have the distinctly largest index among Insectivora, and the lateral geniculate bodies, with the relatively low index of 111 (1.1x AvTn), have one of the highest positions in Insectivora (Fig. 68). The hypophysis (HYP) has a relatively large posterior lobe (Bauchot 1970; Bauchot and Legait 1978).

Various parts of the telencephalon (TEL) have been measured and compared. Data on the olfactory structures are given by Stephan (1966), Stephan et al. (1982), and Baron et al. (1983, 1987); on the septum (SEP) and septal components by Andy and Stephan (1966a) and Stephan (1969); on the amygdala (AMY) by Stephan et al. (1987a); and on the neocortex (NEO) by Frahm et al. (1982). High indices were found for the striatum (STR, 2.7x AvTn), and NEO (2.6x). While the NEO index reaches only a medium position among Insectivora, that of STR is one of the highest.

Relatively high indices (1.6 - 2.1x AvTn) with high positions for Insectivora are reached by several of *Solenodon*'s limbic structures (SCH, SEP, AMY), whereas the hippocampus (HIP) (1.9x AvTn) reaches only a medium position. The highest for Insectivora are the size of the main olfactory bulb (MOB) (1.5 times tenrecine level) and of the olfactory tubercle. In contrast, the accessory olfactory bulb (AOB) is rudimentary or may even be lacking in some individuals. This, however, is not unique in Insectivora, as has been pointed out by Stephan (1967b) and Stephan and Andy (1982), and is also true of *Micropotamogale* spp.

In conclusion, the brain of *Solenodon* is nearly twice the size of that in the average Tenrecinae (AvTn). This is highest after Desmaninae and Echinosoricinae. The indices for the main structures (Tables 12, 45) are all distinctly above the tenrecine level with STR, NEO, and CER the highest i.e., all greater than 2.5x AvTn, and OBL the lowest (1.5x). The main olfactory bulb (MOB) is actually largest, and *Solenodon* may be the most macrosmatic Insectivora. Very high positions were also found for TOL. In clear contrast to the large MOB, there are only slight traces of AOB, a feature *Solenodon* shares with *Micropotamogale*.

Histology

Rehkämper (1981)

Rehkämper (1981) investigated the neocortex (NEO) of *Solenodon* and gave a cortical map containing the four usual areas in Insectivora. Particularities of NEO in *Solenodon* are (1) cell clusters in the dense lamina II in the parietal area 2, (2) sinuosities of lamina II in the caudolateral area 4, and (3) no granular layer. He pointed out similarities with the NEO of *Crocidura* both in laminar and areal differentiation.

Summary of Brain Characteristics in *Solenodon*

Index profiles based on the deviations from the averages of all Insectivora families and subfamilies (AvIF) are given in Figs. 101, 103, 105 and 107. The total brain of *Solenodon* is markedly above AvIF (1.15x). Except DIE the fundamental brain parts are above AvIF; they vary from 0.98x AvIF for DIE to 1.42x for CER. In OBL, the components vary from 0.77x for FCE to 1.45x for FGR, in DIE, from 0.47x for CGM to 1.33x for VET, and in TEL, from 0.87x for NEO to 1.67x for MOB. In contrast to MOB, AOB is extremely small (0.17x). Apart from the main olfactory structures (MOB, PAL), AMY and STR are the largest (1.31 and 1.28x AvIF).

In conclusion, solenodons have well developed brains in which olfactory and motor systems are very well represented, and FCE and CGM are relatively small. The accessory olfactory bulbs are rudimentary.

4.4 Erinaceidae

General Characteristics

Erinaceidae consist of two subfamilies, the Echinosoricinae (moonrats, gymnures) and Erinaceinae (hedgehogs). They vary in size from that of a large mouse to that of a small rabbit. The eyes and pinnae are moderately large and the snout is usually long.

Selected references: Grassé (1955a), Walker (1975), Herter (1979), Anderson and Jones (1984), Vaughan (1985), Fons (1988).

Brain Characteristics

Little is known about the brains of Echinosoricinae, but considerable detailed information is available on brains of Erinaceinae. We will try to give a full review and to complement the knowledge about these animals with detailed quantitative data and a stereotaxic atlas of the brain of the relatively common Algerian hedgehog *Atelerix algirus* (see Atlas, pp. 503 - 553).

4.4.1 Echinosoricinae

General Characteristics

Moonrats and gymnures live in humid forests and mangrove swamps and shelter in hollows in dead logs, under roots of trees and logs or in empty holes. Some are diurnal and some are nocturnal. All members of the Echinosoricinae lack spines but are able to produce obnoxious scents when threatened. The tail may be long or very short. The diet consists mainly of terrestrial invertebrates with aquatic vertebrates and invertebrates as supplementary foods.

Echinosoricinae are generally classified in five monospecific genera: *Echinosorex, Hylomys, Neohylomys, Podogymnura* and *Neotetracus*. Only the first two, the best known, are represented in our brain material.

Brain Characteristics

Little is known about the brains. Most of the information given here is based on original investigations published here for the first time.

Echinosorex gymnurus, Hylomys suillus

Brain Weight, Cranial Capacity, Encephalization

No published data.

Two adult males of *Echinosorex* were caught in May 1987 in Thailand. BoWs were 895 and 750 g; BrWs were 6460 and 5956 mg. The additional brain weight of a subadult male was 5766 mg, and the cranial capacity of the skull of a male was 5941 mm^3, indicating a BrW of 6154 mg. When these additional data on brain size are included, the average brain weight is 6084 mg (Table 5).

From *Hylomys*, only one juvenile specimen was obtained. Its BoW was 31.4 g; its BrW 957 mg. Seven skulls belonging to the Ecological Research Division of the Thailand Institute of Scientific and Technological Research yielded cranial capacities (CrCs in mm^3) of 1115, 1078, 1271, 1120, 1154, 1042, and 1088. The average CrC is 1124 mm^3, indicating an average BrW of about 1200 mg (Table 5).

The brain of the juvenile specimen was measured, and the figures were then converted to the average brain size obtained from adults. BoW data for adults were derived from Miller (1942), Rudd (1965), Harrison (1966), Lim and Heyneman (1968), and Medway (1969). They converge at about 57 grams. This value was used to determine the indices.

The encephalization of Echinosoricinae is 2.1 to 2.4 times greater than that of Tenrecinae and is the second highest of Insectivora (Fig. 53; Table 6). It is surpassed only by that of *Galemys*. Its high level is obvious when *Echinosorex* is compared with the isoponderous *Erinaceus europaeus*. BrWs are 6084 and 3367; i.e., the brain of the moonrat is 1.8 times larger than that of the hedgehog.

Macromorphology

Clark (1932)
Leche (1905, 1907)
Smith (1903)

Smith (1903) described an orbital sulcus in the cerebral sulcus of *Echinosorex*. While Leche (1905, 1907) made mention of having a brain of *Hylomys* for comparison, he gave no details. Clark (1932) presented a detailed description of the brain of *Echinosorex*, pointing out a very conspicuous V-shaped "sulcus orbitalis" 5 mm behind the rostral margin of the hemisphere and a distinct sulcus rhinalis lateralis (ectorhinal fissure) fading close to the caudal margin of the hemisphere.

Clark said that (1) the corpus callosum, measuring 7.5 mm in length and 1.2 mm in thickness, "is rather bigger than might be anticipated in such a primitive type of brain;" (2) the hippocampus is reminiscent "in a moderate degree [of] the hypertrophy of this region in the Macroscelididae," which is not found in other insectivore brains studied; (3) the cerebellum "is much more complex than in most insectivore brains." All the peculiarities given by Clark are confirmed by the measurements discussed in the following section.

Volume Measurements

Most of the data are published here for the first time. Consistent with the large size of the brain, many brain parts have very high indices and high positions in the index scales. The highest index by far was found for AOB of *Hylomys* (810 i.e., 8.1x the tenrecine level, AvTn). In *Echinosorex*, it is less than half as large (349), but it is still the second highest (Fig. 72; Table 50). The highest indices for larger structures (333 - 419) were found for NEO and HIP in both species and SCH in *Hylomys*. Indices between 220 and 280 were obtained for SCH in *Echinosorex*, REL, INO, CER, MES, DIE, DCO, VC, HAM, and TEL. Most other indices were also around 200 (2x AvTn), and only OBL (176) and many of its components, as well as PAL (159) and MOB (126), had smaller indices. FCM was at the tenrecine level. EPI is rudimentary in *Echinosorex*, but relatively large in *Hylomys*.

The relatively low indices for the main olfactory structures (MOB, PAL) do not mean that these structures are small in Echinosoricinae. They are, in fact, the second largest (after *Solenodon*; Fig. 71). Since, however, the Tenrecinae, as the reference group, have a high position in the index scale too, all indices around or above 100 indicate strongly developed olfactory structures.

The highest indices by far in Echinosoricinae without any overlap with other Insectivora were found for MES (Fig. 62), AOB (Fig. 72), HIP (Fig. 87), VC, and AUD with all its components. The highest indices in Echinosoricinae that overlapped with other groups were found for PAL (Fig. 78), SEP (Fig. 85), AMY (Fig. 86), SCH (Fig. 88), INO (Fig. 58), REL, FCE, PRP, MOT, and HAM. Of the larger components, STR was in a relatively low position (Fig. 79).

More quantitative comparisons will be presented on p. 225 in the summary of brain characteristics of Echinosoricinae and its species compared with those of other Erinaceidae.

Histology

Clark (1932)

With regard to histological features, Clark (1932) pointed out that (1) the olfactory regions "are enormously developed;" (2) "the hippocampal formation is convoluted to a degree which is not to be seen in other insectivore brains" examined microscopically; and (3) the thalamus resembles that of *Erinaceus*. In sagittal sections, Clark investigated the position of the "sulcus orbitalis" and found that it "forms a definite sulcus limitans, separating two areas which are quite distinct in cyto-architecture." He described their characteristics and concluded that they "are structurally the equivalent of Brodmann's areas 1 and 4, the general sensory and motor areas." Then "it follows that the "sulcus orbitalis" may be regarded as a homologue (at least in part) of the sulcus centralis of the primate brain. ... Thus it appears that the sulcus centralis is phylogenetically much more ancient than has hitherto been supposed" (Clark 1932, p. 990).

Sense Organs

Ärnbäck-Christie-Linde (1914) | Broom (1915a)

Ärnbäck-Christie-Linde (1914) studied the vomeronasal organ of *Echinosorex* and *Hylomys,* and Broom (1915a) of a young *Echinosorex*. Broom found it to be almost identical to that of *Erinaceus* and to agree equally well with the type found in most Eutherians.

4.4.2 Erinaceinae

General Characteristics

In various parts of their wide range, hedgehogs occupy deciduous woodlands, cultivated lands, and tropical and desert areas. Hedgehogs occupy a variety of shelters, including nests of dry grass or leaves in shallow depressions on the surface, modified burrows of other animals or burrows of their own construction. Many species are known to hibernate, or in the case of desert-adapted species, to aestivate during periods of bad weather or food shortage. Hedgehogs are nocturnal or crepuscular. An obvious specialization is the possession of barbless spines. The sheet of muscle beneath the skin, the panniculus carnosus, is greatly enlarged and is responsible for pulling the skin around the body and the erection of the spines. Most species are omnivorous, but animal matter is usually preferred.

Erinaceinae were classified by Corbet (1988) into four genera *(Hemiechinus, Erinaceus, Atelerix, Paraechinus)* with 14 species. Their classification and thus species names have frequently changed over the course of time.

With regard to the genus *Erinaceus*, *E. europaeus* overlaps with *E. concolor* in Poland, Czechoslovakia, and Austria. *E. concolor* is distributed in eastern Europe and western Asia and includes *E. roumanicus* and *E. europaeus roumanicus* of earlier authors. Thus, it remains uncertain, especially in several papers by eastern European authors, whether *E. europaeus* or *E. concolor* was investigated. The species may not always have been precisely determined for brain studies.

Apart from *Paraechinus*, brain material from all genera was available. The largest number by far of investigations and published papers on Insectivora brains is for hedgehogs, especially *Erinaceus europaeus*.

Brain Characteristics

***Hemiechinus auritus, Erinaceus europaeus, E. concolor, Atelerix algirus, Paraechinus* spp.**

Brain Weight, Encephalization

Bauchot and Stephan (1966)
Bonin (1937)
Cantuel (1943)
Dubois (1897)
Flatau and Jacobsohn (1899)
Haug (1958)
Hrdlicka (1905)
Krompecher and Lipak (1966)
Portmann (1962)
Stephan (1959)
Warncke (1908)
Weber, M. (1896)
Ziehen (1899a)

Pairs of BoW/BrWs for *Erinaceus europaeus* came from several sources (Weber 1896; Dubois 1897; Flatau and Jacobsohn 1899; Warncke 1908; Bonin 1937; Haug 1958; Portmann 1962; Hofer pers. comm.) Every author gave one pair, except Weber, Flatau and Jacobsohn (two pairs), and Warncke (five pairs). Seven pairs were added from our data (Stephan 1959; Bauchot and Stephan 1966). Some additional BoWs were taken from Warncke (1), Cantuel (1943) (8), and our own collection (7), and some additional BrWs from Ziehen (1899a) (2) and Hrdlicka (1905) (1).

The seven pairs of data on each *Atelerix algirus* and *Hemiechinus auritus* are from our own collections. The details have not been published and will be given here. For seven males of *Hemiechinus*, the BoW/BrW data (g/mg) are 233/1650, 222/1920, 216/1990, 191/1980, 218/1910, 193/1670, and 370/2040.

For three *Atelerix algirus* males, the data are 810/3380, 785/3210, and 870/3570, and for 4 females, 574/3230, 730/3180, 770/3160, and 615/3120. The BoW/BrW data on the three *Atelerix algirus* specimens from the Canary Islands which were used for the stereotaxic brain atlas in the appendix of this book were not included in the encephalization data. These animals were held several months in captivity, and two of them were very fat. Thus, their BoWs (on the average 972 grams at the time of preparation) may be much higher than the subspecies average. Their average BrW is much lower than in animals of African origin (2771 versus the 3264 milligrams given in Tables 5 and 6). This indicates a smaller insular subspecies.

Encephalization in hedgehogs is 1.1 - 1.5 times greater than in Tenrecinae, and in *Hemiechinus*, slightly higher than in *Atelerix* and *Erinaceus* (Table 6). Next to Geogalinae, Tenrecinae, and some shrews, Erinaceinae have the lowest encephalization found in Insectivora. It is much lower than that of Echinosoricinae (126 versus 222; Table 7).

Ziehen (1899a) gave the weight of the spinal cord (SPC) of *Erinaceus europaeus* as 900 mg. Brain weight (BrW) is 3.5 times larger than SPC weight, a value adopted by Krompecher and Lipak (1966). For comparison, in man BrW is between 30 and 50 times larger.

Macromorphology

Beccari (1910, olfactory tubercle)
Bischoff (1900)
Bolk (1906, cerebellum)
Bradley (1903, cerebellum)
Brauer (1968, cerebellum)
Brauer and Schober (1970)
Clark (1932)
Dräseke (1901)
Flatau and Jacobsohn (1899)
Flower (1865)
Friant (1955)
Grönberg (1901, ontogenesis)
Hochstetter (1942, meninges)
Igarashi and Kamiya (1972, atlas)
Kapoun et al. (1973, spinal cord, *E. concolor*)
Leche (1905, 1907)
Malinska et al. (1972, 1974a, spinal cord, *E. concolor*)
Malinsky and Vodvarka (1983, ventricle)
Mann (1896)
Retzius (1898)
Schmidt and Lierse (1968, meninges, vessels, *Erinaceus, Atelerix*)
Shtark (1970, 1972, atlas)
Smith (1897, 1902a, b)
Smith (1903, cerebellum)
Stephan (1956a)
Stephan (1963, hippocampus)
Stephan and Andy (1982)
Turner (1891)
Völsch (1906, 1911, amygdala)
Ziehen (1897a)
Zuckerkandl (1887, hippocampus)

Macromorphological details and/or illustrations of the brain of the European hedgehog *(Erinaceus europaeus)* have been given many times since 1865 (Flower, Turner, Mann, Smith, Ziehen, Retzius, Flatau and Jacobsohn, Bischoff, Dräseke, Leche, Völsch, Clark, Friant, Stephan, Brauer and Schober, Igarashi and Kamiya, Stephan and Andy). The topics of more specialized studies are given in the reference list. A roughly drawn atlas with 13 figures was given by Shtark (1970, 1972), and a small atlas with six figures, by Igarashi and Kamiya (1972).

The earliest papers present several informative illustrations. Flower (1865), Zuckerkandl (1887), and Smith (1897) illustrated the medial wall of the cerebral hemisphere, showing the position of the cerebral commissures and the hippocampus. Turner (1891) presented the dorsal surface; Mann (1896) gave dorsal and side views of the telencephalon, indicating the orbital sulcus ("praesylvian fissure"), and Retzius (1898) showed the ventral surface of the telencephalon. Flatau and Jacobsohn (1899) presented a sketch of the midsagittal plane, the brain's position in the skull seen laterally, and a dorsal view of the brain; Smith (1902b) presented lateral, median, and ventral views and (1903) the cerebellar arbor vitae, and Leche (1905, 1907), dorsal and median views.

Most authors compared the hedgehog brain with those of other Insectivora and/or mammalian forms and pointed out that (1) the endbrain is characterized by very large olfactory bulbs, olfactory tubercles, and piriform lobes; (2) the sulcus olfactorius lateralis is well developed and runs relatively high; (3) large parts of the mesencephalic tectum are exposed; and (4) the trigeminal nerves are large while the visual nerves are small, but larger than in most other Insectivora. In shape, hedgehog brains have similarities with those of Tenrecinae and *Solenodon.*

Macromorphological details on the spinal cord (SPC) of the hedgehog were given by Flatau and Jacobsohn (1899), Malinska et al. (1972, 1974a), and Kapoun et al. (1973).

Cell and Area Measurements

Bauchot (1963, 1964)
Bauchot and Stephan (1961)
Bishop et al. (1971)
Brodmann (1913)
Filimonoff (1949, 1965)
Gierlich (1916a, b, c)
Galert (1986)
Grünwald (1903)
Harrison and Irving (1966)
Haug (1958, 1987)
Heffner and Masterton (1983, *Erinaceus, Paraechinus*)
Irving and Harrison (1967)
Malinsky and Sedlak (1981)
Rose, S. (1927)
Sanides and Sanides (1974, *Erinaceus, Hemiechinus, Atelerix*)
Stephan (1956a, b, 1961)
Stephan and Manolescu (1980)
Vysinskaja (1961, ex Blinkov and Glezer 1968)
Wiedemeyer (1974)

Most measurements were made on *Erinaceus europaeus.* In the various nuclei of the diencephalon, Bauchot (1963, 1964) studied cell size, number, and density; Galert (1986) size and cell types in supraoptic and paraventricular nuclei; Vysinskaja (1961), cell number in CGL and visual cortex; and Harrison and Irving (1966) and Irving and Harrison (1967), cell number in the superior olivary complex. Axon sheath thickness and fiber size of pyramids and optic nerves were measured by Bishop et al. (1971). Thickness of cortex, laminae, area of radiating dendrites, and cell number and density were studied by Sanides and Sanides (1974), Wiedemeyer (1974), and Malinsky and Sedlak (1981); grey cell coefficient, cell size, and number by Haug (1958, 1987).

Grünwald (1903) compared the size of the pedunculi cerebellares in 24 mammalian species including *Erinaceus europaeus.* Gierlich (1916a, b, c) measured transverse sections of the pedunculus cerebri, its bundles coming from the prosencephalon, and the pyramids shortly before crossing. In the hedgehog, he found 2.5% of the pedunculus came from the prosencephalon (in man 19.6%); 87.8% of it went to the cerebellum, and 12.2% to the spinal cord (similar in man). The corticospinal tract was measured by Heffner and Masterton (1975, 1983) using photoprints given by Verhaart (1970). The area size of the transversally sectioned tract was discussed with reference to the evolution of dexterity. The areas of sectioned commissures and crossing visual nerves were determined by Bauchot and Stephan (1961). The corpus callosum was found to be relatively small but much larger than in *Tenrec*, the anterior commissure of medium size, and the chiasma opticum the largest in any of the measured Insectivora (13 species).

Area dimensions of cortical surfaces were determined by Brodmann (1913), S. Rose (1927), Filimonoff (1949, 1965), Stephan (1956a, b, 1961), Wiedemeyer (1974), and Stephan and Manolescu (1980). Brodmann found the rhinencephalon to be 67.8% of the total cortex, and the visual cortex and precentral region 4.0% each. The corresponding percentages in man were 2.8, 2.9, and 7.3%, showing that the relative size of man's rhinencephalon is strongly reduced. The primary visual cortex also became proportionally smaller, whereas the motor regions showed a percentage increase. These percentage values are highly influenced by the strong increase in the neocortical association centers in man. Rose, Filimonoff, and Stephan measured olfactory bulb, paleocortex, archicortex (hippocampus), periarchicortex and subdivisions, while Filimonoff and Stephan included the neocortex as well. The areas found by Stephan were in general larger than those measured by the other authors, perhaps because of differences in fixation or conversion to fresh brain.

Whereas Rose gave proportions between the various parts of the allocortex and Filimonoff (1965) presented data on the percentage composition of the cerebral cortex, Stephan (1961) compared allometric indices. In eleven species of Insectivora, Stephan found mostly low or medium positions for the hedgehog.

The areas of the hippocampus (HIP) and several of its parts were measured by Stephan and Manolescu (1980). In the "Basal Insectivora" (Tenrecinae, Erinaceinae, terrestrial Soricidae), the hedgehog position was, in general, slightly above the average with larger indices for the fascia dentata and CA1.

Volume Measurements

Andy and Stephan (1966a, b)
Artjuchina (1952 or 1962, ex Blinkov and Glezer)
Baron (1972, 1974, 1977a, b, 1978, 1979)
Baron et al. (1983, 1987, 1988)
Bauchot (1961, 1963, 1966, 1970, 1979a, b)
Bauchot and Legait (1978)
Bauchot and Stephan (1968)
Etienne (1972)
Frahm et al. (1979, 1982)
Ivlieva (1973)
Kapoun et al. (1973, *E. concolor*)
Kesarev (1964)
Legait et al. (1976)
Malinska et al. (1974a)
Masterton and Skeen (1972)
Masterton et al. (1974)
Matano et al. (1985b)
Stephan (1966, 1967b, 1969)
Stephan and Andy (1962, 1964, 1969, 1970, 1977, 1982)
Stephan and Manolescu (1980)
Stephan et al. (1970, 1981a, 1982, 1984b, 1987a)
West and Schwerdtfeger (1985)
West et al. (1984)
Wiedemeyer (1974)

All volume measurements before 1969 were performed in *Erinaceus europaeus*. Data on *Atelerix (Aethechinus, Erinaceus) algirus* were first given by Stephan and Andy (1969); on *Atelerix albiventris (Paraechinus hindei)* by Masterton and Skeen (1972, prefrontal system); and on *Hemiechinus auritus* by Stephan et al. (1981a). Masterton et al. (1974) gave volumes of visual nuclei based on measurements in *Paraechinus (Atelerix?)* and *Hemiechinus*.

Data on the five **fundamental brain parts** were given by Stephan and Andy (1964, 1969, 1982), Stephan (1967b), Bauchot and Stephan (1968), and Stephan et al. (1970, 1981a). The largest difference in the three species was found for the mesencephalon (MES), which is 1.52 times larger in *Hemiechinus* than in *Erinaceus europaeus* (Table 11). The largest indices were found for DIE, which are on the average 1.56 times larger in the Erinaceinae than in the Tenrecinae (AvTn), and the lowest indices were in OBL, which in hedgehogs are at the tenrecine level (Table 12).

In the **spinal cord** (SPC), Kapoun et al. (1973), and Malinska et al. (1974a) measured volumes of grey matter and the anterior, lateral, and posterior funiculi throughout their length. The authors reported a high proportion of grey matter in the hedgehog, a finding earlier mentioned by Verhaart (1970).

In the **medulla oblongata** (OBL), volumes of vestibular and/or auditory nuclei were given by Baron (1972, 1974, 1977), Ivlieva (1973), Stephan et al. (1981a), and Baron et al. (1988). Data on most other structures are given in Tables 13 - 27 for the first time. The largest index has been found for the nucleus fascicularis cuneatus externus (FCE) which on the average in the three species of Erinaceinae is 1.9x AvTn. Other high values are 1.7x for the nucleus reticularis lateralis (REL), and the nucleus nervi hypoglossi (mot XII), and 1.6x for the inferior olive (INO). The smallest indices (0.9 - 1.0x AvTn) are for the sensory trigeminal complex (TR), the superior vestibular nucleus (VS), and the auditory nuclei (AUD). However, it should be noted that all the auditory nuclei are much larger in the long-eared desert hedgehog *Hemiechinus* than in *Erinaceus europaeus*. The difference is greatest in the superior olive (OLS), which is 2.5 times larger in *Hemiechinus* than in *Erinaceus europaeus,* and smallest in the ventral cochlear nucleus (VCO) (1.6x AvTn). *Atelerix algirus* is always in an intermediate position (Table 26). The nucleus motorius nervi trigemini (mot V), the nucleus fascicularis gracilis (FGR), and the nucleus fascicularis cuneatus medialis (FCM) are near the tenrecine level.

In the **mesencephalon** (MES), the volume of the nucleus ruber (RUB) was measured by Artjuchina (1962) and found to be 1.9% relative to MES. The subcommissural body (SCB) was measured by Stephan (1969). As shown in Table 41, *Erinaceus* and *Atelerix* have a relatively large SCB (av. 3.1x AvTn). Very small nuclei lemnisci lateralis (NLL) (0.5x) and inferior colliculi (INC) (0.7x AvTn) were found in the European hedgehog (Baron 1972).

In the **cerebellum** (CER), the cerebellar nuclei were measured in *Erinaceus europaeus* by Baron (1978) and Matano et al. (1985b). They are somewhat above the tenrecine level, with the medial nucleus showing the highest index.

In the **diencephalon** (DIE), nearly all parts are relatively large in the Erinaceinae. Only the medial habenular nucleus (HAM) was found to be at the tenrecine level (Tables 41, 42). Based on the measurements of Bauchot (1963, 1966, 1979b), the nuclei dorsales (DOT; 2.3x AvTn) and mediales thalami (MLT; 2.1x) are highly progressive (Tables 69, 70) and, based on the measurements of Baron (1972, 1979), the dorsal part of the lateral geniculate body (GLD; 2.45x; Table 38) and the optic tract (TRO; 1.7x; Table 35) are also progressive. The eyes, too, were found to be larger in hedgehogs than in other Insectivora (Stephan et al. 1984b). Volume data on visual nuclei were given by Masterton et al. (1974).

In the subthalamic components, Frahm et al. (1979) measured the pallidum (PALL). Kesarev (1964) measured nine hypothalamic nuclei and found that they constituted 36.6% of the total hypothalamus (23.2% in man). In the diencephalic annexes, the epiphysis (EPI) (Legait et al. 1976) and hypophysis (HYP) (Bauchot 1961, 1970; Bauchot and Legait 1978) of the Erinaceinae were found to be small, particularly the intermediate lobe of HYP, compared with other Insectivora.

In the **telencephalon** (TEL), seven to nine main parts were measured and compared by Stephan and Andy (1964, 1969, 1982), Stephan (1967b), Bauchot and Stephan (1968), and Stephan et al. (1970, 1981a). Additional data were given on cortical structures and striatum by Wiedemeyer (1974); on olfactory structures by Stephan (1966), Stephan and Andy (1970), Etienne (1972), Stephan et al. (1982), and Baron et al. (1983, 1987); on septum (SEP) by Stephan and Andy (1962, 1982), Andy and Stephan (1966a, b), and Stephan (1969, SFB); on amygdala (AMY) by Stephan and Andy (1977) and Stephan et al. (1981a, 1987a); on hippocampus by Stephan and Manolescu (1980), West et al. (1984), and West and Schwerdtfeger (1985); and on neocortex (NEO) by Frahm et al. (1982). A very high index (3.5x AvTn) was found for the visual cortex (derived from GLD data, Frahm et al. 1984; see Table 64), but also for the neocortex in general (2.05x). Erinaceinae have a large accessory olfactory bulb (AOB, 2.6x), relatively well developed schizocortex and striatum (1.6 and 1.5x), and a moderately progressive hippocampus (1.4x; Table 45). Many parts of it were measured and compared with those of rats (West et al. 1984), tree shrews and marmosets (West and Schwerdtfeger 1985), and mice and man (West 1990). The main olfactory bulb and paleocortex are at the tenrecine level, whereas the small subfornical body (SFB, included in SEP) is slightly under this level (0.9x AvTn). Volume estimates of the prefrontal cortex and the caudate nucleus were given by Masterton and Skeen (1972).

The progression in most of the brain structures is similar to that of the total brain (NET 1.24x AvTn). The variation in the three species (*Erinaceus europaeus, Atelerix algirus,* and *Hemiechinus auritus*) is low, except for the auditory nuclei and some small structures, e.g., subfornical body (SFB; Table 41). More quantitative comparisons will be presented on p. 226 in the summary of brain characteristics of Erinaceinae and its species compared with those of other Erinaceidae.

Histology, Histochemistry

Total Brain, General

Antonopoulos et al. (1987, 1989)
Avksentieva (1975)
Bischoff (1900)
Faure and Calas (1977)
Haller (1906)
Papadopoulos et al. (1986a, b)

Schmidt and Lierse (1968)
Shtark (1970, 1972)
Uuspää (1963a, b)
Ziehen (1897a)

General data on the histology of various parts of the brain of the European hedgehog were given by Ziehen (1897a), Bischoff (1900), and Haller (1906). Schmidt and Lierse (1968) described brain vessels, angioarchitecture and capillary density in *Erinaceus europaeus* and *Atelerix algirus*. They found no differences between the two species and no changes during hibernation. Avksentieva (1975) described the chromatin pattern of the nuclei of granular cells and oligodendrogliocytes in cerebellum and cerebral hemispheres. Uuspää (1963a, b) studied the 5-hydroxytryptamine and catecholamine content of the brain. The uptake of tritiated noradrenalin by various parts of the brain in normal active and hibernating hedgehogs was investigated by Faure and Calas (1977).

Somatostatinlike, neurotensinlike, neuropeptide Y-like, VIP-like, and CCK-like immunoreactive neurons in *Erinaceus europaeus* were mapped in the whole brain by Papadopoulos et al. (1986a, b) and Antonopoulos et al. (1987, 1989). Shtark (1970, 1972) gave a stereotaxic atlas of the brain of the European hedgehog and investigated histochemical changes in hibernating animals (see also in Physiology section, p. 223).

Spinal Cord

Bauer (1909b)
Biach (1906)
Curik et al. (1974)
Kotzenberg (1899)
Krajci and Malinsky (1974)
Leszlenyi (1912)
Malinska and Malinsky (1974a, b)
Malinska et al. (1974b)
Malinsky (1973)
Malinsky and Krajci (1974)
Malinsky and Malinska (1973, 1975)
Reich (1909)
Sabbath (1909)
Sano (1909)
Schilder (1910)
Takahashi (1913)
Verhaart (1970)

The **spinal cord** of the hedgehog was studied by Kotzenberg (1899) and, between 1909 and 1913, at the Neurological Institute of the University of Vienna by Bauer (1909b, dorsal roots), Reich (1909, medial zone), Sabbath (1909, ventral roots), Sano (1909, substantia gelatinosa, dorsal column), Schilder (1910, nucleus sacralis), Leszlenyi (1912, dorsal column), and Takahashi (1913, lateral columns).

Between 1973 and 1975, a Czechoslovakian group associated with Malinska and Malinsky (Curik, Kapoun, Krajci, Zrzavy) studied the structures of the spinal cord of *Erinaceus concolor* in hibernating and non-hibernating animals. The group's publications deal with the fine structure of grey matter (Malinsky and Malinska 1973), motor neurons in active hedgehogs (Curik et al. 1974) and during hibernation (Malinska and Malinsky 1974a, b; Malinska et al. 1974b; Malinsky and Malinska 1975), fine structure of synapses (Malinsky 1973), spinal ganglia, normal and during hibernation (Krajci and Malinsky 1974), and enzyme activities, normal and during hibernation (Malinsky and Krajci 1974).

Biach (1906) described the canalis centralis of the hedgehog spinal cord. A detailed description of the spinal cord components was given by Verhaart (1970). He "suggests that propriospinal predominate more than usual and that the cord is much less influenced by the brain stem than in other mammals."

Brain Stem and Cerebellum

Alexander (1931, tractus tegmentalis centralis)
Baron (1972, auditory structures)
Bauer (1909a, substantia nigra)
Berkelbach van der Sprenkel (1924, hypoglossal nerve)
Bishop et al. (1971, pyramidal tract)
Brauer (1968, cerebellar nuclei)
Brunner (1919, cerebellar nuclei)
Campbell, C. (1967, pyramidal tract)
Chauhan et al. (1983, histochemistry)
Earle and Matzke (1974, cerebellar nuclei)
Galert (1985, substantia nigra)
Godlowski (1930, nucleus triangularis vestibularis)
Grünwald (1903, pedunculi cerebellares)
Harrison and Irving (1966, superior olive)
Hatschek (1904, lemniscus medialis, pons)
Herrera Lara (1985, periaqueductal neurons, posterior commissure)
Hofmann (1908, superior olive)
Hulles (1907, nervus trigeminus, radix mesencephalica)
Irving and Harrison (1967, superior olive)
Ishihara (1931, nervus vestibularis)
Kaplan (1913, nucleus vestibularis lateralis and angularis)
Krabbe (1925, subcommissural body)
Marsden and Rowland (1965)
Masterton et al. (1969, 1975, superior olive)
Mohanakumar and Sood (1980, histochemistry)
Pekelsky (1922, raphe)
Petrovicky (1966, reticular formation; 1971 Gudden's tegmental nuclei)
Petrovicky and Kolesarova (1989, parabrachial nuclear complex)
Probst (1900, 1901, pyramidal tract, medial longitudinal fascicle)
Shima (1909, nucleus dorsalis nervi vagi)
Sood and Mohanakumar (1980, histochemistry)
Sood et al. (1982, histochemistry)

Stengel (1924, nucleus commissurae posterioris)
Valeton (1908, inferior colliculi)
VanderVloet (1906, pyramidal tract)
VanValkenburg (1911a, radix spinalis nervi trigemini)
Verhaart (1970, entire OBL)
Victorov (1966, superior colliculi)
Walberg (1952, lateral reticular nucleus)
Werkman (1913, ontogenesis of commissures)
Williams (1909, inferior olive)
Ziehen (1899b, pyramidal tract)

The caudal brain parts of *Erinaceus europaeus* were mentioned and compared mostly with those of other mammalian species between 1903 and 1931 in several papers published by the Neurological Institute of the University of Vienna (Alexander, Bauer, Brunner, Godlowski, Grünwald, Hatschek, Hofmann, Hulles, Ishihara, Kaplan, Pekelsky, Shima, Stengel, Valeton, and Williams). The topics of the publications are given in the reference list. In general, parts of the hedgehog brain were only mentioned or very briefly described. More detailed descriptions of the superior olive were given by Hofmann (1908) and of the raphe by Pekelsky (1922).

A detailed description of OBL components with many figures, tables, and size estimations was given by Verhaart (1970). Probst (1900, 1901) and Verhaart (1970) mention a very large medial longitudinal fascicle in contrast to a very small pyramidal tract. Campbell (1967) summarized his findings by stating that "the fibers of the pyramids became quite sparse at the caudal end of the medulla. No decussation was found in the brain stem. Degenerating fibers in the ipsilateral ventral funiculus of the cervical cord were so fine and in such small numbers that their detection was uncertain in transverse sections. Their presence was revealed in sagittal sections." Using the results of Verhaart's investigations on the hedgehog, Heffner and Masterton (1975, 1983) compared pyramidal tract size with digital dexterity. Bishop et al. (1971) measured axon sheath thickness and fiber size in the pyramids.

Marsden and Rowland (1965) mentioned *Erinaceus europaeus* in a study on the evolution of pons, olive, and pyramid, but gave no details. Walberg (1952), in a careful description of the lateral reticular nucleus (LRE) of *E. europaeus,* subdivided it into three parts. In the superior olivary complex, Harrison and Irving (1966) and Irving and Harrison (1967) did not find a medial olivary nucleus, which they expected to be highly correlated with the VIth nucleus and with eye size. Masterton et al. (1969, 1975) reported that *Paraechinus hypomelas* hear low frequencies poorly and does not localize them. They related this to the low development of the medial superior olive in this species. Victorov (1966) using Nissl and Golgi methods described the differences in the neuronal structure between the three cell layers of the superior colliculus (SUC). Baron (1972), Petrovicky (1971) and Petrovicky and Kolesarova (1989) included *Erinaceus europaeus* in comparative cytoarchitectural studies on brain stem structures.

Histochemical characteristics of cellular and fibrous components of the medulla oblongata and pons of *Paraechinus micropus* were studied by Mohanakumar and Sood (1980), Sood and Mohanakumar (1980), Sood et al. (1982), and Chauhan et al. (1983) based on the distribution of acid phosphatase, acetylcholinesterase, and glycosidases.

Diencephalon

Baron (1972, medial geniculate body)
Bauchot (1963)
Brauer et al. (1978, lateral geniculate body)
Campbell, C. (1969, 1972, visual structures)
Campbell D. (1964, 1966, paraventricular nuclei)
Clark (1929)
Clark (1933, medial geniculate body)
Dinopoulos et al. (1987, visual structures)
Ebbesson et al. (1972)
Ebner (1969, *Paraechinus hypomelas*)
Erickson et al. (1967)
Galert (1986, supraoptic and paraventricular nuclei)
Gil et al. (1985, hypothalamus, hibernation)
Grönberg (1901, epiphysis, ontogenesis)
Hall and Diamond (1968a, visual structures)
Harting et al. (1972, *Paraechinus hypomelas*)
Hochstetter (1923, epiphysis, ontogenesis)
Holmes (1961, 1965, hypothalamus, supraoptic neurons)
Hornet (1933, medial geniculate body)
Környey (1928, lateral geniculate body)
Korf et al. (1985, epiphysis, retinal S-antigen)
Malinska and Malinsky (1985, hypothalamus, hibernation)
Malinska et al. (1985)
Papadopoulos et al. (1985, hypothalamus, OXY and CRF immunoreactivities)
Spiegel and Zweig (1919, tuber cinereum)
Vigh and Vigh-Teichmann (1981, epiphysis, opsin- immunoreactivity, pinealocytes, *Erinaceus concolor*)
Vigh-Teichmann et al. (1986, epiphysis, S-antigen, *E. concolor*)
Warkany (1924, globus pallidus)

The **diencephalon** was investigated by Bauchot (1963; cytoarchitecture) and Malinska et al. (1985; angioarchitecture), and the thalamus by Clark (1929), Erickson et al. (1967), Ebner (1969), Ebbesson et al. (1972) and Harting et al. (1972). The authors of studies on more specific parts of the diencephalon and their respective topics are provided in the reference list.

Clark (1929) emphasized that the cells in the **thalamus** of *Erinaceus* were loosely packed and, as a consequence, the nuclei were more diffuse and less well circumscribed. In particular, identification "of the anterior group of nuclei offers considerable difficulty."

Bauchot (1963) described various thalamic nuclei (e.g., by size and density of the cells), and compared them with other Insectivora, tree-shrews, and galagos. Erickson et al. (1967) described the cytoarchitecture of the posterior dorsal thalamus and noted that there was very little structural difference between the ventroposterior nucleus and the medial geniculate body. Campbell (1969, 1972) and Brauer et al. (1978) described the cytoarchitecture of ventral and dorsal nuclei of the lateral geniculate body in *Paraechinus* and/or *Erinaceus*. They gave a brief review of the comparative anatomy of the lateral geniculate body in mammals and called attention to the lack of lamination in the pars dorsalis of CGL. Ebbesson et al. (1972) compared thalamic organization in a series of vertebrates. Harting et al. (1972) studied the thalamic lateral posterior nucleus of *Paraechinus* and confirmed Clark's statement that it is a homologue of the primate pulvinar which became larger and more complex during mammalian evolution. This tendency is already found in the Insectivora, as shown by the development of DOT within the order relative to its size in Tenrecinae (see Table 70).

In the **hypothalamus**, Galert (1986) described various cell types in the supraoptic and paraventricular nuclei and Gil et al. (1985) found an increase of fusiform cells in these nuclei during hibernation. Holmes studied the esterases in the hypothalamo-hypophyseal system (1961) and the fine structure of supraoptic neurons (1965); Campbell (1964, 1966) described the neurosecretory pathways from the paraventricular nuclei. In their studies of the cyto- and angioarchitectural characteristics of the hypothalamic nuclei in the hedgehog, Malinska and Malinsky (1985) did not find uniform size changes in the cells during hibernation.

Hypophysis

Allara (1958, 1959, hibernation)
Azzali (1955, histophysiology)
Bargmann and Gaudecker (1969, ultrastructure of elementary granula)
Bloom (1960, cytology and histochemistry)
Campbell, D. (1964, 1966, neurosecretory pathway)
Campbell D. and Holmes (1966, neurohypophysis)
Dubois and Girod (1969)
Gerry (1969)
Girod (1971, 1976)
Girod and Curé (1965)
Girod et al. (1965, 1966, 1967, 1982, 1983, 1985, 1986)
Green (1951, blood supply and innervation)
Grönberg (1901, ontogenesis)
Haller (1909b, comparative anatomy)
Hanström (1946, 1952, 1953, 1966, partly *Atelerix frontalis*)
Hochstetter (1924, ontogenesis)
Holmes (1961, esterases)
Holmes and Ball (1974, comparative anatomy)
Holmes and Kiernan (1963, 1964, infundibular process)
Kiernan (1964, carboxylic esterases)
Pokorny (1926, comp. anatomy)
Purves and Bassett (1963, pars intermedia)

Rütten et al. (1988, pars tuberalis, hibernation)
Smit-Vis (1962, hibernation)

The hypophysis of hedgehogs has been studied or mentioned in a great number of papers on active and on hibernating animals. Topics are given in the reference list. The most detailed investigations are those of Hanström and of Girod and co-authors covering morphological, histological, cytological, immunocytochemical, and ultrastructural characteristics.

Non-neocortical Telencephalon

Abbie (1938, hippocampus, subiculum)
Beccari (1910, olfactory tubercle)
Crutcher et al. (1988, hippocampus)
Gaarskjaer et al. (1982, hippocampus)
Hager (1954, main olfactory bulb, paleocortex)
Haller (1906, telencephalon, commissures)
Hewitt (1959, nucleus caudatus)
Kryspin-Exner (1922, substantia perforata anterior)
Lopez-Mascaraque et al. (1986, main olfactory bulb)
Malinsky and Sedlak (1983, olfactory structures, hibernation)
Pines (1927, subfornical body)
Pospisilova and Malinsky (1985, accessory olfactory bulb)
Pospisilova and Malinsky (1986, main olfactory bulb)
Rami et al. (1987, hippocampus)
Rose, M. (1927a, b, allocortex)
Schwerdtfeger et al. (1985, prepiriform region, schizocortex)
Smith (1897, corpus callosum, fornix, hippocampus)
Spiegel (1919, striatum, pallidum, amygdala)
Stengaard-Pedersen et al. (1983, hippocampus)
Stephan (1963, hippocampus)
Stephan (1965, accessory olfactory bulb)
Stephan (1976, allocortex)
Stephan and Andy (1977, amygdala)
Stephan and Andy (1982, septum)
Stephan and Manolescu (1980, hippocampus)
Valverde et al. (1989, retrobulbar region)
Völsch (1906, 1911, amygdala)
West et al. (1984, hippocampus)

In many of the papers, the characteristics of hedgehog telencephalic structures are compared with those in other mammals. Malinsky and Sedlak (1983) found that cell size in the olfactory bulb and tubercle remained the same or increased during hibernation, whereas it diminished in other parts of CNS. Pospisilova and Malinsky (1985) mentioned myelinated neurons in the olfactory glomeruli. Lopez-Mascaraque et al. (1986) described the cytoarchitecture of the olfactory bulb and investigated in detail the cells of the granular layer with the Golgi method. "The mitral cell layer does not stand out as a monolayer as in most mammals; it is arranged as a diffuse stratum with mitral cells displaced into the external plexiform layer." Valverde et al. (1989) investigated in detail the retrobulbar region (nucleus olfactorius anterior) in Nissl-stained and Golgi preparations. They subdivided the region into six parts.

Switzer et al. (1980) included the hedgehog in their comparison of AOB characteristics in mammals. Völsch (1906, 1911) described the amygdala and its relation to other basal telencephalic structures in great detail. Rose (1927a, b) subdivided and mapped the allocortical regions. He described three fields each in PRPI, TOL, and PAM and two fields in ENT. Stephan (1963) and Stephan and Manolescu (1980) gave maps of the allocortex and hippocampus and used the topographical and structural relations in the hedgehog to illustrate evolutionary processes. Schwerdtfeger et al. (1985) studied prepiriform and entorhinal cortices with Timm's sulphide silver and the selenium methods and could clearly separate a medial and a lateral field in the entorhinal cortex (28M and 28L).

Gaarskjaer et al. (1982), Stengaard-Pedersen et al. (1983) and Rami et al. (1987) showed that the hippocampal mossy fibers in *Erinaceus europaeus* are not restricted to the inferior region of the hippocampus (CA 3), as is generally the case, but extend into the superior region (CA 1). In this extension, the mossy fiber terminals have the same large multivesicular, multicontact morphology characteristic of the inferior region. West et al. (1984) studied the staining characteristics of hippocampal components with Timm's heavy metal sulphide technique and argued that the hedgehog hippocampus may be used as a basal reference in comparative studies. Crutcher et al. (1988) found some striking differences in AChE histochemistry and catecholamine histofluorescence from that observed in other mammals, particularly in the dentate area. They found for example uniformly low AChE activity of the molecular layer, and the same density of noradrenergic fibers in the molecular layer as in the hilus.

Neocortex

Abbie (1939, corpus callosum)
Batuev et al. (1980, GOL, ELMI, MAP)
Brodmann (1906, 1909, CTA, MAP)
Demyanenko (1976, GOL, ELMI)
Diamond and Hall (1969, CTA, VIS, MAPM)
Ebner (1969, comparison organization, *Paraechinus hypomelas*)
Ferrer et al. (1986, GOL)
Flores (1911, MYA, MYG, MAP)
Gould and Ebner (1978a, b, CTA, MYA, VIS, *P. hypomelas*)
Gould et al. (1978, VIS, *P. hypomelas*)
Hall and Diamond (1968a, CTA, VIS, *P. hypomelas*)
Haller (1909a, CTA, MAP)
Kaas et al. (1970, CTA, MYA, VIS, MAP)
Malinsky and Sedlak (1981, CTA)
Mott (1907, CTA)
Poliakov (1964, CTA, GOL)
Redlich (1903, cingulum)
Rehkämper (1981, CTA, MAP)
Rose, M. (1928, 1929, CTA, MAP)
Sanides (1969, CTA)
Sanides D. and Sanides F. (1974, CTA, GOL, *Erinaceus, Hemiechinus, Atelerix*)

Sanides F. and Sanides D. (1972, GOL, *Erinaceus, Hemiechinus, Atelerix*)
Sas and Sanides (1970, GOL)
Tolchenova and Demyanenko (1975, GOL, ELMI)
Valverde (1983, 1986, GOL)
Valverde and Facal-Valverde (1986, GOL)
Valverde and Lopez-Mascaraque (1981, GOL)
Valverde et al. (1986, GOL)
Watson (1907, CTA, MAP)

Abbreviations:
CTA = cytoarchitecture
GOL = Golgi stains
ELMI = electron microscopy
MAP = cortical map given
MYA = myeloarchitecture
MYG = myelogenesis
VIS = visual cortex

Watson (1907) differentiated a medial motor area and a lateral general sensory area on the cerebral convexity. Both areas had their main axes parallel to the interhemispheric fissure. Haller (1909a), too, differentiated the cortex in a few longitudinal zones only (inner and outer longitudinal field). In Watson's map, transversal zones of undifferentiated areas were labeled rostrally and caudally, and a small longitudinal band connecting them laterally was noted at the border of the piriform lobe. This areal differentiation was not confirmed by later authors. Brodmann (1909) classified the NEO of the hedgehog in greater detail in about a dozen fields (Fig. 98). He subdivided the convexity into precentral (4+6), postcentral and parietal (5+7), occipital (17), insular (13-16) and temporal (20-22, 36) regions, and the medial surface into cingular (23, 24, 25 and 32) and retrosplenial (26, 29, 30) regions. Flores (1911) classification was nearly identical, but he added the postcentral fields 1-3, and especially a frontal field 8, found by Brodmann in other species. Field 8 may have been included by Brodmann in the frontal insular region. This field is especially early in its myelogenesis and is rich in myelinated fibers. Flores summarized and reviewed the classifications of the earlier authors. Rose investigated the cytoarchitecture of the cingular and retrosplenial regions (1928) and the insular region (1929). He introduced a new nomenclature, but in his infraradiate (cingular, according to Brodmann) and retrosplenial regions, the borders are nearly identical to Brodmann's. The insular region was found to be wholly agranular. It was subdivided by Rose into two fields, the more ventral being narrow and hidden in the rhinal sulcus. A frontal field, described by Flores (field 8), was not mentioned by Rose.

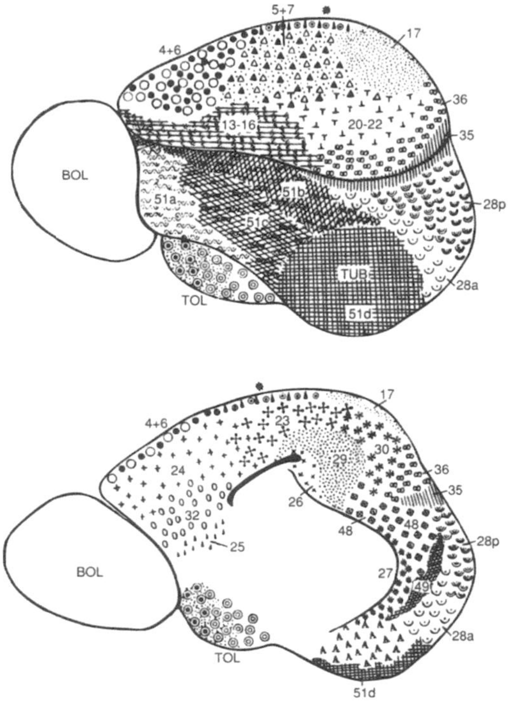

Figure 98

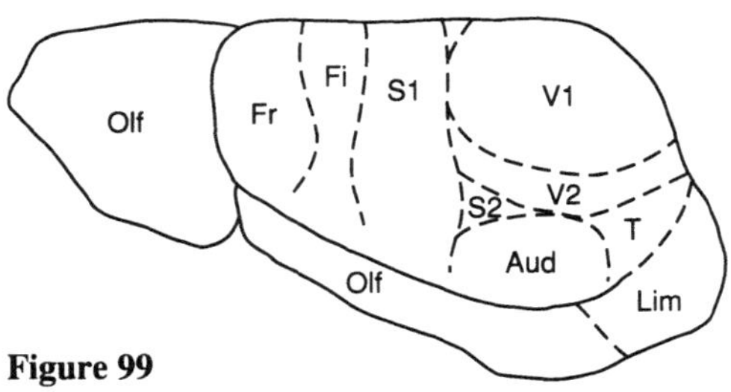

Figure 99

Figure 98: Brodmann's cortical map of the hedgehog brain based on cytoarchitecture (1909)

Figure 99: Map combined from evoked potential studies in the hedgehog brain

Abbreviations:

Aud	auditory
Fi	frontal intermediate
Fr	frontal
Lim	limbic
Olf	olfactory
S1	somatosensory 1
S2	somatosensory 2
T	temporal
V1	visual 1
V2	visual 2

Based on evoked potentials (see Fig. 99 and p. 221), Lende and Sadler (1967) and Lende (1969) determined somatic sensory, somatic motor, visual and auditory fields, with extensive overlap among them. Their somatic sensory field roughly corresponds to Brodmann's postcentral and parietal regions; their somatic motor field to the precentral region; their visual field to the occipital region; and their auditory field to the temporal region. Subsequently, with improved histological and physiological methods, detailed subdivisions were found (Diamond and Hall 1969; Kaas et al. 1970; Gould et al. 1978; Gould and Ebner 1978a, b), especially in the visual cortex (see pp. 218 and 222).

Sanides F. (1969) and Sanides D. and Sanides F. (1974) called attention to the unusual thickness of the first (molecular or zonal) layer, the sharply accentuated layer II, and the very poor cortical lamination with practically no granularization in the NEO of hedgehogs. However, these characteristics are less prominent compared with the NEO of other Insectivora, as shown by Rehkämper (1981). Rehkämper is the only author who has compared the hedgehog NEO with that of eight other Insectivora belonging to eight different subfamilies and/or families. He found a continuous granular layer only in area A 3, which includes the visual cortex, suggesting a greater importance of vision in hedgehogs than in other Insectivora.

Golgi studies were performed by Poliakov (1964), Sas and Sanides (1970), Sanides, F. and Sanides, D. (1972), Sanides, D. and Sanides, F. (1974), Valverde and Lopez-Mascaraque (1981), Valverde (1983, 1986), Valverde and Facal-Valverde (1986), and Ferrer et al. (1986).

Poliakov (1964) investigated the cortical development in hedgehogs and compared it with that in frogs and lizards. In *Erinaceus, Atelerix,* and *Hemiechinus,* Sanides and Sanides (1972) described "extraverted neurons" in layer II all over the neocortex, which in advanced mammals are restricted to the transitional mesocortices. According to Valverde and Facal-Valverde (1986), "layer II contains primitive pyramidal cells representing the most outstanding feature of the neocortex of the hedgehog. Most pyramidal cells in layer II have two, three or more apical dendrites, richly covered by spines predominating over the basal dendrites. These cells resemble pyramidal cells found in the piriform cortex, hippocampus and other olfactory areas. It is suggested that the presence of these neurons reflects the retention of a primitive character in neocortical evolution." Tolchenova and Demyanenko (1975), Demyanenko (1976) and Batuev et al. (1980) investigated the so-called polysensory "assoziation" area described by the Russian authors based on Golgi and electron microscopic methods.

Special interest has been shown in the **visual cortex** (area striata, Brodmann's area 17) of hedgehogs. Mott and Watson (both working in the same institution with the same material) subdivided three small regions on the caudal part of the medial surface, the most ventral of which was determined to be the cortical visual area. That region, however, is the site of the presubicular cortex (Brodmann's area 27, see Figs. 77 and 98). The visual area is on the caudal convexity and its position, cytoarchitecturally determined and mapped by Brodmann (1909) and myeloarchitecturally by Flores (1911), has been confirmed by several studies on visual projections and visual evoked potentials.

Hall and Diamond (1968a, b, *Erinaceus, Hemiechinus, Paraechinus*) and Diamond and Hall (1969) used the relative thickness of layer IV (granular layer) as the chief criterion for defining the hedgehog's visual area, in addition to a decrease in the concentration of neurons in layer VI relative to the adjacent areas. Between striate area (V I) and auditory areas, they found a belt-like area almost free of myelinated fibers which was also found to be visual (V II). Both areas receive thalamic projections from the lateral posterior (LP) and lateral geniculate nucleus (CGL). Gould et al. (1978) added two more thalamic sources, i.e., a transitional zone, adjacent to GLD, and the anterior intralaminar region. Gould and Ebner (1978a) determined four distinct regions of visual cortex from lateral to medial, but "the hedgehog emerges as a mammal that possesses one of the very least differentiated visual systems found in existing mammals." Gould and Ebner (1968b) investigated interlaminar connections and found that both vertical and horizontal connections are important and that horizontal projections originate at three distinct cortical depths.

Fiber connections

Campbell C. (1969, 1972)
Campbell C. et al. (1967)
Cooper et al. (1989)
DeCarlos et al. (1989)
Dinopoulos et al. (1987, 1988)
Earle and Matzke (1974)
Ebbesson et al. (1972)
Ebner (1969)
Ganchrow and Ring (1979)
Gould and Ebner (1978a, b)
Gould et al. (1978)
Hall (1970, 1972)
Hall and Diamond (1968a)
Hall and Ebner (1970)
Jane and Schroeder (1971)
Kiernan (1967)
Killackey (1972)
Killackey and Ebner (1972)
Moore (1973)
Moore et al. (1977)
Ravizza and Diamond (1972)
Reep (1984)
Regidor and Divac (1990)
Ring and Ganchrow (1983)
Schroeder and Jane (1976)
Schroeder et al. (1970)
Tigges and Tigges (1969)
Valverde et al. (1986)
Voneida et al. (1969)

Fiber connections from **the spinal cord and cerebellum to brain stem and within the brain stem** were investigated in *Erinaceus europaeus* by Schroeder et al. (1970), Jane and Schroeder (1971), Ebbesson et al. (1972), Earle and Matzke (1974), Schroeder and Jane (1976), Ganchrow and Ring (1979), and Ring and Ganchrow (1983); in *Hemiechinus auritus* by Earle and Matzke (1974) and Moore et al. (1977); and in *Paraechinus aethiopicus* by Ring and Ganchrow (1983).

Schroeder et al. (1970) mentioned projections from the dorsal column of the spinal cord, cerebellar nuclei, and tectum to the thalamus. Common sites of termination of the pathways are zona incerta, ventral posterior nucleus, and posterior nucleus (PO). Jane and Schroeder (1971) found that spinal hemisections resulted in heavy degeneration in brain stem and mesencephalon but little in the thalamus. In contrast, lesions of the dorsal column nuclei produced stronger degeneration in the thalamus as well. Projections from these nuclei to the inferior colliculi were described by Schroeder and Jane (1976). The projections from the nucleus caudalis (NC) and spinal cord to brain stem and diencephalon were also studied by Ganchrow and Ring (1979) and Ring and Ganchrow (1983). Their results suggest "that somatic input from NC, some of which may be pain-specific, reaches those thalamic areas, a portion of whose neurons are characterized as polymodal and at least partially convergent for somatotopy." Ebbesson et al. (1972) studied the representation of spinal cord and brain stem input in the thalamus and found various degrees of overlap. They found that overlap can be inversely correlated with specialization more readily than with a hypothetical evolutionary position. Earle and Matzke (1974) found that in the hedgehog, the typical mammalian pattern is evident in the distribution of the cerebellofugal fibers.

Moore et al. (1977) investigated the projections of the inferior colliculi and found the only significant difference from other mammals to be a small tecto-thalamic pathway.

In the hedgehog's **visual connections**, the retinal projections to mesencephalic, diencephalic, and telencephalic structures were investigated in *Erinaceus europaeus* by Campbell et al. (1967), Kiernan (1967), Campbell (1969, 1972), Tigges and Tigges (1969), Ebbesson et al. (1972), Dinopoulos et al. (1987) and Cooper et al. (1989); in *Hemiechinus auritus* by Moore (1973); and in *Paraechinus* spp. by Campbell (1969, 1972) and Hall and Ebner (1970). The main results were given by Campbell et al. and confirmed by later authors. Contralateral projections were primarily found to (1) superior colliculus, (2) pretectal complex, (3) pars dorsalis of CGL, except for a few places in which the ipsilateral axons terminate (Campbell 1972), (4) pars ventralis of CGL and (5) accessory optic system. Ipsilateral projections to (1) - (4) exist also, but are more sparse. Those to the dorsal CGL (3) project only to one or a few discrete areas devoid of contralateral terminations (see above).

Campbell (1969) and Moore (1973) found bilateral projections in the hypothalamic suprachiasmatic nuclei. Their existence was supposed earlier by Kiernan (1964), who described cholinesterase-containing fibers possibly passing from the chiasma to the suprachiasmatic nucleus, but was later (1967) denied by the same author. Direct retinal projections to the mediocaudal region of the olfactory tubercle were described by Cooper et al. (1989).

Projections from the tectum opticum to the thalamus were described by Hall and Ebner (1970) in *Paraechinus hypomelas*. The superior colliculus projects mainly to the ventral CGL. The tectum projects to the lateral posterior nucleus as well. Visual projections from the thalamus to the cerebral cortices were investigated in the same species by Hall and Diamond (1968a), Ebner (1969), Hall (1970, 1972), and Gould et al. (1978), and by Hall and Diamond (1968a) in *Erinaceus* and *Hemiechinus*. Hall and Diamond presented figures of sites of retrograde degeneration in the thalamus after cortical lesions of varying size. According to Hall, CGL projects to the striate cortex. The terminations are confined to layer IV and adjacent parts of layer III. The lateral posterior nucleus, receiving dense projections from the optic tectum, projects to striate and peristriate cortex, mainly to layers V and VI. Gould et al. added two more thalamic sources projecting to the visual cortices: a transition zone, adjacent to GLD, and the anterior intralaminar region. Visual cortico-cortical and interlaminar connections were investigated by Gould and Ebner (1978a, b) in *Paraechinus hypomelas.*

Non-visual fiber connections from thalamus and basal forebrain to neocortex were investigated in *Erinaceus europaeus* by Reep (1984), Valverde et al. (1986), and Dinopoulos et al. (1988), in *Paraechinus hypomelas* by Ebner (1969), Killackey (1972), Killackey and Ebner (1972), and Ravizza and Diamond (1972), and in *Atelerix algirus* by Regidor and Divac (1990). Killackey and Killackey and Ebner made small thalamic lesions in the ventral nuclear group (VP and VAL) in *Paraechinus* and found fiber degeneration in layer IV of a limited posterior portion of Lende's somatic sensory-motor (parietal) cortex, which lies rostral to the visual cortex. The central intralaminar nucleus projects diffusely to layers I and VI of the parietal cortex. Ravizza and Diamond found in *Paraechinus* that lesions of the medial geniculate body (CGM) produce terminal degeneration in layers IV and III of auditory cortex with topographic organization and that, following large lesions of the posterior nucleus (PO), degeneration was found only in layer I. Valverde et al. made small injections in layers I and II of the somatosensory cortex in *Erinaceus europaeus* and found labeled cells in nucleus ventralis and nucleus ventromedialis thalami. According to Dinopoulos et al., injections in frontal, parietal and occipital cortices of *Erinaceus europaeus* produce labeled cells throughout the ipsilateral basal forebrain (septum, substantia innominata). "These projections, apparently cholinergic, were not topographically organized." Regidor and Divac made large unilateral cortical deposits of fluorescent somatopetal tracers and found labelled perikarya not only in the ipsilateral but also in the contralateral thalamus, with large numbers in the ventrolateral nucleus. In this strong contralateral projection "the hedgehog brain is so far unique among mammals" and may be a primitive feature in Insectivora.

Reep (1984) investigated the relationship between prefrontal and limbic cortex in a series of mammals including hedgehog, if the prefrontal cortex "is defined as the projection field of the mediodorsal thalamic nucleus (MD)."

Interhemispheric connections were investigated by Voneida et al. (1969, *Erinaceus*), and Gould and Ebner (1978a, *Paraechinus*). They exist in all cytoarchitectonic areas.

Olfactory projections from the MOB and dorsal RB in hedgehogs were investigated by DeCarlos et al. (1989). The authors suggest that the contralateral projection nuclei to the MOB "and the large number of cells with axonal collaterals projecting to both hemispheres, may be a strategy ... to bilaterally integrate brain functions at the expense of its reduced corpus callosum."

Physiology (mainly neocortex), Behavior

Adrian (1942, olfactory centers, EP)
Batuev and Karamyan (1973, EP)
Batuev et al. (1980, EP)
Bretting (1972, olfactory thresholds)
Erickson et al. (1967, MR)
Hall and Diamond (1968b, CA, LE)
Ivanov and Sollertinskaja (1986, LE)
Kaas et al. (1970, EP, MR)
Karamyan et al. (1986, LE)
Konstantinov et al. (1987, EP)
Kristoffersson et al. (1965, 1966, HC)
Lende (1969, EP)
Lende and Sadler (1967, EP)
Malyukova (1984, CA, EC, LE)
Mann (1896, CS)
Masterton and Skeen (1972, LE)
Masterton et al. (1969, 1974, 1975, LE)
Pirogov (1977, MR)
Pirogov and Malyukova (1975, 1976a, b, EP)
Probst (1901, CS, CA)
Putkonen et al. (1964, EP)
Ravizza et al. (1969, audition, LE)
Saarikoski (1967, SL)
Shtark (1970, 1972, EP)
Skeen and Masterton (1982, CA, LE)
Tauber et al. (1968, EP)
Vogt and Vogt (1907, CS)
Weber, E. (1906, CS)
Yellin and Jerison (1980, EP)
Ziehen (1897b, 1899c, CS)

Abbreviations:
CA = cutting and ablation
CS = cortical stimulation
EC = electrocoagulation
EP = evoked potentials
HC = histochemical methods
LE = learning experiments
MR = microelectrode recordings
SL = cortical slices

Many studies have been carried out on the physiology of the hedgehog brain. The earliest were on motor reactions following neocortical stimulation. Such investigations were performed by Mann, Ziehen, Probst, Weber, and Vogt and Vogt (CS in reference list).

In general, the movements obtained after stimulation of medial and lateral parts of the cerebral convexity were described and the results compared with those obtained in other mammalian species. After stimulation of the cortex in front of the "orbital sulcus" Mann observed head movements.

The results were summarized and reviewed by Vogt and Vogt. Later electrophysiological investigations involved recordings of evoked potentials from brain structures (mostly from the neocortical surface) after sensory stimulation (EP in reference list) and/or microelectrode recordings (MR in reference list).

Lende and Sadler (1967) and Lende (1969) gave the first neocortical map based on evoked potential studies. They determined somatic sensory, somatic motor, visual and auditory fields and found extensive overlap of the areas. Approximations to Brodmann's cytoarchitectonic map were found. Lende and Sadler's map was confirmed by later authors, who added more details (see Fig. 99). Russian authors (Batuev and Karamyan 1973; Pirogov 1977; Batuev et al. 1980) were especially interested in localizing zones of intersensory integration. They described an association area in the middle of the lateral NEO.

The electrical activity of the **olfactory** bulb and piriform lobe was studied by Adrian (1942), in animals breathing either normal air or air to which odors had been added, and by Putkonen et al. (1964) of the olfactory bulb and cerebral cortex in hedgehogs arousing from hibernation. Bretting (1972) studied olfactory thresholds for some fatty acids and found similar thresholds in hedgehog and man (see also Section 7.1, pp. 320, 321).

Functional characteristics of the **auditory** system and physical parameters of acoustic signals in hedgehogs were studied by Masterton et al. (1969, 1975; *Paraechinus*) and by Ravizza et al. (1969, *Hemiechinus*) with the behavioral technique of conditioned suppression, and by Konstantinov et al. (1987; presumably *Erinaceus*) with evoked potentials. If the frequency ranges of the two different approaches are at all comparable, they seem to be much broader in *Hemiechinus* (0.25 - 45 kHz) than in *Erinaceus* (1.7 - 7 kHz). This is paralleled by the volume of the auditory nuclei (see p. 206 and Table 26). Erickson et al. (1967) had found that "neither the ventroposterior nucleus nor the medial geniculate body was modality specific." They gave some photoprints and many schematic drawings to show the position of individual neurons responsive to particular sensory stimuli.

Using evoked potentials, Kaas et al. (1970) determined two **visual** fields (V I and V II) which corresponded with the subdivisions of Brodmann and Flores. V I corresponds to area 17. However, the region of the adjacent belt V II was determined by Brodmann (1909) and Flores (1911) to be area 20-22, implying that it was temporal cortex in the hedgehog. Yellin and Jerison (1980) recorded evoked potentials and afterpotentials to visual stimuli in unanesthetized hedgehogs.

Investigations employing methods of **learning** and **behavior** are indicated by LE in the reference list. Ravizza et al. (1969) have already been mentioned in conjunction with the auditory system. Masterton et al. (1974) studied the role of the pulvinar-extrastriate visual system in reversal learning in *Paraechinus* and *Hemiechinus*. Hall and Diamond (1968b) removed both types of visual cortex and investigated pattern discrimination. Masterton and Skeen (1972) and Skeen and Masterton (1982) investigated the role of the prefrontal system in delayed alternation and spatial reversal learning. Malyukova (1984) studied complex motor food-getting conditioned reflexes in normal European hedgehogs and after cortical removal and striatum coagulation. The influence of the oligopeptide tuftsen on hierarchical behavioral interrelationships in the hedgehog was studied by Ivanov and Sollertinskaja (1986). The role of thalamic and hypothalamic formations in regulating conditioned reflex activity in hedgehogs was studied by Karamyan et al. (1986).

Various methods were used to study **sleeping** and **hibernating** hedgehogs by Tauber et al. (1968), Saarikoski (1967), Putkonen et al. (1964), Kristoffersson et al. (1965, 1966), and Shtark (1970, 1972). Tauber et al. investigated electrical activity in the sleep-waking cycle; Saarikoski, oxygen consumption in stimulated and unstimulated cortical slices in hibernating and normal homeothermic hedgehogs; Putkonen et al., electrical activity of the olfactory bulb and cerebral cortex in hedgehogs arousing from hibernation; and Kristoffersson et al., water, GABA, aspartic acid, glutamic acid, and glutamine content of cerebral tissue in hibernating and non-hibernating hedgehogs.

Shtark made a detailed investigation of bioelectric activity and histochemical changes in the hibernating hedgehog.

Sense Organs

Ärnbäck-Christie-Linde (1914, vomeronasal organ)
Boeke (1932, skin)
Broom (1897, vomeronasal organ)
Harvey (1882, vomeronasal organ)
Menner (1929, eye)
Mohr (1936a, nose)
Qayyum et al. (1975, tongue, *Hemiechinus auritus*)
Siemen (1976, eye)
VanVeen et al. (1986, eye, *Erinaceus concolor*)
Wöhrmann-Repenning (1975, nose)

Boeke (1932) described the innervation of the smooth muscle fibers of the spines. The muscle fibers are radially arranged around the broad bases of the spines and can move them in every direction. Mohr (1936a) found clear differences in the shape of the outer border of the rhinarium between *Erinaceus europaeus* and *Atelerix algirus*. In *Atelerix* the border of the rhinarium is more clearly serrated, somewhat reminiscent of that of the mole *Condylura*.

The nasal fossae were compared by Wöhrmann-Repenning (1975) with those of *Talpa, Sorex,* and *Crocidura*. The surface of the maxillo-turbinal has increased enormously in the hedgehog owing to secondary folds. Olfactory glands of Bowman are abundant. The 90 μm thick olfactory epithelium has almost reached the maximum thickness physiologically possible. The vomeronasal organ was studied by Harvey (1882), Broom (1897) and Ärnbäck-Christie-Linde (1914). Broom found it similar to that of *Echinosorex* and typical of that for most eutherian mammals. The self-anointing of hedgehogs (e.g. Herter 1957) was functionally related by Poduschka and Firbas (1968) to the vomeronasal system. The tongue of *Hemiechinus* conforms to the general mammalian pattern and all the usual components are well defined (Qayyum et al. 1975). According to Menner (1929), the retina of *Erinaceus europaeus* is composed of 3.75% cones and 96.25% rods. The nuclei of the rods are smaller and have stronger-staining plasma. Some color vision appears possible (Menner) but improbable (Siemen). Siemen (1976) also found the nuclei and outer parts of the photoreceptors to be rod-like. VanVeen et al. (1986) described a large quantity of rhodopsin in the hedgehog's retina and suggested it was associated with the animal's pronounced nocturnal habits.

4.4.3. Summary of Brain Characteristics in Subfamilies and Species of Erinaceidae

In some macromorphological details of the brains, the two subfamilies of Erinaceidae have common features (see Section 3.1). Similarities in the size indices were found for only a few parts, e.g., the olfactory structures, whereas most other structures are very different (vestibular, auditory, limbic, neocortical). This is clearly shown in the index profiles for Echinosoricinae and Erinaceinae given in Figs. 101, 103, and 107. Given the relatively large eye size in the two subfamilies, similarities may also exist in the size of the visual structures. However, visual brain parts have so far been measured only in Erinaceinae.

Compared with other Insectivora, all Erinaceidae have well developed olfactory structures and seem to have well represented visual structures. In most other brain parts, the two subfamilies differ markedly.

Echinosoricinae

The brains are among the largest in Insectivora and, relative to the average for insectivoran families and subfamilies (AvIF), they are on the average 1.41 times larger (1.41x AvIF). All five fundamental brain parts are clearly above AvIF, varying from 1.27x for OBL to 1.69x for MES (Fig. 101). In OBL, the deviations from the typical Insectivora vary from 0.68x AvIF for FCM to 2.17x for DCO (Fig. 103); in TEL, from 0.95x for STR to 1.83x for HIP, and with an extremely large size, 3.67x AvIF, for AOB (Fig. 107). No measurements are available so far for DIE components. In OBL, all vestibular and auditory components are especially large, whereas only FCM and TR are below the AvIF level. In TEL, all limbic and olfactory structures are well represented. The limbic components vary from 1.34x AvIF for SEP to 1.83x for HIP; the main olfactory centers are 1.39x AvIF for MOB and 1.53x for PAL. The smallest structure in Echinosoricinae, next to STR, is NEO which, however, is still very large in comparison to that of other Insectivora (1.30x AvIF) (see Fig. 107).

In conclusion, Echinosoricinae have large brains with well represented vestibular, auditory, olfactory, and limbic centers. Only some structures in the somatosensory system (FCM and TR) are below the average for families and subfamilies of Insectivora (AvIF).

Species characteristics of the echinosoricine brains are based on direct comparison between the two species so far investigated. It is not worth-while introducing an average echinosoricine level, since the larger size of a brain structure in one species is of course always associated with small size in the other and vice versa. Therefore, the brain characteristics will be described together.

Echinosorex gymnurus and ***Hylomys suillus:*** The total brain, i.e., brain weight or NET volume, is in the two species relatively very large. The five fundamental brain parts, nine TEL parts, about 20 OBL parts, and some periventricular structures were measured. Most brain parts are of similar size in the two species. The largest difference was found in the pineal gland (EPI), which was missing in *Echinosorex*, and, in *Hylomys,* was the second largest of all Insectivora. Large differences were also found in the accessory olfactory bulb (AOB), which is about 2.3 times larger in *Hylomys* than in *Echinosorex*, in the nucleus coeruleus (COE; 1.6x), in the the nucleus vestibularis superior (VS), nucleus fasciculus cuneatus externus (FCE), nucleus mot X, and the schizocortex (SCH) (all 1.4x larger in *Hylomys*). Larger size in *Echisosorex* was found in the dorsal cochlear nucleus (DCO; 1.6x), in the nucleus fascicularis gracilis (FGR; 1.35x), and in the sensory trigeminal complex (TR; 1.3x).

Particularities: *Hylomys* has a large pineal gland and an accessory olfactory bulb (AOB) which is more than twice as large as in *Echinosorex*. *Echinosorex* has no pineal gland.

Erinaceinae

Hedgehogs have smaller brains than the families and subfamilies of the Insectivora on the average (AvIF); NET is on the average 0.78x AvIF. As in Tenrecinae and Crocidurinae, all the five fundamental brain parts are below AvIF (Figs. 101 and 102). The parts are fairly similar, varying in a narrow range from 0.68x AvIF for CER to 0.87x for DIE. In OBL, the relative size of the components varies from 0.53x AvIF for TR to 1.52x for FCE and 1.43x for mot XII (Fig. 103); in DIE, from 0.38x for CGM to 1.43x for CGL (Fig. 105); and in TEL, from 0.68x for NEO to 1.07x for MOB (Fig. 107). Still larger is AOB (1.65x AvIF). Next to the trigeminal structures, in OBL some of the somatosensory centers (FGR and FCM), vestibular centers (VI), and motor nuclei (mot V), are less represented. In TEL, the olfactory structures are well represented whereas, next to NEO, STR (0.70x) and HIP (0.76x AvIF) are least represented. The low position of HIP is in clear contrast to the relations in Echinosoricinae.

In conclusion, hedgehogs have small brains with well represented mot XII, visual, and olfactory (main and accessory) structures. Less well represented are vestibular, auditory, motor, hippocampal, and neocortical centers.

Species characteristics of erinaceine brains are based on deviation from the average erinaceine level (AvEr), which is uniformly set at 1 for each of the structures.

Erinaceus europaeus: Among the three Erinaceinae, *Erinaceus europaeus* has the smallest brain (0.89x AvEr in BrW; 0.90x in NET). All of its five fundamental brain parts are also lowest. They vary from 0.83x AvEr in MES to 0.93x in CER. The OBL structures of *Erinaceus* differ widely in relative size: from 0.58x AvEr in OLS to 1.20x in INO. All the auditory structures (AUD) are lowest (0.68x AvEr); in ascending sequence follow VI (0.87x), REL (0.87x) and FGR (0.89x). The largest OBL components in *Erinaceus* are INO (1.20x AvGr) and mot X (1.15x). In DIE, EPI is large (1.26x). Of the main components, measured by Bauchot (1979b) in *Erinaceus* and *Atelerix*, the anterior thalamic nuclei (ANT) differ the most in the two species and are only two thirds in *Erinaceus* than in *Atelerix*. The TEL structures measured in all three species are all below the erinaceine level in *Erinaceus*. They vary from 0.79x AvEr in AOB to 0.98x in SCH. Lowest next to AOB are NEO (0.84x AvEr), AMY (0.87x), and STR (0.89x). In SEP, a very small SFB (0.66x AvEr) and a small SET (0.89x) are found. The *Erinaceus* brain is the shortest and flattest of erinaceine brains so far measured and it also has the shortest cerebral hemispheres and corpus callosum.

Particularities: *Erinaceus europaeus* has the smallest brain of the three erinaceine species, with very small auditory structures and small accessory olfactory bulb, anterior thalamic nuclei, and neocortex. The pineal gland is large.

Atelerix algirus: In size of brain and many structures, *Atelerix* is intermediate between *Erinaceus* (low) and *Hemiechinus* (high), but it is generally closer to *Erinaceus*. The total brain is 0.95x AvEr in BrW and NET. All five fundamental brain parts are intermediate. They vary from 0.92x AvEr in TEL to 1.00x in OBL. The OBL structures vary from 0.84x AvEr in INO and VL to 1.16x in DCO. VI is large, and this combined with the small VL may be due to uncertain delimitation. INO (0.84x AvEr) is also small, and TR (0.93x) and the MOT nuclei (0.95x) are relatively small. The auditory components are 1.5 times larger in *Atelerix* than in *Erinaceus*, and this is also true for the anterior thalamic nuclei (ANT). Of the diencephalic parts, HAM is relatively small (0.82x AvEr). The TEL parts have a relatively low variation, from 0.89x AvEr in SCH and SEP to 0.99x in HIP. MOB is smaller in *Atelerix* than in the two other species. The *Atelerix* brain is the longest and narrowest (most slender) of erinaceine brains and it also has the longest cerebral hemispheres.

Particularities: In *Atelerix algirus*, the size of the brain and most of its parts is intermediate between *Erinaceus* and *Hemiechinus*. Thus *Atelerix* best represents the erinaceine type in brain characteristics.

Hemiechinus auritus: Of the three Erinaceinae, *Hemiechinus* has the largest brain (1.16x AvEr in BrW and NET). The five fundamental brain parts, all TEL parts, and most parts of OBL are also largest. The greatest deviations are found for the septal SFB (1.59x AvEr), for OLS (1.44x), and for the auditory components as a whole (1.33x), for MES (1.26x), NEO (1.25x), AOB (1.25x), and for AMY (1.23x AvEr). Only a few small structures, such as EPI (0.76x) and the septal SET (0.81x AvEr), are considerably under erinaceine level. The *Hemiechinus* brain is the widest and highest of the erinaceine brains so far measured and it has the longest corpus callosum.

Particularities: *Hemiechinus auritus* has the largest brain in Erinaceinae with most parts largest and especially large auditory components. The pineal gland (EPI) is small.

In conclusion, the three Erinaceinae are characterized by species differences in brain macromorphology (see Section 3.1 and Table 3) and composition. *Atelerix* has the longest and narrowest brain, *Erinaceus* the shortest and flattest, and *Hemiechinus* the widest and highest. The total brain is largest in *Hemiechinus* and smallest in *Erinaceus*. The brain structures show narrow variation and are, in general, below the averages for Insectivora (AvIS and AvIF). Above these averages are the indices for visual and olfactory centers; below are those for auditory, vestibular, trigeminal, and motor structures. The largest differences in Erinaceinae are found in the auditory centers (AUD), which in *Hemiechinus* are on the average 1.94 times larger than in *Erinaceus* and 1.33 times larger than in *Atelerix*.

4.5 Talpidae

General Characteristics

The family of Talpidae includes groups of small rat- or mouse-sized animals which occur in temperate parts of Europe, Asia, and North America. Moles have elongated and cylindrical bodies, short necks, short limbs, minute eyes hidden in the pelage, and no or small external ears. Yates (1984) distinguished three subfamilies: Uropsilinae, Desmaninae, and Talpinae. Brain material is available only from Desmaninae and Talpinae.

Shrew moles (Uropsilinae) are more ambulatory than fossorial (Yates 1984). They lack fossorial or aquatic specializations and resemble shrews in their general form. They are forest and alpine inhabitants, living at elevations of up to 4500 meters. The tail is long. The long scaly snout is formed of two tubular nostrils close together. The ear pinnae extend beyond the fur; the forefeet are small and not adapted for burrowing.

Selected References: Grassé (1955a), Cockrum (1962), Walker (1975), Herter (1979), Yates (1984), Vaughan (1985), Fons (1988).

4.5.1 Desmaninae

General Characteristics

The semiaquatic Desmaninae have webbed feet. In addition, the greatly enlarged hindfeet bear a fringe of stiff hairs that increase their effectiveness as paddles. These animals have flexible snouts with an extremely well developed sense of touch. They live along the banks of lakes, ponds, or streams and feed largely on aquatic invertebrates. Their burrows open beneath the surface of the water and extend upward to a nest chamber above the water level.

Brain Characteristics

Desmana moschata, Galemys pyrenaicus

Brain Weight, Encephalization

Bauchot and Stephan (1966)
Stephan (1961)
Stephan and Bauchot (1959)

Brain and body weights of *Galemys pyrenaicus* have been discussed by Stephan and Bauchot (1959). Based on a large number of BoWs and two BrWs (1270 mg of a male, 1310 mg of a female), the authors arrived at standards of 57.5 g / 1290 mg. Three more brains were prepared in 1960 and the total of five BrWs were published by Stephan (1961). They resulted in an average of 1330 mg. The BoWs were given by Bauchot and Stephan (1966), and in that paper, the three only known pairs of data on *Desmana moschata* were also given (see Table 5).

The average encephalization of Desmaninae is 2.30x AvTn. This is, along with that of Echinosoricinae, the highest of all Insectivora. In *Galemys*, it is higher than in *Desmana* (2.55x versus 2.05x; Table 6).

Macromorphology

Bauchot and Stephan (1967)
Hochstetter (1942)
Kurepina (1974)
Stephan and Andy (1982)
Stephan and Bauchot (1959, 1968b)

Hochstetter (1942) compared the cerebral hemispheres of *Galemys* with those of *Talpa* and found them to be more compact. Detailed macromorphological descriptions with photoprints, drawings, midsagittal reconstructions, and maps of allocortical regions were given by Stephan and Bauchot on *Galemys* (1959) and *Desmana* (1968b).

Bauchot and Stephan (1967) and Stephan and Bauchot (1968b) compared endocast and brain of *Desmana*. Kurepina (1974), in a study on the cytoarchitecture of the neocortex, showed the brain of *Desmana* from dorsal and left side. Stephan and Andy (1982) summarized the results as follows: the border between NEO and piriform lobe is situated relatively far ventrally. The mesencephalic tectum is completely hidden and parts of CER are even covered by the TEL hemispheres. The trigeminal nerves are very large, while the visual nerves are small. Similarities with the brains of Talpinae were pointed out, but the brain of *Galemys* is distinctly higher. The dorsal surface arched upwards with a pronounced convexity. *Desmana* is in an intermediate position between *Talpa* and *Galemys* in this and many other characteristics, e.g., encephalization.

Cell and Area Measurements

Bauchot (1963)
Bauchot and Stephan (1961)
Stephan (1961)
Stephan and Bauchot (1959)
Stephan et al. (1984b)

All measurements were performed in *Galemys*. Bauchot (1963) measured cell size, number, and density in the various components of the diencephalon; Bauchot and Stephan (1961) measured areas of sectioned commissures and chiasma opticum in 13 species of Insectivora. They found in *Galemys* the largest corpus callosum (CC) by far and, next to *Potamogale*, the second smallest anterior commissure (COA). The optic nerves and chiasma were found to be small but distinctly larger than in *Talpa* (Stephan et al. 1984b).

Stephan and Bauchot (1959) gave data on area measurements and found a proportionally larger NEO in *Galemys* than in *Talpa* (41.4% versus 30.9%). In the allocortex, the olfactory centers were proportionally smaller (34.4% versus 43.0%). Calculation and comparison of indices (Stephan 1961) have shown that *Galemys* has the largest NEO of Insectivora (11 species measured) and, in the allocortex, a reduced MOB and PAL but well developed HIP and SCH.

Volume Measurements

Andy and Stephan (1966a)
Baron et al. (1983, 1987, 1988)
Bauchot (1961, 1963, 1966, 1970, 1979a, b)
Bauchot and Legait (1978)
Bauchot and Stephan (1968)
Frahm et al. (1982)
Legait et al. (1976)
Stephan (1966, 1967b, 1969, 1972)
Stephan and Andy (1962, 1964, 1969, 1982)
Stephan and Bauchot (1968a)
Stephan et al. (1970, 1981a, 1982, 1987a)

Many parts of the desman brain have been measured since 1961 by numerous authors and teams (Andy, Baron, Bauchot, Frahm, Legait, Stephan). All measurements before 1966 were performed in *Galemys*. Data on *Desmana* were first given by Stephan (1966) and Andy and Stephan (1966a).

Data on the five **fundamental brain parts** and the main parts of the telencephalon have been given by Stephan and Andy (1964, 1969, 1982), Stephan (1967b, 1972), Stephan and Bauchot (1968a), Bauchot and Stephan (1968), Stephan et al. (1970, 1981a) and Bauchot (1979a). Of the five fundamental brain parts, the largest size was found for DIE, which in Desmaninae is 3.10x AvTn. In descending sequence follow CER (2.74x), TEL (2.27x), MES (2.01x), and OBL (1.84x AvTn). Thus, all the main parts show a clear progression in Desmaninae relative to Tenrecinae (Table 12). Large size was found in Desmaninae for DIE and CER. For DIE, the distance from all other subfamilies and families is large and there is no overlap (Fig. 64), whereas for CER and TEL there is a distinct overlap with other taxonomic units (Figs. 63, 69). In MES, Desmaninae have the second highest position in the index scales next to Echinosoricinae, and in OBL, next to Potamogalinae. The largest differences between the two species were found for CER, which is 1.39 times larger in *Galemys* than in *Desmana* (319 versus 229) and for TEL (1.33 times; 260 versus 195; Table 11).

In the **medulla oblongata**, data on vestibular nuclei have been given by Baron et al. (1988). The size of the various vestibular nuclei is between 1.17x and 2.13x AvTn and is, thus, below the size of the desman NET brain, which is 2.3x AvTn (Table 7). The strongest progression in the OBL components was found for INO (2.93x AvTn) and TR (2.79x). The position of the desmans in the index scales is highest for INO (Fig. 58) and second highest, after Potamogalinae, for TR (Fig. 57). A regression was found for the auditory components (0.85x AvTn), and a low position is shown for DCO in Fig. 60. DCO has the fourth lowest position, next to Erinaceinae, *Talpa* and Soricinae.

In the **mesencephalon**, the subcommissural body (SCB) has been measured by Stephan (1969), and the volume data come from Stephan et al. (1981a). *Desmana* has a large SCB (about 4.4x AvTn). In *Galemys*, SCB is smaller but is still 2.6x AvTn (Table 41).

Measurements on the main parts of the **diencephalon** have been performed by Bauchot (1963, 1979b); on the thalamus (THA) by Bauchot (1966, 1979b); on the epiphysis (EPI) by Stephan (1969), Legait et al. (1976), and Stephan et al. (1981a); and on the hypophysis (HYP) by Bauchot (1961, 1970) and Bauchot and Legait (1978).

Among the main diencephalic parts, Bauchot found an especially large subthalamus (4.94x AvTn) and in the thalamus, very large DOT (7.49x AvTn), CGM (5.15x), MLT (4.20x), and MNT (3.92x). All these components are larger in *Galemys* than in *Desmana*, and the size for DOT in *Galemys* (8.99x AvTn) is the highest found in all the structures and species. EPI is smaller than in Tenrecinae (0.69x AvTn), HYP is much larger in *Galemys* than in *Desmana* and has a large posterior lobe. In *Desmana*, the intermediate lobe is quite large (Bauchot and Legait 1978).

In the **telencephalon**, seven to nine parts have been measured and compared (references: see above under five fundamental brain parts). Additional data were given on olfactory structures by Stephan (1966), Stephan et al. (1982), and Baron et al. (1983, 1987); on septum (SEP) by Stephan and Andy (1962, 1982) and Andy and Stephan (1966a); on amygdala (AMY) by Stephan et al. (1987a); and on neocortex (NEO) by Frahm et al. (1982).

The largest size by far was found for NEO (6.45x AvTn), along with a large size for STR (3.48x AvTn). For NEO and STR, the desmans have the highest positions in the index scales (Figs. 92, 79), in NEO with clear distance and no overlap with the other taxonomic units of Insectivora. The lowest size in the TEL parts of desmans were found for olfactory structures, such as NTO (0.16x AvTn), MOB (0.60x), AOB (0.78x), and PAL (0.82x).

Histology

Herlant (1964)
Kurepina (1974)
Peyre and Herlant (1961)
Rehkämper (1981)
Stephan (1965)
Stephan and Andy (1982)

Many of the papers on volume measurements provide details and/or drawings of histology. Additional information is given by Peyre and Herlant (1961), Herlant (1964), Stephan (1965), Kurepina (1974), Rehkämper (1981), and Stephan and Andy (1982). Peyre and Herlant (1961) and Herlant (1964) described seasonal changes in the hypophysis (HYP) of *Galemys*. In parallel with a long phase of reproductive activity in the two sexes, the γ or LH cells persist longer in a functional state than in other mammals. The characteristics of the γ cells are similar to those of the mole, where they have received much more attention. In *Galemys*, they are strongly reactive to PAS.

Stephan (1965) studied the accessory olfactory bulb (AOB) in *Galemys* and *Desmana*. It was found to be less differentiated in *Desmana*, where it has few internal granular cells and no clear lamination, i.e., diffuse cell distribution. It is better developed in *Galemys*, where the vomeronasal system was supposed by Richard (1985) to be important in the chemoreception of liquids. Stephan and Andy (1982) labeled the various nuclei of the septum in a transverse section through the brain of *Desmana* and showed the well differentiated neocortex (NEO) in *Galemys* at the level of the COA.

Kurepina (1974) compared the cytoarchitecture of NEO in *Desmana* to that of *Talpa* and *Sorex* and found it to be characterized by a dispersed cell arrangement, indicating greater development of interneuronal structures. Areas in central and temporal regions were more differentiated than in *Talpa* and *Sorex*. Other characteristics were better areal differentiation, smaller claustrum and cortical insular region, and better stratification.

Rehkämper (1981) also investigated *Desmana*. In comparison to most other Insectivora (eight species), he found a thinner molecular layer (I), a less dense layer II, which more closely resembles layer III, and a granular layer throughout the NEO. The granular layer was best represented in the lateral field A 4 of NEO, where it is the most prominent layer. A 4 is a very large area thought to be related to the trigeminal system. It is separated from the more medial fields (A 2 rostrally and A 3 caudally) by transitional cortices. All characteristics of NEO suggest progressive differentiation.

Sense Organs

Argaud (1944)
Bauchot et al. (1973)
Richard (1985)

The proboscis of *Galemys* has been investigated by Argaud (1944), Bauchot et al. (1973), and Richard (1985). It contains two types of touch receptors: Eimer's organs around the nostrils and, farther back, on each side, 80 - 100 vibrissae, 4 - 40 mm long. Each of the vibrissae is innervated by three fascicles of trigeminus fibers. More than three quarters of a transversal section through the snout is filled with efferent nerves of the vibrissae. Richard believes the vomeronasal system (VNS) is of great importance for *Galemys*. The size of AOB, which is the first central nervous center of VNS, is above the tenrecine level in *Galemys*, but distinctly smaller in *Desmana* (120 versus 35, Table 50).

4.5.2 Talpinae

General Characteristics

Most true moles (Talpinae) are fossorial or semifossorial. *Condylura* is well adapted for fossorial as well as for aquatic life. In fossorial species, the head is tapered, and the eyes and pinnae are reduced. The forelimbs are short and very broad, have strong claws for digging, and are permanently turned outward from the body. The clavicle and humerus are unusually short and robust. True moles occur in lowland marshes, subalpine forests, meadows, and cultivated areas. All prefer moist and friable soils. Most species construct deep permanent tunnels which include a central nest chamber and shallow subsurface runways. They feed on various small animals including earthworms, adult insects, and larvae.

Brain Characteristics

Talpa europaea, T. micrura, Parascalops breweri, Scalopus aquaticus

Brain Weight, Encephalization

Bauchot and Stephan (1966)
Portmann (1962)
Slifer (1924)
Stephan (1959)
Stephan and Andy (1982)
Stephan et al. (1970)
Welcker and Brandt (1903)
Ziehen (1899a)

Pairs of BoW/BrWs from *Talpa europaea* came from several sources. One pair was obtained from each of the Ziehen, Welcker and Brandt, Slifer, and Portmann reports. Stephan gave six pairs and averages for many additional BoWs. These latter data indicated sex differences, with males heavier than females (84 versus 68 g). Median values of 76 g for BoW and 1020 mg for BrW were obtained and confirmed by Bauchot and Stephan, Stephan et al., and Stephan and Andy. The somewhat different data given in Tables 5 and 6 were obtained by exclusion of two subadult animals and addition of BoW/BrW pairs for two adult animals.

Data from the three other species are so far unpublished and will be given here:

(1) In *Talpa micrura*, only two BoW/BrW pairs were obtained, from animals collected in Thailand. They are 41.4/816 from a male and 34.0/747 from a female. The second animal was subadult and is not included in Table 5, but even the first one may not be fully grown. However, we did not find data on this species in the literature.

(2) In *Parascalops breweri,* six pairs were obtained, from animals collected in Quebec. They are 54.9/826, 53.2/881, 48.0/885, and 55.4/929 for males, 56.9/855 for a female, and 54.1/905 for a specimen of unknown sex. The averages of these values gave the standards (53.8/880) presented in Table 5 .

(3) In *Scalopus aquaticus*, only one pair of BoW/BrW data was obtained. The data are from a full-grown male collected near Jackson (Miss.) with a BoW of 115 g and a BrW of 1310 mg.

The encephalization of moles is on the average 1.76 times larger than in Tenrecinae (Table 7), with no large differences between the species. The one specimen of *Talpa micrura* has somewhat higher values, which also suggests that it was not fully grown. In total, the encephalization of moles is above the averages found for Insectivora which is 1.48 for AvIS and 1.58 for AvIF.

The brain weight relative to spinal cord (SPC) weight was given as 4.7 times by Ziehen (1899a) in *Talpa europaea*, which is similar to that of *Sorex aruneus* (4.9) and distinctly higher than in *Erinaceus europaeus* (3.5). In comparison, man's brain is about 43 times larger than its SPC. The data were republished in a comparative study by Krompecher and Lipak (1966).

Macromorphology

Akert et al. (1961, SFB, *Scalopus*)
Bolk (1906, CER)
Bradley (1903, CER)
Brauer and Schober (1970)
Clark (1932, *Scalopus, Talpa*)
Ganser (1882)
Gudden (1870, optic nerve)
Haller (1909a, TEL)
Heape (1887, morphogenesis)
Hochstetter (1923, EPI morphogenesis; 1924, HYP morphogenesis; 1942, meninges)
Igarashi and Kamiya (1972, *Mogera, Euroscaptor*, small atlas)
Ingvar (1918 CER)
Johnson et al. (1982, *Scalopus, Condylura*)
Krabbe (1942, morphogenesis)
Larsell (1934, CER morphogenesis, *Scapanus*)
Larsell and Dow (1935, CER morphogenesis, *Scapanus*)
Leche (1905, 1907)
Smith (1902b)
Spatz and Stephan (1961)
Stephan (1956a)
Stephan and Andy (1982)
Stephan and Bauchot (1959, 1968b)
Turner (1891)
Ziehen (1897a)
Zuckerkandl (1887, HIP)

The vast majority of papers deal with general brain characteristics of *Talpa europaea*. Other species and the topics in more specialized studies are given in the reference list. Strongly reduced optic nerves were already mentioned by Gudden (1870). Detailed macromorphological descriptions and illustrations of the *Talpa* brain were given in the classical paper of Ganser (1882).

Most authors compared the brains of moles with those of other Insectivora and/or mammalian forms and pointed out that the very flat brains have: (1) a smooth NEO except for a very shallow and ill-defined suppression on the upper surface near the frontal extremity ("orbital sulcus"), (2) a sulcus rhinalis in at least the rostral third of the hemispheres, (3) a long and slender corpus callosum, with SFB far behind, (4) a completely or nearly completely hidden mesencephalic tectum, (5) very small visual nerves and poorly developed visual structures, (6) a large trigeminal nerve, and (7) a small pons. According to Clark (1932), the flocculus is not a definite "lobule" but merely a narrow, compressed folium. With minor differences, the characteristics given are valid for all species so far investigated. Thus, the brains of moles are relatively uniform.

The data on *Talpa micrura, Parascalops breweri* and *Scalopus aquaticus* in this book are new and consistent with the characteristics given. Proportion indices are compared in Section 3.1 and listed in Tables 3 - 4. Based on these data, *Scalopus* has the flattest and broadest brain, which is also long but with relatively short hemispheres.

Macromorphological details of the spinal cord (SPC) were given by Malinska et al. (1976).

Cell and Area Measurements

Bauchot (1963, 1964)
Bauchot and Stephan (1961)
Bishop et al. (1971, *Scalopus*)
Brodmann (1913)
Brouwer (1915, SPC)
Galert (1986)
Gierlich (1916a, b, c)
Grünwald (1903)
Haug (1987)
Linowiecki (1914, *Scalopus*)
Malinska et al. (1976, SPC)
Stephan (1956a, b, 1961)
Stephan and Bauchot (1959)
Stolzenburg et al. (1989)
VanValkenburg (1911b)
Vysinskaja (1961, CGL, Area 17)
Wiedemeyer (1974)

Most measurements were performed on *Talpa europaea*, and only a few on *Scalopus aquaticus* (see reference list).

In the spinal cord, Malinska et al. (1976) measured the transversally sectioned areas of individual segments and their grey and white matters. The grey matter is about 40% in rostral segments and grows to even more than 70% in caudal segments. About 23% of the total white matter of the first cervical segment belongs to the funiculus posterior (Brouwer 1915). For comparison, the figure in man is about 39%.

Quantitative cell and area measurements in the brain may include: (1) fiber thickness and areas of transversally sectioned fiber bundles, (2) cell size, number, and density, (3) cortical thickness and (4) cortical surface areas.

(1) In the pyramidal tract of *Scalopus*, Linowiecki (1914) determined the largest medullated fibers to be 4.84 µm, which is relatively small compared with thickness in other small mammalian species. Linowiecki pointed out that the pyramidal tract is almost completely non-medullated. In it and in the Vth nerve, Bishop et al. (1971, *Scalopus*) measured the sheath thickness of myelinated nerve fibers and plotted them against axon diameters. Grünwald (1903) compared the size of the pedunculi cerebellares in 24 mammalian species including *Talpa europaea*.

Gierlich (1916a, b, c) measured transversally sectioned areas of the cerebral peduncle, its bundles coming from the prosencephalon, and the pyramids shortly before crossing. In *Talpa*, he found 3.9% of the peduncles come from the prosencephalon (for comparison: 2.5% in *Erinaceus*, 19.6% in man); 89.7% of the peduncle goes to the cerebellum and 10.3% to the spinal cord. Bauchot and Stephan (1961) compared midsagittally sectioned areas of the optic chiasma, corpus callosum, and anterior commissure. In 13 species of Insectivora, the chiasma was smallest, and the corpus callosum was larger in *Talpa* than in the Tenrecinae and Erinaceinae.

(2) VanValkenburg (1911b) counted the cells in the mesencephalic trigeminal nucleus, and Vysinskaja (1961), in CGL and area striata. A greater variety of cell characteristics (number, density, size) was determined by Bauchot (1963, 1964) for many diencephalic grisea, and by Wiedemeyer (1974), Haug (1987), and Stolzenburg et al. (1989) for cortical areas. Galert (1986) investigated size and cell types of the supraoptic and paraventricular nuclei of *Talpa*.

(3) Cortical thickness was determined by Stolzenburg et al. (1989).

(4) Cortical surface areas were determined by Brodmann (1913), Stephan (1956a, b, 1961), Stephan and Bauchot (1959), Wiedemeyer (1974), and Haug (1987). Brodmann found a surface area of 307 mm^2 for a hemisphere; Stephan, 306 mm^2; and Wiedemeyer, 263 mm^2. According to Stephan (1956a), the total surface area is composed of 43.2% paleocortex, 25.9% archicortex (hippocampus), and 30.9% neocortex. The percentage of NEO was found to be relatively high and that of PAL relatively low compared with hedgehog and terrestrial shrews. The figures of Wiedemeyer (1974) are 39.1% for PAL, 32.4% for ARCH, and 28.5% for NEO.

Volume Measurements

Andy and Stephan (1966a, SEP)
Artjuchina (1952 or 1962, RUB)
Baron et al. (1983, MOB; 1987, PAL; 1988, VES)
Bauchot (1961, 1970, HYP; 1963, 1966, 1979a, b, DIE)
Bauchot and Legait (1978, HYP)
Bauchot and Stephan (1968)
Frahm et al. (1982, NEO)
Ivlieva (1973, VES)
Legait et al. (1976, EPI)
Malinska et al. (1976, SPC)
Stephan (1966, 1967b, 1972; 1969, PVO)
Stephan and Andy (1962, SEP; 1964, 1969, 1982)
Stephan and Bauchot (1968a)
Stephan et al. (1970, 1981a, 1982 AOB; 1984b VIS; 1987a AMY)
Wiedemeyer (1974 TEL)
Wirz (1950)

All previous measurements on Talpinae were made in *Talpa europaea*. Newly added in this book are so far unpublished data on *Talpa micrura, Parascalops breweri,* and *Scalopus aquaticus*.

Wirz (1950) subdivided the brain macroscopically into five parts (neocortex, basal ganglia, rhinencephalon, cerebellum and "stammrest") and compared their weights by index methods, but did not publish the raw data. Ziehen (1899a) and Krompecher and Lipak (1966) related total brain weight to spinal cord weight.

Volumes of the five **fundamental brain parts** obtained from serial sections of *Talpa europaea* and/or indices of comparison were given by Stephan and Andy (1964, 1969, 1982), Stephan (1967b, 1972), Bauchot and Stephan (1968), Stephan and Bauchot (1968a), and Stephan et al. (1970, 1981a) and compared with other Insectivora and with Primates. In Insectivora, *Talpa europaea* was found to be a slightly progressive species with a large CER and a small OBL. These results were fully confirmed after inclusion of *Talpa micrura, Parascalops breweri,* and *Scalopus aquaticus*. The largest average indices were found for CER, which is 2.37x larger in Talpinae than in the average Tenrecinae (2.37x AvTn). Relatively large indices were also found for DIE (2.14x AvTn). The lowest indices were found for OBL (1.16x AvTn) and MES (1.34x AvTn). In size of the main parts of the brain, all the 4 species were, in general, similar. The largest difference was found in MES, which is 1.46x larger in *Talpa micrura* than in *Parascalops breweri*.

In the **spinal cord** (SPC) Malinska et al. (1976) measured volumes of grey and white matter: 43.3% of SPC was found to be grey matter. This is similar to the percentage in the hedgehog and much higher than in other mammals.

In the **medulla oblongata** (OBL) of *Talpa europaea,* volumes of vestibular nuclei were given by Ivlieva (1973) and Baron et al. (1988). Baron found the vestibular indices to be smaller than in the average Insectivora. Data on other OBL parts of *Talpa europaea* are published here for the first time (Tables 13 - 27). They show that in the mole the funicular nuclei (FUN), especially FCM, are very large (3.20x AvTn); FCE (1.66x) and INO (2.55x) are also large, whereas many other OBL nuclei are relatively small, such as OLS (0.76x AvTn), mot XII (0.72x), VL (0.64x), and VCO (0.56x).

In the **mesencephalon** (MES) comparisons were made by Artjuchina (1952 or 1962) on the nucleus ruber (RUB), and by Stephan (1969) and Stephan et al. (1981a) on the subcommissural body (SCB). Artjuchina found that RUB is 2.6% relative to MES (1.9% in the hedgehog). SCB was measured in *Talpa europaea* and *Scalopus aquaticus* and found to be well differentiated (see Fig. 61) and very large (6.30x AvTn), especially in *Scalopus.*

In the **diencephalon** (DIE), data on main parts are given by Bauchot (1963, 1979b) and Bauchot and Stephan (1968) for *Talpa europaea.* The subthalamus (STH) was found to be relatively large (3.34x AvTn). In the epithalamus (ETH) of the four species (Table 42), the medial habenular nucleus (HAM) is relative large (1.80x), whereas the epiphysis (EPI) is small (0.40x). These results confirm data given for *Talpa* by Stephan (1969), Legait et al. (1976), and Stephan et al. (1981a).

In the thalamus (THA) of *Talpa,* Bauchot (1966, 1979b) found many of the nuclei to be relatively large: DOT (3.51x AvTn), CGM (3.05x), MNT (2.70x), MLT (2.34x), and ANT (1.98x AvTn), but only MNT is markedly larger than the family/subfamily average (1.23x AvIF). VET (1.02x AvTn) and CGL (0.57x) have small indices, which for CGL may be due to the very low size of eyes and optic nerves, which were noted in many previous papers and quantitatively compared by Stephan et al. (1984b). The hypophysis (HYP), especially the anterior lobe, was found to be the smallest in any Insectivora by Bauchot (1961, 1970) and Bauchot and Legait (1978).

In the **telencephalon** (TEL) seven to nine main parts were measured in *Talpa europaea* by Stephan and Andy (1964, 1969, 1982), Stephan (1967b), Bauchot and Stephan (1968), Stephan and Bauchot (1968a), and Stephan et al. (1970, 1981a). Additional data were given for cortex and striatum by Wiedemeyer (1974); for olfactory structures by Stephan (1966), Stephan et al. (1982), and Baron et al. (1983, 1987); for amygdala (AMY) by Stephan et al. (1987a); for septum (SEP) by Stephan and Andy (1962, 1982), Andy and Stephan (1966a), Stephan (1969), and Stephan et al. (1981a); and for neocortex (NEO) by Frahm et al. (1982).

The data on the main parts are complemented in this book by data from three more talpid species. The highest average indices in the four species were found for NEO (3.96x AvTn), STR (2.95x), and SCH (2.37x), and the lowest for AOB (0.80x AvTn) and MOB (0.90x). In the septum, a relatively large triangular nucleus (SET) was found (1.92x) and a relatively small subfornical body (SFB) (0.51x). The variation in the four species is low, except for some very small structures, e.g., AOB, whose index is 3.5 times larger in *Talpa europaea* than in *Scalopus aquaticus*, and SFB, whose index is 2-3 times smaller in *Talpa europaea* than in all other species.

Details on histology are given in many of the papers on volume measurements.

Histology, Histochemistry

Total Brain, General

Avksentieva (1975)
Clark (1932)
Ganser (1882)
Igarashi and Kamiya (1972)
Stieda (1870)
Ziehen (1897a)

As early as 1870, Stieda gave some details of the histology of the brain of *Talpa europaea*. A few years later, a far more detailed report was given by Ganser (1882), who described mesencephalic and diencephalic, but mainly telencephalic structures, such as striatum, hippocampus, amygdala, septum, piriform lobe, olfactory tubercles, tracts and bulbs, and cerebral cortex with white and grey matter. He differentiated five to seven layers in the neocortex, and four layers in the olfactory bulb. An atlas-like illustration of about 20 transverse sections clearly showed the distribution of cells and fibers. A similar illustration of six sections by Igarashi and Kamiya (1972) on the brain of *Mogera wogura* is less informative. Avksentieva (1975) compared the chromatin pattern of the nuclei of granular cells and oligodendrogliocytes in cerebellum and cerebral hemispheres. Some additional histological details can be obtained from the mainly macromorphological studies by Ziehen (1897a) and Clark (1932). Ziehen compared the *Talpa* brain with those of Monotremata and Marsupialia, and Clark pointed out that the thalamus of *Talpa* is more differentiated than that of the Erinaceidae, and that the neocortex is hardly differentiated at all.

Spinal Cord

Biach (1906, canalis centralis)
Dräseke (1904, SPC, pyramids)
Leszlenyi (1912, zona marginalis)
Sabbath (1909, ventral roots)
Sano (1909, substantia gelatinosa)
Schilder (1910, nuclei spinales)
Takahashi (1913, lateral horn)
Verhaart (1970, SPC)

The most detailed descriptions of the spinal cord (SPC) of *Talpa europaea* were given by Dräseke (1904) and Verhaart (1970). Dräseke noted strongly developed ventral horns and relatively small pyramids running in the ventral funiculus. He presented some transverse sections stained for fibers with Weigert methods. According to Verhaart, the pyramidal tract can be followed throughout the cervical cord, where it decreases in size and cannot be recognized more caudally. Sano (1909) described the dorsal substantia gelatinosa and found it well developed with relatively few less differentiated cells. Biach (1906) found a very small lumen in the canalis centralis and a one-layered ependyma. Leszlenyi (1912) reported few fibers in the zona marginalis, and Takahashi mentioned that the lateral horn does not exist in the upper cervical part and is never very clear either in the more caudal parts. Sabbath and Schilder mentioned *Talpa* only in comparison with other mammalian brains and gave no details.

Brain Stem and Cerebellum

Aitkin et al. (1982, auditory structures)
Alexander (1931, tractus tegmentalis centralis)
Bauer (1909a, substantia nigra)
Bauer-Jokl (1919, subcommissural body)
Brunner (1919, cerebellar nuclei)
Forel (1877, tractus tegmentalis, "Haubenregion")
Frankl-Hochwart (1902, colliculus superior)
Fuse (1912, nucleus abducens)
Galert (1985, substantia nigra)
Godlowski (1930, nucleus vestibularis triangularis)
Grünwald (1903, cerebellar peduncles)
Hatschek (1904, medial lemniscus, pons; 1907, nucleus ruber)
Heffner and Masterton (1975, pyramidal tract, *Scalopus*)
Hofmann (1908, superior olive)
Hosaka (1987, facialis)
Hulles (1907, trigeminal root)
Johnson (1954, colliculi, *Scalopus*)
Johnston (1909, mes V)
Kaplan (1913, vestibular nuclei)
Krabbe (1925, subcommiss. body)
Kudo et al. (1989, auditory structures, *Mogera*)
Larsell (1934, cerebellar ontogenesis)
Linowiecki (1914, pyramidal tract, *Scalopus*)
Löwy (1916, flocculus cerebelli)
Marsden and Rowland (1965, pons, olive, pyramid)

Nakamura (1930a, nucleus ambiguus Roller; 1930b, nucleus abducens)
Pekelsky (1922, reticularis, raphe)
Petrovicky (1971, Gudden's tegmental nuclei)
Petrovicky and Kolesarova (1989, parabrachial nuclear complex)
Sato (1977, colliculus superior, *Mogera, Urotrichus*)
Sheehan (1933, nuclei related to mesencephalic aqueduct)
Shima (1909, dorsal vagus nucleus)
Skrzypiec and Jastrzebski (1980, cerebellar nuclei)
Takagi (1925, nucleus vestibularis triangularis)
Tartuferi (1878, colliculus superior)
Valeton (1908, colliculus inferior)
VanValkenburg (1911b, mesV)
Verhaart (1970, OBL)
Warkany (1924, substantia nigra)
Weidenreich (1899, cerebellar nuclei)
Weiss (1916, facialis)
Williams (1909, inferior olive)

The caudal brain parts of *Talpa europaea* were mentioned and compared mainly with those of other mammalian species between 1902 and 1933 in many papers published by the Neurological Institute of the University of Vienna (Alexander, Bauer, Bauer-Jokl, Brunner, Frankl-Hochwart, Godlowski, Grünwald, Hatschek, Hofmann, Hulles, Kaplan, Löwy, Nakamura, Pekelsky, Sheehan, Shima, Takagi, Valeton, Warkany, Weiss, Williams). The topics of the publications are given in the reference list. In general, the structures were merely mentioned or very briefly described. More detailed descriptions were given of the nucleus ruber by Hatschek (1907) and of the superior olive by Hofmann (1908). Verhaart (1970) gave a detailed description of OBL components with many figures, tables, and size estimations. Like many early authors (since Ganser 1882), he stated that there did not seem to be any oculomotor nuclei and nerves.

Johnston (1909) and VanValkenburg (1911b) described mesencephalic trigeminal structures, and Aitkin et al. (1982), and Kudo et al. (1989, *Mogera*) auditory structures. Kudo et al. stressed a bilateral projection from the medial nucleus of the superior olive to the inferior colliculus, and a great development of the nuclei of the lateral lemniscus. Marsden and Rowland (1965) and Heffner and Masterton (1975, *Scalopus*) compared the pyramidal tract of moles with that of other mammals, but gave no details. Petrovicky (1971) and Petrovicky and Kolesarova (1989) included *Talpa europaea* in comparative cytoarchitectural studies on Gudden's tegmental nuclei and parabrachial nuclei.

Hosaka (1987) studied the proprioceptive afferent fibers running in the facial nerve and speculated on the existence of a monosynaptic reflex system between the snout muscles and neck muscles. Galert (1985) gave some details of the substantia nigra, and Johnson (1954, *Scalopus*) gave a detailed description of the mesencephalic tectum. He found a decrease in the visual structures and excellent development of the auditory system. Tartuferi (1878) gave a detailed description of the superior colliculus of *Talpa europaea* and Sato (1977), of *Mogera* and *Urotrichus*.

Larsell (1934) gave a few histological details of the ontogenetic development of the **cerebellum** in *Scapanus townsendi*, and Weidenreich (1899) and Skrzypiec and Jastrzebski (1980) described the cerebellar nuclei in *Talpa*.

Diencephalon

Azzali (1954, HYP)
Bauchot (1959)
Campbell (1972, VIS, *Scalopus*)
Forel (1872, 1877)
Galert (1986)
Girod (1976, HYP)
Green (1951, HYP, *Scalopus*)
Hanström (1946, 1966, HYP)
Herlant (1959a, b, c, 1964, HYP)
Herlant and Klastersky (1961a, b, HYP)
Hochstetter (1923, EPI ontogenesis)
Holmes (1966, HYP)
Johnson (1954, CGL, *Scalopus*)
Kemali (1984, HAB)
Környey (1928, CGL)
Koikegami (1938, HTH, *Mogera*)
Krawczuk (1983, HAB)
Lund and Lund (1965, 1966, VIS)
Pevet et al. (1981, EPI)
Pieters and Herlant (1972, HYP)
Pokorny (1926, HYP)
Sachs (1909, THA)
Sato (1977, VIS, *Mogera, Urotrichus*)
Spiegel and Zweig (1919, tuber cinereum)
Warkany (1924, PALL)
Werkman (1913, ontogenesis of commissural plate)

Nearly all the studies on the **diencephalon** focused on *Talpa europaea*, only a few on *Scalopus, Mogera,* and *Urotrichus*. (These are noted in the reference list). Werkman (1913) gave a detailed description of the ontogenesis of the prosencephalic commissural plate. The most complete description by far of the diencephalic structures was given by Bauchot (1959). His paper includes excellent photoprints of transversal and sagittal sections in which borders between the various structures are given.

Bauchot called attention to the reduction of the optic tract and centers, the weak development of the supraoptic nucleus and the importance and complexity of the ventral nuclear group of the thalamus. This, however, seems to apply to all Insectivora, since in the VET measurements (Tables 69, 70), *Talpa,* with a VET index of 102, is just at the level of the Tenrecinae (100) and slightly below the average of the insectivoran families and subfamilies (AvIF = 110).

In the **epithalamus** (ETH) of *Talpa,* there are studies on the epiphysis (EPI) by Hochstetter (1923, ontogenesis) and Pevet et al. (1981, histochemistry), and on the habenula by Krawczuk (1983, topography and cytoarchitectonics) and Kemali (1984, histological differences between right and left side). In the **thalamus** (THA), there are, besides the fundamental work by Bauchot, some short descriptions given by Sachs (1909), Környey (1928) and Johnson (1954), and more detailed studies on visual structures by Forel (1872, 1877), Lund and Lund (1965, 1966), Campbell (1972), and Sato (1977).

Forel mentioned the absence of a pulvinar and strong reduction of CGL. Sato described and illustrated the CGL, and found its development in the semi-fossorial *Urotrichus* to be intermediate between that in shrews and in the fossorial *Mogera*. Lund and Lund found the pretectum to be the principal primary optic center. Degeneration after eye removal was also found in the ventral CGL, but not in the superior colliculus. There is no accessory optic system. In *Scalopus*, Campbell found entirely crossed projections from the retina to the dorsal CGL. This is in contrast to Lund and Lund's findings in *Talpa*.

In the **subthalamus** (STH), Warkany (1924) briefly mentioned *Talpa* in his study of the pallidum and its relations to the substantia nigra. In the **hypothalamus** (HTH), Spiegel and Zweig (1919) mentioned the tuber cinereum of *Erinaceus* and found similar relations in *Talpa*. A detailed description of the hypothalamic centers in *Mogera* was given by Koikegami (1938). The supraoptic and paraventricular nuclei of *Talpa* were investigated by Galert (1986). The **hypophysis** was studied or mentioned in a great number of papers (see reference list). According to Hanström (1946), the pars anterior of *Talpa* seems to contain a larger number of acidophiles but less colloid formation than in *Sorex* and *Erinaceus*. Studies of HYP during reproductive activity were carried out by Herlant (1959a, b, c, 1964, cytology), Herlant and Klastersky (1961, electron microscopy), Pieters and Herlant (1972, prolactin-containing cells in male moles), and Girod (1976, comparative study on histochemistry of the adenohypophysis). According to Holmes (1966), the mole HYP corresponds to the general mammalian pattern, and adaptive changes in direct response to its underground existence are not apparent. Green (1951), too, in his short description of the HYP in *Scalopus* found no unusual features.

Telencephalon

Akert et al. (1961, SFB, *Scalopus*)
Beccari (1910, TOL)
Brodmann (1906, cortex)
Crosby and Humphrey (1939b, MOB, AOB, RB, *Scalopus*)
Ferrer (1986, NEO)
Ganser (1882)
Hermanides and Köppen (1903, NEO)
Johnson (1957, olf. centers, *Scalopus*)
Johnston (1913, septum, hippocampus, commissures, *Scalopus*)
Kryspin-Exner (1922, substantia perforata anterior)
Kurepina (1974, NEO)
Misek (1989, 1990b, NEO ontogenesis
Mott (1907, NEO)
Rehkämper (1981, NEO)
Rose (1912, 1929, cortex)
Sato (1977, NEO, *Mogera, Urotrichus*)
Spiegel (1919, subcortical grisea)
Stephan (1965, AOB)
Stephan and Andy (1982, SEP)
Switzer et al. (1980, AOB)
Szteyn et al. (1987, striatum, pallidum)

Watson (1907, NEO)

Werkman (1913, commissural plate ontogenesis)

The most extensive description of telencephalic histology was given by Ganser (1882). Additional data on **non-neocortical structures** were given for *Talpa* by Beccari (1910), Spiegel (1919), Kryspin-Exner (1922), Stephan (1965), and Stephan and Andy (1982), and for *Scalopus* by Johnston (1913), Crosby and Humphrey (1939b), Johnson (1957), and Akert et al. (1961).

The accessory olfactory bulb was found to be well developed in *Talpa* (Stephan 1965) but less so in *Scalopus* (Crosby and Humphrey, Johnson). This is fully confirmed by the quantitative data given in Table 50, which show that the AOB in *Talpa europaea* is more than three times larger than in *Scalopus aquaticus*. In the internal granular layer of the *Talpa* AOB, two types of granular cells were found: small diffuse cells in an upper zone and larger, paler cells in a lower zone that are arranged in flat clumps. The septum is well differentiated (Stephan and Andy 1982). The part between corpus callosum and anterior commissure is flat and broad. The septal pars dorsalis extends in contact with the very long corpus callosum far caudally. This corresponds with a far caudally positioned subfornical body (SFB) (Akert et al. 1961).

All investigations on the **neocortex** of moles (except those of Sato 1977) were done in *Talpa europaea*. Areal subdivisions were given by Hermanides and Köppen (1903), Mott (1907), Watson (1907), Rose (1912, 1929), Kurepina (1974), and Rehkämper (1981). Histogenesis and pre- and postnatal development was studied by Misek (1989, 1990b).

Hermanides and Köppen (1903) subdivided the cortex into four areas: a large rostral, a caudodorsal, a caudoventral (facing the mesencephalic tectum) and the piriform lobe. The last one is paleocortex and entorhinal cortex (periarchicortex), the third may be retrosplenial cortex but was assumed by the authors to be visual cortex. Its determination as visual cortex was already doubted by Brodmann (1906), who generally found the mammalian visual cortex in caudodorsal regions. Brodmann believed it unjustified to separate a circumscribed visual cortex in the mole since the tectonic characters are blurred.

Mott (1907) and Watson (1907) also determined a caudoventral area to be visual cortex. The region under consideration was the site of the presubicular cortex (Brodmann's 27 and/or 48) and was separated as such (see Fig. 100) by Rose (1912). Rose was the first to give a complete and detailed map of the cerebral hemispheres of *Talpa* using Brodmann's terminology. Based on cytoarchitectonics, Rose subdivided six types in the neocortex, about nine types in transitional cortices, and seven types in olfactory regions. Most of the regions were well illustrated.

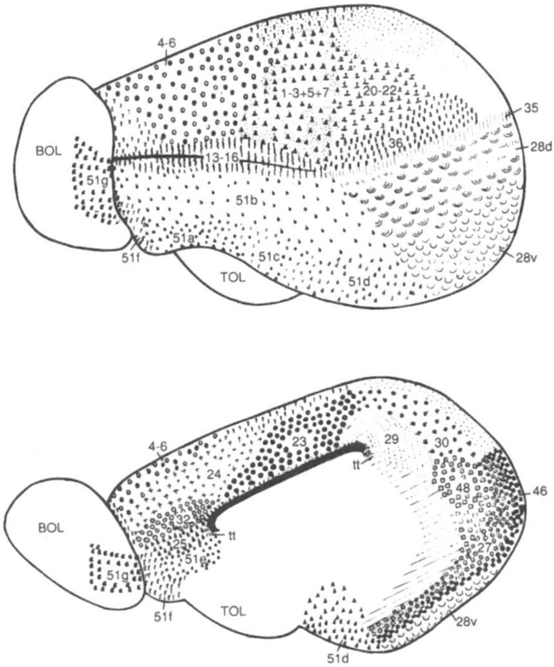

Figure 100: Rose's map of the cerebral cortex of *Talpa europaea* based on cytoarchitecture (1912).

In the true neocortex, Rose combined postcentral + parietal regions (areas 1-3+5+7), gigantopyramidal + frontal agranular regions (areas 4+6), a common insular region (areas 13-16), and a common temporal region (areas 20-22). Furthermore, Rose differentiated an area limbica anterior (area 24) and an area limbica posterior (area 23) in the medial wall. In the transitional cortices, Rose named the following areas: praegenualis (25), praesubicularis (27), entorhinalis interna (28a; 28v in Fig. 100) and externa (28b; 28d in Fig. 100), retrosplenialis granularis (29), retrolimbica agranularis (30), perirhinalis (35), ectorhinalis (36), retrosubicularis (48), and parasubicularis (49) (see Fig. 100). Later, Rose (1929) investigated the insular region in more detail. In contrast to the purely agranular insular cortex of the hedgehog, he described a granular part lying dorsal of the agranular field in the mole. The insular cortex occupies the whole lateral and rostral parts of the frontal lobe and, thus, the frontal pole is made up of insular cortex. Kurepina (1974) also discussed the large extent of the insular region and the claustrum. Rehkämper (1981) classified 4 fundamental areas in which somatosensory and somatomotor (areas 1+2), visual (area 3), and auditory activities (area 4) are localized. In *Talpa* areas 1+2 and 4 show characteristics of progressive differentiation, whereas area 3 is clearly less developed than in other Insectivora. This was related to the minor importance of the talpid visual system. Sato (1977) described and illustrated a visual area in the fossorial *Mogera* and the semifossorial *Urotrichus* and found differences between them.

The Golgi studies by Ferrer (1986) similarly suggest a relatively well-developed neocortex, but the narrow internal granular layer in the occipital region probably reflects a poorly developed visual system in this species which is adapted to life in a subterranean environment. General primitive characteristics of the mole neocortex are the presence of extraverted neurons in the accentuated layer II (Ferrer 1986) and poorly defined stratification (Kurepina 1974).

Physiology, Sense Organs

Aitkin et al. (1982, EP, audition)
Allison and VanTwyver (1970a, EP, *Scalopus;* 1970b, ER, sleep, *Scalopus, Condylura*)
Andreescu-Nitescu (1970, nose)
Armstrong and Quilliam (1961, snout)
Bielschowsky (1907, nose)
Broom (1915b, vomeronasal organ)
Dalquest and Orcutt (1942, sensory physiology, *Neurotrichus*)
Eimer (1871, snout)
Franz (1934, eye)
Graziadei (1966, nose, EM)
Heffner and Masterton (1975, digital dexterity, *Scalopus*)
Henderson (1952, eye)
Henson (1961, ear)
Herzfeld (1889, vomeronasal organ)
Johannesson-Gross (1986, LE, orientation; 1988, LE, visual)
Johannesson-Gross and Gross (1982, LE)
Kohl (1893 - 1895, eye)
Konstantinov et al. (1987, EP, audition)
Kriszat (1940a, SP; 1940b, c, LE, orientation)
Krumbiegel (1961, eye, *Talpa caeca, Urotrichus talpoides*)
Lewis (1983, eye, *Neurotrichus gibbsi*)
Misek (1988, eye, ontogenesis)
Ognev (1962, eye, *Talpa* spp., *Mogera robusta*)
Quilliam (1966a, snout, nose, taste, ear, touch, eye; 1964, 1966b, eye)
Ritter (1898, eye)
Sato (1977, eye, *Mogera, Urotrichus*)
Siemen (1976, eye)
Slonaker (1902, eye, *Scalopus*)
Sokolova (1964, 1965, eye, *Talpa europaea, T. altaica, T. caucasica*)
Suzuki and Kurosumi (1972, snout)
Wöhrmann-Repenning (1975, nose)

Abbreviations:
EM = electron microscopy
EP = evoked potentials
ER = electrical recordings
LE = learning experiments
SP = sensory physiology

In searching for food, the mole's tactile systems seem to play a very important role (e.g., Kriszat 1940a; Dalquest and Orcutt 1942; Quilliam 1966a), whereas the role of vision should be negligible. Auditory and vibratory perception are well developed and seem to be more important than olfaction, which is effective only at short distances (Herter 1957).

In the **olfactory system**, the nose of *Talpa* was investigated by Graziadei (1966), Andreescu-Nitescu (1970) and Wöhrmann-Repenning (1975). Graziadei found the number of olfactory cilia to be relatively low. Andreescu-Nitescu and Wöhrmann-Repenning described a very long cavum nasi, especially at its anterior region.

Wöhrmann-Repenning found characteristics of reduction compared with other Insectivora, e.g., lower cell density and fewer receptor cells. Dalquest and Orcutt also felt "that the sense of smell was very poorly developed" in *Neurotrichus* "and of little or no use in locating food or enemies. The nose is a tactile organ" (1942, p. 398). Such a view is at least partly supported by our measurements on the olfactory brain structures (e.g., Table 47). Details on the vomeronasal system, which projects to the accessory olfactory bulb, were given by Herzfeld (1889) and Broom (1915b). According to Broom, the vomeronasal organ opens through a long narrow duct into the anterior end of the nasopalatine canal.

According to Quilliam (1966a) moles depend on **touch** to a much greater extent than most other mammals. In subterranean passages, the mole is guided almost entirely by tactile and teletactile means. Their organs of touch are: (1) Eimer's organs at the tip of the very elongated snout (tronc), (2) vibrissae which encircle the muzzle and are also present near the manus and on the tail, and (3) Pinkus plates on the hairy skin, especially in the abdominal region (Kriszat 1940a; Quilliam 1966a).

The tronc of the mole contains an enormous number of nerve fibers in a very small area of skin (Bielschowsky 1907), and the largest density of tactile elements ever found (Kriszat 1940a). These elements were first described by Eimer (1871); additional information was given by Bielschowsky (1907), Armstrong and Quilliam (1961), and Suzuki and Kurosomi (1972). Three types of nerve endings were described in Eimer's organs: free nerve endings, Merkel cells near the basal portion, and encapsulated endings in the dermis. Evoked potential studies were conducted by Allison and VanTwyver (1970a) to determine somatic areas. They found that the primary somatic area appears to have shifted posterolaterally, compared to another insectivore, the hedgehog, possibly because of the absence of a visual cortex in the mole.

In the **auditory system**, the evoked potential studies of Allison and VanTwyver (1970a) showed that the location of the auditory area is similar to that in hedgehog and rat. Electrophysiological and behavioral responses to sounds were recorded in *Talpa* by Aitkin et al. (1982) and Konstantinov et al. (1987). Aitkin et al. evoked responses to sounds between 0.1 and 15 kHz, and moles were apparently at least as sensitive to tones of 5 - 8 kHz as they were to low tones. "In general, evidence is presented for the mole having essentially normal mammalian auditory characteristics."

Konstantinov et al. reported, that the auditory system responded to signals within a 0.1 - 22 kHz frequency range, the lowest evoked potential thresholds corresponding to frequencies of 0.3 - 6 kHz, and "most of the acoustic signals are presented by a pulse sequence with an evident spectral maximum at 8.5 - 27.4 kHz in shrews, 0.8 - 7 kHz in moles and 1.7 - 7 kHz in hedgehogs." Size and shape of the auditory ossicles may be suitable only for transmission of sounds of fairly low frequencies (Henson 1961; *Scalopus*), and the cochlea of the mole contains only 1.75 turns and may be the shortest found in mammals (Quilliam 1966a; *Talpa*). Burda et al. (1987) also report on low frequency hearing in subterranean mammals. Such restrictions may explain why the auditory brain centers of the mole are relatively small (Tables 26, 27). In contrast, Dalquest and Orcutt (1942) noted apparent response to vibrations of 8 - 30 kHz in *Neurotrichus*, thus indicating an adaptation to the use of very high-pitched sounds. Ground vibrations may also be more important in the sensory spectrum of moles (Kriszat 1940a; Wöhrmann-Repenning 1975).

Kriszat mentioned that the direction of vibrations is well perceived by *Talpa*. Dalquest and Orcutt (1942) were not able to determine sensitivity to ground vibrations in *Neurotrichus*.

With regard to the **visual system**, there are many studies or remarks on size and/or structure of the eyes and eyelids in moles (e.g., Kohl 1893, 1895; Ritter 1898; Slonaker 1902; Franz 1934; Henderson 1952; Krumbiegel 1961, and according to him, already Aristotle; Ognev 1962; Sokolova 1964, 1965; Quilliam 1964, 1966a, b; Siemen 1976; Sato 1977; Lewis 1983; Misek 1988).

There is agreement that the visual system of moles is very weakly developed and its optical resolution of the retinal image is not very good (Quilliam 1966b), but that it is still a functioning system with the capacity to discriminate light and dark (Quilliam 1966a; Johannesson-Gross 1986; and others). This may be important for conserving the photoreceptive (vegetative) function (Sokolova 1964) and for light/dark cues (i.e., more orientation in time than in space) (Johannesson-Gross 1986). *Neurotrichus* seems to be totally blind (Dalquest and Orcutt 1942). It has, however, a pigmented extension of retina covering the anterior surface of the optic lens (Lewis 1983). Misek (1988) described the ontogenesis of the eye of *Talpa europaea*.

Thus, there seem to be large differences between genera and species. Slonaker (1902) said that the eye of *Scalopus* is much more degenerated than that of *Talpa*. This applies to all its parts: (1) fusion of the lids, (2) reduction in relative size, (3) crowded retina, and (4) shape and size of the lens and its peculiar cell structure. In the evoked potential studies by Allison and VanTwyver, no visual evoked response could be recorded in any animal (*Scalopus*): "In agreement with anatomical studies, we conclude that there is essentially no visual cortex in the mole" (Allison and VanTwyver 1970a, p.554, *Scalopus*). Sato (1977) found the visual structures, including visual cortex, to be better developed in the semifossorial *Urotrichus* than in the fossorial *Mogera*.

4.5.3 Summary of Brain Characteristics in Subfamilies and Species of Talpidae

The brains of the two subfamilies of Talpidae in our material have similar macromorphological features, except for brain height (see Section 3.1). Index profiles are given in Figs. 102, 104, 106, and 108. In terms of index size, all Talpidae have relatively large brains with well represented neocortical and motor structures and less represented olfactory, auditory, and visual centers. The general somatosensory system is well represented, but its orofacial component (TR) is very different: very high in Desmaninae and very low in Talpinae.

Desmaninae

In competition with Echinosoricinae, Desmaninae have the largest brains of all Insectivora, 1.44 times larger than in the average Insectivora (1.44x AvIF). All the five fundamental brain parts are well above AvIF, varying from 1.32x for OBL to 1.72x for DIE (Fig. 102). OBL components vary from 0.67x for VCO to 1.69x for TR (Fig. 104); DIE components, from 1.03x for VET to 2.22x for DOT (Fig. 106); and TEL components, from 0.49x for AOB and 0.65x for MOB to 2.15x AvIF for NEO (Fig. 108). In OBL, after VCO, the lowest indices were found for DCO and OLS, i.e., for all the auditory structures. Although below TR, other somatosensory components, such as FCM and FGR, and some of the vestibular nuclei (VI, VM) show fairly large indices. In DIE, STH is, after DOT, well represented, whereas CGL is, after VET, least represented. In TEL, all olfactory structures are clearly below AvIF, whereas, after NEO, STR and SCH are highest; HIP, SEP, and AMY are also well represented.

In conclusion, Desmaninae have large brains with well developed somatosensory, vestibular, and motor structures, and less well represented olfactory and auditory structures. Compared with the other subfamily with a similarly large brain size, i.e., Echinosoricinae, Desmaninae differ markedly in their large somatosensory centers and small olfactory and auditory components. Thus, brain composition differs widely in the two groups. The brains of Desmaninae may be rated higher since they show the largest concentration of advanced brain characteristics. NEO, STR, and DIE are the largest found in Insectivora, whereas the brains of Echinosoricinae gain in size mainly because of their large olfactory centers.

In most structures, the indices for the desmans deviate markedly from those for the average Insectivora, and the sum of the deviations from the family/subfamily averages (AvIF) is very high. The large OBL and clearly reduced MOB in the Desmaninae are consistent with similar features in other species adapted to searching for food in water.

Species characteristics of desman brains are based on direct comparison between the two species. It is not worth-while to introduce an average desman level, since a larger size of a brain structure in one species is always associated with small size in the other and vice versa. Therefore brain characteristics will be discussed together.

Galemys pyrenaicus and ***Desmana moschata:*** The brain of *Galemys* is 1.24 times larger than that of *Desmana*. In all the comparisons that follow, the direction will be the same, i.e., *Galemys* will be with reference to *Desmana*. Four of the five fundamental brain parts are also larger, with the greatest difference in CER (1.39 times larger). Only MES is somewhat smaller (0.91x). The largest differences in the OBL components are for mot V (1.56x), FCM (1.54x), mot XII (1.36x), and INO (1.25x). COE (0.86x), VS (0.80x), PRP (0.77x), OLS (0.66x), the mesencephalic SCB (0.60x), and the diencephalic EPI (0.50x) are markedly smaller in *Galemys*. The greatest difference in the thalamic nuclei is for ANT, which is 1.45 times larger in *Galemys*. Of the TEL parts, some small structures are much larger in *Galemys* (3.43x for AOB; 2.57x for SFB), while SEP (1.60x), NEO (1.49x), and LAM (1.44x) are moderately larger. The brain of *Galemys* is distinctly higher and slightly longer and narrower than that of *Desmana*; the corpus callosum also is slightly longer.

Particularities: *Galemys* has the largest brain in Insectivora, with very large NEO, sensory trigeminal, and motor (INO, CER, STR) centers. Olfactory, visual, and auditory structures are small. The brain of *Desmana* is similar, but the extremes are less pronounced. NEO and motor centers are smaller; the accessory olfactory bulb is very small. MES and CGL are large, and the CGL of *Desmana* even reaches values characteristic of hedgehogs.

Talpinae

The brain size of moles is only slightly above AvIF (NET = 1.13x). The five fundamental brain parts vary from 0.83x AvIF for OBL to 1.32x for CER (Fig. 102). OBL components vary from 0.50x for VL to 2.18x for FCM (Fig. 104); DIE structures, from 0.59x for CGL to 1.37x for STH and 1.23x for MNT (Fig. 106); and TEL components, from 0.50x for AOB and 0.98x for AMY to 1.40x AvIF for STR (Fig. 108). Of the OBL components, INO is largest (after FCM and FCE), whereas the trigeminal complex (TR), auditory structures (VCO, OLS) and mot XII are among the lowest. In DIE, the large representation of STH and poor development of CGL are shared with the second talpid subfamily, i.e., the Desmaninae, and in TEL, the same is true for AOB; the main olfactory centers of Talpinae are, however, close to AvIF.

In conclusion, the brains of the fossorial Talpinae (*Talpa, Parascalops* and *Scalopus*) are similar in relative size, shape, and composition. They are all flat, broad, and long and have a very long corpus callosum. The mesencephalic tectum is fully or almost fully covered. The average brain size is slightly above the level of the family/subfamily average for the Insectivora (AvIF). Talpinae have very well represented motor (CER, INO, STR) and forelimb components of the general somatosensory system (FCM, FCE), whereas the oro-facial component (TR) is among the least represented. The auditory and especially the visual structures are small.

Species characteristics of brains of Talpinae are based on deviation from the average talpine level (AvTa), which is uniformly set at 1 for each of the structures. The vast majority of investigations on mole brains have been done on *Talpa europaea.*

Talpa europaea: The brain of this species is smaller than in other Talpinae so far investigated: 0.91x AvTa in BrW; 0.90x in NET. The five fundamental brain parts vary from 0.84x AvTa for TEL to 1.03x for CER. Except for AOB, which is very large (1.79x AvTa), all TEL components of *Talpa europaea* are below the talpine level. They vary from 0.79x AvTa for PAL to 0.92x for AMY. Except for AOB, AMY, and SEP all are smaller in *Talpa europaea* than in the other talpine species. In the septum, SFB and SET are also smallest, as are the diencephalic EPI and HAM and the mesencephalic SCB. Other diencephalic and OBL components were measured only in *Talpa europaea* and thus, no comparison with other Talpinae is possible. Relative to other Insectivora, the forelimb components of the general somatosensory system (FCM, FCE) are very well represented, whereas auditory and especially the visual structures (CGL) are small. We do not know whether these results are representative for all Talpinae. Comparisons on the morphology of the eyes suggest that vision is relatively well developed in *Talpa europaea* compared with other

Talpinae. The brain and the cerebral hemispheres are short; the corpus callosum is long.

Particularities: Motor structures (CER, STR, INO) and forelimb components of the general somatosensory system (FCM, FCE) are well represented in *Talpa europaea*; accessory olfactory structures are relatively large; main olfactory, auditory, and visual structures are relatively small.

Talpa micrura: The total brain of *Talpa micrura* is largest in Talpinae: 1.14x AvTa for BrW; 1.13x for NET. Since most brain parts are also largest it is possible that the single individual investigated was not fully grown. Juveniles generally have higher indices. Apart from this reservation, the characteristics are similar to that of *Talpa europaea*. So far, there are no measurements in OBL and DIE.

Particularities: May be similar to *Talpa europaea*, but we have insufficient material and measurements.

Parascalops breweri: The total brain is most typical of Talpinae: 1.03x AvTa for BrW and 1.04x for NET. The five fundamental brain parts vary from 0.87x AvTa for MES to 1.08x for TEL. MES is smaller in *Parascalops* than in other Talpinae; DIE and TEL are above the talpine level, as are all TEL components except AOB (0.59x AvTa) and SCH (0.97x). The largest size is reached by SET (1.34x AvTa), SFB (1.25x), STR (1.16x), and HIP (1.10x). In the diencephalon, EPI is relatively large (1.15x AvTa). So far, the main components of DIE and OBL have not been measured. In shape, the brain of *Parascalops* has the largest height and the shortest corpus callosum of Talpinae.

Particularities: The brain of *Parascalops breweri* has relatively large STR, HIP, and NEO and relatively small accessory olfactory bulbs (AOB).

Scalopus aquaticus: The brain size is below the talpine level: 0.93x AvTa for BrW and 0.94x for NET. All of the five fundamental brain parts are also below this level. They vary from 0.88x AvTa for DIE to 0.95x for OBL. In TEL, AOB is very small (0.51x AvTa) and the other parts vary from 0.85x for SEP to 1.19x for SCH. Besides SCH, only PAL is above the talpine level (1.03x). AOB (0.51x), SEP (0.85x), AMY (0.88x), DIE (0.88x), and CER (0.93x) are smaller in *Scalopus* than in other Talpinae so far measured, SCB (1.36x) and SCH (1.19x) are larger. So far no measurements have been done on the main components of DIE and OBL.The brain of *Scalopus* is the longest in Talpinae; the cerebral hemispheres are the shortest and the flattest, but are very wide.

Particularities: The brain of *Scalopus aquaticus* deviates only slightly from the talpine level, except for AOB, which is clearly smaller, and for SCB, which is larger. In eye characteristics, Slonaker (1902) noted stronger degeneration than that found in *Talpa*.

4.6 Soricidae

General Characteristics

Shrews, which have the widest distribution of any of the Insectivora, are among the smallest and least conspicuous mammals. They are short-legged, mouse-like, with long pointed noses. The limbs are usually generalized plantigrade. In semiaquatic species, the feet are either webbed *(Nectogale)* or fringed with stiff hairs that greatly increase their effectiveness as paddles. The tibia and fibula are fused. The eyes are minute; ear pinnae are present but sometimes reduced. Shrews are mainly terrestrial, but some take to water freely, others burrow a little and a few seem adapted to tree-climbing. Some forms are active day and night, whereas others are active only at night. Their food consists of various invertebrates and small vertebrates. Although the shrew is typically associated with moist conditions, living primarily beneath leaf litter, fallen logs, and grasses, some species inhabit desert areas. Two subfamilies are usually differentiated: the mostly red-toothed Soricinae and the white-toothed Crocidurinae.

Selected References: Grassé (1955a), Pruitt (1957), Cockrum (1962), Kingdon (1974), Walker (1975), Yates (1984), Hutterer (1985), Vaughan (1985), Fons (1988).

Brain Characteristics

4.6.1 Soricinae

Sorex alpinus, S. araneus, S. cinereus, S. fumeus, S. minutus, S. palustris, Microsorex hoyi, Neomys anomalus, N. fodiens, Blarina brevicauda, Cryptotis parva, Anourosorex squamipes

Brain Weight, Cranial Capacity, Encephalization

Andy and Stephan (1961, *Sorex, Neomys*)
Bauchot and Stephan (1966, *Sorex, Neomys, Blarina*)
Bielak and Pucek (1960, *Sorex*)
Brodmann (1913, *Sorex*)
Cabon (1956, *Sorex*)
Crile and Quiring (1940, *Blarina*)
Crowcroft and Ingles (1959, braincase, *Sorex*)
Dapson (1968, skull, *Blarina*)
Dehnel (1949, skull, *Sorex*)
Dubois (1897, *Sorex*)

Fons et al. (1984, *Sorex, Microsorex, Neomys, Blarina*)
Haug (1987, *Sorex*)
Hrdlicka (1905, *Blarina*)
Lapicque (1912, *Sorex*)
Mangold-Wirz (1966, *Sorex, Neomys, Blarina*)
Pucek, M. (1965, *Sorex*)
Pucek, Z. (1955, 1957, 1963, 1964, 1965, 1970, skull, CrC, BrW, *Sorex, Neomys*)
Pucek and Markov (1964, braincase, *Sorex*)
Schubarth (1958, skull, *Sorex*)
Snell (1892, *Sorex*)
Stephan (1959, *Sorex, Neomys*)
Stephan and Andy (1982, *Sorex, Neomys, Blarina*)
Stephan et al. (1970, 1981a, *Sorex, Neomys*)
Welcker and Brandt (1903, *Sorex*)
Wirz (1950, *Sorex, Neomys*)
Yaskin (1984, *Sorex*)
Ziehen (1897a, 1899a, 1901, *Sorex*)

BoW/BrW data on four species of Soricinae were published by Bauchot and Stephan (1966) and on seven species by Fons et al. (1984). Since then, the collection has been enlarged by many new individuals, five of which were from new species, and by new data taken from the literature. Besides the BoW/BrW pairs of data, additional BoWs are available for many species of shrews. They are used to determine the average BoW of a species and to check whether the animals from which the brains were taken were of adult size. The sources of the additional BoWs are given in Table 5 and will not be repeated here. All standard BoWs and BrWs are given in Tables 5 and 6.

(1) *Sorex alpinus:* Four adult males have average BoWs/BrWs (g/mg) of 10.3/259. In addition, (1) BrWs of two more males (269 and 279 mg) and two females (255 and 258 mg), (2) some more BoWs from our collection, and (3) many BoWs given by Spitzenberger (1978) are available. The total of 45 BoWs and eight BrWs resulted in standards of 11.0/262.

(2) *Sorex araneus:* Besides our own vast collection, many data are taken from the literature (details in Fons et al. 1984). The resulting BoW/BrW standards (10.2/216) are well founded.

(3) *Sorex cinereus:* Nine males, with averages of 4.9/174, and 19 females, with averages of 3.9/174, were collected in Canada. One animal with undetermined sex had 3.6/176; 18 subadult animals had relatively small BoWs (3.6 g) and relatively large BrWs (178). Only eleven specimens can be considered to be definite adults. Five adult males had an average of 5.9/167; six adult females averaged 4.5/168. The resulting standards are 5.16/168.

(4) *Sorex fumeus:* No pairs of data, but five BoWs with an average of 8.4 g and two BrWs with an average of 241 mg are known.

(5) *Sorex minutus:* Similar relations as in *Sorex araneus* (details in Fons et al. 1984). The standards are 4.41/115; they are well founded.

(6) *Sorex palustris:* One pair from a male with 14.6/278.

(7) *Microsorex hoyi:* Four males had averages of 2.71/92 and three females averaged 2.53/102. The resulting standards are 2.63/96.

(8) *Neomys anomalus:* Three males had averages of 9.5/285 and two females averaged 9.1/278; 76 additional BoWs are available. The 81 BoWs and five BrWs resulted in averages of 11.6/282, which are taken as standards.

(9) *Neomys fodiens:* Our material includes nine males with averages of 15.2/320 and eight females, with 14.9/336. More than 250 additional BoWs and four BrWs complete the material and result in well founded standards of 15.3/328.

(10) *Blarina brevicauda:* 16 males had averages of 18.9/404 and 24 females averaged 17.6/392. Two animals with undetermined sex had 19.4/394. Only seven animals are found to be definite adults. Their average BoWs/BrWs were 19.7/393, which are taken as standards.

(11) *Cryptotis parva:* Four adult males with averages of 9.3/245 and one adult female, with 10.2/241, were collected near Jackson, Miss. Additional BoWs of four males with an average of 10.3 and one BrW (248) were available. The resulting standards are 9.9/245.

(12) *Anourosorex squamipes:* Four adult males with averages of 20.4/390 and three adult females, with 19.5/388, were collected in Thailand. Two additional BoWs with an average of 20.5 were available. The resulting standards are 20.1/389.

The encephalization of soricine shrews is on the average 1.51 times higher than in Tenrecinae (Table 7), with moderate differences between the species (1.33 - 1.69x AvTn), except for *Sorex minutus*, which has a lower index (1.23x AvTn); 1.33x was found for *Sorex araneus*; 1.69x for *Sorex fumeus* (see Table 6). Relative to all the Insectivora so far measured, the soricine shrews are slightly above the species average (AvIS), which is 1.48x AvTn, and slightly below the family/subfamily average (AvIF), which is 1.58x AvTn.

Seasonal changes occur in body and brain size and are paralleled by a winter depression of the skull (= Dehnel's phenomenon). Changes in brain size have been documented (1) directly, by BrW comparisons in *Sorex minutus* (Cabon 1956; Yaskin 1984) and *Sorex araneus* (Bielak and Pucek 1960; M. Pucek 1965; Z. Pucek 1965, 1970; Yaskin 1984) and (2) indirectly, by comparing cranial capacities (CrCs) in *Sorex* and *Neomys* (Z. Pucek 1955, 1957, 1963, 1964) and by comparing skull measures in *Sorex* (Dehnel 1949; Schubarth 1958; Crowcroft and Ingles 1959; Pucek and Markov 1964) and *Blarina* (Dapson 1968). Seasonal variations, especially the winter decrease (see Section 6.4), were found to be characteristic of all species of the genus *Sorex*, at least in the Palearctic, but have been observed also in *Neomys* and *Blarina* (see Z. Pucek 1970; Lit.).

Ziehen (1899a) related brain weights to spinal cord (SPC) weights. In *Sorex araneus*, BrW/SPC is 4.9, which is similar to that of *Talpa europaea* (4.7) and distinctly higher than in *Erinaceus europaeus* (3.5). In comparison, man's brain is about 43 times heavier than its SPC. Ziehen's data were republished in a comparative work by Krompecher and Lipak (1966).

Macromorphology

Ärnbäck-Christie-Linde (1900, 1907)
Bradley (1903, CER)
Brauer (1969/70, CER)
Brauer and Schober (1970, *Sorex* and *Neomys*)
Clark (1932, *Blarina*)
Fons et al. (1984)
Hochstetter (1942)
Igarashi and Kamiya (1972)
Johnson et al. (1982, *Blarina*)
Stephan (1956a, *Sorex* and *Neomys*)
Stephan and Andy (1982, *Sorex* and *Neomys*)
Ziehen (1897a)

The vast majority of papers deal with general brain characteristics of *Sorex* species (*S. araneus, S. cinereus, S. minutus, S. unguiculatus*). Other species (*Neomys fodiens, Blarina brevicauda*) and the topics of more specialized studies are given in the reference list.

All species have similar brains which may be described as follows: (1) in dorsal view (Fig. 14), the brains are round, with the olfactory bulbs included in the circle; (2) the cerebellar lobulus petrosus is relatively large, but mostly hidden under the cerebellar hemispheres; (3) no true flocculus was found by either Bradley (1903) or Brauer (1969/70); (4) the mesencephalic tectum is fully covered; (5) there is no sulcus in NEO; (6) there is no true rhinal fissure (see also Rose 1929 about totally free insular cortex); (7) an olfactory tubercle (TOL) is very prominent in side view (Fig. 26); it projects strongly rostroventrally (less prominent in *Neomys*); (8) optic nerves, chiasma, and tracts are very small; (9) the trigeminal nerve is large; (10) in the midsagittal plane (Fig. 50), the corpus callosum is relatively thin and long; and (11) CER and its arbor vitae are the simplest found in mammals.

Johnson et al. (1982), in a comparison of several brain characters of *Blarina* with those of many other mammals, found poor optic development. Brauer and Schober (1970, 1976) and Igarashi and Kamiya (1972) labeled a long sulcus rhinalis running over the whole hemisphere. This, however, is mainly a groove formed by a vessel, which, in *Neomys*, does not correspond to the position of the histological border between NEO and piriform lobe (detailed description in Stephan 1956a).

Cell and Area Measurements

Bauchot (1963, 1964, DIE, *Sorex, Neomys*)
Bauchot and Stephan (1961, sectioned areas of CC, COA, CHO, *Sorex, Neomys*)
Brodmann (1913, surface areas of cortex, *Sorex*)
Brouwer (1915, SPC, *Sorex*)
Galert (1986, supraoptic and paraventricular nuclei, *Sorex, Neomys*)
Hyde (1957, motV, eye nerves, *Sorex, Blarina*)
Hyvärinen (1967, pituitary, *Sorex*)
Larochelle (1987, mitral cells, *Sorex, Blarina*)
Larochelle and Baron (1989a, mitral cells, *Sorex, Blarina*)
Ryzen and Campbell (1955, cortex, *Sorex*)
Stephan (1956a, b, 1961, surface areas of cortex, *Sorex, Neomys*)
Stephan and Manolescu (1980, HIP, surface areas, layer thickness, *Sorex*)
Stephan et al. (1984b, VIS, *Sorex, Neomys*)
Stolzenburg et al. (1989, NEO, *Microsorex*)

Measurements of (1) areas of transversally sectioned fiber bundles, (2) size, number and density of fibers and cells, (3) cortex and layer thickness, and (4) surface areas are available.

(1) In the first cervical segment of *Sorex araneus*, Brouwer (1915) measured the transversally sectioned area of the funiculus posterior, which is slightly more than 14% of the white matter. This was the lowest value found in the 20 mammalian species measured. For comparison, the figures are about 23% for the mole and 39% for man. Stephan et al. (1984b) compared transversally sectioned areas of nervus opticus (NOA) and midsagittally sectioned areas of optic chiasma (CHO) of *Sorex araneus, S. minutus,* and *Neomys fodiens*, and Bauchot and Stephan (1961), those of the corpus callosum (CCA), anterior commissure (COA), and CHO. CCA of *Sorex araneus* and *Neomys fodiens* were distinctly above, and that of *Sorex minutus* just at the average for the 13 species measured by Bauchot and Stephan. COA was relatively small in *Neomys* and CHO relatively large in all three soricine species, despite their poorly developed vision. The reason may be that the CHO average is strongly depressed by the Talpidae and Chrysochloridae, which were included in the sample.

(2) Hyde (1957) counted fibers and cells in some trigeminal and visual structures in *Sorex cinereus* and *Blarina brevicauda*. In the trigeminal nerve of *Sorex*, there was only one third as many fibers as were found in *Blarina*, fewer also in the oculomotor nerve, but twice as many in the optic nerve (despite the same area size). In the mesencephalic trigeminal nucleus, he counted 650 cells. Bauchot (1963, 1964) determined various cell characteristics for many parts of the diencephalon in *Sorex araneus, S. minutus,* and *Neomys fodiens.* Galert (1986) investigated size and cell types of supraoptic and paraventicular nuclei in the same three species and Hyvärinen (1967) of the anterior lobe of the pituitary during the life cycle of *Sorex araneus.*

Larochelle (1987) counted the mitral cells of layers 3 and 4 of the main olfactory bulb in *Sorex cinereus, S. fumeus, S. palustris,* and *Blarina brevicauda* and found the relatively largest number in *Blarina,* with an especially large proportion in layer 3. Ryzen and Campbell (1955) considered cell size and density, grey/cell coefficient, and total neuron number in neo- and paleocortical regions of *Sorex pacificus* and constructed a cortical map based on these investigations. Stolzenburg et al. (1989) studied the somatosensory cortex of the very small *Microsorex hoyi.* In comparison to larger Insectivora (*Potamogale, Talpa*), they found small nuclei in neuronal and glial cells, and a large volume density of neuronal cells. Similar relations were found in the still smaller crocidurine *Suncus etruscus.*

(3) The thickness of the cell layer in the hippocampus (HIP) of *Sorex araneus* and *S. minutus* was determined by Stephan and Manolescu (1980). Compared with eight other species of Insectivora, they found relatively thin upper and lower subiculum, and relatively thick cornu ammonis and fascia dentata in both *Sorex* species. Stolzenburg et al. (1989) found a relatively broad molecular layer in the somatosensory cortex of *Microsorex hoyi*. As a rule, they found a decrease in the relative thickness of the molecular layer in larger species of Insectivora.

(4) Brodmann (1913) found a cortical surface area of 76 mm^2 in *Sorex araneus*; while Stephan (1956a, b, 1961), in serial section measurements without correction, found a value of 68 mm^2. After correction for shrinkage, the figure increased to the much higher value of 114 mm^2. According to Stephan, the total surface area in *Sorex araneus* is composed of 47.3% paleocortex (PAL), 26.8% archicortex (ARCH), and 25.9% neocortex (NEO); in *S. minutus*, of 48.5, 26.1, and 25.4%; and in *Neomys fodiens*, of 38.5, 24.8, and 36.7%, respectively. Thus, in *Neomys* the percentage of the mainly olfactory PAL is smaller and that of NEO distinctly larger than in the terrestrial species of *Sorex*. In the hippocampus, the surface areas are smaller in *Sorex minutus* than in *S. araneus* (Stephan and Manolescu 1980). Relative eye size was found to be slightly larger in *Sorex minutus* than in *S. araneus* (Stephan et al. 1984b).

Volume Measurements

Andy and Stephan (1966a, b, 1968, SEP, *Sorex, Neomys*)
Baron (1972, 1974, 1977a, auditory and visual structures, *Sorex*)
Baron (1977b, vestibular nuclei, *Sorex*)
Baron (1978, cerebellar ncl., *Sorex*)
Baron (1979, DIE, *Sorex*)
Baron et al. (1983, MOB, *Sorex, Neomys*)
Baron et al. (1987, PAL, *Sorex, Microsorex, Neomys, Blarina*)
Baron et al. (1988, VES, *Sorex, Microsorex, Neomys, Blarina*)
Bauchot (1961, 1970, HYP, *Sorex, Neomys*)
Bauchot (1963, 1966, 1979b, DIE, THA, *Sorex, Neomys*)
Bauchot (1979a, main parts, *Sorex, Neomys*)
Bauchot and Legait (1978, HYP, *Sorex, Neomys*)
Bauchot and Stephan (1968, main parts, DIE, THA, *Sorex, Neomys*)
Etienne (1972, BOL, *Sorex*)
Frahm et al. (1979, PALL, *Sorex*)
Frahm et al. (1982, NEO, *Sorex, Neomys*)
Larochelle (1987, BOL, *Sorex, Blarina*)
Larochelle and Baron (1989a, BOL, *Sorex, Blarina*)
Legait et al. (1976, EPI, *Sorex, Neomys*)
Mangold-Wirz (1966, weights of main parts, *Neomys*)

Matano et al. (1985b, CER nuclei, *Sorex*)
Stephan (1966, 1967b, main parts, *Sorex, Neomys*)
Stephan (1969, PVO, *Sorex, Neomys*)
Stephan and Andy (1962, 1982, SEP, *Sorex, Neomys*)
Stephan and Andy (1964, 1969, 1982, main parts, *Sorex, Neomys*)
Stephan and Andy (1977, AMY, *Sorex*)
Stephan and Manolescu (1980, HIP, *Sorex*)
Stephan et al. (1970, 1981a, main parts, *Sorex, Neomys*)
Stephan et al. (1982, AOB, *Sorex, Neomys*)
Stephan et al. (1984a, main parts, NEO, *Sorex, Microsorex, Neomys, Blarina*)
Stephan et al. (1984b, CGL, *Sorex*)
Stephan et al. (1987a, AMY, *Sorex, Microsorex, Neomys, Blarina*)
Wirz (1950, weights of main parts, *Sorex, Neomys*)
Yaskin (1984, weights of main parts, *Sorex*)

Wirz (1950) compared **weights of brain parts** in many mammalian species using index methods. She subdivided the brains macroscopically into five parts (neocortex, basal ganglia, rhinencephalon, cerebellum, and "stammrest") but did not publish the raw data (see also Mangold-Wirz 1966). In Soricinae, she investigated *Sorex minutus* and *Neomys*. Seasonal changes in the weights of ten brain parts were investigated by Yaskin (1984) in *Sorex*. Yaskin found that the greatest winter reduction in weight (about 35%) was in the neocortex.

Volumes of the five **fundamental brain parts** obtained from measurements on serial sections of *Sorex araneus, S. minutus,* and *Neomys fodiens* were given by Stephan and Andy (1964, 1969, 1982), Stephan (1966, 1967b), Bauchot and Stephan (1968), Stephan et al. (1970, 1981a), and Bauchot (1979a) and compared with other Insectivora, with Primates and Chiroptera. Stephan et al. (1984a) included four more species: *Sorex cinereus, S. fumeus, Microsorex hoyi,* and *Blarina brevicauda.* Newly added in the present book are *Sorex alpinus, Neomys anomalus, Cryptotis parva,* and *Anourosorex squamipes.*

The largest average indices were found for CER and TEL, which are about 1.7 times larger in Soricinae than in Tenrecinae (1.66 and 1.76x AvTn). The lowest average index was found for MES (1.08x), which is only slightly above the tenrecine level (1.00) and distinctly below AvIF, which is 1.50x AvTn. As in total brain, the relative size of the main parts is similar in all eleven species. The largest difference was found for DIE, which is nearly 1.5 times larger in *Neomys anomalus* than in *Sorex minutus.*

In the **medulla oblongata** (OBL), data on auditory and vestibular structures were given by Baron (1972, 1974, 1977b) and Stephan et al. (1981a) for *Sorex araneus* and *S. minutus*, and vestibular structures for five species by Baron et al. (1988). Many more structures in these five species are published here for the first time (Tables 13 - 27). High average indices in soricine shrews were found for the inferior olive (INO; 1.81x AvTn), nucleus tractus solitarii (TSO; 1.68x), and nucleus dorsalis motorius nervi vagi (mot X; 1.63x), and low indices for the auditory structures (AUD; 0.51x) and nuclei fasciculares cuneatus externus (FCE) and medialis (FCM) (0.69 and 0.79x). The indices for TSO, mot X, and the nucleus nervi hypoglossi (mot XII) are even higher than those for AvIS and AvIF. The soricine indices for auditory, funicular and vestibular nuclei are distinctly lower than AvIS and AvIF. The largest differences among the five species were found for mot X, which is 2.4 times larger in *Blarina* than in *Sorex cinereus*, and for the nucleus vestibularis inferior (VI), which is 2.2 times larger in *Neomys fodiens* than in *Sorex araneus*.

In the **mesencephalon** (MES), volume data were given for the lemniscus lateralis nuclei (NLL) and the colliculi by Baron (1972, 1977a) in *Sorex araneus* and *S. minutus* (see Tables 28 - 30), and for the subcommissural body (SCB) by Stephan (1969) and Stephan et al. (1981a) for these two species and for *Neomys fodiens*. SCB was found to be larger in *Sorex minutus* than in *S. araneus* and *Neomys fodiens*.

In the **cerebellum** (CER) data on the nuclei were given by Baron (1978) and Matano et al. (1985b) for *Sorex araneus* and *S. minutus*. Based on these two species, the cerebellar nuclei are well developed in the soricine shrews and except for the interposed cerebellar nucleus (ICN), are larger in the soricine average than in AvIS and AvIF. The medial cerebellar nucleus (MCN) is especially large, while ICN is relatively small.

In the **diencephalon** (DIE), data on main parts are given by Bauchot (1963, 1966, 1979b) and Bauchot and Stephan (1968) for *Sorex araneus, S. minutus,* and *Neomys fodiens*, and by Baron 1979 and Stephan et al. (1981a) for *Sorex araneus* and *S. minutus*. For the same two species, Baron (1972, 1977a) and Stephan et al. (1984b) gave data for some diencephalic visual structures and Frahm et al. (1979), for pallidum (PALL).

Based on Bauchot's data, all diencephalic structures are small relative to the average Insectivora (AvIS and AvIF). The subthalamus (STH) and, within the thalamus, the medial geniculate body (CGM) are the relatively largest structures. The very high position of CGM, however, is caused by *Neomys*, in which CGM is 2.1 times larger than in the two *Sorex* species. The nuclei dorsales thalami (DOT) are relatively large in *Neomys* but small in *Sorex minutus;* the lateral geniculate body (CGL) is small in all three species measured.

Data on some small diencephalic structures were given for the pineal gland (EPI) and the medial habenular nucleus (HAM) by Stephan (1969) and Stephan et al. (1981a); for EPI, by Legait et al. (1976); and for the hypophysis (HYP) by Bauchot (1961, 1970) and Bauchot and Legait (1978). EPI is very small in all soricine shrews so far investigated. A large intermediate lobe was found in HYP of *Neomys*.

In the **telencephalon** (TEL), data on seven to nine main parts were given by Stephan and Andy (1964, 1969, 1982), Stephan (1966, 1967b), Bauchot and Stephan (1968), and Stephan et al. (1970, 1981a) for *Sorex araneus, S. minutus,* and *Neomys fodiens* and were compared with other Insectivora, and with Primates and Chiroptera. Stephan et al. (1984a) included four more species: which were *S. cinereus, S. fumeus, Microsorex hoyi,* and *Blarina brevicauda*. Newly added in this book are data on *Sorex alpinus, Neomys anomalus, Cryptotis parva,* and *Anourosorex squamipes*. Additional data, mostly from *Sorex araneus, S. minutus,* and *Neomys fodiens*, were given for olfactory structures by Stephan (1966); for MOB, by Etienne (1972), Baron et al. (1983), Larochelle (1987), and Larochelle and Baron (1989a); for AOB, by Stephan et al. (1982); for PAL, by Baron et al. (1987); for SEP, by Stephan and Andy (1962, 1982), Andy and Stephan (1966a, b, 1968), and Stephan (1969); for AMY, by Stephan and Andy (1977) and Stephan et al. (1987a); for HIP, by Stephan and Manolescu (1980); and for NEO, by Frahm et al. (1982) and Stephan et al (1984a). The highest average indices were found for NEO (3.47x AvTn), STR (2.55x), HIP (1.82x), and SCH (1.80x); the lowest, for MOB (0.79x), AOB (0.88x), and PAL (1.03x AvTn). Most of the average indices of Soricinae correspond well with AvIS and AvIF and, thus, are typical for Insectivora. In contrast, the average soricine AOB is only about half the size of that reached in Insectivora.

The largest variation in the indices for the eleven species was found in the small AOB, which is 4.5 times larger in *Sorex araneus* than in *Cryptotis parva*. Of the larger TEL components, variation is highest for NEO, which is twice as large in *Sorex fumeus* as in *Sorex araneus* and for AMY, which is 1.8x larger in *Anourosorex* than in *Sorex minutus*. The lowest variation in the TEL components was found for SEP. In general, TEL composition is similar in all soricine shrews.

Many of the papers mentioned in connection with volume measurements provide details on histology.

Histology

Ärnbäck-Christie-Linde (1900, *Sorex*)
Andy and Stephan (1961, 1966a, SEP, *Sorex, Neomys*)
Barnard (1940, hypoglossal complex, *Blarina*)
Bauchot (1963, DIE, *Sorex, Neomys*)
Bauer (1909a, substantia nigra, *Sorex*)
Brauer (1969/70, CER, *Sorex*)
Campbell and Ryzen (1953, DIE, *Sorex*)
Clark (1932, *Sorex, Blarina*)
Clark (1933, CGM, *Sorex*)
Crosby and Humphrey (1939a, b, MOB, AOB, RB, *Blarina*)
Crosby and Humphrey (1944, AMY, *Blarina*)
Crosby and Woodburne (1943, MES, *Blarina*)
Galert (1985, substantia nigra, *Sorex, Neomys*)
Galert (1986, supraoptic and paraventricular nuclei, *Sorex, Neomys*)
Gillilan (1941, optic route, *Blarina*)
Green (1951, HYP, *Blarina*)
Hanström (1946, 1953, 1966, HYP, *Sorex*)
Huber and Crosby (1943, MES, *Blarina*)
Hyde (1957, trigeminal structures, *Sorex, Blarina*)
Kaplan (1913, vestibular ncl., *Sorex*)
Krabbe (1925, SCB, *Sorex*)
Krasnoshchekova et al. (1987, NEO, *Neomys*)
Krawczuk (1983, HAB, *Sorex, Neomys*)
Kurepina (1974, NEO, *Sorex*)
Larochelle (1987, BOL, *Sorex, Blarina*)
Larochelle and Baron (1989a, BOL, *Sorex, Blarina*)
Misek (1990a, b, NEO, morphogenesis)
Petrovicky and Kolesarova (1989, parabrachial nuclei, *Sorex*)
Pokorny (1926, HYP, *Sorex*)
Rehkämper (1981, NEO, *Neomys*)
Rose (1912, cortex, *Sorex*)
Rose (1929, NEO, *Sorex*)
Ryzen and Campbell (1955, cortex, map, *Sorex*)
Sato (1977, VIS, *Sorex*)
Sigmund et al. (1987, DIE, *Sorex*)
Skrzypiec and Jastrzebski (1980, cerebellar nuclei)
Stephan (1965, AOB, *Sorex, Neomys*)
Stephan and Andy (1982, SEP, *Neomys*)
Stephan and Manolescu (1980, HIP, *Sorex*)
Switzer et al. (1980, AOB, *Blarina*)
Szteyn et al. (1987, corpus striatum, *Sorex, Neomys*)
Valeton (1908, colliculus inferior, *Sorex*)
Verhaart (1970, SPC, OBL, *Sorex*)
Ziehen (1897a, *Sorex*)
Zuckerkandl (1907, indusium griseum, *Sorex*)

Some general remarks on brain histology are found in the paper by Ziehen (1897a), who included *Sorex araneus* as a representative of the Eutheria, in a comparison with Monotremata and Marsupialia.

Verhaart (1970) gave a detailed description of the **spinal cord** (SPC) and **medulla oblongata** (OBL) of the pigmy shrew (*Sorex minutus*). In SPC, the grey matter and ventral horn are relatively large, while dorsal and dorsolateral funiculi are small. Shrews differ from moles and hedgehogs by their large central grey matters, tectospinal systems, and caudal nuclei of the spinal trigeminal tracts. The inferior olive lies more ventrally than usual because the pyramid is so tiny. Heffner and Masterton (1975) compared details of the pyramidal tract (given by Verhaart 1970 for *Sorex minutus*, not *Microsorex hoyi* as assumed by the authors) with those of other mammals. They found it to be very small and its fibers to be extremely thin. Ärnbäck-Christie-Linde (1900) briefly described the area postrema, Kaplan (1913) the vestibular nuclei, and Petrovicky and Kolesarova (1989) the parabrachial nuclear complex of *Sorex araneus*. More detailed descriptions were given by Barnard (1940) on the hypoglossal nucleus of *Blarina brevicauda* and by Hyde (1957) on the trigeminal structures of *Sorex cinereus* and *Blarina brevicauda*. - Brauer (1969/70) and Skrzypiec and Jastrzebski (1980) described and illustrated the **cerebellar nuclei** of *Sorex araneus*.

For the **mesencephalon** (MES), detailed descriptions were given by Huber and Crosby (1943) on the tectum and by Crosby and Woodburne (1943) on the non-tectal portions in *Blarina brevicauda*. In the tectum, Huber and Crosby described, listed, and illustrated the nine strata found in other mammalian species and compared them with those of reptiles. According to Crosby and Woodburne, the differentiation of MES is in line with the reduced optic and large auditory systems. The nuclei associated with the lateral lemniscus are well developed, whereas eye-muscle nuclei and pars lateralis of substantia nigra are poorly represented. The substantia nigra was briefly mentioned by Bauer (1909a) in *Sorex araneus* and described and illustrated by Galert (1985) in *Sorex araneus, S. minutus*, and *Neomys fodiens*. Valeton (1908) mentioned the inferior colliculus of *Sorex araneus*, and Sato (1977) described and illustrated the superior colliculus of *Sorex shinto*. Finally, Krabbe (1925) described the subcommissural body (SCB) of *Sorex araneus* as having an ependym with a single cell layer only.

For the **diencephalon** (DIE), there are descriptions by Campbell and Ryzen (1953) for *Sorex cinereus* and by Bauchot (1963) for *Sorex araneus, S. minutus*, and *Neomys fodiens*. According to Campbell and Ryzen, the differentiation of diencephalic structures in shrews "occurs to nearly the same degree as in the higher members of the insectivore-primate line." The lateral geniculate body and pulvinar are reduced in concert with a poorly developed visual system. In the ventrolateral group of the dorsal thalamus, the lateral portions seem to be poorly developed, and "the conspicuous n. ventralis medialis and possibly the highly developed n. submedius may reflect the extreme development which is attendant upon an exploring snout."

The most complete investigation on structural characteristics was given by Bauchot (1963). In addition to a detailed description of the various diencephalic parts in *Sorex araneus*, Bauchot gave a three-dimensional reconstruction and an atlas with 30 photoprints and drawings. He found small visual structures, well differentiated mamillary bodies, and large periventricular nuclei.

Additional data on non-thalamic diencephalic parts were given by Krawczuk (1983) on the habenular nuclei of *Sorex araneus, S. minutus* and *Neomys fodiens*; by Szteyn et al. (1987), who described and illustrated the globus pallidus in *S. araneus* and *Neomys fodiens* and found no species-specific differences; by Sigmund et al. (1987), who illustrated myelinated axons and the well differentiated hypothalamic nucleus suprachiasmaticus of *Sorex araneus*; and by Galert (1986), who described supraoptic and paraventricular nuclei of *Sorex araneus, Sorex minutus* and *Neomys fodiens.* Pokorny (1926) mentioned the hypophysis of *Sorex araneus*, which was later described in more detail by Hanström (1946, 1953, 1966). Hanström found all the lobes of the HYP to be extremely flat and just one cell layer in the rostral pars tuberalis. The flatness was not found in the HYP of *Blarina*, which Green (1951) presented in a midsagittal section.

Additional data on thalamic structures were given by Clark (1932, 1933), Gillilan (1941), and Sato (1977). According to Clark, the thalamus shows advanced intrinsic differentiation. The lateral geniculate body (CGL) is small, but its dorsal and ventral nuclei are well differentiated. In *Sorex* the medial geniculate body (CGM) is "merely the latero-caudal part of the main sensory nucleus of the thalamus, and is not to be distinguished from the latter except by its topographical position and its fibre connections" (Clark, 1933, p.536). Gillilan (1941) gave a detailed description of the accessory optic tract and its nucleus in *Blarina*. The structures were found to be less reduced than those of the main optic system. Sato (1977) described and illustrated CGL of *Sorex shinto*. The dorsal nucleus shows no lamination and is composed mainly of medium-sized cells.

In the **telencephalon** (TEL), data were given on olfactory structures (MOB, AOB, RB) by Crosby and Humphrey (1939a, b), Stephan (1965), and Switzer et al. (1980); on striatum (STR), by Szteyn et al. (1987); on septum (SEP), by Andy and Stephan (1961); on amygdala (AMY), by Crosby and Humphrey (1944); on hippocampus (HIP), by Zuckerkandl (1907) and Stephan and Manolescu (1980); and on neocortex (NEO), by Rose (1912, 1929), Ryzen and Campbell (1955), Kurepina (1974), Sato (1977), Rehkämper (1981), and Krasnoshchekova et al. (1987).

In *Blarina brevicauda* Crosby and Humphrey (1939b) found the AOB to be extremely minute, and the internal granular layer to be represented only by scattered cells. The scatter was confirmed by Stephan (1965) in *Sorex araneus*, where the granular cells are concentrated in ventro-medial parts of AOB (also in *Neomys fodiens*). Thick bundles of olfactory tract fibers pass above the internal granular layer in *Neomys* (they normally pass below). Their position was assumed to have developed under spatial or mechanical influence. Switzer et al. (1980) used differing positions as a phylogenetic indicator. According to Stephan (1965), *Neomys* has a broad, well circumscribed, and cell-dense layer of AOB mitral cells. Andy and Stephan (1961, 1966a) gave a detailed description of the septum (SEP) and its nuclei in *Sorex araneus, S. minutus,* and *Neomys fodiens* accompanied by many figures and drawings. Subdivision into four major groups of nuclei (dorsal, ventral, medial and posterior) was possible in all species investigated. Some of the nuclear elements appeared better differentiated in Soricinae than in Crocidurinae. Furthermore, SEP of Soricinae was found to be slender, i.e., more elongated in the vertical direction and narrower in the frontocaudal direction. Illustrations of sagittal and transverse sections through SEP of *Neomys* appear in Stephan and Andy (1982). Crosby and Humphrey (1944) described AMY of *Blarina brevicauda* in detail and confirmed the nuclear pattern generally present in mammals. Zuckerkandl (1907) described the supracommissural hippocampus (indusium griseum) of *Sorex araneus*, and Stephan and Manolescu (1980) found and illustrated very clear differences between subiculum and cornu ammonis and within the latter between CA1 and CA2/3 in *Sorex minutus* (see Fig. 89).

Clark (1932) said that the cortical areas in *Sorex araneus* are rather well differentiated. Misek (1990a, b) studied the development of NEO, in particular, the postcentral region. Cortical subdivisions were given by Rose (1912, 1929) for *Sorex araneus*, Ryzen and Campbell (1955) for *Sorex pacificus*, and Rehkämper (1981) for *Neomys fodiens*. The subdivision given by Rose (1912) is similar to that in *Talpa europaea* (see Fig. 100) and was described on p. 246. Slight differences exist in cingular fields 23 and 24, which were combined in *Sorex* since they could not be separated, and in the area striata, which in *Sorex* is not as strongly reduced as in *Talpa*. The insular cortex (Rose 1929) inhabits the whole frontal pole and lies totally on the surface since there is no rhinal sulcus. Ryzen and Campbell (1955) based their map mainly on quantitative investigations, such as total neuronal number, cell size and density, and grey/cell coefficient. They differentiated ten regions, whereas Rehkämper (1981) differentiated four fundamental regions in *Neomys*, as in all other Insectivora. Additional information was given by Kurepina (1974) on projection and association areas in *Sorex araneus*, by Sato (1977) on the visual cortex of *Sorex shinto*, and by Krasnoshchekova et al. (1987) on the spatial arrangement of the neurons in the parietal region of *Neomys fodiens*.

According to Kurepina (1974), the NEO is less developed in *Sorex* than in the Talpidae. In *Sorex* this is based on: poorly defined cortical stratification, large cell-density, large primordial cortex, large insular region and claustrum, absence of a frontal region, and poor development of cortical regions which undergo progressive development in the course of evolution.

Physiology, Sense Organs

Ärnbäck-Christie-Linde (1907, 1914 , vomeronasal organ, *Sorex, Neomys*)
Andreescu-Nitescu (1970, nose, *Sorex, Neomys*)
Braniš (1981, eye, *Sorex, Neomys*)
Braniš (1985, eye ontogenesis, *Sorex*)
Burda (1979a b, ear, *Sorex, Neomys*)
Ganeshina et al. (1957, nose, *Sorex*)
Gould et al. (1964, ear, *Sorex, Blarina*)
Grün and Schwammberger (1980, eye, *Sorex, Neomys*)
Gurtovoi (1966, nose, *Sorex*)
Henson (1961, ear, *Cryptotis*)
Hutterer (1980, nose, *Sorex, Neomys, Nectogale*)
Hyvärinen (1972, snout, vibrissae, *Sorex*)
Konstantinov et al. (1987, ear, *Sorex*)
Larochelle (1987, nose, *Sorex, Blarina*)
Larochelle and Baron (1989a, b, nose, *Sorex, Blarina*)
Sato (1977, eye, *Sorex*)
Schwarz (1935/36, eye, *Sorex*)
Shibkov (1979, sensory systems, *Sorex, Neomys*)
Sigmund (1985, sense organs, *Sorex*)
Sigmund and Claussen (1986, eye, *Sorex*)
Sigmund and Sedlacek (1985, nose, *Sorex*)
Sigmund et al. (1987, eye, *Sorex*)
Söllner and Kraft (1980, nose, *Sorex, Neomys*)
Sokolova (1964, 1965, eye, *Sorex, Neomys*)
Wöhrmann-Repenning (1975, nose, *Sorex*)
Wöhrmann-Repenning and Meinel (1977, nose, *Sorex*)

The structure of the snout and tactile hair system and the distribution of cholinesterase and alkaline phosphatase in it was studied by Hyvärinen (1972) in *Sorex araneus*.

According to Shibkov (1979), the sense of smell predominates for detecting obstacles at a distance in *Sorex araneus* and *Neomys fodiens*. Vision and audition seem not to be used in their detection, and echolocation seems not to exist.

In the **olfactory system**, the vomeronasal organ and the nasal cavity of *Sorex araneus* and *Neomys fodiens* was investigated by Ärnbäck-Christie-Linde (1907, 1914); the nasal cavity by Andreescu-Nitescu (1970); the nasal cavity and the intranasal epithelia of *Sorex araneus* by Ganeshina et al. (1957), Woehrmann-Repenning (1975), and Woehrmann-Repenning and Meinel (1977). The size of the respiratory and olfactory regions of the epithelium was determined by Gurtovoi (1966) in *Sorex araneus*; by Söllner and Kraft (1980) in *Sorex araneus, S. minutus,* and *Neomys fodiens*; by Sigmund (1985) and Sigmund and Sedlacek (1985) in *Sorex araneus*; and by Larochelle (1987) and Larochelle and Baron (1989b) in *Sorex cinereus, S. fumeus, S. palustris,* and *Blarina brevicauda.* Larochelle and Baron reviewed the earlier literature and showed that in terrestrial Soricinae the surface of the olfactory epithelium makes up between 62.6% and 69.3% of the total nasal epithelium; in water-adapted species, such as *Sorex palustris* and *Neomys fodiens*, the corresponding figures are 52.2% and 41.1%, respectively. The total number of olfactory receptors and their density was also reported by Larochelle. Larochelle and Baron (1989a, b) indicated a relationship between adaptation to semiaquatic life and the development of olfactory structures, and confirmed Hutterer's (1985) finding that *Sorex palustris* is less adapted than *Neomys fodiens.* In another paper, Hutterer (1980) described the rhinarium of water-adapted Soricinae, especially of the strongly adapted *Nectogale elegans.* The sensitivity to acetic, proprionic, and butyric acids was examined by Sigmund and Sedlacek (1985). Sensitivity to fatty acids decreases with increasing receptor density, and the size of the olfactory brain centers need not be correlated with this type of sensitivity.

In the **auditory system**, the middle ear of *Cryptotis parva* was investigated in detail by Henson (1961) who found it to be rather primitive and unspecialized. *Cryptotis* was classified, along with some bats, as a high ultrasonic form, "despite the apparent 'primitiveness' of the middle ear, since there is some experimental and anatomical evidence that shrews can hear sounds in the high frequency and ultrasonic ranges" (Henson 1961, p. 237). Advocates are Gould et al. (1964), who studied *Sorex cinereus, S. palustris, S. vagrans,* and *Blarina brevicauda* and found that they emit high pitched sounds while running or exploring an unfamiliar place. Gould et al. said that these facts support the hypothesis that shrews echolocate. Burda (1979b) showed that the auditory organ of soricids is well-developed and anatomically suited to perceiving sounds in the ultrasonic register. Konstantinov et al. (1987) reported that, compared with hedgehogs and moles, the auditory system in shrews exhibits the best adaptation for perception of acoustic signals with a high repetition rate, at an apparent spectral maximum of 8.5 - 27.4 kHz.

Shibkov (1979), however, questions the existence of echolocation (see above), and Sigmund (1985) pointed out that it is still under debate and has not yet been unequivocally proven. Grünwald (1969) denied echolocation in *Crocidura* (two species investigated, see below).

In the **visual system** of Soricinae, the eyes of *Sorex araneus* were investigated by Schwarz (1935/36; species was called *Crocidura aranea*) and Sigmund et al. (1987); of *Sorex araneus* and *Neomys fodiens*, by Sokolova (1964, 1965); of *Sorex shinto*, by Sato (1977); of *Sorex alpinus, S. araneus, S. minutus, Neomys anomalus* and *N. fodiens* by Braniš (1981); and the postnatal development of the eyes in *Sorex araneus* by Braniš (1985). The retina of *Sorex coronatus* and *Neomys fodiens* was ultrastructurally investigated by Grün and Schwammberger (1980) and that of *Sorex araneus*, by Sigmund and Claussen (1986). Braniš (1981) concluded that the eye of shrews is not markedly reduced. The retina is well developed and the wall of the eyeball does not differ from that of other mammals. He found a rod/cone ratio of 5 - 8 : 1 for *Sorex* and of 15 - 17 : 1 for *Neomys*. Grün and Schwammberger (1980) and Sigmund and Claussen (1986) found functioning retinae with well developed photoreceptors as well as inner and outer synaptic layers. Rods and cones are present. An unusual organization of the inner segment of the receptors was found in *Neomys* (and *Crocidura*, see later) by Grün and Schwammberger. They stated that the high number of larger amacrine cells may be an adaptation to semiaquatic life. According to Sigmund et al. (1987), the retina of *Sorex araneus* shows some morphological differences from that of *Crocidura*, which were correlated with the polyphasic pattern of its activity (activity bursts during dark and light period, in contrast to only in the dark period in *Crocidura*).

4.6.2 Crocidurinae

Crocidura attenuata, C. flavescens, C. hildegardeae, C. jacksoni, C. russula, C. suaveolens, Suncus etruscus, S. murinus, Scutisorex somereni, Sylvisorex granti, S. megalura, Ruwenzorisorex suncoides, Myosorex babaulti

Brain Weight, Encephalization

Andy and Stephan (1961, *Crocidura*)
Bauchot and Stephan (1964, 1966, *Crocidura, Suncus, Sylvisorex*)
Fons et al. (1984, *Crocidura, Suncus, Sylvisorex*)
Mangold-Wirz (1966, *Crocidura*)
Schwerdtfeger (1984, *Crocidura*)
Snell (1892, *Crocidura*)
Stephan (1959, *Crocidura, Sylvisorex*)
Stephan and Andy (1982, *Crocidura, Suncus, Sylvisorex*)
Stephan et al. (1970, 1981a, *Crocidura, Suncus*)
Waterlot (1912, *Crocidura*)
Wirz (1950, *Crocidura*)

BoW/BrW data on nine species of Crocidurinae were published by Bauchot and Stephan (1966) and of ten species by Fons et al. (1984). Since then, the collection has been enlarged by many new individuals, 14 from six new species, and by additional data taken from the literature. The sources are given in Table 5 and will not be repeated here.

(1) *Crocidura attenuata:* One male was collected in Thailand. Its body weight (BoW in g) and brain weight (BrW in mg) are 9.3/209.

(2) *Crocidura flavescens:* 41 males have averages of 33.2/432 and eleven females, 25.9/403. There are 58 additional BoWs with an average of 34.3 g and seven additional BrWs with an average of 394 mg. Based on the available data, the males appear to be larger than the females and to have larger brains. Therefore, the median values between males and females (29.3/414) were chosen as standards.

(3) *Crocidura giffardi:* Waterlot (1912) gave one pair of BoW/BrW for this very large species: 82.0/545.

(4) *Crocidura hildegardeae:* Five males have averages of 10.1/226 and four females, 7.8/196. The data seem to indicate sex differences, but additional BoWs do not support this expectation. The available data indicate standards of 10.2/213.

(5) *Crocidura jacksoni:* One pair from a female is 10.3/250. Four additional BoWs have an average of 13.25 g and, thus, indicate a larger species weight. The available data result in BoW/BrW standards of 12.7/250.

(6) *Crocidura leucodon:* One pair from a female with 13.5/190.

(7) *Crocidura russula:* Eleven males have averages of 11.1/199, 13 females, 11.7/192. 61 additional BoWs and twelve additional BrWs result in well founded standards of 11.1/197.

(8) *Crocidura suaveolens:* One pair from a male with 10.3/190.

(9) *Suncus etruscus:* Two males have averages of 2.20/62 and four females, 2.01/62. Nearly 200 additional BoWs, collected by Fons and collaborators, resulted in a slightly lower value. The total data lead to standards of 1.86/62.

(10) *Suncus murinus:* 14 males have averages of 41.4/397 and twelve females, 25.7/366. Additional BoWs and BrWs provide more support for the existence of sex differences. Therefore, the median values between males and females (33.8/383) were chosen as standards.

(11) *Scutisorex somereni:* Four males have averages of 58.9/640. 70 BoWs given by Dieterlen and Heim de Balsac (1979) have an average of 63.7 grams and, thus, indicate a slightly larger size for this species. The resulting standards are 63.4/640.

(12) *Sylvisorex granti:* Three males have averages of 3.83/166, three females, 3.93/164. The average of 25 additional BoWs is 3.90 and, thus, confirm the weight of this species. The resulting, well founded standards are 3.90/165.

(13) *Sylvisorex megalura:* Five males have averages of 5.70/183, eight females, 5.46/191. The average of 40 additional BoWs is 5.47 and, thus, confirm the weight of this species. The resulting, well founded standards are 5.49/188.

(14) *Ruwenzorisorex suncoides:* One pair from a male with 18.2/370. Fons et al. (1984) mentioned a *Sylvisorex* (*lunaris*?) which, based on BoW and BrW (18.5/358), may also have been a *Ruwenzorisorex*.

(15) *Myosorex babaulti:* One pair from a female with 17.0/360. The BoW was exactly confirmed by the average for twelve animals given by Dieterlen and Heim de Balsac (1979).

The encephalization of crocidurine shrews is on the average 1.37 times greater than that of Tenrecinae (Table 7). There are, however, clear differences between two groups of Crocidurinae. Based on brain characteristics, they will be called the *Crocidura* group and *Sylvisorex* group, with no implication that there are closer systematic relationships among all the genera within the groups. The *Crocidura* group includes eleven species belonging to three genera (*Crocidura, Suncus,* and *Scutisorex*). Its average brain size is 1.17x AvTn, with five species even lower than the tenrecine level, which by definition is 1.00. In contrast, the *Sylvisorex*-group, with four species belonging to three genera (*Sylvisorex, Ruwenzorisorex, Myosorex*), has an average brain size of 1.70x AvTn. Relative to all Insectivora, this average is higher than the species average (AvIS), which is 1.48x AvTn, and the family/subfamily average (AvIF), which is 1.58x AvTn. In contrast, the *Crocidura* group average is clearly below AvIS and AvIF.

Macromorphology

Ärnbäck-Christie-Linde (1907, *Crocidura*)
Brauer and Schober (1970, 1976, *Crocidura*)
Clark (1932, *Crocidura*)
Fons et al. (1984, *Suncus*)
Leche (1905, 1907, *Crocidura*)
Schwerdtfeger (1984, HIP, *Crocidura*)
Stephan (1956a, *Crocidura*)
Stephan and Andy (1982, *Crocidura*)

The macromorphology of crocidurine and soricine brains is similar. Therefore, only some of the differences will be pointed out, such as: (1) the olfactory bulbs in the crocidurine brains project farther beyond the cerebral hemispheres; (2) the cerebellar lobulus petrosus is more conspicuous in the dorsal view; (3) small parts of the mesencephalic tectum are visible (Fig. 15); (4) the olfactory tubercle is clearly less prominent (cf. Figs. 26 and 27).

In a midsagittal reconstruction of the smallest brain (*Suncus etruscus*) Fons et al. (1984) found no trace of a fissura secunda in the cerebellar arbor vitae (Fig. 51). It is present in larger shrews (Fig. 50), but always small.

Cell and Area Measurements

Bauchot (1963, DIE, *Crocidura*)
Bauchot and Stephan (1961, sectioned areas of CC, COA, CHO, *Crocidura*)
Schmidt and Nadolski (1979, MOB, *Crocidura*)
Stephan (1956a, b, 1961, surface areas of cortex, *Crocidura*)
Stephan and Manolescu (1980, HIP, surface areas of cortex, layer thickness, *Crocidura, Suncus*)
Stephan et al. (1984b, VIS, *Crocidura, Suncus*)
Stolzenburg et al. (1989, NEO, *Suncus*)

The cell and area measurements may be classified into (1) areas of transversally sectioned fiber bundles, (2) size, number, and density of fibers and cells, (3) cortex and layer thickness, and (4) surface areas.

(1) Stephan et al. (1984b) compared transversally sectioned areas of nervus opticus (NOA) and midsagittally sectioned areas of optic chiasma (CHO) in *Crocidura flavescens, C. russula,* and *Suncus murinus*; Bauchot and Stephan (1961) examined them in corpus callosum (CCA), anterior commissure (COA), and CHO in the two *Crocidura* species. The size is similar to that found in soricine shrews.

(2) Bauchot (1963, 1964) determined several cell characteristics in many parts of the diencephalon of *Crocidura flavescens* and *C. russula* and compared them using allometric methods. Stolzenburg et al. (1989) studied the somatosensory cortex of the very small *Suncus etruscus*. They found similar relations as in *Microsorex hoyi*, which is another very small shrew (see above, Soricinae). In *Crocidura russula* Schmidt and Nadolski (1979) counted about 1600 glomeruli in the MOB with an average diameter of 75μm.

(3) In the hippocampus (HIP) of *Crocidura flavescens, C. russula,* and *Suncus murinus,* the thickness of the cell layer was determined by Stephan and Manolescu (1980). Compared with soricine shrews, the subiculum is distinctly thicker and the cell bands of cornu ammonis and fascia dentata are clearly thinner. Stolzenburg et al. (1989) found a relatively broad molecular layer in the somatosensory cortex of *Suncus etruscus* and relatively thinner molecular layers in larger species of Insectivora.

(4) Based on the measurements by Stephan (1956a, b, 1961), the total cortical surface area in *Crocidura russula* is composed of 50.7% PAL, 22.1% ARCH, and 27.3% NEO, and in *C. flavescens,* of 50.6, 24.0, and 25.4%, respectively. In the hippocampus, the surface areas are larger in *Crocidura flavescens* than in *C. russula* and *Suncus murinus* (Stephan and Manolescu 1980).

Relative eye size was found to be small in all species for which we have data (*Crocidura russula, C. giffardi,* and *C. suaveolens*) (Stephan et al. 1984b).

Volume Measurements

Andy and Stephan (1966a, b, 1968, SEP, *Crocidura, Suncus*)
Baron (1972, 1974, 1977a, auditory and visual structures, *Crocidura, Suncus*)
Baron (1977b, vestibular ncl., *Crocidura, Suncus*)
Baron (1978, cerebellar ncl., *Crocidura, Suncus*)
Baron (1979, DIE, *Crocidura, Suncus*)
Baron et al. (1983, MOB, *Crocidura, Suncus*)
Baron et al. (1987, PAL, *Crocidura, Suncus, Scutisorex, Sylvisorex, Ruwenzorisorex, Myosorex*)
Baron et al. (1988, VES, same genera as 1987)
Bauchot (1961, 1970, HYP, *Crocidura, Suncus*)
Bauchot (1963, 1966, 1979b, DIE, THA, *Crocidura, Suncus*)
Bauchot (1979a, main parts, *Crocidura, Suncus*)
Bauchot and Legait (1978, HYP, *Crocidura, Suncus*)
Bauchot and Stephan (1968, main parts, DIE, THA, *Crocidura, Suncus*)
Etienne (1972, BOL, *Crocidura, Suncus*)
Frahm et al. (1979, PALL, *Crocidura, Suncus*)
Frahm et al. (1982, NEO, *Crocidura, Suncus*)
Legait et al. (1976, EPI, *Crocidura, Suncus*)
Matano et al. (1985b, CER nuclei, *Crocidura, Suncus*)
Schwerdtfeger (1984, HIP, *Crocidura*)
Stephan (1966, 1967b, main parts, *Crocidura, Suncus*)
Stephan (1969, PVO, *Crocidura, Suncus*)
Stephan and Andy (1962, 1982, SEP, *Crocidura, Suncus*)
Stephan and Andy (1964, 1969, 1982, main parts, *Crocidura, Suncus*)
Stephan and Andy (1977, AMY, *Crocidura, Suncus*)
Stephan and Manolescu (1980, HIP, *Crocidura, Suncus*)
Stephan et al. (1970, 1981a, main parts, *Crocidura, Suncus*)
Stephan et al. (1982, AOB, *Crocidura, Suncus*)
Stephan et al. (1984a, main parts, NEO, *Crocidura, Suncus*)
Stephan et al. (1984b, CGL, *Crocidura, Suncus*)
Stephan et al. (1987a, AMY, *Crocidura, Suncus, Scutisorex, Sylvisorex, Ruwenzorisorex, Myosorex*)
Wirz (1950, weights of main parts, *Crocidura*)

Weights of brain parts were compared using index methods by Wirz (1950), who subdivided the brains macroscopically into five parts (neocortex, basal ganglia, rhinencephalon, cerebellum and "stammrest") but did not publish the raw data (see also Mangold-Wirz 1966). In Crocidurinae, she investigated *Crocidura russula*.

Data on the five **fundamental brain parts** obtained from serial sections of *Crocidura flavescens, C. russula,* and *Suncus murinus* were given by Stephan and Andy (1964, 1969, 1982), Stephan (1966, 1967b), Bauchot and Stephan (1968), Stephan et al. (1970, 1981a), and Bauchot (1979a) and compared with other Insectivora, with Primates and Chiroptera. Stephan et al. (1984a) included three more species: *Crocidura hildegardeae, C. jacksoni,* and *Suncus etruscus.* Newly added in this book are *Crocidura attenuata, C. suaveolens, Scutisorex somereni, Sylvisorex granti, S. megalura, Ruwenzorisorex suncoides,* and *Myosorex babaulti.* Some of the new species differed markedly from the others so that two groups were formed (see encephalization, p. 272).

A *Sylvisorex* group, with a clearly larger brain, was separated from a *Crocidura* group. In the *Sylvisorex* group, encephalization was 1.45 times higher and, in the five fundamental brain parts, a similar difference of 1.4x was found for DIE (1.39) and TEL (1.43), while the differences in OBL and MES were smaller (1.24 - 1.29x) and for CER, greater (1.55x). Thus, the largest difference between the *Crocidura* and *Sylvisorex* groups is in CER. The lowest average index in the *Crocidura* group was found for MES (1.20x AvTn) and the highest for CER (1.30x); in the *Sylvisorex* group, the lowest was in OBL (1.52x AvTn) and the highest in CER (2.02x). The size variations within the groups were quite low and the contribution of the five fundamental brain parts to total brain was similar. In the *Crocidura* group, the largest difference was in OBL, which was 1.42 times larger in *Crocidura jacksoni* than in *Suncus murinus*, and in the *Sylvisorex* group, the largest difference was in MES, which was 1.44 times larger in *Sylvisorex megalura* than in *Myosorex babaulti.*

In the **medulla oblongata** (OBL), data on auditory and vestibular structures were given by Baron (1972, 1974, 1977b) and Stephan et al. (1981a) for *Crocidura flavescens, C. russula,* and *Suncus murinus* and vestibular structures for five more species by Baron et al. (1988). Many more OBL structures in eight species are given in this book (Tables 13 - 27): six are from species in the *Crocidura* group and two from species in the *Sylvisorex* group. All structures are on the average larger in the *Sylvisorex* group, but OBL composition is similar in the two groups.

The largest differences in the groups were found for FCE, which in the *Crocidura* group is larger in *Scutisorex* and *Suncus* spp. than in the *Crocidura* spp., with the greatest difference being between *Scutisorex somereni* and *Crocidura russula* (in the former 3.9 times larger). Again in the *Sylvisorex* group, FCE showed large differences in size, and was found to be nearly 1.7 times larger in *Sylvisorex megalura* than in *Ruwenzorisorex suncoides.*

In the **mesencephalon** (MES), volume data were given for the lemniscus lateralis nuclei (NLL) and the colliculi by Baron (1972, 1977a) and for the subcommissural body (SCB) by Stephan (1969) and Stephan et al. (1981a) in *Crocidura flavescens, C. russula,* and *Suncus murinus.* No measurements are available for members of the *Sylvisorex* group. The variation is low in the *Crocidura* group, except for SCB, which was found to be 3.6 timer larger in *Suncus murinus* than in *Crocidura flavescens.*

In the **cerebellum** (CER), data on the nuclei were given by Baron (1978) and Matano et al. (1985b) for *Crocidura flavescens, C. russula,* and *Suncus murinus.* The variation among the three measured species is relatively low. Compared with Tenrecinae, ICN is even smaller (0.83x AvTn), whereas LCN is largest, reaching 1.31x AvTn on the average. No measurements are available for members of the *Sylvisorex* group.

In the **diencephalon** (DIE), data on main parts were given by Bauchot (1963, 1966, 1979b), Bauchot and Stephan (1968), Baron (1979) and Stephan et al. (1981a) for *Crocidura flavescens, C. russula,* and *Suncus murinus.* Baron (1972, 1977a) and Stephan et al. (1984b) gave data on the same species for some diencephalic visual structures and Frahm et al. (1979) provided data on the pallidum (PALL). No measurements are available for members of the *Sylvisorex* group. Based on Bauchot's data, all diencephalic structures are quite small relative to the average Insectivora (AvIS and AvIF). STH (1.51x AvTn) and, within the thalamus, CGM (1.96x) and DOT (1.84x) are largest in relative terms. The variation among the three measured species is relatively low. It is highest for ANT, which is 1.55 times larger in *Crocidura flavescens* than in *Suncus murinus.*

Data on some small diencephalic structures were given for EPI and HAM by Stephan (1969) and Stephan et al. (1981a), for EPI by Legait et al. (1976), and for HYP by Bauchot (1961, 1970) and Bauchot and Legait (1978). HYP was found to be relatively small, but a large intermediate lobe was found in *Crocidura flavescens.* Data on EPI and HAM are added in this book for *Suncus etruscus* and *Scutisorex somereni* and for the four species of the *Sylvisorex* group (*Sylvisorex granti, S. megalura, Ruwenzorisorex suncoides,* and *Myosorex babaulti*).

EPI is very small in all crocidurine shrews so far investigated, and HAM is about average in size for Insectivora, and 1.3 times larger in the *Sylvisorex* group than in the *Crocidura* group. There is, however, large variation in both structures within each of the groups with maximum values for EPI in the *Sylvisorex* group, where it was more than 13 times larger in *Sylvisorex granti* than in *Myosorex babaulti.*

In the **telencephalon** (TEL), data on seven to nine main parts were given by the same authors and in the same species as described for the fundamental parts of the brain (see p. 275). Additional data, mostly from *Crocidura flavescens, C. russula,* and *Suncus murinus*, were given for olfactory structures by Stephan (1966); for MOB by Etienne (1972) and Baron et al. (1983); for AOB by Stephan et al. (1982); for PAL by Baron et al. (1987); for SEP by Stephan and Andy (1962, 1982), Andy and Stephan (1966a, b, 1968), and Stephan (1969); for AMY by Stephan and Andy (1977) and Stephan et al. (1987a); for HIP by Stephan and Manolescu (1980); and for NEO by Frahm et al. (1982) and Stephan et al (1984a). The same species mentioned in conjunction with the fundamental brain parts are newly added in this book, and the differences between the two groups are enormous in TEL. The *Sylvisorex* group, with a 1.43 times larger TEL, is clearly differentiated from the *Crocidura* group. Most fundamental TEL parts were between 1.2 and 1.4 times larger while NEO (1.54 times) and STR (1.67 times) were even larger. Compared with the indices in average Insectivora (AvIS and AvIF), all TEL structures in the *Crocidura* group were clearly lower (lowest in STR, highest in MOB), and those of the *Sylvisorex* group were mostly at or slightly above the AvIS and AvIF levels, except for SCH (below) and MOB and PAL (more strongly above). Compared with the reference base (AvTn), the highest average indices were found for NEO (2.01x AvTn in *Crocidura* group; 3.09x AvTn in *Sylvisorex* group), the lowest for MOB (0.82x and 1.06x AvTn) and PAL (0.91x and 1.27x AvTn, respectively). For MOB and PAL, there is enormous variation within the *Sylvisorex* group. MOB is 2.35 times larger in *Sylvisorex granti* than in *Ruwenzorisorex suncoides*, while PAL is 2.14 times larger. *Sylvisorex megalura* and *Myosorex babaulti* are similar to *S. granti* and, thus, *Ruwenzorisorex* has, in the *Sylvisorex* group, an exceptionally low development of the main olfactory structures.

Details on histology are given in many of the papers on volume measurements.

Histology, Histochemistry, Fiber Connections

Andy and Stephan (1961, 1966b, SEP, *Crocidura*)
Bauchot (1963, DIE, *Crocidura*)
Girod (1976, HYP, *Suncus*)
Hayakawa and Zyo (1983, tegmentum, *Suncus*)
Hayakawa and Zyo (1984, tegmento-mammillar connections, *Suncus*)
Isomura et al. (1986, SPC, *Suncus*)
Kamei et al. (1981, inferior olive, *Suncus*)
Kulshreshtha and Dominic (1971, HTH, *Suncus*)
Mohanty and Kar (1982, biochemistry, *Suncus*)
Naik and Dominic (1971, 1972, 1978, HYP, *Suncus*)
Rehkämper (1981, NEO, *Crocidura*)
Schwerdtfeger (1984, HIP, *Crocidura*)
Sigmund et al. (1984, VIS, *Crocidura*)
Stephan (1965, AOB, *Crocidura, Suncus*)
Stephan and Andy (1982, SEP, *Crocidura, Suncus*)
Stephan and Manolescu (1980, HIP, *Crocidura, Suncus*)
Sugiura and Kitoh (1984, SPC, *Suncus*)

Studies on the histochemistry of the house, or musk, shrew (*Suncus murinus*) were performed by Mohanty and Kar (1982), who investigated some chemical contents as well as acid and alkaline phosphatase activity, and by Kamei et al. (1981), who included the species in their comparison of catecholamine distribution in the inferior olivary complex (INO). INO contained a small number of green fluorescent fibers, while the area just dorsal to it was quite richly innervated by catecholamine neurons.

In the **spinal cord** (SPC) of *Suncus murinus* Isomura et al. (1986) studied the obliteration of the central canal. There is complete obliteration from its rostral beginning ventral to the area postrema until the caudal end of SPC. Sugiura and Kitoh (1984) examined extent and laminar arrangement of the substantia gelatinosa and the axon terminals in it. The occurrence of terminal profiles with flat vesicles was significantly greater in the deeper laminae, unlike the situation in other mammals.

Hayakawa and Zyo (1983, 1984) included *Suncus murinus* in their comparative cytoarchitectural study on Gudden's tegmental nuclei and illustrated its position and subdivision. In contrast to most other mammals, a pars ventralis and dorsalis could not be separated in the dorsal tegmental nucleus of Gudden. Strong connections with the mamillary bodies were confirmed: the dorsal tegmental nucleus of *Suncus* projects to the lateral mamillary nucleus, while the ventral tegmental nucleus projects to the medial and lateral mamillary nuclei.

For the **diencephalon** (DIE) of *Crocidura flavescens* and *C. russula*, Bauchot (1963) provided detailed and comprehensive descriptions. The visual centers are small, as in *Sorex*, Sigmund et al. (1984) described the primary optic pathways in *C. suaveolens*. Degenerated axons were found bilaterally in dorsal and ventral parts of CGL and the pretectal area and extensively in the superficial layers of the colliculus superior. According to Bauchot (1963), the corpora mamillaria are well developed in *Crocidura*. The nucleus paraventricularis, while large in *Sorex*, is small in *Crocidura*. This relation is reversed in the nucleus supraopticus. Some aspects of the hypothalamic neurosecretory system of *Suncus murinus* were studied by Kulshreshtha and Dominic (1971), and of the hypophysis (HYP), by Naik and Dominic (1971, 1972, 1978). Seven cell types were described in the pars anterior of HYP; they were correlated with secretion of various trophic hormones. The results of the first two papers by Naik and Dominic were compared with those on many other species by Girod (1976).

In the **telencephalon** (TEL), the accessory olfactory bulb (AOB) was studied by Stephan (1965) in *Crocidura flavescens, C. russula*, and *Suncus murinus*. In *C. russula*, a very broad and cell-rich layer of mitral cells was found, and in *C. flavescens*, an inner plexiform layer was not identified. Andy and Stephan (1961, 1966b) gave a detailed description of the septum (SEP) and its nuclei in *C. flavescens* and *C. russula*, along with many photoprints. In Crocidurinae, SEP is more compact in the sagittal section and various nuclei are less differentiated than in Soricinae. In contrast, the nucleus preopticus medianus, adjacent to SEP, is better differentiated in *Crocidura*. Illustrations of SEP in *C. russula* and *Suncus murinus* are given in Stephan and Andy (1982). Stephan and Manolescu (1980) found a displacement of cells toward the molecular layer in area CA 1 of the hippocampus (HIP) of *C. flavescens*. This is in contrast to *Suncus murinus* (and *Sorex*), where CA 1 is very compact (see Fig. 89). Schwerdtfeger (1984) gave a detailed and well illustrated description of the HIP in *Crocidura russula* and investigated its structure and fiber connections. He described two circuits of reciprocal connections, one with the anterior thalamus and the second with the dorsal and lateral septum and back to HIP from the medial septum and the diagonal band. Rehkämper (1981) compared the neocortex (NEO) of *C. flavescens* with that of other Insectivora and *Elephantulus*. A very dense second layer and a cell-poor third layer characterize the NEO of *Crocidura*. Regional differences in the laminar structure allowed for subdivision into four neocortical regions apart from the insular cortex.

Physiology, Sense Organs

Ärnbäck-Christie-Linde (1914, vomeronasal organ, *Crocidura*)
Andreescu-Nitescu (1970, nose, *Crocidura*)
Braniš (1981, eye, *Crocidura*)
Burda (1979a, b, ear, *Crocidura*)
Ganeshina et al. (1957, nose, *Crocidura*)
Grün and Schwammberger (1980, eye, *Crocidura*)
Grünwald (1969, senses, *Crocidura*)
Kuramoto et al. (1980, nose, *Suncus*)
Schmidt and Nadolsky (1979, nose, *Crocidura*)
Schwarz (1935/36, eye, *Crocidura*)
Sharma (1958, nose, eye, *Suncus*)
Sigmund (1985, sense organs, *Crocidura*)
Sigmund and Claussen (1986, eye, *Crocidura*)
Sigmund et al. (1987, eye, *Crocidura*)
Söllner and Kraft (1980, nose, *Crocidura*)
Woehrmann-Repenning (1975, nose, *Crocidura*)
Woehrmann-Repenning and Meinel (1977, nose, *Crocidura*)

Grünwald (1969) analyzed the different senses used in orientation by *Crocidura russula* and *C. olivieri.* He pointed out that (1) short-distance orientation is chiefly determined by tactile sensations; (2) vision plays a subordinate role; (3) olfaction is well developed and plays an important role in the search for food; and (4) audition is highly developed, but echolocation does not exist. *Crocidura* rapidly acquires accurate knowledge about its environment. Spatial memory is very important and fixed to such a degree that direct sensorial perception has only subsidiary functions in orientation.

In the **olfactory system**, the vomeronasal organ of *Crocidura russula* was studied by Ärnbäck-Christie-Linde (1914); the nasal cavity of *Suncus murinus* was described by Sharma (1958), and of *Crocidura leucodon* and *C. suaveolens* by Andreescu-Nitescu (1970); the nasal cavity and the intranasal epithelia were studied in *Crocidura suaveolens* by Ganeshina et al. (1957), in *C. russula* by Wöhrmann-Repenning (1975) and Wöhrmann-Repenning and Meinel (1977), and in *Suncus murinus* by Kuramoto et al. (1980). The size of the olfactory epithelium relative to total epithelium was determined in *Crocidura russula* by Schmidt and Nadolsky (1979) and in *C. leucodon* and *C. russula* by Söllner and Kraft (1980). The authors found 58 - 65% to be olfactory epithelium, which is on the average lower than in terrestrial Soricinae, but higher than in semiaquatic Soricinae (reviewed by Larochelle 1987, lit.). Sigmund (1985) found much greater sensitivity to acetic acid in *Crocidura suaveolens* than in *Sorex araneus*.

In the **auditory system**, Burda (1979b) found that all middle as well as inner ear structures were larger in *Crocidura suaveolens* than in the species of the subfamily Soricinae he examined. There may also be differences in the existence of echolocation. Echolocation was denied in *Crocidura* by Grünwald (1969), but was rather convincingly demonstrated in soricine shrews (*Sorex* and *Blarina*) by Gould et al. (1964).

In the **visual system** of Crocidurinae, the eyes of *Crocidura leucodon* were described in detail by Schwarz (1935/36). Further descriptions are for *Suncus murinus* by Sharma (1958) and for *Crocidura leucodon* and *C. suaveolens* by Braniš (1981). Sigmund et al. (1987) studied the photoreceptors of *Crocidura suaveolens* and discussed their characteristics in relation to dark activity and locomotor behavior in crocidurine shrews. Details of the ultrastructure of the eye of *Crocidura russula* were given by Grün and Schwammberger (1980) and of *C. suaveolens* by Sigmund and Claussen (1986). Grün and Schwammberger found that rods and cones are not easily distinguished. Compared with *Sorex* and *Neomys, Crocidura* seems to have the least developed retina. According to Braniš (1981), the rod/cone ratio in *Crocidura* is 11 - 15 : 1, which is intermediate between that of *Sorex* (5 - 8 : 1) and of *Neomys* (15 - 17 : 1).

4.6.3 Summary of Brain Characteristics in Subfamilies and Species of Soricidae and in two Groups within Crocidurinae

Compared with the average Insectivora (AvIF), the brains of the Soricinae are similar, those of the Crocidurinae are smaller. Some macromorphological features are similar for both Soricinae and Crocidurinae (see Section 3.1) and, resulting from the measurements, the somatosensory systems of the forelimbs are poorly represented in both. Otherwise, the two subfamilies differ in many characteristics, including the representation of the main functional systems. Index profiles are given in Figs. 102, 104, 106, and 108.

Soricinae

The NET brain of soricine shrews is only slightly below AvIF (0.98x). Four of the five fundamental brain parts also are below AvIF. They vary from 0.72x AvIF for MES to 0.92x for DIE (Fig. 102). Only TEL is above AvIF (1.09x). In OBL, the components vary from 0.48x for VCO and OLS to 1.18x for mot X (Fig. 104); in DIE, from 0.65x for DOT to 1.09x for CGM; and in TEL, from 0.56x for AOB and 0.86x for MOB to 1.21x for STR (Fig. 108). Of the OBL components, after mot X, only TSO, and mot XII are above AvIF, whereas all the auditory, funicular and vestibular nuclei are less represented.

In conclusion, soricine shrews have small brains with less well represented auditory, olfactory, and vestibular structures. Of the somatosensory structures, those of the forelimbs (FCM) also are less represented. Centers of internal organs (mot X, TSO) are well represented.

Species characteristics of soricine brains are based on the deviation from the average soricine level (AvSn), which is uniformly set at 1 for each of the structures. In *Sorex araneus*, a large body of information found in the literature coincides with our abundant measurements. Therefore, the brain features of this species will be considered first.

Sorex araneus: In Soricinae, *Sorex araneus* has the second smallest brain (0.88x AvSn for BrW; 0.87x for NET). The five fundamental brain parts vary from 0.85x AvSn for CER and TEL to 1.01x for OBL; the eight TEL parts vary from 0.67x AvSn for NEO to 1.03x for MOB and 1.78x for AOB. Only OBL, MOB, and PAL reach the soricine level, whereas NEO, STR, SCH, TEL as a whole, and CER are distinctly below the soricine average. NEO and STR are the smallest found in Soricinae. Of the smaller structures, SFB (0.52x AvSn) and HAM (0.66x) are also relatively small, whereas AOB (1.78x) is the largest found in the eleven species investigated.

Of the structures measured in OBL, mot V was 1.28x AvSn and DCO 1.19x and thus were the largest among the five species so far measured. The nucleus prepositus hypoglossi (PRP; 0.73x) and two of the funicular nuclei (FGR, FCM) are smallest. The third one (FCE) and the nucleus vestibularis inferior (VI) are also small (0.77x and 0.66x). Since, however, the lateral vestibular nucleus (VL) is relatively large, the small size of VI may be explained by uncertain delimitation. The brain of *Sorex araneus* has the longest and the flattest hemispheres (Table 3).

Particularities: *Sorex araneus* has relatively large mot V, auditory DCO, and accessory olfactory structures; general motor structures (CER, STR), funicular nuclei, VI, and PRP are relatively small compared with other soricine species.

Sorex alpinus: All structures so far measured are close to the soricine level (AvSn) and are thus larger than in *S. araneus*. Exceptions are the main olfactory structures (MOB, PAL), which are slightly smaller than in *S. araneus* and in the average soricine shrews. AOB is distinctly smaller than in *S. araneus*. CER is particularly large and among the largest found in Soricinae; STR is also relatively large.

Particularities: Typical soricine brain with relatively large CER and STR.

Sorex cinereus: The total brain is 1.08 times above the soricine average (AvSn), with a relatively small OBL and large TEL. In the telencephalon, STR and PAL are 1.25 and 1.21x AvSn and AOB is distinctly smaller (0.70x). Of the structures measured in OBL, the nucleus reticularis lateralis (REL) was the largest among Soricinae (1.41x AvSn), whereas the motor nuclei (mot V, VII, X and XII) were the smallest (0.61 - 0.85x AvSn).

Particularities: Typical soricine brain, but smaller AOB and motor nuclei.

Sorex fumeus: In BrW and NET, the highest values in soricine shrews were found in *Sorex fumeus* (1.12 and 1.11x AvSn). The small deviation from the soricine average indicates the uniformity of the subfamily. OBL is smallest in *Sorex fumeus* (0.89x AvSn), while the largest figures were found for TEL (1.19x AvSn), and, in the telencephalon, HIP (1.19x), SCH (1.23x), and NEO (1.32x). HIP, SCH, and NEO are the largest in Soricinae, and this is also true of the length of the corpus callosum. OBL parts were not measured.

Particularities: Relatively large brain with large HIP, SCH and NEO.

Sorex minutus: In BrW and NET, the brain of this species has the distinctly lowest position (0.81 and 0.82x AvSn) as it does also for all five fundamental brain parts except OBL, which is slightly smaller in *Microsorex hoyi*. In the telencephalon, SEP, AMY, and HIP are also the smallest, as are most parts of DIE, except VET and CGL, which are close to the soricine average, and CGM, which is lower. The subcommissural body is the largest found in soricine shrews (1.31x AvSn). OBL parts were not measured.

Particularities: Distinctly the smallest brain found in soricine shrews, with most brain parts also the smallest.

Microsorex hoyi: Next to *Sorex minutus* and *S. araneus,* this species has the third smallest brain (BrW and NET) of all the Soricinae so far investigated (0.96x AvSn). Of the five fundamental brains parts, MES is the smallest, and CER and OBL are also very small. These parts are 0.84 - 0.86x AvSn and thus compete with the small size in *S. minutus*. The size of most TEL parts is similarly slightly below AvSn: between 0.91x for MOB and 0.98x for SEP and NEO. Only AOB (1.02x) and STR (1.05x) are slightly above AvSn. OBL parts were measured: the smallest in soricine shrews were TR, REL, INO, the vestibular complex, and the auditory complex (all between 0.71 and 0.83x AvSn). The largest was the tiny nucleus locus coeruleus (COE) (1.47x). The brain of *Microsorex* is the longest and narrowest of all soricine brains.

Particularities: Relatively small brain with very small MES, CER, and OBL and, within the OBL, small REL, INO, trigeminal, and vestibular and auditory structures. Only the tiny nucleus locus coeruleus was found to be relatively large.

Neomys fodiens and ***Neomys anomalus:*** In *Neomys anomalus,* only the five fundamental brain parts and the main telencephalic structures were measured. Since there are no major differences, this species will be considered together with *N. fodiens.* The brains of *Neomys* spp. are relatively large and slightly above AvSn (1.01 - 1.06X). All the fundamental brain parts are the largest found in soricine shrews, except TEL, which is at the soricine level (1.01x AvSn in *N. anomalus*) or slightly below (0.91X in *N. fodiens*). This is mainly because of the small structures in the main olfactory system (MOB, PAL), which are 0.72 - 0.78X AvSn. In contrast, the AOBs are among the largest in Soricinae (1.50 - 1.64X AvSn). A large triangular nucleus was found in the septum. In DIE, measured by Bauchot (1979b) in *N. fodiens*, all nuclear groups except VET are largest in the soricine shrews. Based on our data, EPI and HAM were also the largest. OBL structures were measured in *N. fodiens*, where most of the components were found to be the largest or among the largest. This applies as well to the vestibular and trigeminal structures, the funicular nuclei, TSO, PRP, INO, motor nuclei (except mot X), and the auditory structures. The brain of *N. anomalus* is the shortest and highest of all soricine brains.

Particularities: Nearly all brain structures in *Neomys* spp. are the largest found in Soricinae. Only TEL is at the soricine level, since its olfactory parts are the smallest found in the soricine shrews. The relatively small TEL, attributable to olfactory reduction, reduces BrW and NET and so prevents *Neomys* from being the highest in this overall measure. Nevertheless, the brains of *Neomys* may be regarded as the most progressive in soricine shrews.

Blarina brevicauda: The size of the *Blarina* brain and its five fundamental parts is slightly above the soricine level (1.00 - 1.04X AvSn). In TEL, AOB (0.81X), AMY (0.95X), and STR (0.96X) have small relative size, and HIP (1.18X), SEP (1.15X), and MOB (1.13X AvSn) have large relative size . OBL structures were measured; most were also slightly above or close to AvSn, except for mot X, which was distinctly above (1.46X). In contrast, VL and COE were distinctly below (0.69 and 0.76X AvSn). Since, however, all other vestibular structures are distinctly higher (1.10 - 1.19X AvSn), the explanation may be uncertain delimitation. TR, REL, mot V and mot VII are also lower than AvSn.

Particularities: Relatively large brain, with almost all its parts uniformly large. The olfactory and auditory structures are well developed, whereas the trigeminal structures are relatively small.

Cryptotis parva: *Cryptotis* has one of the most typical brains in soricine shrews. Brain size and four of the five fundamental brain parts are slightly above AvSn (1.02 - 1.06X), and only CER is slightly below (0.95X). Among the TEL parts, AOB is the smallest among all soricine species (0.40X AvSn). STR (0.89X) and NEO (0.88X) are also below AvSn whereas MOB (1.25X) and AMY

(1.19x) are distinctly above. Structures of OBL and DIE have not yet been measured.

Particularities: Close to the type of soricine brains. There are, however, large main olfactory structures, very small accessory olfactory structures, and relatively small motor components (STR, CER) and NEO.

Anourosorex squamipes: This species has brain size and three of the five main parts slightly above AvSn (1.01 - 1.02x). It is similar to *Cryptotis*, but has a slightly smaller brain. Two of the fundamental brain parts are smaller: CER and DIE with 0.97x AvSn. Smallest of the TEL parts are AOB (0.41x), STR (0.86x) and HIP (0.88x). As in *Cryptotis*, the main olfactory structures (MOB, PAL) and AMY are very large - the largest or second largest in soricine shrews. The same applies to the SFB of the septum. The brain of *Anourosorex* has the widest and the shortest hemispheres found in the soricine brains so far studied, and the corpus callosum also is the shortest.

Particularities: Relatively large brains with large main olfactory and amygdaloid structures, very small accessory olfactory structures, and relatively small motor components (STR, CER) and HIP.

In conclusion, the brains of the soricine species differ slightly in shape, relative size, and composition. Compared with the average of Insectivora, they are short and broad and have a long corpus callosum. The mesencephalic tectum is fully or almost fully covered. The average brain size of Soricinae and the averages of most brain parts so far measured, are below the family/subfamily level (AvIF); only STR, SCH, and NEO are above this level, from the OBL components, mot X, mot XII and TSO, and from the DIE components, CGM. Well below the AvIF level are SCB (0.18x), EPI (0.37x), AOB (0.56x), SFB (0.60x), MES (0.72x), and MOB (0.86x), but in some of these small structures, such as AOB, SFB, and EPI, there are very large interspecific differences.

Crocidurinae

The total brain of crocidurine shrews is more distinctly below AvIF (0.89x) than that of soricine shrews, which was 0.98x AvIF. All five fundamental brain parts are at similar levels: between 0.80x AvIF for DIE and 0.95x for OBL (Fig. 102), thus showing narrow variation. Little variation is also found in the components of TEL, reaching from 0.71x AvIF for SCH to 0.99x for PAL (Fig. 108), and of DIE, reaching from 0.54x for MNT to 0.97x for CGL (Fig. 106). The largest variation is found in the OBL components, reaching from 0.56x AvIF for FCE to 1.06x for DCO (Fig. 104). Next to DCO, the other auditory nuclei (OLS, VCO) are also relatively large. In this feature, the crocidurine shrews differ markedly from the soricine shrews. There are differences as well in TEL, and the sequence of the TEL structures is nearly reversed in the two subfamilies. The highest positions in Crocidurinae are found for the olfactory structures, the lowest for SCH, HIP, NEO, and STR (Fig. 108).

In conclusion, Crocidurinae have small brains with well represented auditory and olfactory structures. Somatosensory forelimb centers (FCM) and several of the higher telencephalic components (STR, NEO, SCH) are less well represented. Except for FCM, the relations are the reverse of those found in soricine shrews.

Based on differences in brain size and brain composition, the crocidurine shrews can be separated into two clearly different groups: one includes the genera *Crocidura*, *Suncus*, and *Scutisorex*, with nine species altogether, and the other includes the genera *Sylvisorex*, *Ruwenzorisorex*, and *Myosorex*, with four species altogether. *Myosorex* is in an intermediate position with respect to many of its brain characteristics, but tends to resemble the second group and so will be included in it for comparison. The groups will be called "*Crocidura* group" and "*Sylvisorex* group," respectively. Some index profiles are given in Figs. 102 and 108.

The brains of the representatives of the *Sylvisorex* group are on the average 1.4 times larger than those of the *Crocidura* group. The largest difference was found for FCE, which is nearly 2.4 times larger, and for other structures of the motor systems, such as STR and CER (1.6x) and vestibular centers, such as VC and VM (1.5x). A smaller size in the *Sylvisorex* group was found only in some smaller structures, such as AOB, EPI, and SFB (about 0.8x). Overall, the brains of the *Sylvisorex* group are clearly more progressive than those of the *Crocidura* group.

Crocidura **group**

Species characteristics of the brains of the *Crocidura* group are based on deviations from the average group level (AvCg), which is uniformly set at 1 for each of the structures. Information, given in the literature on brains, is so far limited to *Crocidura* spp. and *Suncus* spp. In *Crocidura russula* a large body of information found in the literature coincides with our abundant measurements. Therefore, brain characteristics of this species will be considered first.

Crocidura russula: The total brain, expressed by BrW or NET volume, is above the tenrecine level (1.15 and 1.18x AvTn, see Table 6). In the *Crocidura* group, however, it is one of the lowest positions and, thus, markedly below the average for this group: for BrW and NET, 0.94x AvCg. The indices for four of the five fundamental brain parts also are under AvCg, and only DIE (1.01x) reaches AvCg. CER (0.88x) is the lowest. In TEL, all the main parts are also slightly below AvCg or just reach it. They vary from 0.91x AvCg for SCH to 1.02x for AMY. Only AOB is smaller (0.53x AvCg) - the smallest in the *Crocidura* group. Predominantly low values are also found in the OBL parts (measured in six species of the group): from 0.42x AvCg for FCE to 1.06x AvCg for COE. The lowest values in the *Crocidura* group were found for REL (0.66x), the group of funicular nuclei (FUN, 0.71x), motor nuclei (MOT, 0.90x), and for TSO (0.88x). Bauchot's DIE measurements in three species of this group for the most part placed *Crocidura russula* in an intermediate position between *Suncus murinus* (lower) and *Crocidura flavescens* (higher). The subthalamus was relatively large (1.14x AvCg). *C. russula* has a relatively wide brain, but differences from the other species are minor.

Particularities: *Crocidura russula* has a relatively small brain in the *Crocidura* group, with most brain parts also small. Especially small are the accessory olfactory bulb (AOB), the nucleus reticularis lateralis (REL), the funicular, and the motor nuclei.

Crocidura attenuata: The largest brain in the group (BrW 1.12x; NET 1.13x AvCg), but this is based on one specimen only. Only the main parts were measured and were found to vary in the five fundamental brain parts from 1.06x AvCg for DIE to 1.15x for CER. CER is the largest in the *Crocidura* group as is TEL (1.13x) as well. The TEL parts vary from 1.00x AvCg for STR to 1.20x for MOB. MOB is the largest found in the *Crocidura* group along with SCH (1.15x), AMY (1.13x), and NEO (1.11x). AOB deviates to a greater extent (1.36x AvCg) and is larger in *Crocidura attenuata* than in any other species of the group. The brain shape is typical for the *Crocidura* group.

Particularities: Relatively large brain with large CER, and main and accessory olfactory structures.

Crocidura flavescens: Relatively large brain (1.04x for BrW; 1.03x for NET) with nearly all main parts above AvCg. Only DIE is slightly below. The five fundamental brain parts vary from 0.98x AvCg for DIE to 1.12x for CER, and the TEL parts, from 1.00x for PAL to 1.14x for HIP and 1.32x for AOB. HIP is the largest found in the *Crocidura* group. OBL parts were measured in six species and were found to be large in *Crocidura flavescens* for REL (1.27x), DCO (1.24x), and among the funicular nuclei, for FGR (1.29x) and FCM (1.19x), but small for FCE (0.54x). Besides FCE, small size was found for COE (0.73x), INO (0.80x), and most vestibular nuclei (0.87 - 0.96x; but VL = 1.18x). Some diencephalic and mesencephalic structures, as well as the cerebellar nuclei were measured in three species of the group and nearly all were found to be larger in *Crocidura flavescens* than in *C. russula* and *Suncus murinus*. In contrast, the mesencephalic SCB was very small in *C. flavescens*. The hemispheres are the longest, competing with those of *Suncus murinus* (see Table 3).

Particularities: Relatively large brain with nearly all measured parts large, especially AOB, CER, HIP, and STR.

Crocidura hildegardeae: Relatively large brain (1.07x AvCg for BrW; 1.08x for NET) with most main parts above AvCg. The five fundamental brain parts vary from 1.02x AvCg for CER to 1.10x for TEL. The TEL parts vary from 0.93x for SCH to 1.23x for PAL. PAL is the largest found in the *Crocidura* group. Besides SCH, HIP, too, is below the AvCg level (0.96x). OBL parts were measured in six species and many of the structures were found to be largest in *C. hildegardeae*. This was the case for COE (1.81x), VI (1.32x; and the whole vestibular group, 1.23x), OLS (1.26x; and the whole auditory group, 1.14x), mot V (1.24x), TSO (1.23x), and TR (1.21x). The smallest were FCE (0.73x) and some of the motor nuclei (mot XII = 0.90x; mot VII = 0.95x). *C. hildegardeae* has the shortest hemispheres, but the differences from the other species are minor.

Particularities: Relatively large brain with nearly all measured parts large, especially PAL, COE, TSO, and vestibular, auditory and trigeminal structures.

Crocidura jacksoni: Relatively large brain (1.09x AvCg for BrW; 1.08x for NET) with most main parts above AvCg. The five fundamental brain parts vary from 1.03x AvCg for TEL to 1.22x for OBL. OBL is the largest found in the *Crocidura* group and this also applies to DIE (1.09x). The TEL parts vary from 0.89x for AOB, 0.95x for SEP, and 0.96x for MOB to 1.08x for AMY. In *Crocidura jacksoni* none of the TEL parts is the largest in the group.

Particularities: Relatively large brain with large brain stem structures (OBL, MES, DIE).

Crocidura suaveolens: Relatively small brain, similar to that of *C. russula*, which is below the average of the *Crocidura* group: 0.95x AvCg for BrW and NET. The indices for all five fundamental brain parts also are below AvCg; the lowest is for OBL (0.89x). In TEL, all the main parts, except SCH and NEO, also are slightly below AvCg or just reach it. They vary from 0.93x AvCg for MOB, PAL, SEP, and HIP to 1.02x for NEO. None of the structures is smallest in *Crocidura suaveolens*. No other structures were measured. *C. suaveolens* has the shortest brain.

Particularities: Relatively small brain with nearly all measured parts slightly below the level of the *Crocidura* group.

Suncus etruscus: The brain of this tiny species is below the average of the *Crocidura* group: 0.96x AvCg for BrW and NET, and most main parts are also small. Exceptions are OBL and DIE, which are slightly above. The five fundamental brain parts vary from 0.84x AvCg for CER to 1.03x for OBL and DIE, and the TEL parts from 0.78x for AMY to 1.05x for STR. Among the periventricular structures (five species measured), the subfornical body also is the smallest (0.83x AvCg), whereas the medial habenular nucleus is the largest (1.54x). OBL parts were measured and INO was found to be largest in *Suncus etruscus* (1.23x AvCg). Also large are FCE (1.16x), VM and mot VII (1.15x). Especially small were COE (0.81x), VL (0.73x), and the auditory structures (0.71x). VL and the auditory structures are smaller in *Suncus etruscus* than in any other member of the *Crocidura* group. The brain is the flattest competing with that of *Suncus murinus*.

Particularities: The smallest brain of the group with most main parts also the smallest. Only some structures of OBL and DIE, such as FCE, VM, INO, PRP, and HAM, are well developed.

Suncus murinus: The brain of *Suncus murinus* is the smallest (0.87x AvCg for BrW and NET) except for BrW in *Crocidura giffardi* and *C. leucodon* (Tables 5 and 6). The five fundamental brain parts (except CER) also are the smallest in the *Crocidura* group (0.85 - 0.88x) and from TEL parts this is true for STR (0.84x), NEO (0.87x), and SEP (0.93x). Small (but not the smallest in the group) are AMY (0.88x), PAL (0.89x), and HIP (0.91x AvCg). Only AOB is above AvCg level (1.13x). Of the structures measured in OBL, MES, CER, and DIE, many are in *Suncus murinus* the smallest of the group. The smallest structures are COE (0.68x), HAM (0.70x), FGR (0.76x), PRP (0.79x), VS (0.83x; and the vestibular structures as a whole, 0.89x), as well as OLS (0.85x), TR (0.85x), and the triangular nucleus of the septum (SET = 0.84x). Some of the tiny periventricular structures and certain motor nuclei are larger in *Suncus murinus*, however, than in any other member of the *Crocidura* group. Examples are SCB (1.52x), EPI (1.43x), mot XII (1.31x), and mot VII (1.17x). The brain is much longer than in the other members of the *Crocidura* group. The hemispheres are long and flat, and the corpus callosum is also long.

Particularities: Long, relatively small brain with small vestibular and trigeminal structures and large motor nuclei.

Scutisorex somereni: Relatively small brain (0.96x AvCg for BrW; 0.97x for NET) with most main parts also slightly below the AvCg level. The five fundamental brain parts vary from 0.89x AvCg for OBL to 1.08x for CER. CER is the only of the five main structures above the AvCg level. The TEL parts vary from 0.85x for AOB, and 0.95x for STR and NEO to 1.14x for SEP. SEP is largest when compared with the other species of the group. Besides SEP, HIP (1.10x), MOB and PAL (1.05x) are above the AvCg level. OBL parts and periventricular structures were measured: the auditory components and the funicular nuclei, as well as the subfornical body are large; the epiphysis and the motor nuclei are small. EPI and mot VII are the smallest found in any member of the *Crocidura* group so far measured. The brain is slender and high. The corpus callosum is short.

Particularities: *Scutisorex somereni* has a relatively large brain with large olfactory and auditory components and funicular nuclei, and small motor nuclei.

In conclusion, the *Crocidura* group is quite uniform, except for a few generic differences, such as: (1) the two *Suncus* species have a particularly small brain size and most of their brain parts also are among the smallest, and (2) all *Crocidura* species measured so far have a very small ncl. fascicularis cuneatus externus (FCE), which in *Suncus* species and in *Scutisorex* is considerably larger.

Sylvisorex group

Species characteristics of the brains of the *Sylvisorex* group are based on deviations from the average group level (AvSg), which is uniformly set at 1 for each of the structures.

Sylvisorex granti: *Sylvisorex granti* has the largest brain in the *Sylvisorex* group (1.13x AvSg for BrW and NET). All main structures are above the AvSg level. The five fundamental brain parts vary from 1.03x for OBL and CER to 1.19x for TEL. TEL is by far the largest found in the *Sylvisorex* group. The high position of TEL is primarily due to the main olfactory structures, which are 1.26x AvSg for MOB and 1.24x for PAL. The other structures vary from 1.08x for AMY to 1.12x for HIP and NEO. MOB, PAL, SEP, AMY, and NEO are in *Sylvisorex granti* the largest in the *Sylvisorex* group, but there is little variation. Only AOB is below the AvSg level (0.95x). Among the periventricular structures, a very large EPI was found (1.73x AvSg). The brain shape is typical for the *Sylvisorex* group.

Particularities: Largest brain of the *Sylvisorex* group, with large main olfactory structures and epiphysis.

Sylvisorex megalura: The most typical of the group, with BrW and NET 1.03x AvSg. The five fundamental parts vary from 0.99x AvSg for TEL to 1.12x for MES. OBL, MES, and CER are the largest found in the members of the *Sylvisorex* group. The TEL parts vary from 0.70x for HIP to 1.17x for AOB. AOB is the largest in the group, followed by STR (1.07x); the smallest, after HIP, are SCH (0.80x), AMY and SEP (0.98x). The subfornical body (SFB) is the largest in the group (1.12x), and the pineal gland (EPI, 1.31x) is the second largest, after that of *Sylvisorex granti* (see above). OBL parts were measured in only two species of the group *(Sylvisorex megalura* and *Ruwenzorisorex)* and, thus, AvSg was determined from these species and the size comparisons are limited. *S. megalura* has the larger sizes in nearly all OBL structures, arriving at values of 1.25x for FCE and 1.38x for VL. Since, however, VI is especially small (0.83x), the explanation may lie in uncertain delimitation of VL and VI. Apart from VI, small size was also found for COE (0.80x) and TR (0.90x). The auditory structures were markedly larger in *Sylvisorex megalura* than in *Ruwenzorisorex* (1.12x versus 0.88x), as were the funicular nuclei (1.13x versus 0.87x). Very small differences were found for the motor nuclei. *Sylvisorex megalura* has the longest brain in the group, with the shortest hemispheres and corpus callosum. The hemispheres are the narrowest.

Particularities: In size and composition *Sylvisorex megalura* has the most typical brain of the group with small limbic structures (HIP, SCH) and, presumably, large auditory structures and funicular nuclei.

Ruwenzorisorex suncoides: *Ruwenzorisorex* has the smallest brain of the group (0.91x AvSg for BrW and NET), a fact that is totally accounted for by its small TEL (0.85x), whereas the other fundamental brain parts vary between 0.99x AvSg for MES and 1.03x for OBL. The small TEL is mainly caused by the main olfactory structures, which are 0.54x AvSg for MOB and 0.58x for PAL and, thus, are by far the smallest in the group. AMY (0.87x) and STR (0.88x) are also the smallest, whereas SCH (1.25x) and HIP (1.16x) are the largest in the group. The triangular nucleus (SET) of the septum (1.18x) and the epithalamic medial habenular nucleus (HAM, 1.21x) also are the largest. OBL parts were measured in two species only and the results were the opposite of those given for *Sylvisorex megalura*, i.e., small auditory structures and funicular nuclei and relatively large trigeminal structures. *Ruwenzorisorex* has the widest and flattest brain of the group.

Particularities: The small olfactory structures of *Ruwenzorisorex* reduce the size of TEL and that of the brain as a whole. The limbic structures (SCH, HIP) and, presumably, the trigeminal structures as well are large, whereas the auditory and funicular structures seem to be small.

Myosorex babaulti: *Myosorex babaulti* has a relatively small brain (0.93x AvSg for BrW and NET), with all the five fundamental brain parts below the group level (ranging from 0.78x for MES to 0.97x for TEL). In TEL, the main olfactory structures (1.10x for MOB and 1.09x for PAL) and AMY (1.07x) are well developed. SCH and NEO (0.83x) are relatively small. NEO is by far the smallest found in the group. EPI is extremely small (0.13x) and HAM (0.74x), SFB (0.82x), and SET (0.89x) are also small. OBL parts have not yet been measured. *Myosorex* has the shortest brain, with long hemispheres and a long corpus callosum. The brain is also heigh.

Particularities: Relatively small brain with small MES and NEO and especially small EPI. The main olfactory structures (MOB and PAL) are quite large.

In conclusion, the *Sylvisorex* group is fairly uniform in the relative size of the non-telencephalic fundamental brain parts, whereas there are large differences in TEL, such as: (1) strongly reduced main olfactory components in *Ruwenzorisorex* in contrast to all other species of the group, and (2) large differences in the limbic structures (HIP and SCH), which are particularly small in *Sylvisorex megalura* and large in *Ruwenzorisorex*. The epiphysis is extremely small in *Myosorex* and large in *Sylvisorex granti*.

Based on special features of skull and teeth *Ruwenzorisorex suncoides* was recently separated by Hutterer (1986) from the genus *Sylvisorex*. This separation is corroborated by some brain characteristics that are very deviant from other brains in the group: (1) Main olfactory bulbs (MOB) and paleocortex (PAL) are only half as large in *Ruwenzorisorex* as in the other species. (2) Distinctly larger structures are COE (in *Ruwenzorisorex* 1.5 times larger than in *Sylvisorex*), HIP (1.6 times), SCH (1.3 times) and TR (1.2 times). Index profiles are given in Figs. 102 and 108.

A strong decrease in the MOB with a simultaneous increase in TR was also found when comparing *Neomys* with *Sorex*. Such a brain composition is highly characteristic of animals adapted to hunting for prey in water. Thus, based on its brain composition, such an adaptation is likely for *Ruwenzorisorex* but, so far, has not been confirmed by ecoethological data.

4.7 Index Profiles of Taxonomic Units Relative to the Average of Insectivora

To show the special brain characteristics of the taxonomic units (mainly families and subfamilies) in the context of the Insectivora as a whole, their deviations from the average ("typical") Insectivora were calculated and are illustrated in Figs. 101 - 108. As a measure of the average size of a brain structure in Insectivora, the average value of the family and subfamily averages (AvIF) is used. AvIF values are given for each structure in the tables and are summarized in Table 74. They are indicated by dotted lines in the diagrams and scales. In order to obtain the deviations of the families and subfamilies from the typical Insectivora, the averages of the various families and subfamilies were divided by the AvIF values, or, which is the same, the AvIF values are set at 1 for each of the structures. In the diagrams (Figs. 52, 54, 70, 91), this corresponds to a shifting of the reference lines into the AvIF position. The averages of the family and subfamily averages (AvIF) were chosen instead of the averages between species (AvIS), because the shrews are strongly over-represented in AvIS. For Crocidurinae, the strong differences between the two groups of genera and species (called *Crocidura* and *Sylvisorex* groups) are shown in Figs. 102 and 108.

Abbreviations for Figures 101 - 108 (on the following pages):

AMY	amygdala	OLS	superior olive
ANT	nuclei anteriores thalami	PAL	paleocortex
AOB	accessory olfactory bulb	PRP	nucleus prepositus hypoglossi
CER	cerebellum	REL	nucleus reticularis lateralis
CGL	corpus geniculatum laterale	SCH	schizocortex
CGM	corpus geniculatum mediale	SEP	septum telencephali
COE	locus coeruleus	STR	striatum
DCO	dorsal cochlear nucleus	TEL	telencephalon
DIE	diencephalon	TR	complexus sensorius nervi trigemini
DOT	nuclei dorsales thalami	TSO	nucleus tractus solitarii
FCE	nucleus fascicularis cuneatus externus	VCO	nucleus cochlearis ventralis
FCM	nucleus fascicularis cuneatus medialis	VET	nuclei ventrales thalami
FGR	nucleus fascicularis gracilis	VI	nucleus vestibularis inferior
HIP	hippocampus	VL	nucleus vestibularis lateralis
INO	inferior olive	VM	nucleus vestibularis medialis
MES	mesencephalon	VS	nucleus vestibularis superior
MLT	nuclei mediales thalami	V	nucleus motorius nervi trigemini
MNT	nuclei mediani thalami	VII	nucleus nervi facialis
MOB	main olfactory bulb	X	nucleus dorsalis motorius nervi vagi
NEO	neocortex	XII	nucleus nervi hypoglossi
OBL	medulla oblongata		

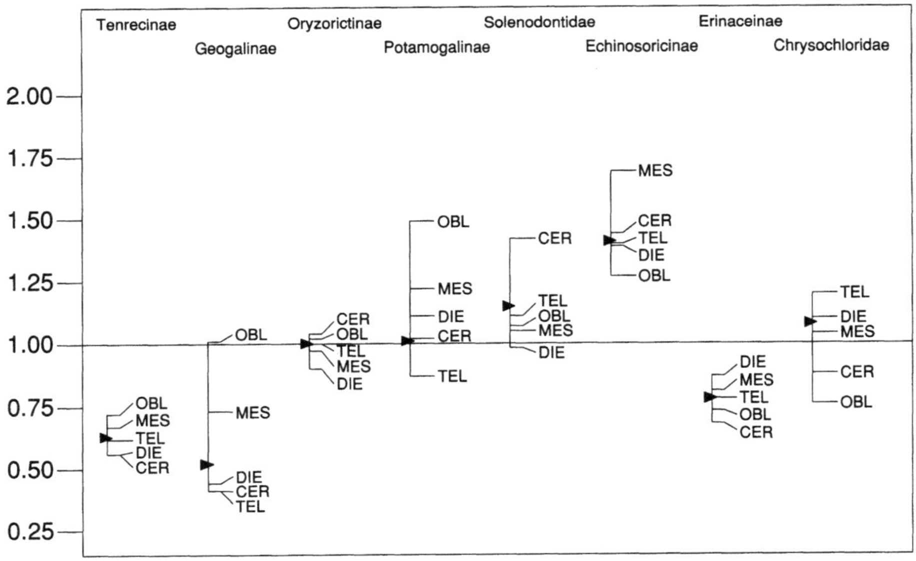

Figure 101

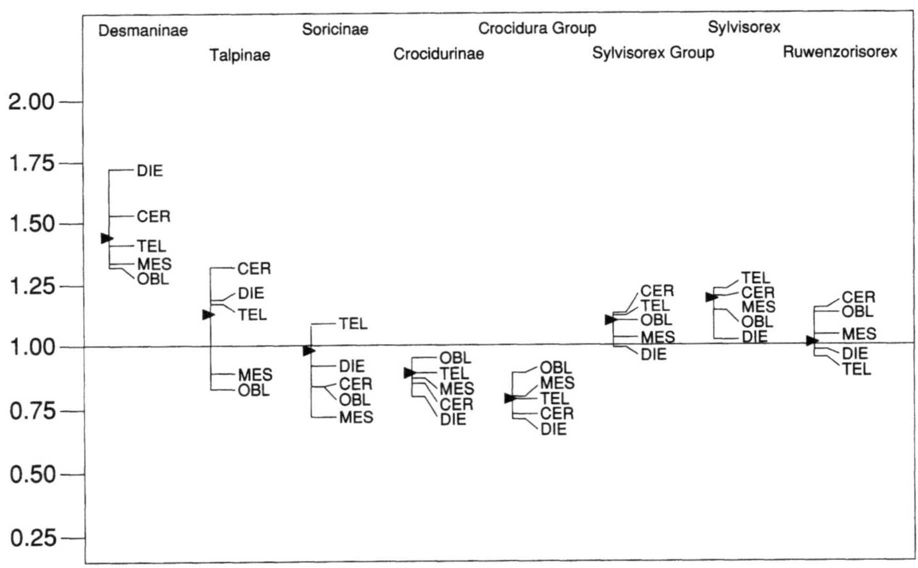

Figure 102

Figures 101 - 108: Deviation of taxonomic units (mostly subfamilies) from the typical Insectivora for some of the measured brain parts. The typical Insectivora is represented for each structure by the average index of families and subfamilies (AvIF) summarized in Table 74. AvIF for all structures is uniformly set = 1. The scales indicate the direction and size of the deviation from the typical Insectivora (= average index of taxonomic units / AvIF). Abbreviations on p. 293.

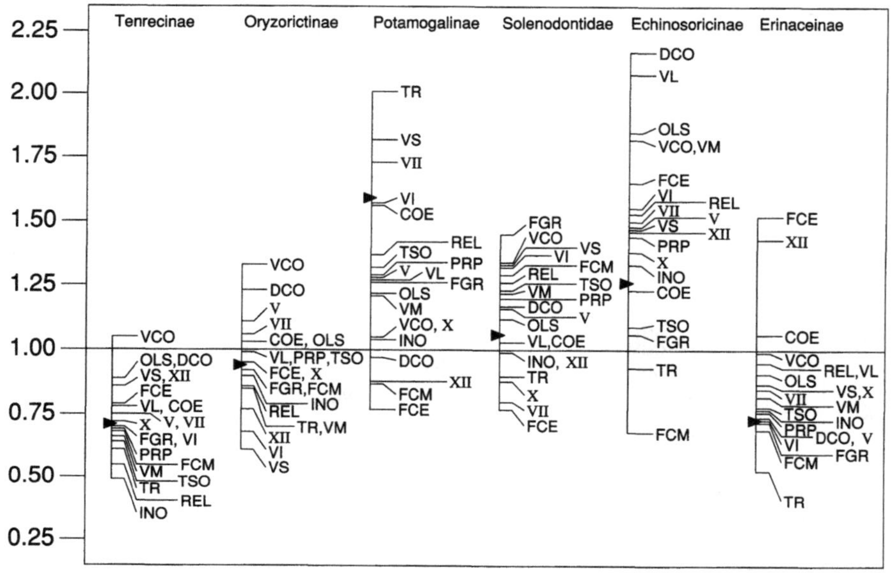

Figure 103

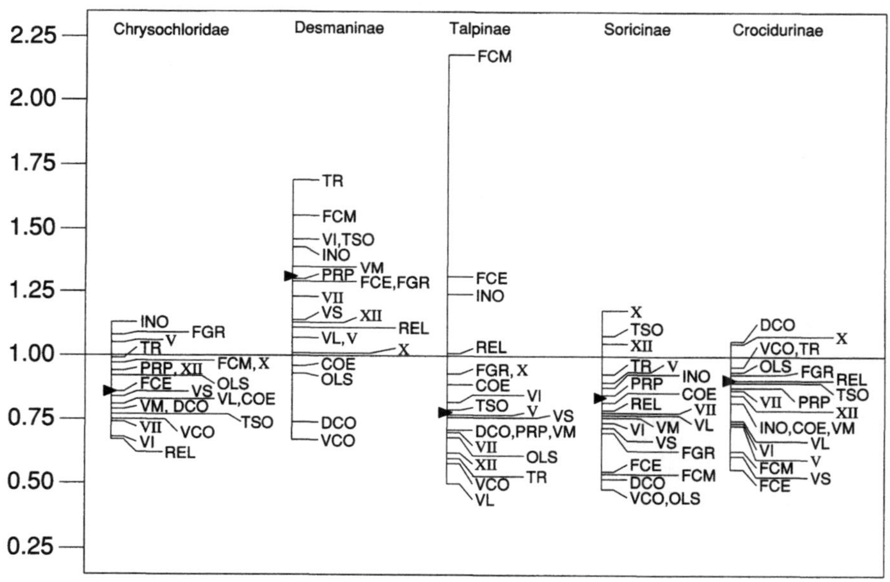

Figure 104

Figures 103 - 104: 20 parts of the medulla oblongata. Arrow heads give the position of the total OBL calculated from the indices and averages given in Tables 17 and 18. Abbreviations on p. 293.

Figures 101 - 102: The five fundamental brain parts. Arrow heads give the position of the total NET brain. Abbreviations on p. 293. (on facing page)

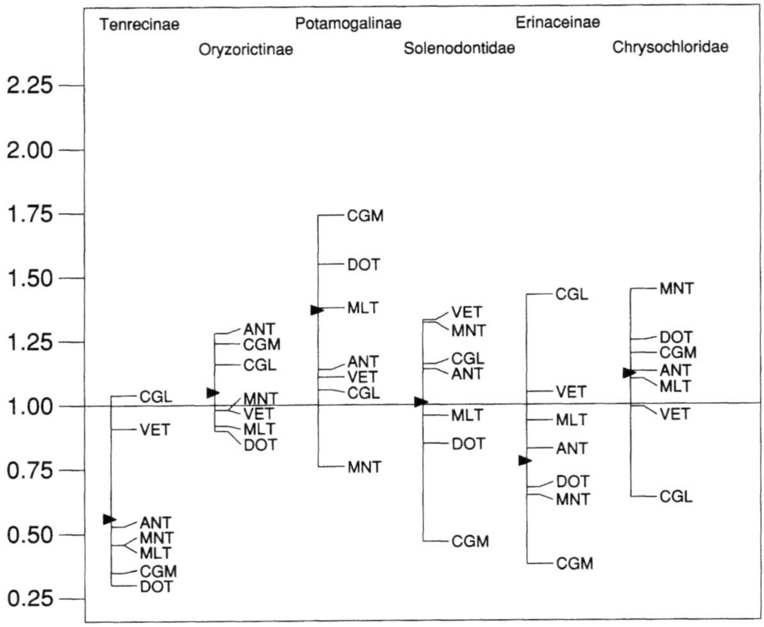

Figure 105

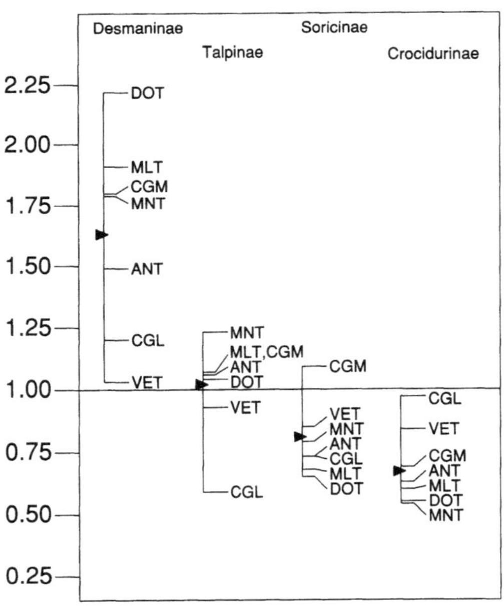

Figure 106

Figures 105 - 106: Seven parts of the thalamus based on measurements by Bauchot (1979b). Arrow heads give the position of the total THA. Abbreviations on p. 293.

(on facing page)
Figures 107-108: Nine parts of the telencephalon. Arrow heads give the position of the total TEL as shown in Figs. 101 and 102. Abbreviations on p. 293.

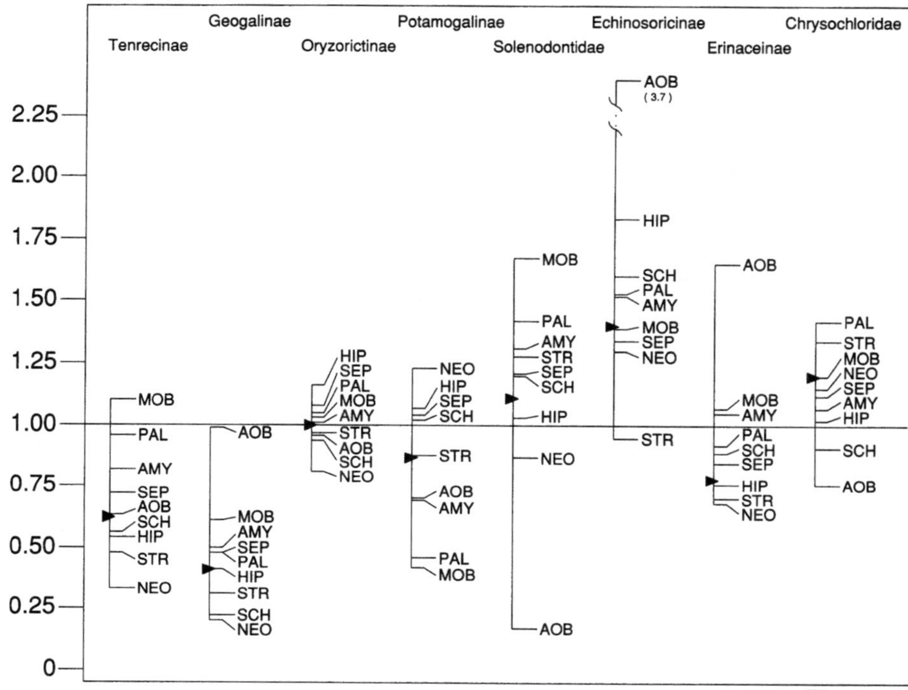

Figure 107

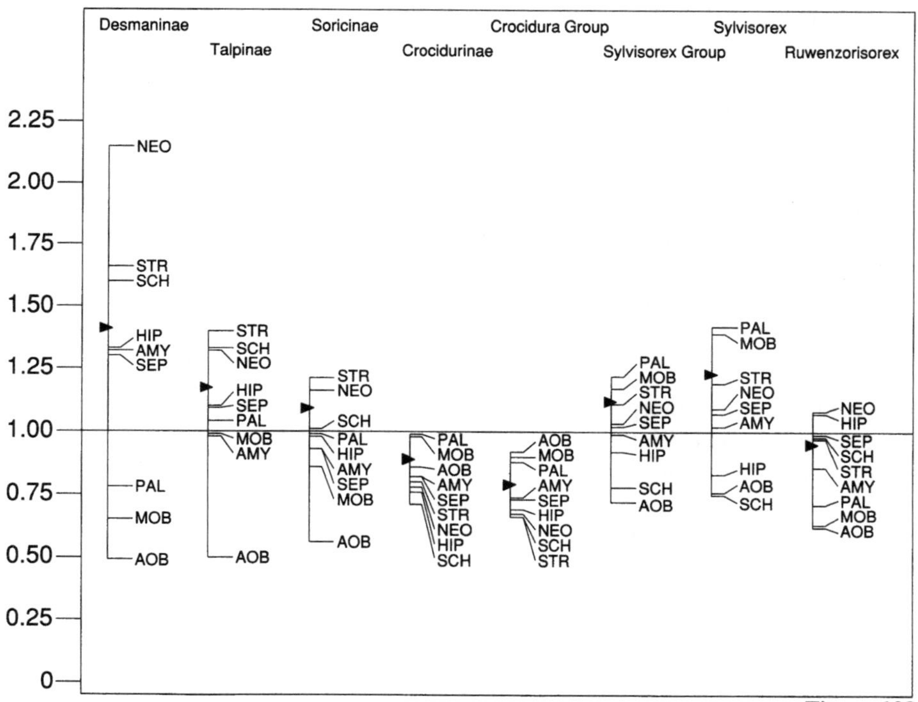

Figure 108

5 Adaptive Radiation

While Insectivora have invaded all continents except Australia, they are rather rare in the neotropics. Only one genus, *Cryptotis* (Soricidae), has reached South America and another, *Solenodon* (Solenodontidae), has survived in the Greater Antilles (Cuba, Haiti).

Insectivora form a very heterogeneous group having as a common characteristic a relative abundance of conservative traits, such as abdominal testes, ring-shaped tympanic bone, a clavicle, paired anterior vena cava, unconvoluted cerebral hemispheres and many others. None of the primitive characters is universal in the Insectivora, however, and all occur in other orders as well. Moreover, derived characters are widespread in the Insectivora (Grassé 1955b; Thenius and Hofer 1960; Thenius 1969, 1972; Butler 1972). Since the earliest investigations on brain size and brain composition by Stephan and collaborators (1956), it became evident that various insectivorous species have reached very different evolutionary levels, sometimes surpassing species from other orders. Not only does their brain size vary enormously, but their brain composition, too, shows extensive differences.

The vast heterogeneity makes classification of the Insectivora one of the most controversial problems of mammalian taxonomy (Butler 1972). The great number of unsolved problems still remaining make it unlikely that a satisfactory and generally accepted solution is imminent. Quantitative brain composition has little taxonomic value. Brain size and composition in closely related species may be very different, while distantly related species may be very similar in these traits. Although brain size and quantitative brain composition are unsuitable for establishing similarities as a basis for phylogenetic and taxonomic relationships, they are very useful in revealing the diversity and extent of radiation within a taxon. A shift into a new niche requires, almost without exception, a change in behavior (Mayr 1976) and hence a structural reorganization of the brain. Quantitative brain structure may therefore be considered an accurate measure for evaluating niche diversity and specialization.

Based on molar morphology, the Insectivora sensu stricto (or Lipotyphla) have been subdivided into Zalambdodonta and Dilambdodonta (Gill 1883). The living Zalambdodonta comprise three families: the Tenrecidae, Chrysochloridae, and Solenodontidae. These families are restricted to Madagascar (Tenrecidae, except Potamogalinae), West and Central Africa (Potamogalinae), South and Central Africa (Chrysochloridae), and the Greater Antilles (Solenodontidae). Given the profound differences, it seems doubtful that there is any close phylogenetic relationship between these families. The recent Dilambdodonta also comprise three families, the Erinaceidae, Talpidae, and Soricidae.

Tenrecidae

The Tenrecidae have radiated into a variety of ecological niches. They comprise four subfamilies: the Tenrecinae, Oryzorictinae, Geogalinae, and Potamogalinae. With the exception of the Potamogalinae, which occupy the wet regions of Central and West Africa, the living Tenrecidae are confined exclusively to Madagascar. The Potamogalinae are the only survivors of an earlier radiation on the African continent (Thenius 1969). The exceptional diversity of the Tenrecidae on Madagascar represents a more recent radiation which was favored by the isolation of the island and, consequently, an absence of ecologic competitors (Thenius 1980). The founder stock for the Tenrecidae must have been isolated on Madagascar at approximately the Paleocene (Eisenberg 1981). The Tenrecinae, which include only terrestrial forms, that may be to some extent scansorial, radiated into opossum-like types (*Tenrec* and *Hemicentetes*) and hedgehog-like types (*Setifer, Echinops*). The ecologic variety is greater among Oryzorictinae, which include shrew-like terrestrial and partly scansorial forms (*Microgale*), talpoid subterranean (*Oryzorictes*), and semiaquatic forms (*Limnogale*).

The taxonomic position of *Geogale* has long been disputed. Brain characteristics, as shown recently by Rehkämper et al. (1986), support the classification of *Geogale* into a separate subfamily (**Geogalinae**), as already suggested by Dobson (1882), Starck (1974, 1978), and Thenius (1980). According to Thenius (1980), *Geogale aletris* found in the East African Miocene cannot be separated generically from *Geogale aurita*. This genus may have split from the other Tenrecidae very early and should have invaded Madagascar independently. Thenius' statement, however, will be opposed by Poduschka and Poduschka (1985) and Butler (1984) who regard *Geogale aletris* (*Parageogale aletris*, Butler 1984) not to be congeneric with *Geogale aurita*. With regard to the size of the brain and its parts, especially of the neocortex, *Geogale aurita* is perhaps the most conservative extant mammalian species and must be placed at the bottom of the mammalian brain scale, even lower than the Tenrecinae (Figs. 101, 107). This means that the brain of *Geogale* may be, in its size and quantitative composition, close to the starting point of mammalian brain evolution.

If *Geogale* is excluded **Tenrecinae** have the most conservative brain. The most progressive brain structures in evolutionary terms, the neocortex and the striatum, are far below the insectivoran average (Table 45, Fig. 107). In fact, they are the smallest found in any Insectivora except *Geogale*. Of the sensory structures, the olfactory (Table 45, Fig. 107), auditory (Table 27, Fig. 103) and visual centers (Table 70, Fig. 105) are fairly well developed, being close to the insectivoran average, but the somatosensory structures are below the average (Table 15, Fig. 103). The orofacial motor nuclei (V, VII) are among the smallest found in Insectivora (Table 21).

Tenrecinae show conservative as well as advanced characters which are not shared equally by all species. Like other Tenrecidae, they have a cloaca and the males lack a scrotum. The position of the testicles, however, is more conservative in Tenrecinae than in the other Tenrecidae in that they maintain the original contact with the kidneys (Grassé 1955a). Reduction of the tail and possession of a spiny pelage, which is particularly well developed in *Setifer* and *Echinops*, is a specialized character. With respect to litter size, *Tenrec*, which may have up to 30 young, is more conservative than the other species, which have fewer young. The prolonged period of somatic growth in *Tenrec* may also be considered a plesiomorphic character (Heim de Balsac 1972). *Echinops* exhibits the most profound reduction in tooth number, not only within Tenrecinae but within the entire family. *Hemicentetes* has the most teeth (40) but they are reduced in size when compared with the other genera (Eisenberg and Gould 1970). Encephalization and neocorticalization are most advanced in *Setifer* and most reduced in *Tenrec*, followed by *Echinops*.

This incomplete enumeration of plesiomorphic and apomorphic traits in various species clearly shows that brain evolution does not follow the same evolutionary trend as other structures and systems.

Although all Tenrecinae can be considered terrestrial surface dwellers, two foraging types may be distinguished: the almost completely terrestrial forms with digging ability, including *Tenrec* and *Hemicentetes*; and terrestrial forms with some climbing ability, including *Setifer* and *Echinops* (Eisenberg and Gould 1970).

Olfactory, auditory, and visual structures are similarly developed in *Tenrec* and *Hemicentetes*. Niche differentiation is mainly revealed by the fact that the somatosensory structures of the orofacial (TR) and limb regions (FUN) are better developed in *Hemicentetes* than in *Tenrec* (Table 14). Among the orofacial motor structures, the trigeminal motor nucleus (V) is smaller in *Hemicentetes* than in *Tenrec*. The two differ in their specialization to a particular biotope and in their foraging behavior. *Tenrec* is ubiquitous with respect to biotope selection and is rather omnivorous while *Hemicentetes*, which characteristically inhabits rain forest areas, is specialized for soft food such as worms and coleopteran larvae, a finding consistent with its reduced teeth and its motor centers for the mastication muscles (Herter 1962a; Eisenberg and Gould 1970; Heim de Balsac 1972). The involvement of limbs in foraging behavior seems to be more important in *Hemicentetes*, which also has better developed funicular nuclei than *Tenrec*.

Setifer and *Echinops* are sufficiently different in brain size and quantitative brain composition to consider them two different adaptive trends despite a number of morphological and behavioral similarities and their tendency to become generalized omnivores like *Tenrec*. In addition to the differences in encephalization and neocorticalization already mentioned, there are differences in the development of their sensory structures. The orofacial somatosensory structures are better developed in *Setifer* than in *Echinops*, while those of the limb regions are bigger in *Echinops* than in *Setifer* (Table 14). *Setifer* also has a better developed olfactory system (Table 44), but *Echinops* surpasses *Setifer* in the size of its auditory structures (Table 26) while the visual structures are similar in the two (Table 69). The differences in the sensory structures are partly reflected by differences in the corresponding morphological features. *Echinops* has shorter vibrissae but larger ear pinnae than *Setifer* (Herter 1979). According to Herter (1962a, 1963), olfaction and audition are the predominant senses and are used in prey detection. Morphometric data on the sensory centers, however, suggest that olfaction is slightly predominant in *Setifer* but that audition is so in *Echinops*. *Echinops* also seems to be more scansorial than *Setifer* (Herter 1962a, 1963, 1979, Eisenberg and Gould 1970) and this might explain the somewhat better development of the somatosensory system of its limb region.

Among the **Oryzorictinae**, the terrestrial *Microgale* have the least apomorphic characters and represent in many respects the most primitive Insectivora (Thenius 1979). The teeth of *Microgale* show the most primitive types found in Insectivora (Heim de Balsac 1972). However, according to brain characteristics, they must be considered more evolved than the terrestrial Tenrecinae, although the latter in general have more specialized features. Brain size, which is slightly larger in *Microgale* than the mean for Oryzorictinae, is increased nearly 70% compared with Tenrecinae (Table 7). The neocortex and hippocampus have progressed the most (Table 45). Visual (Table 69), acoustic (Table 26), somatosensory structures of the orofacial and limb regions (Table 14), and the olfactory bulb (Table 44) show similar moderate increase. Motor centers of the orofacial region (mot V and VII, Table 20) exhibit the same increase as the sensory trigeminal complex.

The fossorial *Oryzorictes* is, relative to the generalized *Microgale*, a more modified genus. The regression or disappearance of the thumb, unique within the Tenrecidae, the development of claws on the enlarged hands, and the talpoid aspect of the body are derived characters. The teeth are almost as primitive as in *Microgale* (Heim de Balsac 1972). *Oryzorictes* exhibits brain characteristics which differ from those of the other oryzorictine ecotypes. The brain and neocortex are the smallest in the subfamily. The olfactory structures are smaller than in *Microgale* (Table 44). Compared to the other Oryzorictinae, the medulla oblongata is slightly reduced, a tendency generally observed in fossorial forms.

The semiaquatic *Limnogale*, which resembles *Micropotamogale*, was classified by Frechkop (1957) as Potamogalinae. However, osteological features which are not adaptive, such as the ringshaped tympanic bone, characterize *Limnogale* as an aquatic "*Microgale*" (Guth et al. 1959). The brain shows characteristics typical of the semiaquatic Potamogalinae: very well developed neocortex and strongly reduced olfactory structures (Table 44). Visual structures are in the tenrecine range of variation (Table 69). No data are available for the other sensory and motor structures. However, the large medulla, which is characteristic of the semiaquatic Insectivora, combined with the extremely well developed vibrissae, strongly indicates large somatosensory structures of the orofacial region.

The taxonomic position of the **Potamogalinae** has often been questioned. Early authors considered them a subfamily of the Tenrecidae (Cabrera 1925; Weber 1928; Frechkop 1957). In support of these authors but in opposition to Simpson (1945), who classified *Potamogale* as a separate family, the discovery of the *Micropotamogale* induced Heim de Balsac (1954) and Heim de Balsac and Bourlière (1955) to reclassify the two genera as a subfamily. Later authors have supported this classification (Guth et al. 1959, 1960; Verheyen 1961a, b; Rahm 1961) and emphasized the intermediate position of *Micropotamogale* between Oryzorictinae and *Potamogale*. More recently, Heim de Balsac (1972) classified *Micropotamogale* and *Potamogale* as genera of the Oryzorictinae using osteological criteria of the tail. Because of its primitive molar pattern, Butler (1985) considers the otter shrews a subfamily with an independent evolution going back to the Oligocene.

Micropotamogale and *Potamogale* have common characteristics which differentiate them from the other subfamilies. The Potamogalinae are the only Insectivora without clavicles. In mammalian phylogeny, this is generally regarded as an apomorphic character. Another distinctive feature is the syndactylous second and third toes. Kingdon (1974) considers the Potamogalinae to be the last relics in Africa of a very ancient group of mammals which have survived because of exploration of a very peculiar niche.

The general evolutionary level, as expressed by relative brain size of the Potamogalinae, is considered to be in the upper half of the total range found in Insectivora. The neocortex reaches the greatest development by far among Tenrecidae (Table 45). Olfactory structures are strongly reduced (Table 44, Fig. 107), while the somatosensory structures of the orofacial region are strongly increased relative to the Tenrecinae (Table 15, Fig. 103). The other sensory structures are not uniformly developed in all genera, indicating diverging trends. The orofacial motor nuclei, particularly the facial nucleus (VII) are among the best developed in any Insectivora (Table 21, Fig. 103). The most striking feature of the otter shrews head is the enormous muzzle formed by an important mass of labial muscles and erector muscles of the vibrissae. These muscles are innervated by the facial nerve.

According to brain characteristics, *Potamogale* is the most progressive, and *Micropotamogale lamottei*, the least progressive species of the subfamily. With respect to water adaptation, the brain of *Potamogale* is recognizably more specialized than that of *Micropotamogale*. The olfactory bulb is clearly smaller, and the sensory trigeminal complex and the cerebellum are distinctly larger than in *Micropotamogale*.

Although measurements of sensory and motor nuclei are not available for *Micropotamogale lamottei*, there are strong indications that this species is the least specialized for life in water. All brain structures which tend to increase in water-adapted species show the least progression, and those which tend to decrease show the least regression in *M. lamottei*. Brain characteristics are in full agreement with morphological and behavioral data suggesting that *M. lamottei* is clearly less adapted to water than the other otter shrews (Heim de Balsac 1954, 1956, 1957; Heim de Balsac and Bourlière 1955; DeWitte and Frechkop 1955; Guth et al. 1959, 1960; Rahm 1961; Dubost 1965). General morphology and brain characteristics also support the hypothesis that *M. lamottei* is the most primitive form of the three species, and is close to the basic form from which *Potamogale velox* and *Micropotamogale ruwenzorii* developed as two divergent trends (Guth et al. 1960).

Chrysochloridae

Since the Chrysochloridae being highly specialized to the subterranean ecotope, they, together with the semiaquatic Potamogalinae, are the only Zalambdodonta which have survived on the African continent. Among Insectivora, their brain can be considered relatively advanced. They have on the average a slightly larger relative brain size as the semiaquatic Potamogalinae (Table 7), which have the best developed brains of all the Tenrecidae. Progressive brain parts, such as neocortex and striatum, are well developed (Table 45, Fig. 107). Olfactory structures (Table 45) have barely changed, while somatosensory structures of the orofacial and the limb regions (Table 15, Fig. 104) are clearly increased, but acoustic structues (Table 27) are slightly reduced, and visual structures (Table 69, Fig. 105) are strongly reduced relative to Tenrecinae. Of the orofacial motor nuclei, the trigeminal motor nucleus (mot V) is clearly increased, suggesting an increase in the functional importance of the masticatory muscles. The facial motor nucleus (mot VII), however, has not increased relative to Tenrecinae (Table 21). Among the muscles innervated are the orbital, auricular, and proboscis (Kahle 1976; Travers 1985), which are strongly reduced in the golden mole.

Solenodontidae

There is no unanimity among authors as to the classification and origin of the Solenodontidae. VanValen (1967), for example, has grouped the Solenodontidae with the Tenrecidae in the suborder Zalambdodonta following Simpson (1945), who grouped the two families in the superfamily Tenrecoidea. He added, however, that the view that *Solenodon* is related to *Nesophontes* and therefore to the Soricidae, is about as defensible as its opposite. *Nesophontes*, which lived in the Greater Antilles till the Holocene (Thenius 1980), may represent, according to Patterson (1957), the structural ancestry of the shrews. The solenodons indeed show morphological and behavioral affinities with the Soricidae (McDowell 1958).

The encephalization index is higher than that of *Potamogale* which is the highest among Tenrecidae. It is also higher than in the Soricidae. *Solenodon* differs from all the other Insectivora in that the conservative and the progressive brain structures are well developed. Its olfactory structures have the highest size indices, while the striatum is above, and the neocortex is slightly under the insectivoran average (Fig. 107). The somatosensory centers of the orofacial region have the same relative size as in terrestrial Soricidae and Oryzorictinae (Fig. 103). The somatosensory centers of the limb region are the best developed, except in Talpidae. The size of the visual and acoustic nuclei are close to the average of Oryzorictinae.

Erinaceidae

Erinaceidae are divided into the spiny hedgehogs, or Erinaceinae, and the hairy hedgehogs, or Echinosoricinae, which differ in their geographic distribution, degree of conservatism and the evolutionary level of their brains. The Echinosoricinae are confined to Southeast Asia, the Erinaceinae, to Eurasia and Africa (with the exception of Japan, Sri Lanka, and Madagascar). In comparison to the Erinaceinae, the Echinosoricinae have retained numerous plesiomorphic characters, such as a complete dental formula in most cases, a long facial cranium, a hairy pelage, and a long naked tail except in *Hylomys*. Echinosoricinae more closely resemble the ancestral form. The genera of the Echinosoricinae differ from each other in the specialization of the teeth and skull structure. These diverging evolutionary tendencies reveal, according to Thenius (1969), the mosaicism of phylogenetic development. *Hylomys* is more conservative with respect to skull structure, but *Echinosorex* has the more primitive teeth (Leche 1902).

The advanced characters in the Erinaceinae compared with the Echinosoricinae are short facial cranium, differentiation of the anterior teeth, spiny fur with development of a specialized cutaneous muscle, an orbicularis panniculi and reduction of the tail. The genera and species differ among themselves in the degree of specialization of particular morphologic traits. *Erinaceus europaeus* has the best developed orbicularis panniculi. With respect to skull structure, however, *Erinaceus* appears to be the most plesiomorphic and *Paraechinus*, the most apomorphic species (Verma 1962). *Hemiechinus* has the most progressive limbs, which are longer than in the other species, and larger ear pinnae. According to Thenius and Hofer (1960), *Erinaceus* and *Atelerix* are, in general, the most specialized forms of the Erinaceinae.

With respect to brain size and brain composition, the Echinosoricinae are on a higher evolutionary level than the Erinaceinae. The brain in Echinosoricinae is almost as large as in Desmaninae, which have the best developed brain of all Insectivora, whereas the brain in Erinaceinae is smaller than the insectivoran average and is close to the mean size for Soricidae (Table 7). The neocortex in Echinosoricinae is the third largest after the Desmaninae and Talpinae; that in Erinaceinae is much smaller (Table 45, Fig. 107). The most pronounced difference is in the size of the hippocampus: more than twice as large in the Echinosoricinae as in the Erinaceinae. While the difference in the size of the olfactory structures is less pronounced in the two subfamilies (Table 45), there are important differences in the size of the somatosensory and the auditory systems: both are better developed in Echinosoricinae (Fig. 103). No comparative data are available for the visual centers.

The differences between the two subfamilies in brain evolution are certainly related to fundamental differences in their antipredator strategies. Selective pressure favoring an active antipredator strategy also favors an increase in brain size, particularly in the neocortex and hippocampus. Active defence and flight involve not only more complex motor coordination, but also more accurate evaluation of the danger and more careful decision-making than stereotypical defence reactions. Because active defence demands more energy than passive defence, precise evaluation of the situation and making the right decision are critical for survival.

Among the **Echinosoricinae**, *Echinosorex* and *Hylomys* differ in development of the somatosensory centers. The structures of the orofacial and the posterior limb regions are better developed in *Echinosorex* than in *Hylomys*, whereas the reverse is true of the somatosensory structures of the anterior limb region. Although both species are broadly parapatric, *Hylomys suillus* occupies higher ground, but there is probably wide overlap (Corbet 1988). Differences in body size and sensory adaptations may prevent complete competitive exclusion and contribute to niche diversification. *Hylomys* lives mainly in primary montane forest, where it forages on the floor and occasionally climbs in low bushes. *Echinosorex* inhabits mangrove swamps, primary and secondary forest, and cultivated land (Lekagul and McNeely 1977). The diet of *Hylomys* is mainly made up of arthropods and earthworms, while *Echinosorex* prefers terrestrial and aquatic invertebrates and vertebrates (Davis 1962; Lim 1967; Medway 1969; Lekagul and McNeely 1977). The better developed vibrissae and somatosensory centers of the orofacial region in *Echinosorex* may make for greater efficiency in prey capture in an aquatic environment.

Among **Erinaceinae**, *Hemiechinus auritus* has the best developed, and *Erinaceus europaeus*, the least developed brain and neocortex. *Atelerix algirus, Erinaceus europaeus*, and *Hemiechinus auritus* differ particularly in development of the auditory structures, while somatosensory center and olfactory structure development are similar in all three species. The well developed acoustic system and long ears *Hemiechinus* shares with *Paraechinus* are characters commonly associated with a desert habitat. In fact, *Hemiechinus auritus* inhabits arid grassland and scrub, although it is less tolerant of desert conditions than *Paraechinus* (Niethammer 1973). There is no overlap in the distribution of *Erinaceus europaeus* and *Atelerix algirus*. The similarity in habitat, diet and especially brain size strongly suggests similarity in ecological niche.

Soricidae

The Soricidae, which are among the smallest living mammals, are distributed in the Holarctic, Africa, Southeast Asia, and the northern regions of South America. In Australia, their niche is occupied by small marsupials and, in South America, which was successfully invaded by only one genus (*Cryptotis*), the soricid niche is occupied by cricetid rodents and small marsupials (Eisenberg 1981). Recent Soricidae include two subfamilies, the Soricinae and Crocidurinae. The Soricinae have a Holarctic and northern Neotropical distribution, whereas the Crocidurinae have a Palearctic, Oriental, and Ethiopian distribution (George 1986). Shrews are found in a wide range of habitats, including tropical rain forests and deserts, mangrove swamps and montane streams, alpine meadows, and sandy plains. They have adapted to terrestrial, fossorial, semiaquatic, and occasionally to scansorial life-styles (Thenius 1969; Hutterer 1985).

It is noteworthy that the genera are characterized either by a remarkable rate of speciation and extensive distribution (*Crocidura, Sorex, Suncus, Cryptotis*) or by very few species (*Anourosorex, Blarina*) often restricted to a small area (Thenius 1969). More than 70% of the species belong to two genera only: the *Crocidura* (about 125 species) and *Sorex* (about 64 species). These genera also have the widest geographic distribution. Soricinae and Crocidurinae show conservative along with evolved characteristics. Apart from a few exceptions, the postcranial skeleton did not undergo any major changes. The skull and teeth, however, show several apomorphic features. The middle incisors are enlarged and have acquired a peculiar prominence that serves as a sort of pincers. Most of the premolars have disappeared along with the lower canines (Kingdon

1974). In Crocidurinae, the mandibular condyle conformation and the primitive triangular P4 have remained conservative (Repenning 1967; Thenius 1969), whereas they show evolved features in Soricinae. On the other hand, the tooth pigmentation found in Soricinae represents a conservative trait (Repenning 1967; Thenius 1969). Furthermore, the large litter size and extreme altricial state at birth that is found in Soricinae also represent plesiomorphic traits (Vogel 1980).

Based on their brain characteristics, Crocidurinae appear to be generally more primitive than Soricinae. In fact, despite a small overlap between the two groups and with few obvious exceptions, Soricinae are more highly encephalized and have better developed progressive brain structures, such as neocortex and striatum, than Crocidurinae.

Among the **Soricinae**, differences in relative brain size and brain composition are only partly related to taxonomic affiliation. A more likely explanation lies in biotopical and zoogeographical aspects, which to some extent reflect differences in environmental complexity. Despite the scarcity of detailed studies on specific differences in behavioral adaptations, the data available indicate trends in brain evolution which correlate with niche diversification. Regardless of their taxonomic categories, Soricinae may be grouped into terrestrial, semifossorial, semiaquatic, and scansorial forms, into ground and underground dwellers, into New World and Old World species and so forth.

Of the three tribes distinguished in Soricinae (Repenning 1967; George 1986), the Soricini with *Sorex* as the most representative genus, are characterized by exceptional variability in distribution, niche diversity, and brain characteristics, although the group exhibits a generalized appearance, that has changed little from the early soricids (George 1986).

Of the four terrestrial *Sorex* species, the two New World forms, *Sorex fumeus* and *S. cinereus*, have a larger brain and neocortex than the Old World forms, *S. minutus* and *S. araneus*. The Palearctic *S. alpinus*, however, which exhibits scansorial capabilities (Hutterer 1982), has a brain and neocortex almost as large as the Nearctic species, while the semiaquatic *Sorex palustris*, surprisingly, has a brain weight similar to that of the Palearctic forms (Tables 6 and 44). The data are, however, from poor material. No measurements of neocortex are available for *S. palustris*. The tribe Blarini, comprising two living genera, *Blarina* and *Cryptotis*, is restricted to the New World and exhibits more specialized characters than the Soricini. With respect to size of the brain and neocortex, the Blarini follow the evolutionary trend of the two American *Sorex* species. The third tribe, Neomyini, to which *Neomys* and *Anourosorex* belong, are Holarctic in distribution.

The Neomyini are the most specialized Soricinae and include semiaquatic and fossorial forms. Although *Neomys* and *Anourosorex* have the same relative brain size, they differ considerably in brain composition. Like other semiaquatic forms, the two *Neomys* species have a relatively large medulla and neocortex but relatively small olfactory bulbs. *Anourosorex*, with morphological features for fossorial life, has larger bulbs but a smaller medulla, similar to the fossorial Talpidae (Table 44). Other brain structures, such as the cerebellum and sensory trigeminal complex, are larger in *Neomys* than in other Soricinae. No data are available for *Anourosorex*.

Soricine shrews live in very varied habitats, such as grassland, forest, wooded steppes, river banks, marshy areas, and mountain taiga (Aitchison 1987). At least part of the variation in brain size and quantitative brain structure in the *Sorex* species seems to have ecological correlates. In comparison to *Sorex araneus* and *S. minutus*, New World species living in the northern regions occupy larger home ranges (Buckner 1966, 1969; Michielsen 1966). *S. araneus* and *S. minutus*, which differ to a certain extent in brain composition, also differ ecologically. Although both are sympatric and show similar habitat preferences, they are partially segregated: *Sorex araneus* spends most of its active time underground, whereas *S. minutus* is active mainly on the ground (Michielsen 1966). Niche separation results at least partially from different feeding strategies as well. Pernetta (1977) reports a large component of Lumbricidae in the diet of *Sorex araneus*, which also digs extensively; this prey is absent in the diet of *S. minutus*, which does little digging. The species also differ in the differential development of sensory structures. Olfactory structures are better developed in *S. araneus*. The differences between *Sorex alpinus* and the two other Old World species are striking. The cerebellum, striatum, and neocortex are considerably larger in *S. alpinus*. Behavioral and morphological adaptations characterize this species as an inhabitant of rock and soil fissures with scansorial abilities (Spitzenberger 1966; Hutterer 1982).

All Nearctic soricines investigated are sympatric in the area where they were captured (except *Cryptotis*). Although there is considerable overlap in diets, niche separation is established by differences in size, habitat preferences, feeding strategies, and/or brain structures. *Blarina*, the biggest and most fossorial of the American shrews, consumes the greatest variety of prey (Aitchison 1987), and its preferences are related to prey size, rather than abundance (Buckner 1964). It hunts underground in its own tunnels or in tunnels of other species and on the surface. Climbing abilities, reported by Carter (1936), may be performed rarely. The relative size of *Blarina*'s brain and neocortex is slightly above the soricine average. *Microsorex hoyi* has the smallest brain and neocortex of the Nearctic species. Its diet is less varied than in *Blarina*. It does not eat any large invertebrates (Banfield 1974). *Sorex cinereus*, which is about the same size as *Microsorex* but has a larger brain and neocortex, is one of the most successful and widely distributed shrews in Canada. It considers most invertebrates greater than 3 mm long as potential prey (Platt and Blakley 1973; Aitchison 1987). *Sorex fumeus*, which has the best developed brain and neocortex, feeds mainly on insects, especially beetles, but only rarely lumbricids (Aitchison 1987). Sensory structures differ somewhat in *Blarina* and *Sorex cinereus*. Olfactory structures are slightly better developed in *Blarina*, but the somatosensory centers of the orofacial region are better developed in *S. cinereus*. The long facial vibrissae of *S. cinereus* are an important sensory organ. When hunting, the shrew frequently holds its snout in the air, with vibrissae twitching, to investigate the environment (Banfield 1974).

Crocidurinae are largely Paleotropical forms. With respect to brain evolution, Crocidurinae can be divided into two groups. The *Crocidura, Suncus,* and *Scutisorex* group has small brains, and the other group, including *Sylvisorex, Ruwenzorisorex,* and *Myosorex*, is characterized by larger brains. *Crocidura,* which has the most reduced dentition, is the most advanced and, at the same time, the most adaptable and generalized genus (Kingdon 1974). However, its brain and neocortex are among the smallest in Soricidae. The smallest brain and neocortex is found in *Suncus,* which differs from *Crocidura* by the presence of a vestigial third premolar (Grassé 1955a). *Scutisorex* shows the same dental formula as *Suncus* and a similar level of brain evolution, but it also has the most specialized backbone. No satisfactory explanation has been offered of the possible function of its backbone. Although brain size is similar in the first group, there are marked differences in brain composition even within the same genus.

Myosorex, which has the largest number of teeth and is regarded as the most archaic genus of the Soricidae (Grassé 1955a), nevertheless has a well developed brain and neocortex. According to Heim de Balsac and Lamotte (1956, 1957), it shares a number of characters with Soricinae. The genus is almost exclusively confined to high mountains. *Sylvisorex* resembles the ancient stock of *Crocidura* and *Suncus* with respect to dental structure. Some species, however, differ by specialized features, such as an elongated tail and feet, which are related to scansorial abilities (Vogel 1974). Brain and neocortex size reach the highest values found in Crocidurinae. *Ruwenzorisorex suncoides* (Hutterer 1986) differs only slightly from *Sylvisorex* in that its brain is smaller. Brain composition, however, is different, particularly with respect to sensory equipment, where the characteristics are similar to those found in semiaquatic forms. The olfactory bulbs are less than half the size of those in *Sylvisorex*, while the sensory trigeminal complex is significantly larger than in *Sylvisorex* (Tables 14, 44, Fig. 108).

The conspicuous differences in brain size and composition between the *Crocidura* group and the *Sylvisorex* group, including *Sylvisorex, Ruwenzorisorex,* and *Myosorex* (Figs. 102, 108), raise question about the taxonomic unity of the Crocidurinae. Heim de Balsac and Lamotte (1956) claimed that the Crocidurinae constitute an artificial group of forms with no direct phyletic relation among them. Based on cranial and dental features, *Myosorex* appears to represent a separate lineage from the Crocidurinae. *Sylvisorex*, on the other hand, is close to *Suncus* and *Crocidura* (Heim de Balsac and Lamotte 1957; Butler and Greenwood 1979), but, based on the number of conservative traits, it is more primitive and ancestral to the two other genera. According to brain evolution, the *Sylvisorex* group is clearly more progressive than the *Crocidura* group and is characterized by a relatively uniform size in the five fundamental brain parts. In the TEL, however, there are considerable differences. Differential brain evolution corroborates the heterogeneity of the Crocidurinae. The taxonomic position of *Sylvisorex* and *Ruwenzorisorex,* in addition to that of *Myosorex,* should therefore be reconsidered.

Talpidae

The Talpidae are generally considered specialized descendants of soricid ancestors (Thenius and Hofer 1960). They are confined to Eurasia (including Japan) and North America. In comparison with the Soricidae, the teeth of moles are more conservative but the forelimbs and the scapular girdle may be highly apomorphic in connection with specialized habits.

Recent Talpidae have differentiated into two adaptive types: the fossorial and the semiaquatic forms. Among the fossorial forms, Yates (1984) distinguished the Uropsilinae, which are more ambulatory than fossorial, and the Talpinae. *Uropsilus,* from East Asia, has the most specialized teeth, but its postcranial skeleton makes it the most conservative genus, while *Talpa,* which is the most specialized fossorial form, has the complete dental formula (Thenius 1969). Based on the characteristics of the postcranial skeleton (for example the humerus in Talpinae) it is possible to establish an evolutionary series from the most conservative to the most specialized form (cf. Campbell 1939). The recent Desmaninae, which are the semiaquatic forms, are confined to two disjunctive relic areas, one on the Iberian Peninsula (*Galemys*), and the other to eastern Europe and western Central Asia (*Desmana*). As in *Talpa,* the teeth are conservative, but unlike *Talpa*, the hind limbs show stronger modifications than the forelimbs. According to Thenius (1969), *Galemys* is generally less specialized than *Desmana.*

Brain size is distinctly larger in the Desmaninae than in the Talpinae. The difference between the subfamilies is even greater with respect to neocortex size (Fig. 108). The marked increase in brain size and of progressive brain structures is related to adaptation to an aquatic environment. The development of the sensory structures reflects the special requirement for successful substrate utilization. The olfactory structures are strongly reduced in Desmaninae (Table 45, Fig. 108) and the visual structures are small in Talpinae (Table 70,

Fig. 106). As in all semiaquatic Insectivora, the somatosensory center of the orofacial region is strongly developed in the Desmaninae, that of the forelimb region is large in the fossorial Talpinae (Table 15). Auditory structures are relatively small in both subfamilies (Table 27, Fig. 104).

In conclusion, there is no parallel between the evolutionary level of the somatic structure and of the brain. Species with a large number of conservative traits may have a very progressive brain, while species with many derived features may be at a low level with respect to brain evolution.

Although Echinosoricinae are more conservative than Erinaceinae (Leche 1897, 1902; Viret 1955), they have a much better developed brain and neocortex. *Sylvisorex* is more conservative than *Suncus* and *Crocidura* but has the best developed brain and the most progressive neocortex of all the Crocidurinae. *Myosorex* has the largest number of primitive characters and its teeth are the most conservative among extant Soricidae. However, its brain is much more progressive than that of *Suncus* and *Crocidura,* which are considered to be the most advanced forms of the Crocidurinae (Heim de Balsac 1958; Heim de Balsac and Lamotte 1956, 1957; Butler and Greenwood 1979; Hutterer 1985).

In contrast, among Potamogalinae, *Potamogale velox* is the most apomorphic with respect to brain morphology and somatic morphology, while *Microgale lamottei* is the most plesiomorphic with respect to both features (Heim de Balsac 1956).

Although the fact of mosaic evolution is well established, sparse quantitative data on detailed brain structure have limited our understanding of the various phylogenetic changes occuring in Insectivora. It appears now beyond any doubt that somatic and brain evolution can progress at different rates. A species with a conservative "Bauplan" may reach a high level of brain evolution, while a specialized and progressive species may be characterized by a very conservative brain structure.

6 Brain Characteristics Related to Ecoethological Adaptations

Many of the derived brain characters are clearly related to specific adaptations to particular environments and are therefore discussed with respect to ecological niches.

6.1 Surface Dwellers

Adaptation to terrestrial surface dwelling involves the greatest variation in body and brain morphology. Although their basic body plan comes closest to a conservative insectivore, terrestrial species show large variations in body size, limbs, external appendages, and fur structure (Baron and Stephan in press). Similarly, brain size and quantitative brain composition show extreme variation.

When terrestrial Insectivora are compared with semiaquatic forms and Primates, the most conspicuous brain structure is the main olfactory bulb (MOB), suggesting that olfaction is the predominant information channel. However, there are clear differences in the development of MOB among the terrestrial forms. *Solenodon* not only has the best developed MOB of the Insectivora, but olfaction also seems to be its predominant sensory channel. *Solenodon* hunts arthropods and other invertebrates in leaf litter and other debris of vegetation, rooting and probing the ground with its snout or using its claws (Cockrum 1962; Eisenberg and Gould 1966). Eisenberg and Gould stated that the prey is apparently located by tactile and auditory or olfactory stimuli. The sensory trigeminal complex (TR) is also well developed in *Solenodon* but, compared with the other terrestrial Insectivora, it is somewhat less developed than MOB. Although the exact role of the different senses in foraging behavior is not known, it is likely that an animal, when adopting the typical foraging position with its nose to the ground, is searching for olfactory cues.

Echinosoricinae have a better developed MOB than Erinaceinae. Nevertheless, olfaction seems to be the predominant sensory channel in hedgehogs, whereas in moonrats, the auditory nuclei (AUD) reaches the highest development found in Insectivora (Fig. 103). In Tenrecinae, MOB is also better developed than AUD and TR. Erinaceinae (Herter 1957; Kingdon 1974) and Tenrecinae (Eisenberg and Gould 1970) display similar foraging behavior to *Solenodon*. Probing with the nose in the mat of vegetation certainly indicates olfactory investigation.

Of all terrestrial Insectivora, Soricidae have the smallest MOB. Furthermore, the Soricinae and Crocidurinae differ in their respective sensory equipment. In the Crocidurinae, which have better developed auditory structures than Soricinae, audition seems to play a major role, whereas in the Soricinae, the somatosensory system of the orofacial region seems to be the predominant sensory channel. Observers disagree about the sensory cues which are important in finding food. While some authors maintain that shrews hunt almost entirely by smell (Middleton 1931; Holling 1958; Grünwald 1969; Maser and Hooven 1974), others believe that olfactory cues are of minor importance, and that food seems to be located mainly by means of the long vibrissae (Blossom 1932; Hamilton 1940; Crowcroft 1957; Rudge 1968; Kingdon 1974). For many Crocidurinae, hearing seems to be important in hunting prey (Ansell 1964; Fons 1974; Burda and Bauerova 1985) (see p. 331). Although optimization of feeding behavior implies that all sensory channels are involved in hunting prey, comparative volumetric brain data suggest that even closely related species may differ in their predominant information channel.

The great variation in the progressive brain parts, especially the neocortex (NEO), indicates major differences in niche complexity. This is particularly striking in the Erinaceidae. The Echinosoricinae, which display more complex behavior, have a better developed NEO than the Erinaceinae. While the Erinaceinae use a highly stereotyped passive defence reaction, the Echinosoricinae adopt an active strategy (see pp. 337, 339). *Echinosorex* may be very aggressive when threatened and dash against a potential aggressor with open mouth (Lim 1967; pers. obs.) or flee when in danger (Gould 1978). Moonrats also exploit a great variety of feeding niches. They feed on terrestrial, limnetic, and aquatic invertebrates and vertebrates (Lim 1967; Medway 1969).

Of the two soricid subfamilies, the terrestrial Soricinae have a larger NEO on the average than the terrestrial Crocidurinae. The two groups show vast differences in their geographic distribution, spatial organization, population density, feeding strategy, and activity rhythm (see pp. 339, 340). In general, for most Crocidurinae, which live mainly in tropical environments, food supplies are more predictable than for the Soricinae, which live in the temperate climatic zone. The increase in progressive brain parts may be primarily related to the characteristics of the niche occupied. Of the Soricinae, New World species have by far the best developed NEO. In comparison to *Sorex araneus* and *S. minutus*, New World species usually occupy larger home ranges (Buckner 1966; Newman 1976). The extreme climatic conditions found on the North American continent increase the heterogeneity of the biotope and require considerable behavioral plasticity.

6.2 Subterranean or Fossorial Species

Among the Insectivora, burrowing adaptations evolved independently in the Talpidae and Chrysochloridae. Since moving and digging underground demand more energy than locomotion on the surface, subterranean Talpidae and Chrysochloridae exhibit structural adaptations for burrowing and underground life which should reduce energetic costs. The animals are characterized by a cylindrical body and several structural reductions and modifications, the most remarkable involving the structure of the pectoral girdle and the forelimbs (Slonaker 1920; Edwards 1937; Reed 1951; Ellerman 1956; Yalden 1966; Puttick and Jarvis 1977; Nevo 1979; Gasc et al. 1986). Differences between the Talpidae and Chrysochloridae in the forelimb modifications are related to differences in digging technique.

There are also large differences in brain shape. Talpinae have flat and broad brains, whereas Chrysochloridae have very high, antero-posteriorly compressed brains. This difference, too, may be related to differences in burrowing. Unlike the lateral stroke burrowing of talpids, golden moles burrow by means of the leathery snout (ref. in Spatz and Stephan 1961) and powerful thrusts of the forepaws resembling cursorial progression. The relative size of the brain and neocortex (NEO) is above the insectivoran average. There is a marked tendency for the striatum (STR) to increase and the medulla oblongata (OBL) to decrease in size, while the olfactory structures (MOB, PAL) and cerebellum (CER) show divergent tendencies. Olfactory structures are close to the subfamily average in the Talpinae, but clearly above AvIF in the Chrysochloridae (Figs. 107, 108). In contrast, CER is under the subfamily average in the Chrysochloridae, but above, in the Talpinae (Figs. 101, 102). The three sensory complexes (main olfactory bulb, MOB; auditory nuclei, DCO, VCO, OLS; and sensory trigeminal structures, TR) are in all cases better developed in golden moles than in talpids (Figs. 104, 107, 108). Talpids differ from the chrysochlorids in their foraging strategy as well. According to Mellanby (1966) talpids collect prey in the tunnels, which serve as pitfall traps. It seems, therefore, that they feed rather opportunistically on prey encountered in the course of regular excursions, just as do subterranean shrews. Chrysochloridae seem to detect worms by smell. For food other than worms, however, they rely on contact (Kingdon 1974) or, perhaps, on vibration produced by potential prey. Differences in the development of the sensory brain structures are consistent with differences in feeding strategies.

The high STR indices in talpids and chrysochlorids should be interpreted as an adaptation to fossorial life. However, the differences in the relative size of the CER suggest marked differences in motor abilities. In fact, as already mentioned, the mode of digging and the structures associated with this behavior are fundamentally different in the two groups. In golden moles, digging movements are similar to running motions (Campbell 1938; Puttick and Jarvis 1977); in talpids, the forelimb functions more as a spade when burrowing.

6.3 Semiaquatic Species

Adaptation to semiaquatic life evolved in three insectivoran families: the Tenrecidae, Talpidae, and Soricidae. In adaptation to the very specialized habit of hunting for prey in water, several morphologic changes occured, that enable the animals to move freely in water. Species, within a family, often show a series of progressively stronger structural modifications for semiaquatic life. The best adapted have a long and streamlined body, like *Potamogale velox* among the Tenrecidae, *Desmana moschata* among the Talpidae, and *Nectogale elegans* among the Soricidae. All semiaquatic species show a reduction of the ear pinnae and an elongation of the tail.

There are two trends in adaptation to an aquatic environment and the structural modifications vary according to their involvement in swimming. Animals propelled by the axial muscles, the trunk-tail swimmers, have a hypertrophied flattened tail; those propelled mainly by limb movements, the limb-tail swimmers, tend to have webbed digits.

In contrast to the great differences in the adaptations in external morphology, there is relative uniformity in the adaptations of the brains compared with the terrestrial forms, although some differences related to the type of swimming and degree of adaptation are found. Of the 50 species of Insectivora in which we have investigated brain composition, eight are predators in limnetic ecosystems: *Limnogale mergulus, Potamogale velox, Micropotamogale lamottei, M. ruwenzorii, Neomys anomalus, N. fodiens, Desmana moschata,* and *Galemys pyrenaicus.* In all semiaquatic species, the medulla oblongata (OBL) is much larger than in closely related non-aquatic forms, while the main olfactory bulb (MOB) is much smaller. Reduced olfaction is compensated by an increase in the trigeminal system (TR). In subfamilies containing both terrestrial and semiaquatic forms, such as the Oryzorictinae and Soricinae, the cerebellum (CER), diencephalon (DIE), and striatum (STR) are all larger in the semiaquatic forms. In *Limnogale,* the neocortex (NEO), too, is larger than in the terrestrial Oryzorictinae. Although *Neomys* does not have the best developed NEO in the Soricinae, its size index is still above the subfamily average. Similar tendencies are observed in the other semiaquatic forms. Desmaninae and Potamogalinae have the highest NEO indices of their respective families. The indices of MES and DIE also tend to be higher than in closely related terrestrial forms. Although motor structures, such as CER and STR, show a tendency to increase in size over related terrestrial or subterranean species, the increase is more pronounced in Desmaninae, which are limb-tail swimmers.

The high size indices of two sensory brain complexes, i.e., of the vestibular nuclei complex (VC) and sensory trigeminal complex (TR), are characteristic of the semiaquatic forms. Movements in three-dimensional space demand more precision in equilibrium than movements confined to the ground. It is therefore not surprising that Insectivora moving in water have well developed vestibular nuclei. The tremendous size of TR is mainly responsible for the high OBL index. Since mammalian olfaction, relying on respiration, is inefficient in water, semiaquatic Insectivora depend on their trigeminal system for detecting prey. The extremely long vibrissae, innervated by the trigeminal nerves, may act as teletactile or vibration receptors (Stephan and Bauchot 1959; Stephan and Spatz 1962; Bauchot and Stephan 1968).

Like body morphology, brain morphology may indicate the degree of adaptation to aquatic life. This is best seen in the case of Potamogalinae. All brain structures which tend to increase in semiaquatic forms are better developed in *Potamogale velox* than in *Micropotamogle lamottei.* On the other hand, structures which tend to decrease, such as MOB, are more reduced in *Potamogale velox* than in *Micropotamogale lamottei.* The papers by Heim de Balsac (1954, 1956, 1957), Heim de Balsac and Bourlière (1955), DeWitte and Frechkop (1955), Guth et al. (1959, 1960), Rahm (1961), and Dubost (1965) suggest that *Micropotamogale lamottei* is clearly less adapted to water than the other otter shrews. It does not have the laterally compressed tail found in *Potamogale velox*, nor does it have webbed hands and feet like *Micropotamogale ruwenzorii. Micropotamogale lamottei* swims by paddling and lateral movements of the body (Vogel 1983) and is able to catch prey in water (Guth et al. 1959).

6.4 Morphological Adaptations for Winter Survival

In response to cold temperatures, Insectivora inhabiting north-temperate and boreal regions have evolved morphological adjustments as well as physiological and behavioral ones. Like many small mammals inhabiting cold regions, Insectivora such as *Sorex araneus* have been reported to show a general decline in body mass during winter (Adams 1912, Dehnel 1949, Mezhzherin 1964, Mezhzherin and Melnikova 1966), sometimes by about 35%, and a reduction in body length of about 7% (Siivonen 1954, Hyvärinen and Heikura 1971, Hyvärinen 1984). The shortening of the body is caused by flattening of the intervertebral discs and by resorption of cartilage in the discs (Hyvärinen 1969, 1984). Winter remodeling of the skull in shrews was first described by Dehnel 1949 and later confirmed by Z.Pucek (1955, 1957), Schubarth (1958), and Crowcroft and Ingles (1959). The decreased height of the skull is due to resorption of the parietal and occipital bones at the edges of the sutura sagittalis and sutura lambdoides (Z.Pucek 1955, Hyvärinen 1968, 1969, 1984). Many internal organs, such as kidneys, liver, and spleen, also lose weight in winter (Z.Pucek

1965, 1970). Only the weight of the heart (Z.Pucek 1965, 1970) and the stomach (Myrcha 1967) do not decrease. The most surprising phenomenon, perhaps, is the reduction in brain weight first noted by Cabon (1956) in *Sorex minutus* and later by several authors in *Sorex araneus* (Bielak and Pucek 1960, M.Pucek 1965, and Z.Pucek 1965). Yaskin (1984) reported a differential reduction of brain parts in *Sorex araneus* and *Sorex minutus*. The most pronounced decreases were observed in the neocortex and striatum.

There is adequate evidence that the size regression is followed by an increase again as spring and breeding season approach. This is clearly shown in the length and weight of the body and the skull dimensions in *Sorex minutus*. However, the skull does not attain its dimensions of the previous summer (Grainger and Fairly 1978). Z.Pucek (1970) maintained that the phenomenon was characteristic of shrews of northern regions. In fact, representatives of the genus *Crocidura* show only minor changes. *Blarina brevicauda*, however, which is not reported to nest communally nor to undergo any form of physiological heterothermy but rather remains active throughout the winter months, shows a significant and steady increase of 39% in body mass between September and April unlike *Sorex araneus* (Merrit 1986).

Another morphological change seen in shrews from cold climates is an increase in mass of interscapular brown adipose tissue and in liver lipids (Hyvärinen 1984). Brown adipose tissue has been shown to be an important site of non-shivering thermogenesis (Foster and Frydman 1978, Feist 1980).

Non-hibernating Insectivora may minimize thermal conductance and reduce heat loss by increasing pelage density and length during winter. In the only study of seasonal pelage changes in shrews, Borowski (1958) found an inverse relationship between pelage density and environmental temperature in *Sorex araneus*.

7 Brain Characteristics Related to Functional Systems

7.1 Main Olfactory System

Despite the importance of olfaction as a predominant sensory channel in Insectivora, numerous reports on the morphology of the insectivoran nasal cavity and intranasal epithelia suggest the existence of enormous interspecific differences (Ganeshina et al. 1957; Negus 1958 ex Quilliam 1966b; Gurtovoi 1966; Graziadei 1966; Andreescu-Nitescu 1970; Wöhrmann-Repenning 1975; Wöhrmann-Repenning and Meinel 1977; Schmidt and Nadolski 1979; Kuramoto et al. 1980; Larochelle 1987; Larochelle and Baron 1989a, b).

Subterranean, semiaquatic, and terrestrial Insectivora differ in the relative size of the various regions of the nasal cavity. Compared to terrestrial Insectivora, in *Talpa europaea,* the rostral regio vestibularis is enlarged, providing better protection for the very sensitive mucosa in a subterranean environment (Wöhrmann-Repenning 1975). On the other hand, the regio respiratoria, which, according to Negus (1965), serves to warm and moisten the air, is reduced in the subterranean *Talpa*, which lives in a relatively humid environment with fairly constant temperatures (Wöhrmann-Repenning 1975). The diameter of the lumina between the ethmoturbinals was found to be larger in the semiaquatic *Neomys* than in terrestrial Soricidae (Söllner and Kraft 1980). These "bottlenecks" help bring odorous particles in the air into close contact with the receptor cells.

The size of the olfactory epithelium and the number of receptor cells indicate that olfaction is highly developed in all Insectivora. There are, however, major variations among species living in different biotopes. Both are clearly reduced in semiaquatic forms when compared with terrestrial species. In *Sorex palustris*, the olfactory epithelium covers about 52.2% (Larochelle and Baron 1989a), and in *Neomys fodiens*, as little as about 41% (Söllner and Kraft 1980) of the total inner surface of the cavum nasi. In terrestrial shrews, it has been found to vary from 61.7% in *Sorex minutus* (Söllner and Kraft 1980) to 69.3% in *Sorex cinereus* (Larochelle and Baron 1989a). The difference between the two water-adapted species is paralleled by differences in morphological adaptations. According to Hutterer (1985), *Neomys* is better adapted to semiaquatic life than is *Sorex palustris*. Similarly, J. Niethammer (1978) found that 74 to 95% of the diet of *Neomys* is of aquatic origin, whereas in *Sorex palustris*, the figure is only 49% (Conaway 1952).

From the observation that receptor density is lower in the subterranean *Talpa* than in three terrestrial species *(Sorex araneus, Crocidura russula,* and *Eri-*

naceus europaeus) Wöhrmann-Repenning (1975) and Wöhrmann-Repenning and Meinel (1977) concluded that the mole has poorer olfaction. However, the relationship between receptor density and olfactory capacity is not clear. Comparative studies on the sensitivity of the olfactory organ seem to indicate that threshold values increase with increasing receptor density (Sigmund 1985; Sigmund and Sedlacek 1985). Morphological studies on the olfactory mucosa, however, have shown that receptor density in semiaquatic species is similar to that in terrestrial shrews (Söllner and Kraft 1980; Larochelle 1987; Larochelle and Baron 1989b), despite poorer olfactory capacity as indicated by reduction of the peripheral and central olfactory structures.

Differences similar to those found in the olfactory membrane have been reported in the size of the olfactory brain centers. Larochelle (1987), investigating olfactory structures in shrews, has found a strong positive correlation between the size and number of cells in the olfactory epithelium and the size of the main olfactory bulb (MOB) and the number of mitral cells.

Compared with other mammalian groups, the main olfactory bulbs (MOB) in Insectivora constitute a major part of the telencephalon (TEL). In fact, the MOB account for an average of about 14% of the TEL (number of species = N = 50; see p. 100). In Tenrecinae (N = 4), even more than 1/5 of the TEL (22.1%) is MOB, but only 6.3%, in Potamogalinae (N = 3). In Primates the average values are 2.4% in Prosimians (N = 18) and less than 0.2% in Simians (N = 27). In Microchiroptera (N = 125), the MOB occupy about 5% and in Megachiroptera (N = 36), close to 8% of the TEL.

The size indices for MOB and the paleocortex (PAL) and its components indicate vast differences in the development of insectivoran olfactory structures, all clearly related to ecological niche (Baron et al. 1983, 1987). Terrestrial or ground-dwelling Insectivora, including *Solenodon*, Erinaceidae, and several species of Tenrecidae and Soricidae, have the best developed MOB of this order. The size indices for the fossorial Insectivora are within the range of the terrestrial forms, although the MOB in the Talpinae is somewhat smaller than in the Chrysochloridae. Therefore, total olfactory information appears to play a lesser role in the Talpinae than in the Chrysochloridae. In the Oryzorictinae, the semifossorial *Oryzorictes* also has a relatively small MOB. Most of the terrestrial Soricidae have even smaller MOBs than the fossorial Talpinae. It is not yet known to what extent olfactory cues are used by shrews to detect food. However, based on MOB size indices, it is likely that olfaction plays a lesser role in terrestrial shrews than in other terrestrial Insectivora. There are some interesting differences in Crocidurinae. *Sylvisorex* and *Myosorex* have very large olfactory centers, whereas *Ruwenzorisorex* has indices similar to that of the semiaquatic *Neomys*. With a small MOB, *Ruwenzorisorex* is clearly different from the genus *Sylvisorex* from which it was recently separated (Hutterer 1986).

The most reduced MOB, as indicated in many publications (Stephan and Bauchot 1959; Bauchot and Stephan 1968; Stephan and Kuhn 1982; Baron et al. 1983; Frahm 1985), is found in semiaquatic Insectivora regardless of taxonomic affiliation. In Potamogalinae, the MOB is only 38%, and in Desmaninae, 60% of the average size in Tenrecinae. Water shrews and *Limnogale* have the lowest indices of the Soricinae and Oryzorictinae, respectively. Since olfaction is associated with breathing in mammals, it is of no use in detecting prey under water. Therefore, reduced bulbs in species that obtain a considerable portion of their food in the water is not surprising. Heim de Balsac (1954, 1956, 1957), Heim de Balsac and Bourlière (1955), DeWitte and Frechkop (1955), Guth et al. (1959, 1960), Rahm (1961), and Dubost (1965) suggest that *Micropotamogale lamottei* is clearly less adapted to water than the other otter-shrews. Correspondingly, it has the least reduced, i.e., the highest size index for MOB, of the Potamogalinae.

In parallel with MOB, the following paleocortical components were reduced in the eight semiaquatic Insectivora under investigation: retrobulbar region (RB; = anterior olfactory nucleus), prepiriform cortex (PRPI), olfactory tubercle (TOL), lateral olfactory tract (TRL), nucleus of the lateral olfactory tract (NTO), and the fiber complex of the anterior commissure (COA). The reduction in MOB and all the structures mentioned is nearly 1/2 in semiaquatic as compared with terrestrial Insectivora, but is clearly stronger in NTO (more than 2/3) (Baron et al. 1987). Thus, the olfactory paleocortex (PAL) shows a similar pattern to MOB, with NTO even more strongly reduced in semiaquatic Insectivora than MOB. Since a strong NTO reduction occurs also in simians and man, it cannot be related to water adaptation, but must be a general phenomenon which occurs when olfaction is no longer one of the main systems for detecting food.

Perhaps the best known insectivoran surface foragers are the Erinaceidae. The size development of their MOB is similar to that of the other terrestrial Insectivora. Oddly, there is no agreement among authors about their olfactory capacity and the precise role of olfaction in their foraging behavior. According to Herter (1957), hedgehogs move along with their noses only a few millimeters from the soil when foraging and produce audible sniffing sounds while inhaling and exhaling. Objects with strong odors can be detected from a distance of eleven meters. Furthermore, it has been shown that they ignore prey hidden behind a glass cover but can detect food buried three or four centimeters in the ground (Herter 1979). On the other hand, Bretting (1972), who determined threshold levels of perception of some fatty acids, considered that the olfactory capacity of *Erinaceus europaeus* was mediocre and comparable to that of man. Such comparisons may be misleading since they do not take into account specialized sensitivity to odorants which may be biologically significant for one group but not for the other.

With the exception of some semiaquatic forms, most of the Soricidae may be considered terrestrial and subterranean but not fossorial forms, even if they occasionally dig burrows. They inhabit terrains covered by vegetation offering good hiding places and shelter under stones or in burrows of moles and small rodents. Subterranean shrews collect food by tunneling through the loose accumulation of plant detritus (Rudge 1968), in narrow spaces and crevices beneath rocks and logs, or by patrolling in surface runways and burrow systems of small mammals. We do not yet know to what extent olfactory cues are used by shrews in food detection. Observers disagree about the role of their sense of smell in finding food (see 6.1). Although shrews can detect buried or openly available food items by olfaction (Grünwald 1969; Vlasak 1972; Schmidt 1979) and may distinguish odors of different kinds of hidden prey, such as healthy and fungusinfected cocoons (Holling 1958), it seems that they feed rather opportunistically on available prey encountered more or less accidentally during their regular excursions at various levels of the substratum. Our results indicate that shrews have generally less developed olfactory centers than the other terrestrial Insectivora. This supports the view that olfaction is less important in the foraging strategy of Soricidae.

There are two types of true fossorial Insectivora which differ in their digging techniques and in the relative size of their olfactory bulbs. The size indices for *Talpa* and *Oryzorictes* are within the range of variation of the size indices for subterranean shrews, whereas those of the fossorial Chrysochloridae are similar to the indices for the ground foragers. During their regular excursions through complicated tunnel systems, which essentially serve as pitfall traps (Mellanby 1966), talpids apparently collect prey in a manner similar to subterranean shrews. One may therefore suppose that, as in shrews, olfactory cues do not play an important role in the detection of food in fossorial talpids. In fact, they ignore prey only a few centimeters away. Among Tenrecidae, *Oryzorictes* is the most pronounced subterranean type. Like common moles, it possesses powerful digging claws and has a similar MOB size indicating that, in contrast to terrestrial Tenrecidae, olfactory stimuli certainly play a less important role in the detection of food. The second fossorial group, the Chrysochloridae, dig tunnels in soft soil and vegetable debris just under the surface of the ground as they search for prey, usually worms, which seem to be detected by smell. However, for food other than worms, Chrysochloridae rely on actual contact (Kingdon 1974).

The functional significance of the quantitative development of the olfactory centers has to be considered not only in relation to feeding behavior but also in relation to social behavior. Most Insectivora possess scent glands producing pheromones used in intra- and interspecific communication (for references see Poduschka 1977a; Baxter and Meester 1980; Hutterer 1985). According to Thiessen (1977), animals with a high basal metabolic rate are able to produce an enormous amount of glandular secretions. One may assume that, for the Insectivora, which are known to have a high basal metabolism, pheromones are a very economical means of communication. Furthermore, the particular characteristics of their terrestrial and subterranean ecotopes, which favor population dispersal (Nevo 1979), render olfaction the most efficient channel for communication. In fact, marking behavior has been observed in representatives of most insectivoran families and ecotypes (Poduschka 1977a; Baxter and Meester 1980; Jannett and Jannett 1981). Behavioral studies suggest an important role for olfactory cues in intraspecific and interspecific interaction in shrews (Hawes 1976; Hausser and Juhlin-Dannfeldt 1979; Larochelle and Baron 1986), tenrecs (Eisenberg and Gould 1970; Poduschka 1974), moonrats (Gould 1978), hedgehogs (Poduschka 1977b), and moles (Stone 1986).

Given the present state of knowledge, it is difficult to correlate the relative importance of the olfactory channel with differential development of the olfactory centers in the various types of Insectivora. The available data, however, strongly suggest that, regardless of its precise role in feeding behavior, olfaction is the predominant sensory channel in intraspecific and interspecific interaction in the Insectivora, even in species with apparently reduced olfactory structures, such as the semiaquatic forms.

7.2 Accessory Olfactory (Vomeronasal) System

Accessory olfactory bulbs (AOB) have been found in all Insectivora so far investigated, but they have been very small or rudimentary and sometimes difficult to distinguish in *Micropotamogale, Solenodon,* some Soricinae, and some Talpidae (indices between 6 and 47). The Echinosoricinae, with a mean size index of 580 and a maximum of 810 in *Hylomys* (Tables 50, 51), have by far the best developed AOB of all Insectivora. The average indices are twice as large as in Erinaceinae (261), which also have very well developed AOB. The mean indices for the other families and subfamilies vary from 152 in Oryzorictinae to 78 in Desmaninae. No clear trends could be found with respect to known ecological adaptations or taxonomic affiliations, although marked differences in accessory bulb development among species of the same family had been found in various Insectivora families (Stephan 1965, 1975; Stephan et al. 1982).

Unlike the main olfactory bulb (MOB), the accessory olfactory bulb seems to be unaffected by adaptation to water. This small bulb, which is well developed in *Potamogale velox* (index 288), is even more reduced in *Micropotamogale* than the main bulb (indices 22 versus 44). Similarly, *Limnogale, Neomys,* and the Desmaninae, which are also predators in limnetic ecosystems, show no consistency in AOB size, but rather show a variation covering the complete range found in Insectivora. Furthermore, there is no obvious relation between the size of AOB and the preferred diet of a species, the variety of the food eaten, or the feeding strategy adopted. The role of the accessory olfactory (vomeronasal) system in sexual behavior and other social interactions has been discussed by many authors (for ref., see Stephan et al. 1982).

The vomeronasal system may be considered a parallel olfactory system specialized for the reception of particular odors. It differs from the main olfactory system in operational and functional characteristics. Like the specialized receptor cells of the cochlear membrane in some bats, which are finely tuned to receive a specific band of ultrasound frequencies (Bruns 1976a, b), the vomeronasal organ could be considered an olfactory organ finely "tuned" for reception of specific odorant molecules which may or may not be perceived by the main olfactory system.

Several factors could explain the adaptive value of an accessory olfactory system that operates as a peripheral filter. Sexual pheromones are highly specific and serve as a sexual barrier. To play this role efficiently, the pheromones must be captured by a receptor of high selective sensitivity. The fact that sexual pheromones function as primer pheromones affecting the endocrine system also favors the existence of a second and independent olfactory system. The morpho-functional organization of the accessory system should be such that its receptors, unlike those in the main olfactory system, could be activated by more persistent stimuli which do not interfere with other often rapidly changing chemical stimuli in the environment.

This very specific role in the perception of pheromones could also explain the absence of a functional vomeronasal system in a number of mammals. In the predominantly visual Old World simians, for example, static visual signals (sexual dimorphism, swellings and color changes of the sexual skin) and dynamic visual signals (body postures, gestures, facial expressions) play a major role in sexual behavior. Besides their releaser effect (immediate modification in the behavior of the receiver), these signals also have a primer effect on sexual motivation by modifying the endocrine level. Without overstating the comparison, it is interesting to note that birds, which are often characterized by marked sexual dimorphism and make use of very elaborate visual displays, also lack a vomeronasal system. Another question concerns the degree of selectivity of the vomeronasal filter. Is the vomeronasal organ merely a specialized receptor for sexual pheromones or does it respond to a broader array of chemical signals?

Flehmen has been associated with the perception of sexual pheromones by the vomeronasal organ (Knappe 1964; Estes 1972). It has been observed in the European hedgehog (Poduschka and Firbas 1968), but it does not seem to be exclusively related to sexual behavior in hedgehogs.

It is well known that the same pheromones may have different functions in different situations and at different periods (e.g., maintenance of pair-bonds, aggressive behavior etc.). It is therefore possible that the vomeronasal system is involved not only in sexual but also in general social interaction. The extent of the involvement remains, however, an open question. In Primates, chemical signals may be contained and discharged by several body products. In our present state of knowledge, it would be impossible to find a clear-cut correlation between the variety of pheromones produced and AOB development. However, there is sufficient evidence that Primates disposing of a great variety of chemical signals, for example, have a better developed vomeronasal system than Primates with a predominantly visual communication system. In this context, a divergent development of the main and the accessory olfactory systems becomes more plausible. Despite the same or even, in many cases, smaller relative size of their MOB (Stephan 1975; Baron et al. 1983), prosimians in general have a much better developed AOB than basal Insectivora. The increase in size of the AOB relative to the MOB in prosimians can be explained by their more complex and sophisticated communication system.

Thus, there does not appear to be any simple relation between the size of the AOB and any known ecological, behavioral, or phylogenetic factor. Species belonging to the same taxonomic group may be as divergent in terms of AOB size as species with similar behavior patterns and similar ecological adaptations. The size and, hence, functional importance and even precise functional role of the AOB seem to depend upon a variety of interacting factors, including general sensory equipment, the ecological niche occupied, social organization, complexity of behavior, genotype etc. These complex relationships may explain both the enormous size variation in the AOB and the differential development of its components. Therefore, some of the attributes previously assigned to the vomeronasal system may be purely species-specific.

A detailed review of the vomeronasal system in social odors was given by Von Holst (1985) for Insectivora, Scandentia and Macroscelidea.

7.3 Visual System

The visual capacities of Insectivora are generally considered to be poorly developed compared to those of most other mammals. Nevertheless, many studies indicate that the insectivoran eye, though reduced in some species, contains all the structures usually regarded as essential for vision (Kohl 1895; Slonaker 1902; Schwarz 1935/36; Verrier 1935; Kolmer 1936; Rochon-Duvigneaud 1943, 1972; Cei 1946; Sharma 1958; Sokolova 1965; Quilliam 1966b; Dubost 1968a, b; Siemen 1976; Grün and Schwammberger 1980; Braniš 1981). There are, however, clear differences in size and in the differentiation of the eye among species living in different environments.

The available data indicate that, in the terrestrial Insectivora, the eyeball is largest in Erinaceidae. The size indices for half the surface of the eye (**HSE**) are about 12 times larger in hedgehogs than in shrews (Stephan et al. 1984b). Other visual structures follow a similar trend. Thus, the relative size of the optic nerves is about 5.6 times, and that of the lateral geniculate bodies 5.5 times larger in hedgehogs than in isoponderous shrews. The area striata of the hedgehog is about 12% of the total neocortex (Brodmann 1913); its granular layer is better differentiated than in other Insectivora (Rehkämper 1981). In terrestrial Tenrecinae, for which no eye measurements are available, the relative size of the visual structures based on the indices for the optic nerves and lateral geniculate bodies is intermediate between that of shrews and hedgehogs (Stephan et al. 1984b). Although shrews have the smallest eyes and visual structures of terrestrial Insectivora, they have well developed functional retinae containing two types of receptors: rods and cones (Schwarz 1935/36; Kolmer 1936; Braniš 1981; Sigmund and Claussen 1986), or rods and rod-like cones (Grün and Schwammberger 1980).

The eyes of the water-adapted Insectivora do not seem to differ from those of terrestrial shrews. The semiaquatic Desmaninae and *Neomys* have optic nerve indices similar to those found in terrestrial shrews (Stephan et al. 1984b). In *Neomys*, the morphologic characteristics of the retina resemble the terrestrial shrews, differing only by a higher proportion of amacrine cells (Grün and Schwammberger 1980; Braniš 1981). No differences between terrestrial and semiaquatic species are found in the visual centers of the diencephalon (Bauchot 1963).

The ratio of HSE to the transversally sectioned area of the nervus opticus, which may be considered an estimator of convergency, is similar in Soricinae to that in diurnal prosimians. In Erinaceidae and in *Crocidura*, it comes close to the value found for nocturnal primates. Hedgehogs (Kristiansson and Erlinge 1977) and Crocidurinae (Baxter et al 1979; Genoud and Vogel 1981; Sigmund et al. 1987) are active mainly at night, while Soricinae have a polyphasic activity pattern (Crowcroft 1957; Siegmund and Kapischke 1983; pers. obs.).

Investigations on the ultrastructure of the retina in Soricidae have revealed differences between the Soricinae and Crocidurinae. The shrew retina has all the anatomical structures known to be involved in light reception and impulse-transmission usually observed in other mammals (Grün and Schwammberger 1980). Nevertheless, there are differences in the cytoarchitecture and ultrastructure of the retina, resulting in different visual properties. The retina in *Sorex* and, particularly, *Neomys* seems to perform more information processing than in *Crocidura*. Furthermore, in contrast to *Sorex* and *Neomys*, the rods and cones are less well differentiated and perhaps also less abundant in *Crocidura*.

The smallest visual structures are found in subterranean Insectivora. Their eyes are extremely small. In *Talpa*, for instance, the HSE index is only 0.6, compared to 14.5 in *Erinaceus europaeus* and 22.5 in *Atelerix algirus* (Stephan et al. 1984b). Furthermore, the palpebral fissure may be a minute hole, as in *Talpa europaea* (Kohl 1895; Quilliam 1964), or completely obturated, as in *Talpa wogura* (Sato 1977). Moles are by no means blind, as suggested by several early observers (e.g., Lee 1870, cited by Quilliam 1966b; Slonaker 1902). The eyes of all moles so far investigated, though considerably reduced, contain all the structures considered typical of the mammalian eye (Quilliam 1966b; Rochon-Duvigneaud 1972; Sato 1977). Siemen (1976) mentions a lack of contact between the microvilli of the pigment epithelium and the outer segment of the receptor cells in the retina of *Talpa*. Since contact seems to be essential for rhodopsin production (Bloom and Fawcett 1970, ex Siemen 1976), Siemen considered this an argument for inefficiency in the *Talpa* eye. Quilliam (1966b), however, mentions some features that may be considered specific adaptations to a subterranean environment. The eyes may be protruded, as usually occurs during phases of hyperactivity, or withdrawn through a tightly adapted cuff of eyelid skin. The dermal roots of the fine, short eye-lashes around the edges of the palpebral folds are almost as richly innervated as vibrissae. Their role may be to signal the necessity for withdrawing the eye or closing the palpebral fissure. The mole's lens is entirely cellular and comparable to those found in prenatal developmental stages (Ritter 1898; Quilliam 1966b; Sato 1977). The precise role of such a lens is unknown. Several possible functions have been suggested by Quilliam (1966b), such as diffusing light, transmitting polarized light, acting as a device for improving image contrast or simply as a diffuser to increase the image area on the retina.

Visual brain centers of Insectivora with subterranean life habits are not only smaller (Bauchot 1963; Stephan et al. 1984b) but also less differentiated. In a comparative study, Sato (1977) showed that the cytoarchitecture of the visual system in the semifossorial *Urotrichus* exhibits patterns intermediate between the terrestrial *Sorex shinto* and the purely fossorial *Talpa wogura*. Furthermore, retinal projections in the terrestrial European hedgehog are essentially the same as in *Tupaia* (Campbell et al. 1967), while *Talpa* lacks components of the accessory optic system and retinal projections to the superior colliculus (Lund and Lund 1965).

Behavioral observations confirm the results of the morphological studies. Rudimentary eyes enable moles to perceive moving objects, distinguish day and night and cross rivers under visual guidance (Rochon-Duvigneaud 1943; Quilliam 1966b). Responses to visual stimuli have been noted by Lund and Lund (1965), but nearly none was found by Kriszat (1940a). Of all terrestrial Insectivora, hedgehogs have the best vision. They are able to differentiate shapes and colors (Herter 1933, 1935) and to discriminate between patterns (Hall and Diamond 1968b).

7.4 Somatosensory System

The funicular-trigeminal complex relays somatosensory input to higher brain centers. The funicular or dorsal column nuclei receive general somatosensory information from the body below the head, while the trigeminal complex receives input from the orofacial region. Fibers from the hindlimb region go via the gracile fasciculus to the nucleus fascicularis gracilis (FGR); fibers from the forelimb region go via the cuneate fasciculus to the ncl. fascicularis cuneatus medialis (FCM) and ncl. fascicularis cuneatus externus (FCE).

In the terrestrial Tenrecinae *Microgale* (Oryzorictinae), *Solenodon* (Solenodontidae), and Erinaceinae, the FGR and FCM have very similar size indices in spite of marked interspecific differences. Although nothing can be said about the relative importance of the different somatosensory modalities such as touch, pain, temperature, and proprioceptive information from joints and muscles, the similarity in locomotor behavior and in FGR and FCM indices suggests a similarity in the functional importance of both nuclei (and thus fore and hind limbs).

The terrestrial Echinosoricinae and Soricidae differ from the other terrestrial forms in that their FGR index is higher than their FCM index. In two species, *Sylvisorex megalura* and *Echinosorex gymnurus,* the FGR index is particularly high relative to both the FCM and the FGR index for the other terrestrial species. *Sylvisorex*, which has a long prehensile tail, is known to be scansorial (Vogel 1974), and *Echinosorex* may be aquatic to some extent (Lim 1967; Walker 1975; Gould 1978).

In the semiaquatic Potamogalinae, the FGR index is also much higher than the FCM index. *Potamogale velox* is a trunk-tail swimmer which is propelled by a sinuous movement of the back and tail (Dubost 1965). *Micropotamogale lamottei*, which swims by paddling and lateral movement of the body (Vogel 1983), has the best developed FCM of the Tenrecidae. *Desmana moschata* and *Galemys pyrenaicus* differ in the size relation between FGR and FCM and between the FCM indices. The FCM indices indicate that the somatosensory capacity of the forelimbs is greater in *Galemys* than in *Desmana.* The two differ in morphology, behavior and ecology. *Desmana* uses the compressed tail as a propeller as well as using the feet (Walker 1975). *Galemys* uses its feet as paddles and its tail as a rudder control (Richard and Viallard 1969; G. Niethammer 1970). While *Desmana* lives in quiet freshwater ponds and streams (Walker 1975), *Galemys* inhabits sometimes very turbulent montane streams (Richard and Viallard 1969; Stone 1987; pers. obs.). *Galemys* has been observed using its hands when eating a prey.

The best developed FCM is found in the fossorial *Talpa*, whereas FGR is in the range of variation of the terrestrial forms. The highly enlarged manus executes lateral stroke movements that demand finely tuned muscular contractions in which sensory feedback certainly plays an important role. Sensory hairs comparable to facial whiskers have been found on the back of each forefoot (Slonaker 1920; Godet 1951; Godfrey and Crowcroft 1960). *Chrysochloris*, which differs from *Talpa* by its digging technique and the anatomy of its pectoral girdle and forelimbs (Puttick and Jarvis 1977), has a FCM index only half of that found for *Talpa*, although it is higher than in all terrestrial forms except *Solenodon*.

The FCE receives muscle spindle and tendon organ afferents from the upper body and limbs via the cuneate fasciculus. The nucleus is part of the spinocerebellar system. Like the spinocerebellar system, the cuneocerebellar tract is concerned with unconscious proprioception, yet it also seems to project to the thalamus (Fukushima and Kerr 1979). Neurons in the FCE were recently shown to receive fibers from the glossopharyngeal and vagus nerves, as well as from the vasopressor and cardioacceleratory areas of the hypothalamus. Stimulation of the FCE seems to produce brachycardia and hypotension. The functional complexity of this nucleus is reflected in the great variation of the size indices among closely related species. The highest index is reached by *Hylomys suillus* a species known to have scansorial habits; similar habits are reported for *Sylvisorex megalura*, which has by far the largest FCE in shrews.

General somatic sensory information from the orofacial region to the brain is mediated by the trigeminal nerves. This type of information is more important to an animal than that transmitted from other parts of the body for at least two reasons. (1) The orofacial region is the most complex area of the body surface and includes such diversified areas as the mouth, tongue, teeth, nostrils, cornea, and facial skin, which in Insectivora, contains specialized tactile organs, the facial vibrissae (Darian-Smith 1973). (2) The oro-facial region is the first part of the body to enter a new environment and thus provides the animal with precise information about the changes in its immediate surroundings. The importance of trigeminal input, which supplements information obtained from other sensory organs, should, therefore, vary with the characteristics of the ecological niche.

In fact, the differences in development of the sensory trigeminal complex (TR) among insectivoran species can be at least partly explained by differences in the role of the orofacial region in exploratory behavior. Facial whiskers are very prominent in all Insectivora, although there may be differences in the basic pattern as well as in the size. With the development of the vibrissae, Insectivora cannot only actively explore the environment well beyond their facial profile but may also be capable of effective telereception, particularly in an aquatic environment. Eisenberg and Gould (1970) describe mystacial, genal, superorbital, and mental vibrissae in the Tenrecinae, Grassé (1955a) in *Eri-*

naceus. Hutterer (1985), however, mentions only mystacial and submental vibrissae in shrews. There is no positive correlation between the variety of whiskers and size of the trigeminal complex. Although shrews seem to have fewer types of whiskers, they have higher size indices than the Tenrecinae and Erinaceinae. The salient feature common to all Insectivora are the mystacial vibrissae, which are complex tactile exploratory organs capable of active, rapid, and accurate movement in space (Darian-Smith 1973).

The best developed TR by far was found in semiaquatic Insectivora (mean index: 283). Consistent with the huge TR, the vibrissae reach their greatest development in semiaquatic species. In addition to the very abundant and long mystacial vibrissae arranged in 11 to 13 lines, many other facial tactile hairs have been described (Guth et al. 1959; Verheyen 1961a, b; Kuhn 1964). The proboscis of water moles such as *Galemys* contains two types of touch organs. The proximal part is covered on each side with 80 to 100 vibrissae (Bauchot et al. 1973). Around the nostrils are Eimer's organs (Argaud 1944; Richard and Viallard 1969; Bauchot et al. 1973), as many as about 100 000 over a few mm^2 (Richard 1985). A teletactile function has been suggested for the long whiskers in *Galemys* and other semiaquatic Insectivora in previous studies (Stephan and Bauchot 1959; Stephan and Spatz 1962; Bauchot and Stephan 1968).

In the water shrew *Neomys*, which is the least adapted semiaquatic form of the Insectivora in this study, the morphologic modifications for aquatic life are less pronounced than in the other semiaquatic species. Consistent with these morphologic characteristics, *Neomys* has the least developed TR of the semiaquatic Insectivora.

The TR size index for the terrestrial forms is very variable even within the same family. In Tenrecidae, *Microgale* (140) has a better developed TR than Tenrecinae. The shrew-like *Microgale* occupies a similar ecological niche in tropical Madagascar to the shrews in the temperate boreal zone with which it shares a similar TR size index. It has been observed inhabiting moist meadows covered with dense grass and reeds. The vegetation covering the runways was so dense and matted that sunlight did not penetrate to the ground (Walker 1975). It is, therefore, reasonable to suspect that they rely much more on tactile information than do the mainly surface dwelling Tenrecinae.

Despite the fact that all Erinaceidae are terrestrial forms, there are significant differences in the size indices for the Echinosoricinae and Erinaceinae. The two differ in their morphological and ecoethological adaptations. Although we lack detailed information on the exploratory behavior of gymnures, there are indications that the complexity of the gymnures' environment may be greater than hedgehogs'. Gymnures live in humid forests with dense undergrowth. *Echinosorex* may be aquatic to some extent, and *Hylomys* has been observed to climb in trees (Herter 1938, 1979; Lim 1967; Gould 1978). The better developed vibrissae in *Echinosorex* may increase the efficiency of prey capture in the aquatic milieu. True hedgehogs avoid humid regions.

Although the Soricinae and Crocidurinae differ in biotope and feeding ecology, there are no clear diverging trends in their TR development. The Soricinae generally live in a densely covered biotope and spend a considerable amount of their active time in underground tunnels (Hutterer 1976). The Crocidurinae are predominantly surface dwellers. *Sylvisorex megalura* and, particularly, *Ruwenzorisorex suncoides*, however, have higher indices than the other terrestrial Soricidae. *S. megalura* is scansorial. The TR index for *R. suncoides* falls in the range of variation of semiaquatic species.

The two fossorial species in this study differ in development of the TR. While *Chrysochloris stuhlmanni* has an index above the mean for the terrestrial Insectivora, that of *Talpa europaea*, surprisingly, is far below, although its whiskers and Eimer's organ (Eimer 1871; Bielschowsky 1907; Slonaker 1920; Quilliam 1966a) are well developed and its tactile organs seem to play an important role in its orientation in subterranean tunnels (Kriszat 1940c; Godet 1951; Quilliam and Armstrong 1963). The hands and the snout, however, are sensitive tactile organs (Kriszat 1940a; Quilliam 1966b). Furthermore, the fur plays an important role in tactile sensation. Moles also have, especially in the abdominal region, Pinkus plates, small skin elevations containing several Merkel disk-type receptors (Quilliam 1966a). It is, therefore, possible that *Talpa* relies less on the orofacial region alone for tactile information.

7.5 Auditory System

The importance of the auditory senses for all Insectivora has rarely been questioned, despite the fact that the size of the ear pinnae varies widely across species. Size, form, and position of the external ears are certainly related to the features of the biotope. The external ears are largest in terrestrial Insectivora, reduced in semifossorial forms and very small or even absent in semiaquatic and subterranean species.

The morphology of the pinnae in several species has been described by Boas (1907, 1912, 1934) and Burda (1979a). Of the six shrews studied by Burda, the two semiaquatic species *Neomys fodiens* and *Neomys anomalus* had the smallest, and the terrestrial *Crocidura suaveolens*, the largest pinnae. Although the differences in the relative size of the pinnae in the three species of Soricinae are not very great, it is interesting to note that *Sorex alpinus*, apparently with scansorial abilities (Hutterer 1982), has larger pinnae than *Sorex araneus*, with pronounced semifossorial habits. Hutterer (1985) showed how position and form of the pinnae change in shrews relative to the particular substrate utilized. The superior part of the helix of the large pinnae of scansorial and terrestrial shrews is often extended rostrally, as in *Sylvisorex megalura* and *Crocidura bottegi*. In the semifossorial *Blarina brevicauda*, reduction of the rostro-dorsal part of the helix gives the pinnae a backward position. In the semiaquatic

Neomys fodiens and particularly in *Nectogale elegans*, as well as in the semifossorial *Anourosorex*, this tendency is even stronger. According to Hutterer (1985), the rearward position of the pinnae in water-adapted species might reduce water resistance. Similar trends can be observed in the Tenrecidae. The terrestrial Tenrecinae and Oryzorictinae have larger pinnae than the semiaquatic *Limnogale*, Potamogalinae, and, especially, the semifossorial *Oryzorictes*.

The attributes of the middle and inner ear in Insectivora have been described by several authors (Henson 1961; Platzer and Firbas 1966; Firbas and Platzer 1969; Fleischer 1973; Burda 1979b; Burda and Bauerova 1985; Bruns 1985, 1987; Walther and Bruns 1987). Comparative studies have shown that, despite a generally uniform organization, the structure of the ear is adapted to the species' particular acoustic environment. Bruns (1987) mentioned two structural features that may be considered as peripheral filters and which are correlated with ecoethological parameters: (1) the acoustic fovea, which improves frequency analysis, and (2) macro- and micromechanical resonators, which increase sensitivity to frequencies biologically relevant to the species. Dilatations in the fluid spaces, functioning as a resonator, have been found in subterranean species such as *Talpa europaea* (Bruns 1985). A comparison of the inner ear in certain species of Soricidae revealed that the morphological features in *Sorex araneus* are typical of small echolocating mammals that use high frequencies. In *Crocidura russula*, the structure of the basal membrane favors perception of lower frequencies (Walther and Bruns 1987). Important structural differences have also been found by Burda and Bauerova (1985) in the middle ear of *Sorex araneus* and *Crocidura suaveolens*. The eardrum and the footplate of the stapes are larger in *Crocidura suaveolens* than in *Sorex araneus*. The higher transformation ratio of the middle ear apparatus in *Crocidura suaveolens* indicates greater capacity to perceive low sound intensities.

Ultrasonic signals have been recorded from many shrews (Gould et al. 1964; Gould 1969; Grünwald 1969; Sales and Pye 1974; Hutterer 1976; Hutterer and Vogel 1977), and it has been shown that different species have different frequency ranges. *Sorex* is able to produce sounds of up to 77 kHz (Hutterer 1976) and *Blarina* to about 107 kHz (Gould 1969), whereas African Crocidurinae utter sounds no higher than 32 kHz (Hutterer and Vogel 1977), and *Crocidura russula* and *Crocidura oliviera* of 32 - 34 kHz (Grünwald 1969).

The existence of echolocation has been demonstrated in several species of Soricinae (Gould et al. 1964; Buchler 1976), but has been denied for Crocidurinae (Grünwald 1969). It is possible that, in Soricinae, audition is involved mainly in orientation by echolocation and in communication.

Behavioral studies have shown that insectivoran species other than shrews may respond to high frequency sounds (Sales and Pye 1974) or use them for echolocation (Gould 1965). Chang (1936) and Sales and Pye (1974) found that hedgehogs are sensitive to frequencies of up to 84 kHz, with an optimum frequency of 20 kHz. Poduschka (1968) claimed that hedgehogs can perceive sounds higher than 24 kHz. The anatomy of the middle ear of *Erinaceus europaeus* has been studied by Henson (1961), who concluded that although they may be able to hear sounds approaching the ultrasonic range, it does not appear likely that hedgehogs would have any special acuity or use for such sounds.

Consistent with the great variation in ear pinnae and middle ear apparatus, the size of the auditory nuclei varies according to ecological adaptations. In the terrestrial Insectivora are found the species with the best developed *(Echinosorex* and *Hylomys)* and the species with the least developed auditory nuclei *(Microsorex)*. In the terrestrial Tenrecinae, *Echinops* and *Setifer* have better developed subcortical acoustic nuclei than *Hemicentetes* and *Tenrec*. They differ not only morphologically but also behaviorally, in addition to living in different habitats. *Echinops*, and to a certain extent *Setifer* as well, prefer dry forests and bushlands or semiarid regions, while *Tenrec* and *Hemicentetes* seem to prefer humid regions, although *Tenrec* also occurs in arid plains (Gould 1965; Gould and Eisenberg 1966; Eisenberg and Gould 1970; Herter 1979). *Tenrec* may den in hollow trees and is somewhat arboreal (Gould and Eisenberg 1966); *Tenrec* and *Hemicentetes* feed on various prey, such as worms, snails and arthropods found in the soil or on the ground. *Setifer* and *Echinops* hunt insects, particularly grasshoppers, which they detect by sound (Herter 1979). Audition seems therefore more important to *Setifer* and *Echinops* than to *Tenrec* and *Hemicentetes*, although *Tenrec* and *Hemicentetes* rely on the perception of stridulation sounds used in communication (Gould 1965). Perception of sounds produced by a potential prey apparently requires greater auditory capacity than perception of sounds produced by conspecifics, such as stridulation.

In the Erinaceidae, the Echinosoricinae have a much better developed auditory system than the Erinaceinae. Besides habitat, the two differ in antipredator strategy. Although, according to Gould (1978), compared to many insectivoran species (shrews, moles and most tenrecs), *Echinosorex* seems more alert because of its apparent visual acuity, the development of its acoustic nuclei indicates the importance of audition in detecting changes in the environment. In the Erinaceinae, Masterton et al. (1975) reported from *Paraechinus hypomelas* that animals of this species hear low frequencies poorly and does not localize them. Since, however, the long-eared desert hedgehog *(Hemiechinus)* has higher size indices in the acoustic structures than *Erinaceus* and *Atelerix*, there may exist larger interspecific differences.

In the Soricidae, the auditory nuclei are twice as large in Crocidurinae as in Soricinae. Members of the Soricinae and Crocidurinae differ considerably in their behavior, preferred biotope, and feeding ecology. *Sorex araneus*, for example, lives in a densely covered biotope and spends about 80% of its time in underground tunnels (Hutterer 1976) where it feeds on fleshy, slow moving hypogeic prey, such as earthworms. In contrast, most crocidurine shrews are terrestrial species, hunting faster moving epigeal prey. *Crocidura suaveolens*, for instance, hunts at dusk and at night and thus relies more on hearing to detect food than *Sorex araneus* (Burda and Bauerova 1985).

The size indices for the semiaquatic and fossorial forms fall in the range of variation of the terrestrial Insectivora, suggesting a lack of any particular adaptation of the auditory structures to the aquatic and subterranean environment. However, in subterranean mammals, low frequency hearing appears to be well developed (e.g. Burda et al. 1987), and Kudo et al. (1989) described for *Mogera* a well developed medial nucleus of the superior olive, compared with that in other similar body-sized mammals and, most strikingly, a bilateral projection from this nucleus to the inferior colliculus.

7.6 Vestibular System

Maintenance of equilibrium during locomotion requires delicate muscular coordination. The size of the vestibular complex (VC), concerned with regulation of the spatial orientation of the body and its parts, may be considered an indicator of its efficiency. Since movements in three-dimensional space make higher demands on vestibular functions than movements confined to the ground, one would expect animals living mainly on the ground to have a less developed vestibular complex than animals exploiting a more complicated environment. This is clearly reflected in the size differences between terrestrial and semiaquatic Insectivora. The semiaquatic forms have a larger VC, in general, than the terrestrial forms. It is almost twice as large in the semiaquatic Potamogalinae than in the terrestrial Tenrecinae (Baron et al. 1988). Similarly, the vestibular index of water-adapted Desmaninae is 1.8 times larger than in the fossorial *Talpa*. In Soricidae, *Neomys* is at the top of the index scale. It is interesting that *Sylvisorex megalura* and *Ruwenzorisorex* have similar size indices as the water shrews, and it has been shown that at least *Sylvisorex megalura* has scansorial abilities (Vogel 1974).

The best developed VC is found in the Echinosoricinae. It seems that *Echinosorex*, for instance, is quite agile in water and may even dive (Gould 1978). However, it is doubtful that moonrats have the same swimming abilities as the semiaquatic Tenrecidae, Talpidae, and even Soricidae. The higher VC indices therefore remain puzzling for the time being.

Estimates of the size of the flocculonodular lobe and related cerebellar structures (vestibulocerebellum) may be obtained from the size of the vestibular nuclei. Considering the large size of the vestibular nuclei in *Potamogale*, the vestibulocerebellar parts may be quite large too. This is also indicated by the very large and highly visible dorsal paraflocculus in *Potamogale* (Fig. 19).

7.7 Motor Systems

Although the cerebellum (CER) and striatum (STR) tend to be better developed in semiaquatic forms than in terrestrial species, there are major differences among the aquatic forms themselves. Desmaninae have the best developed CER and STR (Figs. 63, 79). The difference between Desmaninae and Talpinae is less pronounced, but it is more marked between Desmaninae and Potamogalinae. The size index for CER in Potamogalinae is close to the mean index for all insectivoran subfamilies, while that of the STR is even below the subfamily average. The two groups differ in swimming behavior and general locomotory repertoire. *Potamogale velox*, in particular, uses a more primitive lateral movement of the back and tail, similar to lower vertebrates. Its movement on land is rather clumsy (Kingdon 1974). G. Niethammer (1970) described the swimming movement in *Galemys*: the hindfeet may execute alternating or synchronous paddle movements, and the forelimbs are mainly used as rudders. In Desmaninae, locomotion seems to be more versatile than in Potamogalinae. It has already been shown in birds, primates, and bats that species capable of more diversified and finely tuned movements have better developed motor centers than species with less complex locomotive abilities, which, however, may be highly specialized (Portmann 1947; Stephan et al. 1974, 1987b). *Limnogale* and *Neomys* follow the trend of the semiaquatic species.

The rather high size indices for STR in the fossorial Talpinae and Chrysochloridae (Fig. 79) should be interpreted in the same manner as for Primates and semiaquatic Insectivora. However, unlike Talpinae, Chrysochloridae have a relatively small CER (Fig. 63). The mode of digging and the structures associated with this behavior are fundamentally different in the two groups. In Chrysochloridae, the forelimbs execute antero-posterior digging movements parallel to the sagittal plane of the body, similar to running motions (Campbell 1938; Puttick and Jarvis 1977). In Talpidae, the forelimbs are placed in an extreme position of pronation and execute lateral stroke movements.

The extreme variation in the relative size of the CER and STR suggests profound differences in the locomotor abilities of terrestrial Insectivora. The relative size of CER in Echinosoricinae is more than twice that in Erinaceinae. The difference in relative size of STR is less pronounced, but the size index is still 1.4 times greater in Echinosoricinae than in Erinaceinae. It is obvious, then, that gymnures have a wider locomotor spectrum than hedgehogs. Although terrestrial, lesser gymnures *(Hylomys)* have been observed to climb around in branches of trees (Walker 1975), and moonrats, to swim; moonrats are, in fact, considered aquatic to some extent (Gould 1978).

Solenodon has, next to some Soricinae, the best developed STR and, next to *Hylomys*, the best developed CER of all terrestrial Insectivora. The significance of this is not immediately obvious because *Solenodon* seems to be a very typical and conservative terrestrial Insectivore with many archaic morphological characters (Eisenberg 1981). There are, however, a few behavioral observations which may furnish some tentative explanations. Solenodons are able to climb, dig holes and tear up rotten logs. They may use the forepaws in capturing and manipulating prey (Mohr 1936b, c, d, 1938; Eisenberg and Gould 1966).

With regard to the size of CER and STR in terrestrial Soricidae, two points stand out: the larger mean index of STR in Soricinae than in Crocidurinae and the size increase of the two motor centers in *Sylvisorex, Ruwenzorisorex,* and *Myosorex* compared to the other Crocidurinae. The Soricinae show a stronger tendency to occupy the subterranean biotope than the Crocidurinae. They occasionally dig burrows, although they usually use burrows of moles and small rodents. On the other hand, some Soricinae, such as *Sorex alpinus*, are scansorial. In contrast, most Crocidurinae, particularly of the *Crocidura* group, are typical ground-dwellers. Some Crocidurinae, however, diverge from the rule. This is particularly true of *Sylvisorex* and *Ruwenzorisorex. Sylvisorex megalura* has morphological adaptations to arboreal life, and indeed, observations indicate extensive arboreal behavior (Vogel 1974).

7.8 Limbic System

There is great confusion among authors about the structures composing the limbic system. Brodal (1981, p. 690) stated that "the limbic system appears to be on its way to including all brain regions and functions. As this process continues, the value of the term as a useful concept is correspondingly reduced." Of the TEL structures differentiated, measured, and compared in the tables and figures in this book, those generally included in the limbic system are hippocampus (HIP), septum (SEP), and schizocortex (SCH). The HIP is regarded as the core area of the limbic system. Those parts of the amygdala (AMY) which are not mainly olfactory (Stephan and Andy 1977; Stephan et al. 1987a) must also be included as well.

Still greater confusion exists as to the system's possible functions, which include motivation, emotion, memory and learning, and other complex cognitive functions, whatever that means. In most general terms, the limbic system may be considered an interface between the neocortex (NEO) and the hypothalamus (HTH) and brain stem, passing cortical information about the conditions in the external environment to subcortical centers and passing information about the internal state to the NEO (Horel 1988).

Limbic structures, particularly the HIP, are best developed in the Echinosoricinae, whereas in the Erinaceinae size indices, except for the AMY, are far below the subfamily average. In the Oryzorictinae, which have relatively large limbic structures, the HIP is best developed in *Microgale*. Other groups with indices above the subfamily average are the Talpidae and *Solenodon*. In the Soricidae, the Soricinae have on the average better developed limbic structures than the Crocidurinae, particularly the genera *Crocidura* and *Suncus*. All the species with small limbic structures belong to the group which we have called "Basal Insectivora" since they have the largest concentration of conservative characters.

The most conspicuous difference within the same family is found between Echinosoricinae and Erinaceinae. The species of the two subfamilies have fundamentally different antipredator strategies. The defence strategy of hedgehogs is purely passive: they erect their spines when they roll into a tight ball to protect their vulnerable furry parts. Echinosoricinae have an active antipredation strategy: they flee or attack when threatened. In this context it is interesting to note that all Insectivora equipped with spines (*Setifer, Echinops*, Erinaceinae) also have relatively small limbic structures. This does not imply, however, that species which completely lack spines always have large limbic structures. In the Tenrecinae, the spiny *Setifer* has larger limbic structures than the less spiny *Tenrec*. Hairy species, such as *Geogale* and the Crocidurinae, also have small limbic structures. There are, however, no species with spines among those species with a well developed limbic system. It may, thus, well be that there is a relation between passive antipredation and small limbic structures.

Comparative studies in many mammalian groups raise interesting questions about the functional significance of size differences in HIP and other limbic structures. The elephant shrews (Macroscelidea) have extremely large HIP as well as large SCH, SEP, and AMY, larger even than in most Primates (Stephan and Spatz 1962; Stephan and Andy 1964, 1969, 1982). In Chiroptera, the Megachiroptera have a larger HIP than the Microchiroptera (Stephan et al. 1974), and in the Microchiroptera, HIP is strongly progressive in *Kerivoula papillosa* (Stephan et al. 1987c). In contrast, HIP is strongly regressive in *Tylonycteris* (Stephan et al. 1987b). Finally, in a comparison of brains of European wild boars and domestic pigs, it was shown that HIP was about 43% smaller in the domestic animals (Kruska and Stephan 1973). The strong reduction in the pigs was explained by the special environment of the domestic forms (entspanntes Feld, Lorenz 1959), in which there is no need for continuous attentiveness in order to search for food or to avoid enemies. The safety of *Tylonycteris* in its roosting place, the bamboo culms, may have allowed for a HIP reduction similar to that found in domestication. Outside its roost sites, *Tylonycteris* seems to be fairly easy prey for enemies, as e.g., *Megaderma lyra* (Medway 1967; pers. obs.). On the other hand, *Kerivoula papillosa* roosts in relatively exposed sites, such as abandoned hanging bird nests, foliage and flowers (Brosset 1966; Walker 1975). The necessity for a high degree of alertness is enhanced by its monogamous social structure. In fact, *Kerivoula* lacks the advantages offered by social life. An obvious advantage of living in a large group is that there are more ears and eyes to detect potential predators. In comparison to most Microchiroptera, Megachiroptera resting in uncovered trees are less protected and need to be more alert (Stephan et al. 1974). In behavioral experiments, Möhres and Kulzer (1956a, b) observed that Megachiroptera are maximally sensitive to disturbances even if they are kept in captivity a long time. Microchiroptera, however, become extremely indifferent to such disturbances when they are accustomed to them. In fact, most Microchiroptera roost in dark caves in large groups and are therefore better protected against a surprise attack by potential enemies. In elephant shrews, the extremely large HIP may also be functionally related to high attentiveness, which, together with well developed vision and high agility, allows them to avoid enemies and to survive.

According to Gould (1978), compared to many Insectivora (shrews, moles, and most tenrecs), *Echinosorex* seems more alert because of its apparent visual acuity. The importance of vision and audition agrees also with the great development of the mesencephalon, which is made up to a large extent of the inferior (acoustic) and the superior (visual) colliculi. As already mentioned, *Echinosorex* and the closely related hedgehogs have fundamentally different antipredator strategies. Protected by their spiny fur, hedgehogs can afford to be less attentive to the environment than the Echinosoricinae. This may be one of the reasons that HIP is 2.4 times larger in the Echinosoricinae than in the Erinaceinae (Table 45).

The functions of attentiveness were discussed by Hassler (1964) for the limbic system. This particular role would also be consistent with Smythies' model (1967), in which the hippocampus transmits information about environmental conditions to other components of the limbic system.

As to the question of learning and memory, Stephan and Manolescu (1980) showed that comparative data do not corroborate limbic learning theories. The very marked development of HIP in forms belonging to very different evolutionary levels, such as Macroscelidea, the prosimian *Daubentonia,* and the simian *Hylobates*, as well as the very weak development in several whales, contradict such theories. The HIP seems to be of secondary importance in learning and may be related to functions of activation, attentiveness, and vigilance (see e.g., Stephan 1975), which are without doubt an essential requirement for efficiency in learning.

Warrington and Weiskrantz's theory, which seems to imply that conscious awareness of what is learned is related to the hippocampus (Horel 1988), would be consistent with our interpretation.

7.9 Neocortex

Of all brain parts, the size and differentiation of the NEO have proved to be the most reliable indicators of niche complexity (see e.g., Stephan et al. 1987a). For the Insectivora, there is a strong tendency for semiaquatic species to have a better developed NEO than closely related subterranean and terrestrial species. Desmaninae have the highest NEO index of all Insectivora so far investigated (Fig. 92): it is 1.6 times larger than in the fossorial Talpinae. The NEO of the Potamogalinae is actually 3.7 times larger than that of the terrestrial Tenrecinae. In the Oryzorictinae, the water-adapted *Limnogale* has the best developed NEO by far of the group, with a size index of 3.3x. Similarly, in the Soricidae, the water and swamp shrews *Neomys* are found in high positions in the NEO size scale. According to Bauchot and Stephan (1968), a similar trend is observed in other mammalian orders. The Cetacea are known to have a well developed NEO (Schwerdtfeger et al. 1984); Mangold-Wirz (1966) reports similar findings for pinnipeds.

The semiaquatic niche is more complex than the terrestrial and subterranean. Semiaquatic species have to cope, not only with aquatic, but also with terrestrial and sometimes even semifossorial requirements. In the aquatic environment, food distribution is usually less dense, more patchy or less ubiquitous and hence less predictable in space and time than in the terrestrial habitat. The prey are mobile and often exhibit complex antipredatory behavior. Consequently, the home range is quite large. The large home range combined with an absence of underwater hiding places and the special physiological requirements (respiration) increase vulnerability to predation. As a result, semiaquatic species require extensive information storage for efficient exploitation of their food supply, greater mobility for both avoiding predators and reaching prey, and greater plasticity to cope with the continuously changing environment; they therefore require a more complex nervous system.

The Echinosoricinae have the best developed NEO of terrestrial surface dwellers: it is 1.9 times larger than in Erinaceinae and 3.9 times larger than in Tenrecinae. Several converging traits suggest that the niche of Echinosoricinae might be more complex than that of the Erinaceinae. Indeed, Echinosoricinae differ from Erinaceinae in a variety of morphological and ecoethological features. Lacking spines, Echinosoricinae prefer humid, covered areas. The lesser gymnure *Hylomys*, which has the largest NEO of all Erinaceidae, lives in humid forests with dense undergrowth. Although mostly terrestrial, it climbs on shrubs and trees (Herter 1938, 1979). The moonrat *Echinosorex*, with a NEO only slightly smaller than that of *Hylomys*, inhabits mangrove swamps, primary and secondary forests, often near streams, and may be aquatic to some extent (Lim 1967; Walker 1975; Gould 1978). Moonrats feed on terrestrial, limnetic, and aquatic invertebrates and vertebrates (Lim 1967; Medway 1969; Lekagul and McNeely 1977). True hedgehogs avoid humid regions and are not very agile climbers.

The two erinaceid subfamilies differ, as already mentioned, in their antipredator strategies. While Erinaceinae use passive defence by erecting their spines when curling up, the Echinosoricinae adopt an active strategy in addition to producing obnoxious scents. *Echinosorex* may be very aggressive when threatened and may dash forward with open mouth if attacked (Lim 1967), but it usually flees with considerable speed when in danger (Gould 1978), as does *Hylomys*. The highly stereotyped defence reaction of the true hedgehogs is usually released by acoustic or olfactory stimuli (Herter 1938; Lindemann 1951) and certainly involves less complex motor coordination than the very plastic flight and defence behavior of *Echinosorex*.

The terrestrial Soricinae have on the average a larger NEO than the terrestrial Crocidurinae. The two subfamilies not only show vast differences in their geographic distribution, but have adopted rather different biological strategies. An indicator of differences in ecological niche is the spatial distribution of a population: Crocidurine shrews usually show higher population densities with less territoriality than soricine shrews (Vogel 1980). It is difficult to relate these differences to differences in brain size or brain composition, since both conditions may necessitate more information about the environment. According to Clutton-Brock and Harvey (1980), there is a close positive relation in Primates between increase in brain size for different families and mean home range size. They explain this relation by a need for more spatial information when the home range is large. Based on this hypothesis, the more developed brains of the Soricinae compared to the Crocidurinae would be in agreement with their larger home range. On the other hand, high population density might increase social contacts and, hence, the amount of social information to be processed, but in this case, there would be a negative correlation between brain development and social behavior, which is highly improbable. Therefore, the high population density in Crocidurinae does not seem to result in increased social behavior but, at best, in greater mutual tolerance (cf. Genoud 1981).

Of the Soricinae, New World species have by far the best developed NEO. In comparison to *Sorex araneus* and *Sorex minutus*, New World species usually occupy larger home ranges (Buckner 1966; Newman 1976). The extreme climatic conditions found on the American continent increase the heterogeneity of the biotope and require considerable behavioral plasticity.

Since morphological traits express ecological adaptations, even slight differences in form should reflect differences in the ecological niche. It is commonly accepted that jaw morphology can be used as an indicator of feeding habits. Dötsch (1983a, b, 1985) conducted detailed studies of mandibular articulation in shrews and confirmed differences in the condylar structure between Soricinae and Crocidurinae mentioned in earlier studies (for ref. see Repenning 1967). Dötsch (1982) also found equally large differences in the masticatory musculature, leading her to conclude that Soricinae can execute more differentiated mandibular movements than Crocidurinae. Increasing complexity in movement requires more and more control units in the nervous system. While these differences should not be overemphasized, they do suggest the existence of a trend towards a more complex ecological niche in Soricinae than in Crocidurinae, particularly of the *Crocidura* group.

According to Vogel (1976, 1980), crocidurine shrews have much lower basal metabolic rates than most soricine shrews. Their relative brain size is also lower. The increase in diencephalon and non-olfactory telencephalon accompanying the increment in metabolic rate in Soricidae demands an ecological interpretation. One of the most important factors influencing energy budgets is the thermal regime in the environment. In shrews, the differences in the metabolic rate may be a consequence of soricine evolution in a temperate climate and of crocidurine's in a tropical climate (Vogel 1976, 1980). Living in a physically more favorable environment than the Soricinae, the Crocidurinae have lower metabolic heat production. Furthermore, variations in food availability may be fairly unimportant for Crocidurinae: since they live mostly in tropical environments, food supplies may well be continuously available despite seasonal fluctuations, whereas for Soricinae in temperate regions, this is highly unlikely. The ecological constraints characteristic of the temperate climatic zone make the trophic niche less predictable, and hence more complex. Thus, it seems clear that the increase in progressive brain parts may be primarily related to the characteristics of the niche occupied. Greater energy requirements combined with higher variability in the environmental conditions demand greater behavioral plasticity.

8 Conclusions

The diversity in brain evolution in Insectivora can only be fully understood if ecological diversity is adequately taken into account. To cope successfully with the biotic and abiotic requirements of the environment, an animal must have, in addition to a special set of morphological and physiological features, particular brain characteristics. Insectivora are adapted to a great variety of ecological niches. They are indeed one of the most diversified eutherian orders. From the fact that adaptive radiation can be observed in different taxonomic levels, it follows that convergent brain evolution has occurred several times in this order (see also Pruitt 1957; Hutterer 1985; Baron and Stephan, in press).

Adaptation to semiaquatic life, for instance, does not only occur in different families and subfamilies, but has evolved several times independently in different genera of the same subfamily. It is found in the Talpidae, Tenrecidae, and Soricidae. In Tenrecidae, adaptation to aquatic life developed independently in Potamogalinae and Oryzorictinae and, in Soricinae, in four different genera. Similarly, adaptation to fossorial or semifossorial life occurs in Talpidae, Chrysochloridae, and Soricidae. Soricinae and Crocidurinae comprise several genera with fossorial or semifossorial species. The ecological resemblance and diversity raise several questions concerning brain evolution in Insectivora. (1) Which ecological factors exert the strongest selective pressure to increase the size of the brain and particularly the neocortex? (2) To what extent does utilization of the same environmental substrate correlate with brain structures? (3) How can the differential brain organization in closely related species belonging to the same ecotype be explained?

(1) The relative size of the brain and the brain parts shows strong interspecific differences. The most progressive part is the neocortex. Enlarged brain and neocortex appeared independently in several families, subfamilies, and genera. The strongest selective pressure for increase in the neocortex is related to adaptation to aquatic life. In all taxonomic groups, there is a strong tendency for semiaquatic species to have a better developed neocortex than closely related fossorial and terrestrial species. Selective forces which acted to increase size of the brain and neocortex can, however, be observed in terrestrial forms as well. It was shown, for instance, that the two erinaceid subfamilies, which differ in antipredator strategies, differ also in neocortical development. Echinosoricinae, which adopt an active strategy, have a larger brain and neocortex than Erinaceinae, which have a highly stereotyped passive defence reaction. Precise coordination of movement and accurate evaluation of the situation are critical for a successful defence reaction and demand a greater problem-solving capacity and hence a larger brain and neocortex. Differential development of brain and neocortex, although on a lower scale, can be observed in closely

related species, e.g., within genera of shrews, even if the species use the same substrate. In this case too, increase of brain and neocortex has, as it was shown, ecological correlates, such as home range and niche complexity in general.

One of the common denominators in the three cases, just mentioned, is a relatively high degree of uncertainty and unpredictability of the ecological conditions, and this renders the decision-making process particularly complex. An animal's decision on how to allocate time and energy to alternative behaviors and its assessment of the risks and benefits attached to them are considered, to a large extent, outcomes of learning. The ability to learn emerged early in evolutionary history, long before the neocortex evolved. However, the neocortex provides the basis for the integration and retention of complex information. It enables an animal to modify its behavior to novel events and, thus, to adjust to fluctuations in the environment (Konorski 1967; Oakley 1979; Jolicoeur et al. 1984). One may therefore argue that selective pressure favors bigger brains and neocortex in animals living in situations where changing conditions are highly unpredictable.

(2) Quantitative comparison of brain structures in Insectivora reveals the existence of several ecomorphological trends within the order, regardless of the taxonomic affiliation of the species. However, species and genera often form a series characterized by progressively greater adaptation to a particular environmental substrate.

Adaptation to terrestrial surface dwelling involves the greatest variation not only in body but also in brain morphology. Brain size and quantitative brain composition show extreme variation. The only clear trend is the increase in the olfactory structures.

Fossorial species are characterized by a number of morphological features, the most typical being the highly modified thoracic girdle and forelimbs. There are two trends in forelimb modification in the most specialized forms and they are related to differences in digging technique. The relative size of the brain and neocortex is only slightly above the insectivoran average. Based on brain development, the niche seems to be less complex than that of aquatic species. There is a marked tendency for the striatum to increase in size, while the olfactory structures show a divergent tendency in the Talpidae and Chrysochloridae. Visual structures are strongly reduced.

Two trends have emerged in adaptation to the aquatic milieu: the trunk-tail swimmers and the limb-tail swimmers. In both groups, the size of the whole brain, medulla oblongata, and neocortex is strongly increased. The reduction in olfaction is compensated by an increase in the trigeminal system. The motor structures of the brain show a tendency to increase in size when compared to related terrestrial and fossorial species; this increase is more pronounced in limb-tail swimmers. The progressively stronger structural modifications for semiaquatic life are paralleled by stronger size modifications in the brain structures.

(3) The sometimes extensive differences in brain structure, especially in sensory and motor centers, found at the generic level may be interpreted as character displacements (Brown and Wilson 1956; Grant 1972) to prevent competitive exclusion in related sympatric species which, by physiological and ecological requirements, are forced to share the same habitat. If food availability is optimal and competition is reduced to a minimum, food choice may be based solely on energy assessment. In this case, animals may indiscriminately share the same prey without affecting their foraging effectiveness. During periods of food scarcity, however, niche partitioning would be a better survival strategy. Differential brain structure gives sympatric species the ability to exploit different parts of the otherwise same feeding niche preferentially. Species with a better developed acoustic system are better equipped to react to acoustic stimuli, while species with a better developed olfactory system may benefit more from olfactory detection of prey. The selective advantage of the evolution of differential brain structure in sympatric species is therefore obvious: it tends to minimize interspecific competition and favor stable coexistence over an appreciable area.

From the preceeding considerations, it is clear that any attempt to establish precise phylogenetic relationships based on analysis of quantitative similarities in brains must fail. In fact, two studies showing dendrograms based on quantitative analyses of brain structure show different results and, moreover, contradict traditional taxonomic relationships (Douglas and Marcellus 1975; Bauchot 1982). In an earlier study of interspecific distances within Insectivora and Primates, Bauchot (1979a) stressed the impossibility of distinguishing between convergent adaptations and phylogenetic inertia based solely on volume measurements.

Identical brain size found in different species may be the result of differential growth in different brain parts. This means, for example, that the same brain size found in the Desmaninae and Echinosoricinae does not have the same ecological valence. Although there are several interesting studies relating brain size to ecological aspects or niche complexity (e.g. Bauchot and Stephan 1966; Pirlot and Stephan 1970; Eisenberg and Wilson 1978; Clutton-Brock and Harvey 1980), brain size alone is insufficient for assessing the ecological significance of brain morphology. This can only be satisfactorily with sufficiently detailed knowledge about quantitative brain structure. Diversity of brain structure, which is often more easily measured than behavior, may therefore serve as an indicator of niche diversity.

9 References

Abbie AA (1938) The relations of the fascia dentata, hippocampus and neocortex, and the nature of the subiculum. J. comp. Neurol. 68: 307-333

Abbie AA (1939) The origin of the corpus callosum and the fate of the structures related to it. J. comp. Neurol. 70: 9-44

Adams LE (1912) The duration of life of the common and the lesser shrew, with some notes on their habits. Mem. Proc. Manch. Lit. Phil. Soc. 56: 1- 10

Adrian ED (1942) Olfactory reactions in the brain of the hedgehog. J. Physiol. 100: 459-473

Ärnbäck-Christie-Linde A (1900) Zur Anatomie des Gehirnes niederer Säugetiere. Anat. Anz. 18: 8-16

Ärnbäck-Christie-Linde A (1907) Der Bau der Soriciden und ihre Beziehungen zu anderen Säugetieren. Morphol. Jb. 36: 463-514

Ärnbäck-Christie-Linde A (1914) On the cartilago palatina and the organ of Jacobson in some mammals. Morph. Jb. 48: 343-364

Aitchison CW (1987) Review of winter trophic relations of soricine shrews. Mammal Review 17: 1-24

Aitkin LM, Horseman BG, Bush BMH (1982) Some aspects of the auditory pathway and audition in the european mole, *Talpa europaea*. Brain Behav. Evol. 21: 49-59

Akert K, Potter HD, Anderson JW (1961) The subfornical organ in mammals. I. Comparative and topographical anatomy. J. comp. Neurol. 116: 1-13

Alexander A (1931) Untersuchungen über die zentrale Haubenbahn. Arb. Neurol. Inst. Wien 33: 261-288

Allara E (1958) La struttura dell'ipofisi del riccio durante l'ibernazione. Monit. Zool. ital. (suppl. Atti Soc. ital. Anat.) 67: 72-74

Allara E (1959) La struttura dell'ipofisi del riccio durante l'ibernazione e nelle varie fasi del risveglio. Boll. Soc. ital. Biol. sper. 35: 160-161

Allen GM (1910) *Solenodon paradoxus.* Cambridge Mass. Mem. Mus. Comp. Zool. Harvard Coll. 40: 1-54 + 9 plates

Allison T, VanTwyver H (1970a) Sensory representation in the neocortex of the mole, *Scalopus aquaticus*. Exp. Neurol. 27: 554-563

Allison T, VanTwyver H (1970b) Sleep in the moles, *Scalopus aquaticus* and *Condylura cristata*. Exp. Neurol. 27: 564-578

Anderson S, Jones JK (1984) Orders and Families of Recent Mammals of the World. John Wiley and Sons, New York

Andreescu-Nitescu I (1970) Étude comparative des cornets nasaux chez: *Talpa europaea* L., *Crocidura leucodon* Herm., *C. suaveolens* Pall., *Sorex araneus* L. et *Neomys fodiens* Schreb. (Ord. Insectivora) de Roumanie. Trav. Mus. Hist. Nat. Grig. Antipa 10: 359-363

Andy OJ, Stephan H (1959) The nuclear configuration of the septum of *Galago demidovii*. J. comp. Neurol. 111: 503-545

Andy OJ, Stephan H (1961) Septal nuclei in the Soricidae (insectivors). Cytoarchitectonic study. J. comp. Neurol. 117: 251-274

Andy OJ, Stephan H (1966a) Phylogeny of the primate septum telencephali. In: Hassler R, Stephan H (eds) Evolution of the Forebrain. Thieme, Stuttgart, pp 389-399

Andy OJ, Stephan H (1966b) Septal nuclei in primate phylogeny. A quantitative investigation. J. comp. Neurol. 126: 157-170

Andy OJ, Stephan H (1968) The septum in the human brain. J. comp. Neurol. 133: 383-409

Ansell WFH (1964) Captivity behaviour and post-natal development of the shrew *Crocidura bicolor*. Proc. Zool. Soc., London 142: 123-127

Antonopoulos J, Papadopoulos GC, Karamanlidis AN, Parnavelas JG, Dinopoulos A, Michaloudi H (1987) VIP- and CCK-like-immunoreactive neurons in the hedgehog (*Erinaceus europaeus*) and sheep (*Ovis aries*) brain. J. comp. Neurol. 263: 290-307

Antonopoulos J, Karamanlidis AN, Papadopoulos GC, Michaloudi H, Dinopoulos A, Parnavelas JG (1989) Neuropeptide Y-like immunoreactive neurons in the hedgehog (*Erinaceus europeus*) and the sheep (*Ovis aries*) brain. J. Hirnforsch. 30: 349-360

Argaud R (1944) Signification anatomique de la trompe du Desman des Pyrénées (*Galemys pyrenaicus*). Mammalia 8: 1-6

Armstrong E (1982) A look at relative brain size in mammals. Neurosci. Lett. 34: 101-104

Armstrong E (1983) Relative brain size and metabolism in mammals. Science 220: 1302-1304

Armstrong J, Quilliam TA (1961) Nerve endings in the mole's snout. Nature 191: 1379-1380

Artjuchina NJ (1952 or 1962) Vergleichende Untersuchungen am Nucleus ruber der Säugetiere. Autoref. Diss. Moskau. (cited from Blinkov und Glezer 1968)

Arvidsson J (1982) Somatotopic organization of vibrissae afferents in the trigeminal sensory nuclei of the rat studied by transganglionic transport of HRP. J. comp. Neurol. 211: 84-92

Aström KE (1953) On the central course of afferent fibres in the trigeminal, facial, glossopharyngeal, and vagal nerves and their nuclei in the mouse. Acta physiol. scand. (Suppl. 106) 29: 209-320

Avksentieva LI (1975) Structure of nuclei of granular cells and oligodendrocytes in the mammal brain. Zool. Zh. 54: 734-740 (russ)

Azzali G (1954) Aspetti morfologici dell'apparato neurosecretorio diencefalo-ipofisario della Talpa (Talpa europaea) e dello Scoiattolo (*Sciurus vulgaris*). Boll. Soc. ital. Biol. sper. 30: 1013-1015

Azzali G (1955) Aspetti istofisiologici dell'apparato neurosecretorio diencefalico nel riccio (*Erinaceus europaeus*) in condizioni normali e sperimentali. Z. Zellforsch. 41: 391-406

Banfield AWF (1974) The Mammals of Canada. University Toronto Press, Toronto-Buffalo

Bang BG, Cobb S (1968) The size of the olfactory bulb in 108 species of birds. Auk 85: 55-61

Bargmann W, Gaudecker B (1969) Über die Ultrastruktur neurosekretorischer Elementargranula. Z. Zellforsch. 96: 495-504

Barnard JW (1940) The hypoglossal complex of vertebrates. J. comp. Neurol. 72: 489-524

Baron G (1970) Étude comparative de quelques noyaux moteurs encéphaliques chez des chiroptères néo-tropicaux. II. Nucleus nervi facialis et nucleus motorius nervi trigemini. Rev. Can. Biol. 29: 115-128

Baron G (1972) Morphologie comparative des relais auditifs chez les Chiroptères. Thèse, Univ. Montreal

Baron G (1974) Differential phylogenetic development of the acoustic nuclei among Chiroptera. Brain Behav. Evol. 9: 7-40

Baron G (1977a) Signification éco-éthologique du développement allométrique des deux systèmes visuels chez les chiroptères. Rev. Can. Biol. 36: 1-16

Baron G (1977b) Allometric development of the vestibular nuclei in relation to flight behavior among Chiroptera. Zool. Anz. (Jena) 199: 227-250

Baron G (1978) Volumetric analysis of cerebellar nuclei among Chiroptera. In: Olembo RJ, Castelino JB, Mutere FA (eds) Proc. 4th Internat. Bat Res. Conf. Nairobi 1975. Kenya Nat. Acad. Adv. Arts Sci., Nairobi, pp 115-126

Baron G (1979) Quantitative changes in the fundamental structural pattern of the diencephalon among primates and insectivores. Folia primatol. 31: 74-105

Baron G, Pirlot P (1978) Phylogenetic development of diencephalic components among insectivores, bats and primates. Anat. Rec. 190: 332

Baron G, Stephan H (in press) Anatomical adaptations of the Insectivora in relation to their ecology. In: Stone (ed).,

Baron G, Frahm HD, Bhatnagar KP, Stephan H (1983) Comparison of brain structure volumes in Insectivora and Primates. III. Main olfactory bulb (MOB). J. Hirnforsch. 24: 551-568

Baron G, Frahm HD, Stephan H (1988) Comparison of brain structure volumes in Insectivora and Primates. VIII. Vestibular complex. J. Hirnforsch. 29: 509-523

Baron G, Stephan H, Frahm HD (1987) Comparison of brain structure volumes in Insectivora and Primates. VI. Paleocortical components. J. Hirnforsch. 28: 463-477

Baron G, Stephan H, Frahm HD (1990) Comparison of brain structure volumes in Insectivora and Primates. IX. Trigeminal complex. J. Hirnforsch. 31: 193-200

Bartelmez GW (1960) Neural crest from the forebrain in mammals. Anat. Rec. 138: 269-281

Bates CA, Killackey HP (1983) Pattern of trigeminal afferents in the brainstem trigeminal complex of the rat. Soc. Neurosci. Abstr. 9: 245

Batuev AS, Karamyan AI, Pirogov AA, Demianenko GP, Maliukova IV (1980) Structural and functional characteristics of hedgehog polysensory cortical zone. Int. J. Neurosci. 10: 69-83

Batuev AS, Karamyan AI (1973) Sensory projections in hedgehog neocortex. Dokl. Akad. Nauk USSR, Otd. Biol. 211: 350-352

Bauchot R (1959) Étude des structures cytoarchitectoniques du diencéphale de *Talpa europaea* (Insectivora Talpidae). Acta anat. 39: 90-140

Bauchot R (1961) Le volume hypophysaire chez les Insectivores. Mammalia 25: 162-183

Bauchot R (1963) L'architectonique comparée, qualitative et quantitative, du diencéphale des insectivores. Mammalia (Suppl. 1) 27: 1-400

Bauchot R (1964) La densité en neurones des noyaux gris diencéphaliques. J. Hirnforsch. 6: 327-330

Bauchot R (1966) Le développement phylogénétique du thalamus chez les Insectivores. In: Hassler R, Stephan H (eds) Evolution of the Forebrain. Thieme, Stuttgart, pp 346-355

Bauchot R (1970) Le volume hypophysaire chez les Insectivores et les Lémuriens. Étude des corrélations liant les volumes de l'hypophyse et de ses parties constituantes au poids somatique et au poids encéphalique chez les Insectivores et les Lémuriens. J. Hirnforsch. 11: 419-432

Bauchot R (1979a) Indices encéphaliques et distances interspécifiques chez les Insectivores et les Primates. 1. Encéphale et telencéphale. Mammalia 43: 173-189

Bauchot R (1979b) Indices encéphaliques et distances interspécifiques chez les Insectivores et les Primates. II. Diencéphale et thalamus. Mammalia 43: 407-426

Bauchot R (1982) Brain organization and taxonomic relationships in Insectivora and Primates. In: Armstrong E, Falk D (eds) Primate Brain Evolution, Methods and Concepts. Plenum Press, New York-London, pp 163-175

Bauchot R, Buisseret C, Leroy Y, Richard PB (1973) L'equipement sensoriel de la trompe du desman des pyrénées (*Galemys pyrenaicus*, Insectivora, Talpidae). Mammalia 37: 17-24

Bauchot R, Diagne M (1973) La croissance encéphalique chez *Hemicentetes semispinosus* (Insectivora Tenrecidae). Mammalia 37: 468-477

Bauchot R, Legait H (1978) Le volume de l'hypophyse et des lobes hypophysaires chez les Mammifères. Corrélations et allométries. Mammalia 42: 235-253

Bauchot R, Stephan H (1959) Le cerveau de *Setifer setosus* (Schreber) (Insectivora Tenrecidae). Mémoires de l'Institut Scientifique de Madagascar, Série A 13: 139-148

Bauchot R, Stephan H (1961) Étude quantitative de quelques structures commissurales du cerveau des insectivores. Mammalia 25: 314-341

Bauchot R, Stephan H (1964) Le poids encéphalique chez les insectivores Malgaches. Acta Zool. 45: 63-75

Bauchot R, Stephan H (1966) Données nouvelles sur l'encéphalisation des Insectivores et des Prosimiens. Mammalia 30: 160-196

Bauchot R, Stephan H (1967) Encéphales et moulages endocraniens de quelques Insectivores et Primates actuels. In: Colloques Internationaux du Centre National de la Recherche Scientifique 163. Problèmes actuels de Paléontologie (Evolution des Vertébrés)., Paris, pp 575-587

Bauchot R, Stephan H (1968) Étude des modifications encéphaliques observées chez les insectivores adaptés à la recherche de nourriture en milieu aquatique. Mammalia 32: 228-275

Bauchot R, Stephan H (1969) Encéphalisation et niveau évolutif chez les simiens. Mammalia 33: 225-275

Bauchot R, Stephan H (1970) Morphologie comparée de l'encéphale des insectivores Tenrecidae. Mammalia 34: 514-541

Bauer J (1909a) Die Substantia nigra Soemmeringii. Arb. Neurol. Inst. Wien 17: 435-512

Bauer J (1909b) Vergleichend anatomische Untersuchung der hinteren Rückenmarkswurzeln der Säugetiere nebst Bemerkungen zur tabischen Hinterstrangserkrankung. Arb. Neurol. Inst. Wien 17: 98-117

Bauer-Jokl M (1919) Über das sogenannte Subkommissuralorgan. Arb. Neurol. Inst. Wien 22: 41-79

Baxter RM, Goulden EA, Meester J (1979) The activity patterns of some southern African *Crocidura* in captivity. Acta theriol. 24: 61-68

Baxter RM, Meester J (1980) Notes on the captive behaviour of five species of southern African shrews. Säugetierkdl. Mitt. 28: 55-62

Bayer SA (1985) Hippocampal region. In: Paxinos G (ed) The Rat Nervous System. Vol 1, Forebrain and midbrain. Academic Press, Sydney, pp 335- 352

Beccari N (1910) Il lobo paraolfattorio nei mammiferi. Arch. ital. Anat. Embriol. 9: 173-220

Beddard FE (1901) Some notes upon the brain and other structures of *Centetes*. Novitates Zoologicae 8: 89-92

Belford GR, Killackey HP (1979) Vibrissae representation in subcortical trigeminal centers of the neonatal rat. J. comp. Neurol. 183: 305- 321

Berkelbach van der Sprenkel H (1924) The hypoglossal nerve in an embryo of *Erinaceus europaeus*. J. comp. Neurol. 36: 219-270

Beyerl BD (1978) Afferent projections to the central nucleus of the inferior colliculus in the rat. Brain Res. 145: 209-223

Biach P (1906) Vergleichend-anatomische Untersuchungen über den Bau des Zentralkanales bei den Säugetieren. Arb. Neurol. Inst. Wien 13: 399-454

Bielak T, Pucek Z (1960) Seasonal changes in the brain weight of the common shrew (*Sorex araneus araneus* Linnaeus, 1758). Acta theriol. 3: 297- 300

Bielschowsky M (1907) Über sensible Nervenendigungen in der Haut zweier Insectivoren (*Talpa europaea* und *Centetes ecaudatus*). Anat. Anz. 31: 187- 194

Bischoff E (1900) Beitrag zur Anatomie des Igelgehirns. Anat. Anz. 18: 348-358

Bishop GH, Clare MH, Landau WM (1971) The relation of axon sheath thickness to fiber size in the central nervous system of vertebrates. Int. J. Neurosci. 2: 69-77

Blanks RHI, Precht W, Torigoe Y (1983) Afferent projections to the cerebellar flocculus in the pigmented rat demonstrated by retrograde transport of horseradish peroxidase. Exp. Brain Res. 52: 293-306

Bleier R, Byne W (1985) Septum and hypothalamus. In: Paxinos G (ed) The Rat Nervous System. Vol 1, Forebrain and midbrain. Academic Press, Sydney, pp 87-118

Blinkov SM, Glezer II (1968) Das Zentralnervensystem in Zahlen und Tabellen. Fischer, Jena

Bloom RS (1960) Cytological and histochemical study of the hypothalamo- hypophysial system of the European hedgehog (*Erinaceus europaeus*). Thesis Univ. Birmingham

Bloom W, Fawcett DW (1970) A Textbook of Histology (tenth edition 1975). Saunders, Philadelphia-London-Toronto

Blossom PM (1932) A pair of long-tailed shrews (*Sorex cinereus cinereus*) in captivity. J. Mammal. 13: 136-143

Boas JEV (1907) Zur vergleichenden Anatomie des Ohrknorpels der Säugetiere. Anat. Anz. 30: 434-442

Boas JEV (1912) Ohrknorpel und äusseres Ohr der Säugetiere. Nielsen & Lydiche, Kopenhagen

Boas JEV (1934) Äusseres Ohr. In: Bolk L, Göppert E, Kallius E, Lubosch W (eds) Handbuch der vergleichenden Anatomie der Wirbeltiere, Vol 2. Urban & Schwarzenberg, Berlin, pp 1433-1444

Boeke J (1932) Some observations on the structure and innervation of smooth-muscle fibers (arrectores spinarum of the hedgehog, and blood vessels). J. comp. Neurol. 56: 27-48

Böker H (1924) Begründung einer biologischen Morphologie. Z. Morphol. Anthropol. 24: 1-22

Böker H (1935) Einführung in die vergleichende biologische Anatomie der Wirbeltiere. Vol 1. Fischer, Jena

Böker H (1937) Einführung in die vergleichende biologische Anatomie der Wirbeltiere. Vol 2, Biologische Anatomie der Ernährung. Fischer, Jena

Boire D (1989) Comparaison quantitative de l'encéphale, de ses grandes subdivisions et de relais visuels, trijumeaux et acoustiques chez 28 espèces d'oiseaux. Thèse, Montréal

Bolk L (1906) Das Cerebellum der Säugetiere. Eine vergleichend anatomische Untersuchung. Fischer, Haarlem-Jena

Boller N (1969) Untersuchungen am Gehirn von *Solenodon paradoxus*, Brandt 1833 (Insectivora, Solenodontidae). Hirnform und Bestimmung des Neocortex-Index. Morph. Jb. 113: 346-374

Bonin G (1937) Brain-weight and body-weight of mammals. J. General Psychol. 16: 379-389

Borowski S (1958) Variations in density of coat during the life cycle of *Sorex araneus araneus* L. Acta theriol. 2: 286-289

Borowsky S, Dehnel A (1953) Materialy do biologii Soricidae (Angaben zur Biologie der Soriciden). Ann. Univ. M. Curie-Sklodowska. Sect. C. 7: 305-448

Braak H (1972) Zur Pigmentarchitektonik der Großhirnrinde des Menschen. I. Regio entorhinalis. Z. Zellforsch. 127: 407-438

Bradley OC (1903) On the development and homology of the mammalian cerebellar fissures. J. Anat. Physiol. 37: 112-130

Braniš M (1981) Morphology of the eye of shrews (Soricidae, Insectivora). Acta Univ. Carolinae-Biologica 1979: 409-445

Braniš M (1985) Postnatal development of the eye of *Sorex araneus*. Acta Zool. Fennica 173: 247-248

Brauer K (1968) Vergleichend-anatomische Untersuchungen am Kleinhirn der Insektivoren. I. Das Kleinhirn von *Erinaceus europaeus*. J. Hirnforsch. 10: 89-100

Brauer K (1969) Untersuchungen über den Paraflocculus der Rodentier. Gedanken zur Paraflocculusevolution. Anat. Anz. 125: 128-142

Brauer K (1969/70) Vergleichend-anatomische Untersuchungen am Kleinhirn der Insektivoren. II. Das Kleinhirn von *Sorex araneus* und *Elephantulus intufi*. J. Hirnforsch. 11: 537-548

Brauer K, Schober W (1970, 1976) Katalog der Säugetiergehirne I, II. Fischer, Jena

Brauer K, Schober W, Winkelmann E (1978) Phylogenetical changes and functional specializations in the dorsal lateral geniculate nucleus (dLGN) of mammals. J. Hirnforsch. 19: 177-187

Bretting H (1972) Die Bestimmung der Riechschwellen bei Igeln (*Erinaceus europaeus* L.) für einige Fettsäuren. Z. Säugetierkd. 37: 286-311

Brodal A (1940) Experimentelle Untersuchungen über die olivo-cerebellare Lokalisation. Z. ges. Neurol. Psychiat. 169: 1-153

Brodal A (1981) Neurological Anatomy in Relation to Clinical Medicine (third edition). Oxford University Press, NewYork-Oxford

Brodal A, Pompeiano O (1957) The vestibular nuclei in the cat. J. Anat. 91: 438-454

Brodal A, Pompeiano O (1958) The origin of ascending fibres of the medial longitudinal fasciculus from the vestibular nuclei. Acta morph. neerl.- scand. 1: 306-328

Brodmann K (1906) Beiträge zur histologischen Lokalisation der Grosshirnrinde. Fünfte Mitteilung: Über den allgemeinen Bauplan des Cortex pallii bei den Mammaliern und zwei homologe Rindenfelder im besonderen. Zugleich ein Beitrag zur Furchenlehre. J. Psychol. Neurol. 6: 275-400

Brodmann K (1909) Vergleichende Lokalisationslehre der Großhirnrinde. Barth, Leipzig

Brodmann K (1913) Neue Forschungsergebnisse der Großhirnrindenanatomie mit besonderer Berücksichtigung anthropologischer Fragen. 85. Verh. Ges. dtsch. Naturf. Ärzte Wien Teil 1: 200-240

Broom R (1897) A contribution to the comparative anatomy of the mammalian organ of Jacobson. Trans. roy. Soc. Edinburgh 39: 231-255

Broom R (1915a) On the organ of Jacobson and its relations in the "Insectivora" - Part I. *Tupaia* and *Gymnura*. Proc. Zool. Soc., London 25: 157-162

Broom R (1915b) On the organ of Jacobson and its relations in the "Insectivora" - Part II. *Talpa*, *Centetes*, and *Chrysochloris*. Proc. Zool. Soc., London 25: 347-354

Brosset A (1966) Les chiroptères du Haut-Ivindo (Gabon). Biol. Gabonica 2: 47-86

Brouwer B (1915) Die biologische Bedeutung der Dermatomerie. Beitrag zur Kenntnis der Segmentalanatomie und der Sensibilitätsleitung im Rückenmark und in der Medulla oblongata. Folia neuro-biol. 9: 225-336

Brown WL, Wilson EO (1956) Character displacement. Syst. Zool. 5: 49-64

Brunner H (1919) Die zentralen Kleinhirnkerne bei den Säugetieren. Arb. Neurol. Inst. Wien 22: 200-277

Bruns V (1976a) Peripheral auditory tuning for fine frequency analysis by the CF-FM bat, *Rhinolophus ferrumequinum*. I. Mechanical specializations of the cochlea. J. comp. Physiol. 106: 77-86

Bruns V (1976b) Peripheral auditory tuning for fine frequency analysis by the CF-FM bat, *Rhinolophus ferrumequinum.* II. Frequency mapping in the cochlea. J. comp. Physiol. 106: 87-97

Bruns V (1985) Adaptions of the inner ear of mammals. In: Duncker HR, Fleischer G (eds) Functional Morphology in Vertebrates. Fortschritte der Zoologie 30. Fischer, Stuttgart, pp 653-656

Bruns V (1987) Vergleichende und funktionelle Morphologie des Innenohrs der Säugetiere. 61. Vers. Deutsche Gesellschaft für Säugetierkunde Berlin, p 10

Buchler ER (1976) The use of echolocation by the wandering shrew (*Sorex vagrans*). Anim. Behav. 24: 858-873

Buckner CH (1964) Metabolism, food capacity, and feeding behavior in four species of shrews. Can. J. Zool. 42: 259-279

Buckner CH (1966) Populations and ecological relationships of shrews in tamarack bogs of south-eastern Manitoba. J. Mammal. 47: 181-194

Buckner CH (1969) Some aspects of the population ecology of the common shrew, *Sorex araneus*, near Oxford, England. J. Mammal. 50: 326-332

Burda H (1979a) Morphologie des äusseren Ohres der einheimischen Arten der Familie Soricidae (Insectivora). Vest. cs. Spolec. zool. 43: 1-15

Burda H (1979b) Morphology of the middle and inner ear in some species of shrews (Insectivora, Soricidae). Acta Scientiarum Naturalium Academiae Scientiarum Bohemoslovacae Brno 13: 1-46

Burda H, Bauerova Z (1985) Hearing adaptations and feeding ecology in *Sorex araneus* and *Crocidura suaveolens* (Soricidae). Acta Zool. Fennica 173: 253-254

Burda H, Müller M, Bruns V, Dannhof BJ (1987) Hörbiologie subterraner Säugetiere. 61. Vers. Deutsche Gesellschaft für Säugetierkunde Berlin, pp. 11-12

Burkitt AN (1938) The external morphology of the brain of *Notoryctes typhlops*. Proc. kon. ned. Akad. Wet. 41: 921-933

Butler PM (1972) The problem of insectivore classification. In: Joysey KA, Kemp TS (eds) Studies in Vertebrate Evolution. Oliver & Boyd, Edinburgh, pp 253-265

Butler PM (1984) Macroscelidea, Insectivora and Chiroptera from the Miocene of East Africa. Palaeovertebrata 14: 117-198

Butler PM (1985) The history of African insectivores. Acta Zool. Fennica 173: 215-217

Butler PM, Greenwood M (1979) Soricidae (Mammalia) from the Early Pleistocene of Olduvai Gorge, Tanzania. Zool. J. Linn. Soc. 67: 329-379

Bystrzycka EK, Nail BS (1985) Brain stem nuclei associated with respiratory, cardiovascular and other autonomic functions. In: Paxinos G (ed) The Rat Nervous System. Vol 2, Hindbrain and spinal cord. Academic Press, Sydney, pp 95-110

Cabon K (1956) Untersuchungen über die saisonale Veränderlichkeit des Gehirnes bei der kleinen Spitzmaus (*Sorex minutus minutus* L.). Ann. Univ. M. Curie-Sklodowska. Sect. C. 10: 93-115

Cabrera A (1925) Genera mammalium: Insectivora, Galeopithecia. Mus. Nacion. Cienc. Natur., Madrid

Cajal SR (1893) Beiträge zur feineren Anatomie des großen Hirns. I. Über die feinere Struktur des Ammonshornes. Z. wiss. Zool. 56: 615-663

Cajal SR (1911) Histologie du système nerveux, part 2. Maloine, Paris (Reprint:1951, Inst. Ramon y Cajal, Madrid)

Campbell B (1938) A reconsideration of the shoulder musculature of the Cape golden mole. J. Mammal. 19: 234-240

Campbell B (1939) The shoulder anatomy of the moles. A study in phylogeny and adaptation. Am. J. Anat. 64: 1-39

Campbell B, Ryzen M (1953) The nuclear anatomy of the diencephalon of *Sorex cinereus*. J. comp. Neurol. 99: 1-22

Campbell CBG (1967) Pyramidal tracts in primate taxonomy. Diss. Abstr. 27: 2574-B

Campbell CBG (1969) The visual system of insectivores and primates. Ann. N. Y. Acad. Sci. 167: 388-403

Campbell CBG (1972) Evolutionary patterns in mammalian diencephalic visual nuclei and their fiber connections. Brain Behav. Evol. 6: 218-236

Campbell CBG, Jane JA, Yashon D (1967) The retinal projections of the tree shrew and hedgehog. Brain Res. 5: 406-418

Campbell DJ (1964) An experimental investigation of the paraventriculo-neurohypophysial neurosecretory pathway in the hedgehog (*Erinaceus europaeus*). Thesis University Birmingham

Campbell DJ (1966) Neurosecretory pathways from the paraventricular nuclei of the hedgehog. J. Anat. 100: 702-703

Campbell DJ, Holmes RL (1966) Further studies on the neurohyhophysis of the of the hedgehog (*Erinaceus europaeus*). Z. Zellforsch. 75: 35-46

Cantuel P (1943) Poids moyens de quelques micromammiferes de la faune française. Mammalia 4: 113-117

Carleton SC, Carpenter MB (1983) Afferent and efferent connections of the medial, inferior and lateral vestibular nuclei in the cat and monkey. Brain Res. 278: 29-51

Carpenter MB, Hanna GR (1961) Fiber projections from the spinal trigeminal nucleus in the cat. J. comp. Neurol. 117: 117-131

Carpenter MB, Nakano K, Kim R (1976) Nigrothalamic projections in the monkey demonstrated by autoradiographic technics. J. comp. Neurol. 165: 401- 416

Carter T (1936) The short-tailed shrew as a tree climber. J. Mammal. 17: 285

Catsman-Berrevoets CE, Kuypers HGJM (1978) Differential laminar distribution of corticothalamic neurons projecting to the VL and the center median. An HRP study in the cynomolgus monkey. Brain Res. 154: 359-365

Cazin L, Magnin M, Lannou J (1982) Non-cerebellar visual afferents to the vestibular nuclei involving the prepositus hypoglossal complex: An autoradiographic study in the rat. Exp. Brain Res. 48: 309-313

Cei G (1946) Morfologia degli organi della vista negli insettivori. Arch. Ital. Anat. Embriol. 52: 1-42

Chan-Palay V, Palay SL, Brown JT, VanItallie C (1977) Sagittal organization of olivo-cerebellar and reticulocerebellar projections: Autoradiographic studies with 35 S-methionine. Exp. Brain Res. 30: 561-576

Chang HT (1936) An auditory reflex of the hedgehog. Chinese J. Physiol. 10: 119-124

Chapman RN (1919) A study of the correlation of the pelvic structure and its habits of certain burrowing mammals. Am. J. Anat. 25: 185-219

Chauhan AK, Sood PP, Mohanakumar KP (1983) Functional significance of ß-galactosidase and ß-glucuronidase in lipid metabolism in the medulla oblongata and pons of hedgehog (*Paraechinus micropus*). Acta morph. neerl.-scand. 21: 229-238

Chibuzo GA, Cummings JF (1982) An enzyme tracer study of the organization of the somatic motor center for the innervation of different muscles of the tongue: Evidence for two sources. J. comp. Neurol. 205: 273-281

Clark WE le Gros (1929) Studies on the optic thalamus of the Insectivora. - The anterior nuclei. Brain 52: 334-358

Clark WE le Gros (1932) The brain of the Insectivora. Proc. Zool. Soc., London 1932: 975-1013

Clark WE le Gros (1933) The medial geniculate body and the nucleus isthmi. J. Anat. 67: 536-548

Clutton-Brock TH, Harvey PH (1980) Primates, brains and ecology. J. Zool., London 190: 309-323

Cockrum EL (1962) Introduction to Mammalogy. Ronald Press, New York

Conaway CH (1952) Life history of the water shrew (*Sorex palustris navigator*). Am. Midl. Nat. 48: 219-248

Cooper HM, Mick G, Magnin M (1989) Retinal projection to mammalian telencephalon. Brain Res. 477: 350-357

Corbet GB (1988) The family Erinaceidae: a synthesis of its taxonomy, phylogeny, ecology and zoogeography. Mammal Review 18: 117-172

Crile G, Quiring DP (1940) A record of the body weight and certain organ and gland weights of 3690 animals. Ohio J. Science 40: 219-260

Crosby EC, Humphrey T (1939a) A comparison of the olfactory and the accessory olfactory bulbs in certain representative vertebrates. Papers Mich. Acad. Sci., Arts and Letters. 24: 95-104

Crosby EC, Humphrey T (1939b) Studies of the vertebrate telencephalon. I. The nuclear configuration of the olfactory and accessory olfactory formations and of the nucleus olfactorius anterior of certain reptiles, birds and mammals. J. comp. Neurol. 71: 121-213

Crosby EC, Humphrey T (1944) Studies of the vertebrate telencephalon. III. The amygdaloid complex in the shrew (*Blarina brevicauda*). J. comp. Neurol. 81: 285-305

Crosby EC, Woodburne RT (1940) The comparative anatomy of the preoptic area and the hypothalamus. Res. Pub. Ass. Nerv. Ment. Dis. 20: 51-169

Crosby EC, Woodburne RT (1943) The nuclear pattern of the non- tectal portions of the midbrain and isthmus in the shrew and the bat. J. comp. Neurol. 78: 253-288

Crowcroft P (1957) The Life of the Shrew. Reinhardt, London

Crowcroft P, Ingles JM (1959) Seasonal changes in the brain-case of the common shrew (*Sorex araneus* L.). Nature 183: 907-908

Crutcher KA, Danscher G, Geneser FA (1988) Hippocampus and dentate area of the european hedgehog. Comparative histochemical study. Brain Behav. Evol. 32: 269-276

Curik R, Malinska J, Malinsky J (1974) Light and electron microscopical structure of motor nerve cells in the spinal cord of normal hedgehog. Acta Univ. Palackianae Olomucensis, Fac. med. 71: 355-362

Dalquest WW, Orcutt DR (1942) The biology of the least shrew-mole, *Neurotrichus gibbsii minor*. Am. Midl. Nat. 27: 387-401

Dapson RW (1968) Growth patterns in a post-juvenile population of short-tailed shrews (*Blarina brevicauda*). Am. Midl. Nat. 79: 118-129

Darian-Smith L (1973) The Trigeminal System. In: Iggo A (ed) Handbook of Sensory Physiology. Vol 2, Somatosensory System. Springer, Berlin-New York, pp 271-314

Davis DD (1962) Mammals of the lowland rain-forest of North Borneo. Bull. Singapore Nat. Mus. 31: 1-129

Deacon TW (1988) Human brain evolution: II. Embryology and brain allometry. In: Jerison HJ, Jerison I (eds) Intelligence and Evolutionary Biology. NATO ASI Vol G 17. Springer, Berlin-Heidelberg, pp 383-415

Dean P, Redgrave P, Lewis G (1982) Locomotor activity of rats in open field after microinjection of procaine into superior colliculus or underlying reticular formation. Behav. Brain Res. 5: 175-187

DeCarlos JA, López-Mascaraque L, Valverde F (1989) Connections of the olfactory bulb and nucleus olfactorius anterior in the hedgehog (*Erinaceus europaeus*): fluorescent tracers and HRP study. J. comp. Neurol. 279: 601-618

Dehnel A (1949) Studies on the genus *Sorex* L. Ann. Univ. M. Curie-Sklodowska. Sect. C. 4: 17-102

Demyanenko GP (1976) Structural organization of the "associative" area of the hedgehog neocortex. Biol. Nauki (Mosc.) 10: 54-60

DeWitte GF, Frechkop S (1955) Sur une espèce encore inconnue de mammifère africain, *Potamogale ruwenzorii*, sp.n. Bull. Inst. Roy. Sci. Nat. Belgique 31: 1-11

Diamond IT, Hall WC (1969) Evolution of neocortex. Science 164: 251-262

DiChiara G, Morelli M, Imperato A, Porceddu ML (1982) A re-evaluation of the role of superior colliculus in turning behaviour. Brain Res. 237: 61-77

Dieterlen F, Heim de Balsac H (1979) Zur Ökologie and Taxonomie der Spitzmäuse (Soricidae) des Kivu-Gebietes. Säugetierkdl. Mitt. 27: 241-287

Dinopoulos A, Karamanlidis AN, Michaloudi H, Antonopoulos J, Papadopoulos G (1987) Retinal projections in the hedgehog (*Erinaceus europaeus*). Anat. Embryol. 176: 65-70

Dinopoulos A, Michaloudi H, Karamanlidis AN, Antonopoulos J, Parnavelas JG (1988) Basal forebrain neurons project to the cortical mantle of the European hedgehog (*Erinaceus europaeus*). Neuroscience Letters 86: 127-132

Dobson GE (1882) Monograph of the Insectivora, systematic and anatomical. Part I. Including the families Erinaceidae, Centetidae, and Solenodontidae. Van Voorst, London

Dobson GE (1883) Monograph of the Insectivora, systematic and anatomical. Part II. Including the families Potamogalidae, Chrysochloridae, and Talpidae. Van Voorst, London

Dötsch C (1982) Der Kauapparat der Soricidae (Mammalia, Insectivora). Funktionsmorphologische Untersuchungen zur Kaufunktion bei Spitzmäusen der Gattungen *Sorex* Linnaeus, *Neomys* Kaup und *Crocidura* Wagler. Zool. Jb. (Anat.) 108: 421-484

Dötsch C (1983a) Das Kiefergelenk der Soricidae (Mammalia, Insectivora). Z. Säugetierkd. 48: 65-77

Dötsch C (1983b) Morphologische Untersuchungen am Kauapparat der Spitzmäuse *Suncus murinus* (L.), *Soriculus nigrescens* (Gray) und *Soriculus caudatus* (Horsfield) (Soricidae). Säugetierkdl. Mitt. 31: 27-46

Dötsch C (1985) Masticatory function in shrews (Soricidae). Acta Zool. Fennica 173: 231-235

Douglas RJ, Marcellus D (1975) The ascent of man: Deductions based on a multivariate analysis of the brain. Brain Behav. Evol. 11: 179-213

Dräseke J (1901) *Centetes ecaudatus*. Ein Beitrag zur vergleichenden makroskopischen Anatomie des Centralnervensystems der Wirbeltiere, mit besonderer Berücksichtigung der Insektivoren. Mschr. Psychiat. Neurol. 10: 413-431

Dräseke J (1904) Zur Kenntnis des Rückenmarks und der Pyramidenbahnen von *Talpa europea*. Mschr. Psychiat. Neurol. 15: 401-409

Dräseke J (1930) Zur makroskopischen Hirnanatomie der Chrysochlorida. Festschrift zum 50jährigen Bestehen des Akademisch-wissenschaftlichen Vereins zu Jena

Dubois E (1897) Über die Abhängigkeit des Hirngewichtes von der Körpergrösse bei den Säugethieren. Arch. Anthropol. 25: 1-28

Dubois P, Girod C (1969) Aspects ultrastructuraux des cellules limitant les formations colloidales dans l'antéhypophyse du Hérisson. C. R. Soc. Biol. (Paris) 163: 1390-1393

Dubost G (1965) Quelques renseignements biologiques sur *Potamogale velox*. Biol. Gabonica 1: 257-272

Dubost G (1968a) Les mammifères souterrains, I. Rev. Ecol. Biol. Sol. 5: 99-133

Dubost G (1968b) Les mammifères souterrains, II. Rev. Ecol. Biol. Sol. 5: 135-197

Earle AM, Matzke HA (1974) Efferent fibers of the deep cerebellar nuclei in hedgehogs. J. comp. Neurol. 154: 117-131

Ebbesson SOE, Jane JA, Schroeder DM (1972) A general overview of major interspecific variations in thalamic organization. Brain Behav. Evol. 6: 92-130

Ebner FF (1969) A comparison of primitive forebrain organization in metatherian and eutherian mammals. Ann. N.Y. Acad. Sci. 167: 241-257

Eccles JC (1982) The future of studies on the cerebellum. In: Palay SL, Chan-Palay V (eds) The Cerebellum - New Vistas. Exp. Brain Res. Suppl. 6. Springer, New York, pp 607-620

Eccles JC, Ito M, Szentágothai J (1967) The Cerebellum as a Neuronal Machine. Springer, Berlin-Heidelberg-New York

Edwards LF (1937) Morphology of the forelimb of the mole (*Scalops aquaticus* L.) in relation to its fossorial habits. Ohio J. Sci. 37: 20-41

Eimer T (1871) Die Schnautze des Maulwurfs als Tastwerkzeug. Arch. mikr. Anat. 7: 181-191

Eisenberg JF (1981) The Mammalian Radiations. An Analysis of Trends in Evolution, Adaptation, and Behavior. University Chicago Press, Chicago

Eisenberg JF, Gould E (1966) The behavior of *Solenodon paradoxus* in captivity with comments on the behavior of other insectivora. Zoologica 51: 49-60

Eisenberg JF, Gould E (1970) The tenrecs: A study in mammalian behavior and evolution. Smithson. Contrib. Zool. 27: 1-137

Eisenberg JF, Wilson DE (1978) Relative brain size and feeding strategies in the chiroptera. Evolution 32: 740-751

Ellerman JR (1956) The subterranean mammals of the world. Trans. Roy. Soc. S. Afr. 35: 11-20

Erickson RP, Hall WC, Jane JA, Snyder M, Diamond IT (1967) Organization of the posterior dorsal thalamus of the hedgehog. J. comp. Neurol. 131: 103-130

Erzurumlu RS, Killackey HP (1979) Efferent connections of the brainstem trigeminal complex with the facial nucleus of the rat. J. comp. Neurol. 188: 75-86

Erzurumlu RS, Killackey HP (1980) Diencephalic projections of the subnucleus interpolaris of the brainstem trigeminal complex in the rat. Neuroscience 5: 1891-1901

Estes, RD (1972) The role of the vomeronasal organ in mammalian reproduction. Mammalia 36: 315-341

Étienne A (1972) Premiers résultats sur l'histologie quantitative comparée du bulbe olfactif dans la lignée des Insectivores aux Primates. Acta oto-rhino-laryng. belg. 26: 493-505

Faull RLM, Mehler WR (1985) Thalamus. In: Paxinos G (ed) The Rat Nervous System. Vol 1, Forebrain and midbrain. Academic Press, Sydney, pp 129- 168

Faure A, Calas A (1977) Étude radioautographique de l'incorporation in vitro de noradrénaline tritiée dans des fibres catécholaminergiques centrales chez le hérisson actif et en hibernation. C. R. Soc. Biol. (Paris) 171: 136-141

Feist DD (1980) Norepinephrine turnover in brown fat, skeletal muscle and spleen of cold exposed and cold acclimated Alaskan red-backed voles. J. Therm. Biol. 5: 89-94

Ferrer I (1986) Golgi study of the isocortex in an Insectivore: the common European mole (*Talpa europaea*). Brain Behav. Evol. 29: 105-114

Ferrer I, Fabregues I, Condom E (1986) A Golgi study of the sixth layer of the cerebral cortex. I. The lissencephalic brain of Rodentia, Lagomorpha, Insectivora and Chiroptera. J. Anat. 145: 217-234

Filimonoff IN (1947) A rational subdivision of the cerebral cortex. Arch. Neurol. Psychiat. 58: 296-311

Filimonoff IN (1949) Vergleichende Anatomie der Großhirnrinde der Säugetiere (russ.). Moskau

Filimonoff IN (1965) On the so-called rhinencephalon in the dolphin. J. Hirnforsch. 8: 1-23

Findley JS (1967) Insectivores and dermopterans. In: Anderson S, Jones JK (eds) Recent Mammals of the World. Ronald Press, New York, pp 87-108

Findley JS, Wilson DE (1982) Ecological significance of chiropteran morphology. In: Kunz TH (ed) Ecology of Bats. Plenum, New York-London, pp 243-260

Firbas W, Platzer W (1969) Über die Cochlea der Insectivoren. Anat. Anz. 124: 233-243

Flatau E, Jacobsohn L (1899) Handbuch der Anatomie und vergleichenden Anatomie des Centralnervensystems der Säugetiere. Karger, Berlin

Fleischer G (1973) Studien am Skelett des Gehörorgans der Säugetiere, einschließlich des Menschen. Säugetierkdl. Mitt. 21: 131-239

Flores A (1911) Die Myeloarchitektonik und die Myelogenie des Cortex Cerebri beim Igel (*Erinaceus europaeus*). J. Psychol. Neurol. 17: 215-247

Flower WH (1865) On the commissures of the cerebral hemispheres of the Marsupialia and Monotremata as compared with those of the placental mammals. Roy. Soc. Phil. Trans., London 155: 633-651

Flumerfelt BA, Hrycyshyn AW (1985) Precerebellar nuclei and red nucleus. In: Paxinos G (ed) The Rat Nervous System. Vol 2, Hindbrain and spinal cord. Academic Press, Sydney, pp 221-250

Fons R (1974) Le repertoire comportemental de la pachyure étrusque, *Suncus etruscus* (Savi, 1822). Terre et Vie 1: 131-157

Fons R (1979) Durée de vie chez la Pachyure étrusque, *Suncus etruscus* (Savi, 1822) en captivité (Insectivora, Soricidae). Z. Säugetierkd. 44: 241-248

Fons R (1988) Insektenesser. In: Grzimeks Enzyklopädie, Säugetiere, Band 1. Kindler, München, pp. 418-419, 425-519

Fons R, Saint Girons MC (1975) Notes sur les mammifères de France. XIV. - Données morphologiques concernant la Pachyure étrusque, *Suncus etruscus* (Savi, 1822). Mammalia 39: 685-688

Fons R, Stephan H, Baron G (1984) Brains of Soricidae. I. Encephalization and macromorphology, with special reference to *Suncus etruscus*. Z. zool. Syst. Evol.-forsch. 22: 145-158

Forel A (1872) Beiträge zur Kenntniss des Thalamus opticus und der ihn umgebenden Gebilde bei den Säugethieren. Sitzungsber. kaiserl. Akademie Wissenschaften 3.Abt. : 25-58

Forel A (1877) Untersuchungen über die Haubenregion und ihre oberen Verknüpfungen im Gehirne des Menschen und einiger Säugethiere, mit Beiträgen zu den Methoden der Gehirnuntersuchung. Arch. Psychiat. 7: 393-495

Foster DO, Frydman ML (1978) Nonshivering thermogenesis in the rat. II. Measurements of blood flow with microspheres point to brown adipose tissue as the dominant site of the calorigenesis induced by noradrenaline. Can. J. Physiol. Pharmacol. 56: 110-122

Frahm HD (1985) Comparison of main olfactory bulb size in mammals. In: Duncker HR, Fleischer G (eds) Functional Morphology in Vertebrates. Fortschritte der Zoologie 30. Fischer, Stuttgart, pp 691-693

Frahm H, Hassler R, Graisarn S (1979) Comparative volumetric studies on pallidum in Insectivora and Primates. Appl. Neurophysiol. 42: 91-94

Frahm HD, Stephan H, Baron G (1984) Comparison of brain structure volumes in Insectivora and Primates. V. Area striata (AS). J. Hirnforsch. 25: 537-557

Frahm HD, Stephan H, Stephan M (1982) Comparison of brain structure volumes in Insectivora and Primates. I. Neocortex. J. Hirnforsch. 23: 375-389

Frankl-Hochwart L (1902) Zur Kenntnis der Anatomie des Gehirns der Blindmaus (*Spalax typhlus*). Arb. Neurol. Inst. Wien 8: 190-220

Franz V (1934) Vergleichende Anatomie des Wirbeltierauges. In: Bolk L, Göppert E, Kallius E, Lubosch W (eds) Handbuch der vergleichenden Anatomie der Wirbeltiere. Urban & Schwarzenberg, Berlin, pp 989-1292

Frechkop S (1957) A propos de nouvelles espèces de Potamogalinés. Mammalia 21: 226-234

Friant M (1947) Le cerveau du Centetes, Insectivore Malgache. C. R. Ass. Anat. (Paris) 34: 196-200

Friant M (1955) Morphologie et développement du cerveau des Mammifères euthériens. I. Séries des Insectivores et des Carnassiers. Ann. Soc. Roy. Zool. Belgique 86: 249-279

Fukushima T, Kerr FWL (1979) Organization of trigeminothalamic tracts and other thalamic afferent systems of the brainstem in the rat: presence of gelatinosa neurons with thalamic connections. J. comp. Neurol. 183: 169-184

Fuse G (1912) Über den Abduzenskern der Säuger. Arbeiten hirnanatomisches Inst. Univ. Zürich. 6: 401-447

Gaarskjaer FB, Danscher G, West MJ (1982) Hippocampal mossy fibers in the regio superior of the european hedgehog. Brain Res. 237: 79-90

Gacek RR (1969) The course and central termination of first order neurons supplying vestibular endorgans in the cat. Acta oto-laryng. (Suppl.) 254: 1-66

Galert D (1985) Morphology and topography of substantia nigra in Insectivora. Folia morphol. (Warszaw) 44: 186-193

Galert D (1986) The supraoptic and paraventricular nuclei in insectivores. Folia morphol. (Warszaw) 45: 182-191

Ganchrow D, Ring G (1979) Efferent projections of n. caudalis in hedgehog (*Erinaceus europaeus*). Anat. Rec. 193: 742

Ganeshina LV, Vorontsov NN, Chabovsky VI (1957) Comparative morphological study of the nasal cavity structure in certain representatives of the order Insectivora (russ.). Zoologichesky Zhurnal 36: 122-138

Ganser S (1882) Vergleichend-anatomische Studien über das Gehirn des Maulwurfs. Morph. Jb. 7: 591-725

Gasc JP, Jouffroy FK, Renous S (1986) Morphofunctional study of the digging system of the Namib Desert Golden mole (*Eremitalpa granti namibensis*): cinefluorographical and anatomical analysis. J. Zool., London 208: 9-35

Gebczynska Z, Gebczynski M (1965) Oxygen consumption in two species of water shrews. Acta theriol. 10: 209-214

Gebczynski M (1965) Seasonal and age changes in the metabolism and activity of *Sorex araneus* (Linnaeus 1758). Acta theriol. 10: 303-331

Gebczynski M (1971a) Oxygen consumption in starving shrews. Acta theriol. 16: 288-292

Gebczynski M (1971b) The rate of metabolism of the lesser shrew. Acta theriol. 16: 329-339

Gebczynski M (1977) Body temperature in five species of shrews. Acta theriol. 22: 521-530

Genoud M (1981) Contribution a l'étude de la stratégie énergétique et de la distribution écologique de *Crocidura russula* (Soricidae, Insectivora) en zone tempérée. Thèse, Lausanne

Genoud M, Vogel P (1981) The activity of *Crocidura russula* (Insectivora, Soricidae) in the field and in captivity. Z. Säugetierkd. 46: 222-232

George SB (1986) Evolution and historical biogeography of soricine shrews. Syst. Zool. 35: 153-162

Gerry H (1969) Contribution a l'étude de la cytologie antéhypophysaire (Recherches chez un Mammifère hivernant). Thèse, Lyon

Gierlich N (1916a) Neuere Untersuchungen über die Ausbildung der Großhirnbahnen bei Mensch und Tier. Neurologisches Centralblatt 16: 2-7

Gierlich N (1916b) Zur vergleichenden Anatomie der aus dem Großhirn stammenden Faserung. 1. Der Anteil des Pes pedunculi am Pedunculusquerschnitte bei verschiedenen Säugetieren. Anat. Anz. 49: 24-28

Gierlich N (1916c) Zur vergleichenden Anatomie der aus dem Großhirn stammenden Faserung. 2. Der Anteil des Kleinhirns an den im Pes pedunculus herabziehenden Gehirnbahnen bei verschiedenen Säugetieren. Anat. Anz. 49: 123- 128

Gil E, Machin C, Abella G (1985) Quantitative study of the supraoptic and paraventricular nuclei of the hedgehog in two different physiological situations. Folia morphol. (Prague) 33: 166-174

Gill TN (1883) On the classification of the insectivorous mammals. Bull. Philos. Soc. Washington 5: 118-120

Gillilan LA (1941) The connections of the basal optic root (posterior accessory optic tract) and its nucleus in various animals. J. comp. Neurol. 74: 367-408

Girod C (1971) Recherches sur les cellules gonadotropes antéhypophysaires du Hérisson: Étude cytologique, ultrastructurale et cytophysiologique. 3es Entretiens de Chizé, Fonction gonadotrope et rapports hypothalamo-hypophysaires chez les animaux sauvages, Série physiolog. no. 2: 57-81

Girod C (1976) Histochemistry of the Adenohypophysis. In: Graumann W, Neumann K (eds) Handbuch der Histochemie. Suppl. Part 4. Fischer, Stuttgart

Girod C, Curé M (1965) Étude des corrélations hypophyso-testiculaires, au cours du cycle annuel, chez le Herisson (*Erinaceus europaeus* L.). C. R. Acad. Sci. (Paris) 261: 257-260

Girod C, Curé M, Dubois P (1965) Mise en évidence, au microscope optique et au microscope électronique, de trois catégories de cellules gonadotropes antéhypophysaires chez le Hérisson (*Erinaceus europaeus* L.). C. R. Soc. Biol. (Paris) 159: 2202-2205

Girod C, Dubois P, Curé M (1966) Recherches sur la cytologie antéhypophysaire du Hérisson. C. R. Ass. Anat. 135: 436-444

Girod C, Dubois P, Curé M (1967) Recherches sur les corrélations hypophyso-génitales chez la femelle de Herisson (*Erinaceus europaeus* L.). Ann. Endocrin. (Paris) 28: 581-610

Girod C, Lhéritier M, Trouillas J, Dubois MP (1982) Cell types of the pars distalis of the hedgehog (*Erinaceus europaeus* L.) adenohypophysis: cytological, immunocytochemical and ultrastructural studies. I. Somatotropic cells. Acta anat. 114: 248-258

Girod C, Lhéritier M, Trouillas J, Dubois MP (1983) Cell types of the pars distalis of the hedgehog (*Erinaceus europaeus* L.) adenohypophysis: cytological, immunocytochemical and ultrastructural studies. 2. Prolactin cells. Acta anat. 117: 102-111

Girod C, Lhéritier M, Trouillas J, Dubois MP (1985) Cell types of the pars distalis of the hedgehog (*Erinaceus europaeus* L.) adenohypophysis: cytological, immunocytochemical and ultrastructural studies. 3. Opiocorticomelanotropic cells. Acta anat. 123: 67-71

Girod C, Lhéritier M, Trouillas J, Dubois MP (1986) Cell types of the pars distalis of the hedgehog (*Erinaceus europaeus* L.) adenohypophysis: cytological, immunocytochemical and ultrastructural studies. 4. Thyrotropic cells. Acta anat. 127: 48-52

Godet R (1951) Contribution à l'éthologie de la taupe (*Talpa europaea* L.). Bull. Soc. Zool. France 76: 107-128

Godfrey G, Crowcroft P (1960) The life of the mole (*Talpa europaea* Linnaeus). Museum Press, London

Godlowski WJ (1930) Über den Nucleus triangularis. Arb. Neurol. Inst. Wien 32: 289-340

Gould E (1965) Evidence for echolocation in the Tenrecidae of Madagascar. Proc. Am. Phil. Soc. 109: 352-360

Gould E (1969) Communication in three genera of shrews (Soricidae): *Suncus, Blarina & Cryptotis*. Comm. Behav. Biol., Part A 3: 11-31

Gould E (1978) The behavior of the moonrat, *Echinosorex gymnurus* (Erinaceidae) and the pentail shrew. *Ptilocercus lowi* (Tupaiidae) with comments on the behavior of other Insectivora. Z. Tierpsychol. 48: 1-27

Gould E, Eisenberg JF (1966) Notes on the biology of the Tenrecidae. J. Mammal. 47: 660-686

Gould E, Negus NC, Novick A (1964) Evidence for echolocation in shrews. J. exp. Zool. 156: 19-37

Gould HJ, Ebner FF (1978a) Connections of the visual cortex in the hedgehog (*Paraechinus hypomelas*). II. Corticocortical projections. J. comp. Neurol. 177: 473-502

Gould HJ, Ebner FF (1978b) Interlaminar connections of the visual cortex in the hedgehog (*Paraechinus hypomelas*). J. comp. Neurol. 177: 503-517

Gould HJ, Hall WC, Ebner FF (1978) Connections of the visual cortex in the hedgehog (*Paraechinus hypomelas*). I. Thalamocortical projections. J. comp. Neurol. 177: 445-471

Grainger JP, Fairley JS (1978) Studies on the biology of the pygmy shrew *Sorex minutus* in the west of Ireland. J. Zool., London 186: 109-141

Grandidier G, Petit G (1930) Étude d'un mammifère insectivore malgache le *Geogale aurita* A.M.-E. et A.G. Faune des Colonies Francaises 4: 441-493

Grandidier G, Petit G (1932) Zoologie de Madagascar. Soc. Edit. Géographiques, Paris

Grant PR (1972) Convergent and divergent character displacement. Biol. J. Linn. Soc. 4: 39-68

Grassé PP (1955a) Ordre des Insectivores. Anatomie et reproduction. In: Grassé PP (ed) Traité de Zoologie. Vol 17, Anatomie-Systématique-Biologie. Masson, Paris, pp 1574-1641

Grassé PP (1955b) Ordre des Insectivores. Affinités zoologiques des diverses familles entre elles et avec les autres ordres de mammifères. In: Grassé PP (ed) Traité de Zoologie. Vol 17, Anatomie-Systématique-Biologie. Masson, Paris, pp 1642-1653

Graybiel AM, Hartwieg EA (1974) Some afferent connections of the oculomotor complex in the cat: an experimental study with tracer techniques. Brain Res. 81: 543-551

Graziadei P (1966) Electron microscopic observations of the olfactory mucosa of the mole. J. Zool., London 149: 89-94

Green JD (1951) The comparative anatomy of the hypophysis, with special reference to its blood supply and innervation. Am. J. Anat. 88: 225-311

Grönberg G (1901) Die Ontogenese eines niedern Säugergehirns nach Untersuchungen an *Erinaceus europaeus*. Zool. Jb. (Anat.) 15: 261-384

Grün G, Schwammberger KH (1980) Ultrastructure of the retina in the shrew (Insectivora: Soricidae). Z. Säugetierkd. 45: 207-216

Grünwald A (1969) Untersuchungen zur Orientierung der Weißzahnspitzmäuse (Soricidae-Crocidurinae). Z. vergl. Physiol. 65: 191-217

Grünwald HF (1903) Zur vergleichenden Anatomie der Kleinhirnarme. Arb. Neurol. Inst. Wien 10: 368-377

Gudden B (1870) Über einen bisher nicht beschriebenen Nervenfaserstrang im Gehirne der Säugethiere und des Menschen. Arch. Psychiat. 2: 364-366

Gurdjian ES (1925) Olfactory connections in the albino rat, with special reference to the stria medullaris and the anterior commissure. J. comp. Neurol. 38: 127-163

Gurtovoi NN (1966) Ecological-morphological differences in the structure of the nasal cavity in the representatives of the orders Insectivora, Chiroptera and Rodentia (russ.). Zoologichesky Zhurnal 45: 1536-1551

Guth C, Heim de Balsac H, Lamotte M (1959) Recherches sur la morphologie de *Micropotamogale lamottei* et l'évolution des Potamogalinae. I. - Écologie, denture, anatomie crânienne. Mammalia 23: 423-447

Guth C, Heim de Balsac H, Lamotte M (1960) Recherches sur la morphologie de *Micropotamogale lamottei* et l'évolution des Potamogalinae. II. Rachis, viscères, position systématique. Mammalia 24: 190-217

Haberly LB, Price JL (1978a) Association and commissural fiber systems of the olfactory cortex of the rat. I. Systems originating in the piriform cortex and adjacent areas. J. comp. Neurol. 178: 711-740

Haberly LB, Price JL (1978b) Association and commissural fiber systems of the olfactory cortex of the rat. II. Systems originating in the olfactory peduncle. J. comp. Neurol. 181: 781-807

Hager H (1954) Die Cytoarchtitektonik des Bulbus olfactorius des Igels, *Erinaceus e. europaeus* Linne, 1758. Säugetierkdl. Mitt. 2: 8-15

Haines DE (1977) Cerebellar corticonuclear and corticovestibular fibers of the flocculonodular lobe in a prosimian primate (*Galago senegalensis*). J. comp. Neurol. 174: 607-629

Hall WC (1970) Visual pathways from the thalamus to the telencephalon in the turtle and hedgehog. Anat. Rec. 166: 313

Hall WC (1972) Visual pathways to the telencephalon in reptiles and mammals. Brain Behav. Evol. 5: 95-113

Hall WC, Diamond IT (1968a) Organization and function of the visual cortex in hedgehog: I. Cortical cytoarchitecture and thalamic retrograde degeneration. Brain Behav. Evol. 1: 181-214

Hall WC, Diamond IT (1968b) Organization and function of the visual cortex in hedgehog. II. An ablation study of pattern discrimination. Brain Behav. Evol. 1: 215-243

Hall WC, Ebner FF (1970) Parallels in the visual afferent projections of the thalamus in the hedgehog (*Paraechinus hypomelas*) and the turtle (*Pseudemys scripta*). Brain Behav. Evol. 3: 135-154

Haller B (1906) Beiträge zur Phylogenese des Großhirns der Säugetiere. Arch. mikr. Anat. 69: 117-222

Haller B (1909a) Die phyletische Stellung der Grosshirnrinde der Insektivoren. Jena. Z. Med. Naturw. 45: 279-298

Haller B (1909b) Über die Hypophyse niederer Placentalier und den Saccus vasculosus der urodelen Amphibien. Arch. mikr. Anat. 74: 812-843

Hamilton WJ (1940) The biology of the smoky shrew (*Sorex fumeus fumeus* Miller). Zoologica 25: 473-492

Hanström B (1946) The pituitary in Swedish Insectivora. Arkiv för Zoologi 38A (7): 1-20

Hanström B (1952) The hypophysis in some South-African Insectivora, Carnivora, Hyracoidea, Proboscidea, Artiodactyla, and Primates. Arkiv för Zoologi 4: 187-294

Hanström B (1953) The neurohypophysis in the series of mammals. Z. Zellforsch. 39: 241-259

Hanström B (1966) Gross anatomy of the hypophysis in mammals. In: Harris GW, Donovan BT (eds) The Pituitary Gland, Vol 1. Butterworths, London, pp 1-57

Harrison JL (1966) An Introduction to Mammals of Singapore and Malaya. Malayan Nature Society, Singapore

Harrison JM, Irving R (1966) Visual and nonvisual auditory systems in mammals. Science 154: 738-743

Harting JK, Hall WC, Diamond IT (1972) Evolution of the pulvinar. Brain Behav. Evol. 6: 424-452

Harvey PH (1988) Allometric analysis and brain size. In: Jerison HJ, Jerison I (eds) Intelligence and Evolutionary Biology. NATO ASI Series G17. Springer, Berlin-Heidelberg, pp 199-210

Harvey RT (1882) Note on the organ of Jacobson. Quart. Journ. Micr. Sci. 22: 50-52

Hassler R (1964) Zur funktionellen Anatomie des limbischen Systems. Nervenarzt 35: 386-396

Hatschek R (1904) Bemerkungen über das ventrale Haubenfeld, die mediale Schleife und den Aufbau der Brücke. Arb. Neurol. Inst. Wien 11: 128-155

Hatschek R (1907) Zur vergleichenden Anatomie des Nucleus ruber tegmenti. Arb. Neurol. Inst. Wien 15: 89-136

Haug H (1958) Über die Beziehungen des Hirngewichtes zum Grauzellkoeffizienten der Sehrinde bei den Primaten und einigen primitiven Säugern. J. Hirnforsch. 4: 189-204

Haug H (1987) Brain sizes, surfaces, and neuronal sizes of the cortex cerebri: A stereological investigation of man and his variability and a comparison with some mammals (primates, whales, marsupials, insectivores, and one elephant). Am. J. Anat. 180: 126-142

Hausser J, Juhlin-Dannfelt B (1979) Réponse de la musaraigne musette *Crocidura russula*, aux marquages odorants d'individus de son espèce (Insectivora, Soricidae). Terre et Vie 33: 555-562

Hawes ML (1976) Odor as a possible isolating mechanism in sympatric species of shrews (*Sorex vagrans* and *Sorex obscurus*). J. Mammal. 57: 404-406

Hawkins AE, Jewell PA (1962) Food consumption and energy requirements of captive British shrews and the mole. Proc. Zool. Soc., London 138: 137-155

Hawkins AE, Jewell PA, Tomlinson G (1960) The metabolism of some British shrews. Proc. Zool. Soc., London 135: 99-103

Hayakawa T, Zyo K (1983) Comparative cytoarchitectonic study of Gudden's tegmental nuclei in some mammals. J. comp. Neurol. 216: 233-244

Hayakawa T, Zyo K (1984) Comparative anatomical study of the tegmentomammillary projections in some mammals: a horseradish peroxidase study. Brain Res. 300: 335-349

Heape W (1887) The development of the mole (*Talpa europaea*). Stages E to J. Part 2. Quart. J. Microsc. Sci. 27: 123-163

Heffner RS, Masterton RB (1975) Variation in form of the pyramidal tract and its relationship to digital dexterity. Brain Behav. Evol. 12: 161-200

Heffner RS, Masterton RB (1983) The role of the corticospinal tract in the evolution of human digital dexterity. Brain Behav. Evol. 23: 165-183

Heidary H, Tomasch J (1969) Neuron numbers and perikaryon areas in the human cerebellar nuclei. Acta anat. 74: 290-296

Heim de Balsac H (1954) Un genre inédit et inattendu de mammifère (Insectivore Tenrecidae) d'Afrique Occidentale. C. R. Acad. Sci. (Paris) 239: 102-104

Heim de Balsac H (1956) Morphologie divergente des Potamogalinae (Mammifères Insectivores) en milieu aquatique. C. R. Acad. Sci. (Paris) 242: 2257-2258

Heim de Balsac H (1957) Évolution des vertèbres caudales chez les Potamogalinae (Mammifères Insectivores Tenrecidae). C. R. Acad. Sci. (Paris) 245: 562-564

Heim de Balsac H (1958) Persistance inattendue de certains caractères ancestraux dans la denture de Soricidae (Mammifères Insectivores). C. R. Acad. Sci. (Paris) 247: 1499-1501

Heim de Balsac H (1972) Insectivores. In: Battistini R, Richard-Vindard G (eds) Biogeography and Ecology in Madagascar. Junk, The Hague, pp 629- 660

Heim de Balsac H, Bourlière F (1955) Ordre des Insectivores. Systématique. In: Grassé PP (ed) Traité de Zoologie. Vol 17, Anatomie-Systématique-Biologie. Masson, Paris, pp 1653-1697

Heim de Balsac H, Lamotte M (1956) Évolution et phylogénie des Soricidés Africains. Mammalia 20: 140-167

Heim de Balsac H, Lamotte M (1957) Évolution et phylogénie des Soricidés Africains. II. La lignée *Sylvisorex-Suncus-Crocidura*. Mammalia 21: 15-49

Heimer L (1978) The olfactory cortex and the ventral striatum. In: Livingston KE, Hornykiewicz O (eds) Limbic Mechanisms. The continuing Evolution of the Limbic System Concept. Plenum Press, New York-London, pp 95-187

Heimer L, Alheid GF, Zaborszky L (1985) Basal ganglia. In: Paxinos G (ed) The Rat Nervous System. Vol 1, Forebrain and midbrain. Academic Press, Sydney, pp 37-86

Henderson T (1952) The eye of the mole. Brit. J. Ophthal. 36: 637

Henson OW (1961) Some morphological and functional aspects of certain structures of the middle ear in bats and insectivores. Univ. Kansas Sci. Bull. 42: 151-255

Herkenham M, Nauta WJH (1977) Afferent connections of the habenular nuclei in the rat. A horseradish peroxidase study, with a note on the fiber-of-passage problem. J. comp. Neurol. 173: 123-145

Herlant M (1959a) Les modifications hypophysaires chez la Taupe au cours de la phase d'activité sexuelle. C. R. Ass. Anat. 107: 347-354

Herlant M (1959b) Contribution a l'étude des deux cellules gonadotropes de l'hypophyse. C. R. Ass. Anat. 101: 417-422

Herlant M (1959c) L'hypophyse de la Taupe au cours de la phase d'activité sexuelle. C. R. Acad. Sci. (Paris) 248: 1033-1037

Herlant M (1964) The cells of the adenohypophysis and their functional significance. Int. Rev. Cytol. 17: 299-382

Herlant M, Klastersky J (1961a) Étude preliminaire au microscope électronique de l'hypophyse antérieure de la chauve-souris *Myotis myotis* et de la Taupe. C. R. Ass. Anat. 110: 382-391

Herlant M, Klastersky J (1961b) Étude au microscope électronique des cellules chez la Taupe. C. R. Acad. Sci. (Paris) 253: 2415-2417

Hermanides SR, Köppen M (1903) Über die Furchen und über den Bau der Grosshirnrinde bei den Lissencephalen insbesondere über die Localisation des motorischen Centrums und der Sehregion. Arch. Psychiat. Nervenkr. 37: 616-634

Herre W, Stephan H (1955) Zur postnatalen Morphogenese des Hirnes verschiedener Haushundrassen. Morph. Jb. 96: 210-264

Herrera Lara M (1985) Periaqueductal neurons associated to the posterior commissure: A morphological study in the hedgehog, rat and cat. Anat. Anz. 159: 195-201

Herter K (1933) Dressurversuche mit Igeln. (I. Orts-, Helligkeits- und Farbendressuren). Z. vergl. Physiol. 18: 481-515

Herter K (1935) Dressurversuche mit Igeln. (II. Form-, Helligkeitsdressuren, Farbenunterscheidung, Labyrinthversuche, Rhythmus- und Selbstdressuren). Z. vergl. Physiol. 21: 450-462

Herter K (1938) Die Biologie der europäischen Igel. Monographien der Wildsäugetiere, Vol 5. Schöps, Leipzig

Herter K (1957) Das Verhalten der Insektivoren. In: Helmcke JG, Lengerken H, Starck D (eds) Handbuch der Zoologie 8/10. DeGruyter, Berlin, pp 1-50

Herter K (1962a) Über die Borstenigel von Madagaskar (Tenrecinae). Sitzungsberichte Ges. Naturf. Freunde Berlin N.F. 2: 5-37

Herter K (1962b) Untersuchungen an lebenden Borstenigeln (Tenrecinae), 1. Über Temperaturregulierung und Aktivitätsrhythmik bei dem Igeltanrek *Echinops telfairi* Martin. Zool. Beiträge 7: 239-292

Herter K (1963) Untersuchungen an lebenden Borstenigeln (Tenrecinae). 2. Über das Verhalten und die Lebensweise des Igeltanreks *Echinops telfairi* Martin in Gefangenschaft. Zool. Beiträge 8: 125-165

Herter K (1964) Untersuchungen an lebenden Borstenigeln (Tenrecinae). 4. Über das Verhalten und die Aktivitätsrhythmik eines *Setifer setosus* (Schreber) in Gefangenschaft. Zool. Beiträge 10: 161-187

Herter K (1979) Die Insektenesser. In: Grzimek B (ed) Grzimeks Tierleben. Vol 10, Säugetiere 1. Deutscher Taschenbuchverlag dtv, München, pp 169-232

Herzfeld P (1889) Ueber das Jacobson'sche Organ des Menschen und der Säugethiere. Zool. Jb. (Anat. Ontog.) 3: 551-574

Hewitt W (1959) The mammalian caudate nucleus. J. Anat. 93: 169-176

Hochstetter F (1923) Beiträge zur Entwicklungsgeschichte des menschlichen Gehirns. Vol 2, Part 1, Die Entwicklung der Zirbeldrüse. Deuticke, Wien-Leipzig

Hochstetter F (1924) Beiträge zur Entwicklungsgeschichte des menschlichen Gehirns. Vol 2, Part 2, Die Entwicklung des Hirnanhanges. Deuticke, Wien-Leipzig, pp 73-79

Hochstetter F (1942) Über die harte Hirnhaut und ihre Fortsätze bei den Säugetieren nebst Angaben über die Lagebeziehungen der einzelnen Hirnteile dieser Tiere zueinander, zu den Fortsätzen der harten Hirnhaut und zur Schädelkapsel. Denkschrift Akad. Wiss. Wien, math.-naturw. Kl. 106: 1-114

Hofman MA (1982) Encephalization in mammals in relation to the size of the cerebral cortex. Brain, Behav. Evol. 20: 84-96

Hofmann F (1908) Die obere Olive der Säugetiere nebst Bemerkungen über die Lage der Cochlearisendkerne. Arb. Neurol. Inst. Wien 14: 76-328

Holling CS (1958) Sensory stimuli involved in the location and selection of sawfly cocoons by small mammals. Can. J. Zool. 36: 633-653

Holloway RL, Post DG (1982) The relativity of relative brain measures and hominid mosaic evolution. In: Armstrong E, Falk D (eds) Primate Brain Evolution. Methods and Concepts. Plenum Press, New York, pp 57-76

Holmes RL (1961) Esterases of the hypothalamo-hypophysial system. In: The Cytology of Nervous Tissue. Proc. Anat. Soc. G.B. and I. pp 1-4. Taylor & Francis, London

Holmes RL (1965) The fine structure of supraoptic neurons of hedgehogs. Z. Zellforsch. 66: 685-689

Holmes RL (1966) The pituitary gland of the mole in relation to that of other insectivores. J. Zool., London 149: 71-75

Holmes RL, Ball JN (1974) The pituitary gland. A comparative account. In: Harrison RJ, McMinn RMH, Treherne JE (eds) Biological Structure and Function, Vol 4. University Press, Cambridge

Holmes RL, Kiernan JA (1963) Nerve fibres of the infundibular process of the hedgehog. J. Anat. 97: 613-615

Holmes RL, Kiernan JA (1964) The fine structure of the infundibular process of the hedgehog. Z. Zellforsch. 61: 894-912

Holst D von (1985) The primitive eutherians I: orders Insectivora, Macroscelidea, and Scandentia. In: Brown RE, MacDonald DW (eds) Social odours in mammals. Oxford University Press, Oxford, pp 105-154

Horel JA (1988) Limbic neocortical interrelations. In: Steklis HD, Erwin J (eds) Comparative Primate Biology. Vol 4, Neurosciences. Alan R. Liss, New York, pp 81-97

Hornet T (1933) Vergleichend anatomische Untersuchungen über das Corpus geniculatum mediale (c.g.m.). Arb. Neurol. Inst. Wien 35: 76-92

Hosaka K (1987) Experimental morphological studies on central projection of snout proprioceptive afferents in moles. J. Stomatol. Soc. Japan 54: 208-232

Hrdlicka A (1905) Brain weight in vertebrates. Proc. Smithson. Misc. Coll. 48: 89-112

Huber GC, Crosby EC (1943) A comparison of the mammalian and reptilian tecta. J. comp. Neurol. 78: 133-168

Huerta MF, Frankfurter A, Harting JK (1983) Studies of the principal sensory and spinal trigeminal nuclei of the rat: projections to the superior colliculus, inferior olive, and cerebellum. J. comp. Neurol. 220: 147-167

Hulles E (1907) Zur vergleichenden Anatomie der cerebralen Trigeminuswurzel. Arb. Neurol. Inst. Wien 16: 469- 486

Hutterer R (1976) Deskriptive und vergleichende Verhaltensstudien an der Zwergspitzmaus, *Sorex minutus* L., und der Waldspitzmaus, *Sorex araneus* L. (Soricidae-Insectivora-Mammalia). Dissertation, Wien

Hutterer R (1980) Das Rhinarium von *Nectogale elegans* und anderen Wasserspitzmäusen (Mammalia, Insectivora). Z. Säugetierkd. 45: 126-127

Hutterer R (1982) Biologische und morphologische Beobachtungen an Alpenspitzmäusen (*Sorex alpinus*). Bonner zool. Beitr. 33: 3-18

Hutterer R (1985) Anatomical adaptations of shrews. Mammal Review 15: 43-55

Hutterer R (1986) Eine neue Soricidengattung aus Zentralafrika (Mammalia: Soricidae). Z. Säugetierkd. 51: 257-266

Hutterer R, Hürter T (1981) Adaptive Haarstrukturen bei Wasserspitzmäusen (Insectivora, Soricinae). Z. Säugetierkd. 46: 1-11

Hutterer R, Vogel P (1977) Abwehrlaute afrikanischer Spitzmäuse der Gattung *Crocidura* Wagler, 1832 und ihre systematische Bedeutung. Bonner zool. Beitr. 28: 218-227

Hyde JB (1957) A comparative study of certain trigeminal components in two soricid shrews, *Blarina brevicauda* and *Sorex cinereus*. J. comp. Neurol. 107: 339-351

Hyvärinen H (1967) Variation of the size and cell types of the anterior lobe of the pituitary during the life cycle of the common shrew (*Sorex araneus* L.). Aquilo, Ser. Zool. 5: 35-40

Hyvärinen H (1968) On the seasonal variation of the activity of alkaline phosphatase in the kidney of the bank vole (*Clethrionomys glareolus* Schr.) and the common shrew (*Sorex araneus* L.). Aquilo, Ser. Zool. 6: 7-11

Hyvärinen H (1969) On the seasonal changes in the skeleton of the common shrew (*Sorex araneus* L.) and their physiological background. Aquilo, Ser. Zool. 7: 1-32

Hyvärinen H (1972) On the histology and histochemistry of the snout and vibrissae of the common shrew (*Sorex araneus* L.). Z. Zellforsch. 124: 445-453

Hyvärinen H (1984) Wintering strategy of voles and shrews in Finland. In: Merritt JF (ed) Winter Ecology of Small Mammals. Carnegie Mus. Nat. Hist. Publ. pp 139-148

Hyvärinen H, Heikura K (1971) Effects of age and seasonal rhythm on the growth patterns of some small mammals in Finland and in Kirkenes, Norway. J. Zool., London 165: 545-556

Igarashi S, Kamiya T (1972) Atlas of the Vertebrate Brain. University of Tokyo Press, Tokyo

Ingvar S (1918) Zur Phylo- und Ontogenese des Kleinhirns nebst ein Versuch zu einheitlicher Erklärung der zerebellaren Funktion und Lokalisation. De Erven F. Bohn, Haarlem

Irving R, Harrison JM (1967) The superior olivary complex and audition: a comparative study. J. comp. Neurol. 130: 77-86

Ishihara K (1931) Zur vergleichenden Anatomie des Nervus vestibularis. Arb. Neurol. Inst. Wien 33: 233-247

Isomura G, Kozasa T, Tanaka S (1986) Absence of the central canal and its obliterative process in the house shrew spinal cord (*Suncus murinus*). Anat. Anz. 161: 285-296

Ivanov EI, Sollertinskaya TN (1986) Influence of tuftsen on hierarchical behavioral interrelationships in the hedgehog *Erinaceus auritus*. J. evol. Biochem. Physiol. 22: 137-145

Ivlieva LF (1973) Vestibular nuclei of brain stem in mammals with different locomotion. Zool. Zh. 52: 1104-1107 (russ)

Jane JA, Schroeder DM (1971) A comparison of dorsal column nuclei and spinal afferents in the European hedgehog (*Erinaceus europaeus*). Exp. Neurol. 30: 1-17

Jannett FJ, Jannett JA (1981) Convergent evolution in the flank gland marking behavior of a rodent and a shrew. Mammalia 45: 473-481

Jansen J, Brodal A (1940) Experimental studies on the intrinsic fibers of the cerebellum. II. The cortico-nuclear projection. J. comp. Neurol. 73: 267-321

Jerge CR (1963a) Organization and function of the trigeminal mesencephalic nucleus. J. Neurophysiol. 26: 379-392

Jerge CR (1963b) The function of the nucleus supratrigeminalis. J. Neurophysiol. 26: 393-402

Jerison HJ (1973) Evolution of the Brain and Intelligence. Academic Press, New York

Johannesson-Gross K (1986) Zur taktilen Orientierung des Maulwurfs (*Talpa europaea* L): Lernversuche in einem Y-förmigen Röhrenlabyrinth. Verhandl. dtsch. zool. Ges. 79: 216-217

Johannesson-Gross K (1988) Lernversuche in einer Zweifachwahlapparatur zum Hell-Dunkel-Sehen des Maulwurfs (*Talpa europaea* L.). Z. Säugetierkd. 53: 193-201

Johannesson-Gross K, Gross H (1982) Lernversuche mit Maulwürfen (*Talpa europaea* L.) unter Anwendung einer speziellen Labyrinthmethode. Z. Säugetierkd. 47: 277-282

Johnson JI, Switzer RC, Kirsch JAW (1982) Phylogeny through brain traits: the distribution of categorizing characters in contemporary mammals. Brain Behav. Evol. 20: 97-117

Johnson TN (1954) The superior and inferior colliculi of the mole (*Scalopus aquaticus machrinus*). J. comp. Neurol. 101: 765-799

Johnson TN (1957) The olfactory centers and connections in the cerebral hemisphere of the mole (*Scalopus aquaticus machrinus*). J. comp. Neurol. 107: 379-425

Johnston JB (1909) The radix mesencephalica trigemini. J. comp. Neurol. 19: 593-644

Johnston JB (1913) The morphology of the septum, hippocampus, and pallial commissures in reptiles and mammals. J. comp. Neurol. 23: 371-478

Jolicoeur P, Baron G (1980) Brain center correlations among Chiroptera. Brain Behav. Evol. 17: 419-431

Jolicoeur P, Pirlot P, Baron G, Stephan H (1984) Brain structure and correlation patterns in Insectivora, Chiroptera, and Primates. Syst. Zool. 33: 14-29

Kaas J, Hall WC, Diamond IT (1970) Cortical visual areas I and II in the hedgehog: relation between evoked potential maps and architectonic subdivisions. J. Neurophysiol. 33: 595-615

Kahle W (1976) Nervensystem und Sinnesorgane. In: Kahle W, Leonhardt H, Platzer W (eds) Taschenatlas der Anatomie, Vol 3. Thieme, Stuttgart

Kamei I, Shiosaka S, Senba E et al. (1981) Comparative anatomy of the distribution of catecholamines within the inferior olivary complex from teleosts to primates. J. comp. Neurol. 202: 125-133

Kaplan M (1913) Die spinale Acusticuswurzel und die in ihr eingelagerten Zellsysteme. Nucleus Deiters - Nucleus Bechterew. Arb. Neurol. Inst. Wien 20: 375-559

Kapoun S, Malinska J, Zrzavy J (1973) Anatomical peculiarities of the spinal cord in some Insectivora and Chiroptera. Folia morphol. (Prague) 21: 136-138

Karamyan AI, Sollertinskaya TN, Dustov SB (1986) Role of thalamic and hypothalamic formations in regulating conditioned-reflex activity in insectivores (hedgehogs). Dokl. Akad. Nauk USSR Biol. 285: 816-819

Kemali M (1984) Morphological asymmetry of the habenulae of a macrosmatic mammal, the mole. Z. mikr.-anat. Forsch. 98: 951-954

Kesarev VS (1964) Vergleichende Morphologie des Hypothalamus des Menschen und einiger Tiere (russ.). Dissertation, Moskau (ex Blinkow and Glezer 1968)

Kiernan JA (1964) Carboxylic esterases of the hypothalamus and neurohypophysis of the hedgehog. J. roy. micr. Soc. 83: 297-306

Kiernan JA (1967) On the probable absence of retino-hypothalamic connections in five mammals and an amphibian. J. comp. Neurol. 131: 405-408

Killackey HP (1972) Projections of the ventral nucleus to neocortex in the hedgehog. Anat. Rec. 172: 345

Killackey HP, Ebner FF (1972) Two different types of thalamocortical projections to a single cortical area in mammals. Brain Behav. Evol. 6: 141-169

Kingdon J (1974) East African Mammals. An Atlas of Evolution in Africa. Vol 2, Part A (Insectivores and Bats). Academic Press, London-New York

Knappe H (1964) Zur Funktion des Jacobsonschen Organs (Organon vomeronasale Jacobsoni). Zool. Garten, N.F. 28: 188-194

Kölliker A (1896) Handbuch der Gewebelehre des Menschen. Vol 2, Nervensystem des Menschen und der Thiere (6. Aufl.). Engelmann, Leipzig

Köppen M (1915) Über das Gehirn eines Blindtieres *Chrysochloris*. Mschr. Psychiat. Neurol. 38: 201-216

Környey S (1928) Zur vergleichenden Morphologie des lateralen Kniehöckers der Säugetiere. Arb. Neurol. Inst. Wien 30: 93-120

Kohl C (1893-1895) Rudimentäre Wirbelthieraugen. Bibliotheca Zoologica, Vol 5. Part 2: 1-180; Part 3: 181-274. Nägele, Stuttgart

Koikegami, H (1938) Beiträge zur Kenntnis der Kerne des Hypothalamus bei Säugetieren. Arch. Psychiat. Nervenkrankh. 107: 742-774

Kolmer W (1936) Die Netzhaut (Retina). In: Möllendorff W (ed) Handbuch der mikroskopischen Anatomie des Menschen. Vol 3, Haut- und Sinnesorgane. Teil 2 Auge. Springer, Berlin, pp 295-468

Konorski, J (1967) Integrative activity of the brain. Univ. Chicago Press, Chicago

Konstantinov AI, Movchan VN, Shibkov AA (1987) Functional properties of the auditory system and acoustic signalling in insectivores. J. evol. Biochem. Physiol. 23: 321-328

Kooy FH (1916) The inferior olive in vertebrates. Folia neuro-biol. 10: 205-369

Korf HW, Moller M, Gery I, Zigler JS, Klein DC (1985) Immunocytochemical demonstration of retinal S-antigen in the pineal organ of four mammalian species. Cell Tissue Res. 239: 81-85

Kotzenberg W (1899) Untersuchungen über das Rückenmark des Igels. Bergmann, Wiesbaden

Krabbe KH (1925) L'organe sous-commissural du cerveau chez les mammifères. Det. Kgl. Danske Videnskabernes Selskab. Biologiske Meddelelser 4: 3-83

Krabbe KH (1942) Studies on the morphogenesis of the brain in lower mammals (Morphogenesis of the vertebrate brain II). Munksgaard, Copenhagen

Krajci D, Malinsky J (1974) Morphology, ultrastructure and histochemistry of the spinal ganglia of normal and hibernating hedgehogs. Folia morphol. (Prague) 22: 344-347

Krasnoshchekova EI, Figurina II, Burikova NV (1987) Characteristics of the spatial organization of neurons in the neocortex of certain mammals. Vestn. Leningr. Univ. Biol. 3: 57-63

Krawczuk A (1983) Topography and structure of habenular nuclei in four species of Insectivora. Acta theriol. 28: 351-356

Kretschmann HJ, Weinrich W (1986) Neuroanatomy and Cranial Computed Tomography. Thieme, Stuttgart-New York

Kristiansson H, Erlinge S (1977) Rörelsen och aktivitetsomrade hos igelkotten. (Movements and home range of the hedgehog). Fauna och Flora 72: 149-155

Kristoffersson R, Ahlström A, Suomalainen P (1966) Studies on the physiology of the hibernating hedgehog. 4. Cerebral free amino acids in hibernating and non-hibernating animals. Ann. Acad. Sci. Fenn. A: IV. Biologica 104: 3-9

Kristoffersson R, Soivio A, Suomalainen P (1965) Studies on the physiology of the hibernating hedgehog. 3. Changes in the water content of various tissues in relation to seasonal and hibernation cycles. Ann. Acad. Sci. Fenn. A: IV. Biologica 92: 1-17

Kriszat G (1940a) Untersuchungen zur Sinnesphysiologie, Biologie und Umwelt des Maulwurfs (*Talpa europaea* L.). Z. Morphol. Ökol. Tiere 36: 446- 511

Kriszat G (1940b) Die Orientierung im Raume bei *Talpa europaea* L. Z. Morphol. Ökol. Tiere 36: 512-556

Kriszat G (1940c) Wie orientiert sich der Maulwurf in seinen Gängen. Umschau 44: 561-565

Krompecher S, Lipak J (1966) A simple method for determining cerebralization. Brain weight and intelligence. J. comp. Neurol. 127: 113-120

Krumbiegel I (1961) Rückbildungserscheinungen im Tierreich. Neue Brehm Bücherei. Ziemsen, Wittenberg

Kruska D, Stephan H (1973) Volumenvergleich allokortikaler Hirnzentren bei Wild- und Hausschweinen. Acta anat. 84: 387-415

Kryspin-Exner W (1922) Vergleichend-anatomische Studien über die substantia perforata anterior der Säugetiere. Arb. Neurol. Inst. Wien 23: 148-187

Kudo M, Nakamura Y, Tokuno H, Kitao Y, Moriizumi T (1989) Auditory brainstem in the mole (*Mogera*): Anatomical basis for possible hearing specialization of the underground dweller. Soc. Neurosci. Abstr. 15: 744

Künzle H (1976) Thalamic projections from the precentral motor cortex in *Macaca fascicularis*. Brain Res. 105: 253-267

Künzle H (1988) Retinofugal projections in hedgehog-tenrecs (*Echinops telfairi* and *Setifer setosus*). Anat. Embryol. 178: 77-93

Kuhlenbeck H (1975) The Central Nervous System of Vertebrates. Vol 4, Spinal Cord and Deuterencephalon. Karger, Basel

Kuhn HJ (1964) Zur Kenntnis von *Micropotamogale lamottei* Heim de Balsac, 1954. I. Z. Säugetierkd. 29: 152-173

Kulshreshtha A, Dominic CJ (1971) Inhibition by reserpine of histological changes induced by hypertonic saline in the hypothalamic neurosecretory system of the musk shrew, *Suncus murinus* L. J. Endocr. 50: 707- 708

Kuramoto K, Nishida T, Mochizuki M (1980) Morphological studies on the nasal turbinates of the musk shrew (*Suncus murinus riukiuanus*). Jap. J. vet. Sci. 42: 377-380

Kurepina MM (1974) Einige Angaben zur Cytoarchitektonik von Projektions- und Assoziationsformationen des Neocortex bei Insektivoren. J. Hirnforsch. 15: 283-298

Lande R (1979) Quantitative genetic analysis of multivariate evolution applied to brain: body size allometry. Evolution 33: 402-416

Lapicque ML (1912) Remarques sur la série de pesées encéphaliques recueilles au Dahomey par M. Waterlot. Bull. Mus. Hist. Nat., Paris 18: 495-497

Larochelle R (1987) Aspects morphologiques et éthologiques de l'olfaction chez quatre espèces de musaraignes (Soricidae) nord-américaines. Thèse, Montreal

Larochelle R, Baron G (1986) Discrimination des odeurs d'espèces différentes chez la musaraigne à queue courte *Blarina brevicauda* (Say). Naturaliste can. (Rev. Écol. Syst.) 113: 251-256

Larochelle R, Baron G (1989a) Morphometric study of the olfactory system in red-toothed shrews. In: Splechtna H, Hilgers H (eds) Fortschritte der Zoologie, Vol 35. Fischer, Stuttgart, pp. 360-362

Larochelle R, Baron G (1989b) Comparative morphology and morphometry of the nasal fossae of four species of North American shrews (Soricinae). Amer. J. Anat. 186: 306-314

Larsell O (1934) Morphogenesis and evolution of the cerebellum. Arch. Neurol. Psychiat. 31: 373-395

Larsell O, Dow RS (1935) The development of the cerebellum in the bat (*Corynorhinus* sp.) and certain other mammals. J. comp. Neurol. 62: 443-468

Leche W (1897) Zur Morphologie des Zahnsystems der Insectivoren. Anat. Anz. 13: 1-11

Leche W (1902) Zur Entwicklungsgeschichte des Zahnsystems der Säugethiere. Theil 2: Phylogenie, Heft 1: Die Familie der Erinaceidae. Bibliotheca Zoologica 15. Nägele, Stuttgart

Leche W (1905) Ein eigenartiges Säugetierhirn, nebst Bemerkungen über den Hirnbau der Insectivora. Anat. Anz. 26: 577-589

Leche W (1907) Zur Entwicklungsgeschichte des Zahnsystems der Säugetiere. Teil 2: Phylogenie. Heft 2: Die Familien der Centetidae, Solenodontidae und Crysochloridae. Bibliotheca Zoologica 20: 1-158

Lee RJ (1870) On the organs of vision in the common mole. Proc. Roy. Soc., London. 18: 322-327

Legait H, Bauchot R, Stephan H, Contet-Audonneau JL (1976) Étude des corrélations liant le volume de l'épiphyse aux poids somatique et encéphalique chez les rongeurs, les insectivores, les chiroptères, les prosimiens et les simiens. Mammalia 40: 327-337

Lekagul B, McNeely JA (1977) Mammals of Thailand. Assoc. Conserv. Wildlife, Bangkok

Lende RA (1969) A comparative approach to the neocortex: Localization in monotremes, marsupials and insectivores. Ann. N.Y. Acad. Sci. 167: 262-276

Lende RA, Sadler KM (1967) Sensory and motor areas in neocortex of hedgehog (*Erinaceus*). Brain Res. 5: 390-405

Leszlenyi O (1912) Vergleichend-anatomische Studie über die Lissauersche Randzone des Hinterhorns. Arb. Neurol. Inst. Wien 19: 252-304

Lewis TH (1983) The anatomy and histology of the rudimentary eye of *Neurotrichus*. Northwest Science 57: 8-15

Lim BL (1967) Note on the food habits of *Ptilocercus lowii* Gray (Pentail tree-shrew) and *Echinosorex gymnurus* (Raffles) (Moonrat) in Malaya with remarks on "ecological labelling" by parasite patterns. J. Zool., London 152: 375-379

Lim BL, Heyneman D (1968) A collection of small mammals from Tuaran and the southwest face of Mt. Kinabalu, Sabah. Sarawak Mus. J. 16: 257- 276

Lindemann W (1951) Zur Psychologie des Igels. Z. Tierpsychol. 8: 224-251

Linowiecki AJ (1914) The comparative anatomy of the pyramidal tract. J. comp. Neurol. 24: 509-530

Löwy R (1916) Über die Faseranatomie und Physiologie der Formatio vermicularis cerebelli. Arb. Neurol. Inst. Wien 21: 359- 382

Lopez-Mascaraque L, DeCarlos JA, Valverde F (1986) Structure of the olfactory bulb of the hedgehog (*Erinaceus europaeus*): Description of cell types in the granular layer. J. comp. Neurol. 253: 135-152

Lorente de Nó R (1922) Contribution al conocimento del nervio trigemino. Libro en honor Ramon y Cajal 2: 13

Lorente de Nó R (1934) Studies on the structure of the cerebral cortex. II. Continuation of the study of the ammonic system. J. Psychol. Neurol. 46: 113-177

Lorenz K (1959) Psychologie und Stammesgeschichte. In: Heberer G (ed) Die Evolution der Organismen. Fischer, Stuttgart, pp 131-172

Loughlin SE, Fallon JH (1985) Locus coeruleus. In: Paxinos G (ed) The Rat Nervous System. Vol 2, Hindbrain and spinal cord. Academic Press, Sydney, pp 79-93

Lund RD, Lund JS (1965) The visual system of the mole, *Talpa europaea*. Exp. Neurol. 13: 302-316

Lund RD, Lund JS (1966) The central visual pathways and their functional significance in the mole (*Talpa europaea*). J. Zool., London 149: 95-101

Lund RD, Webster KE (1967) Thalamic afferents from the spinal cord and trigeminal nuclei. An experimental anatomical study in the rat. J. comp. Neurol. 130: 313-327

Mace GM, Harvey PH, Clutton-Brock TH (1981) Brain size and ecology in small mammals. J. Zool. (London) 193: 333-354

Malinska J, Malinsky J (1974a) Quantitative changes in the spinal cord motor neurons of the hibernating hedgehog. Folia morphol. (Prague) 22: 341- 343

Malinska J, Malinsky J (1974b) Changes of motor neurons in the spinal cord of hedgehog during hibernation as seen under the light and electron microscope. Acta Univ. Palackianae Olomucensis, Fac. med. 71: 347-354

Malinska J, Malinsky J (1985) Changes in the hypothalamus of the hibernating hedgehog. Folia morphol. (Prague) 33: 209-212

Malinska et al. in cronological order:

Malinska J, Kapoun S, Malinsky J (1972) Topography of the spinal cord in the east central European hedgehog (*Erinaceus roumanicus centroeuropaeus*). Folia morphol. (Prague) 20: 182-184

Malinska J, Kapoun S, Malinsky J (1974a) A topographical and quantitative anatomical study of the spinal cord in the hedgehog. Acta Univ. Palackianae Olomucensis, Fac. med. 71: 331-338

Malinska J, Malinsky J, Krajci D (1974b) Light, electron microscopical, and histochemical study of motor nerve cells in spinal cord of hibernating and non-hibernating hedgehog. Activitas Nervosa Superior (Praha) 16: 105

Malinska J, Hubackova E, Malinsky J (1976) A topographical and quantitative anatomical study of the spinal cord in the mole. Acta Univ. Palackianae Olomucensis, Fac. med. 79: 169-178

Malinska J, Malinsky J, Gregor R (1985) Angioarchitectonics of hedgehog diencephalon. Acta Univ. Palackianae Olomucensis, Fac. med. 108: 81-91

Malinsky J (1973) Fine structure of synapses in the spinal cord of Insectivora and Chiroptera. Folia morphol. (Prague) 21: 130-131

Malinsky J, Krajci D (1974) Histochemistry of enzymes in the spinal cord of the hedgehog. Folia morphol. (Prague) 22: 348-349

Malinsky J, Malinska J (1973) A comparative anatomical and ultrastructural study of the grey matter of the spinal cord in hedgehog and bat. Activitas Nervosa Superior (Praha) 15: 30-31

Malinsky J, Malinska J (1975) Changes of the motoneurons and synapses in the spinal cord of the hedgehog during hibernation. Proc. Xth Internat. Congress Anatomists, Tokyo : 151

Malinsky J, Sedlak C (1981) Quantitative assessment of cytoarchitectonics of hedgehog cerebral cortex. Acta Univ. Palackianae Olomucensis, Fac. med. 97: 111-120

Malinsky J, Sedlak C (1983) Changes of the rhinencephalic neurons in the hibernating hedgehog. Folia morphol. (Prague) 31: 407-409

Malinsky J, Vodvarka T (1983) Cerebral ventricles of the hedgehog brain. Acta Univ. Palackianae Olomoucensis., Fac. med. 104: 195-203

Malyukova IV (1984) Basic principles of structural-functional organization of complex forms of behavior in some vertebrates. J. evol. Biochem. Physiol. 19: 336-341

Mangold-Wirz K (1966) Cerebralisation und Ontogenesemodus bei Eutherien. Acta anat. 63: 449-508

Mann G (1896) On the homoplasty of the brain of Rodents, Insectivores and Carnivores. J. Anat. Physiol., New Series X, London 30: 1-35

Marsden CD, Rowland R (1965) The mammalian pons, olive and pyramid. J. comp. Neurol. 124: 175-187

Martin RD (1981) Relative brain size and basal metabolic rate in terrestrial vertebrates. Nature 293: 57-60

Martin RD (1982) Allometric approaches to the evolution of the primate nervous system. In: Armstrong E, Falk D (eds) Primate Brain Evolution, Methods and Concepts. Plenum Press, New York-London, pp 39-56

Maser C, Hooven EF (1974) Notes on the behavior and food habits of captive pacific shrews, *Sorex pacificus pacificus*. Northwest Sci. 48: 81-95

Masterton B, Heffner H, Ravizza R (1969) The evolution of human hearing. J. Acoustic Soc. Am. 45: 966-985

Masterton B, Skeen LC (1972) Origins of anthropoid intelligence: I. Prefrontal system and delayed alternation in hedgehog, tree shrew, and bush baby. J. comp. physiol. Psychol. 81: 423-433

Masterton B, Skeen LC, RoBards MJ (1974) Origins of anthropoid intelligence. II. Pulvinar-extrastriate system and visual reversal learning. Brain Behav. Evol. 10: 322-353

Masterton B, Thompson GC, Bechtold JK, RoBards MJ (1975) Neuroanatomical basis of binaural phase-difference analysis for sound localization: A comparative study. J. comp. physiol. Psychol. 89: 379-386

Matano S, Stephan H, Baron G (1985a) Volume comparisons in the cerebellar complex of Primates. I. Ventral pons. Folia primatol. 44: 171-181

Matano S, Baron G, Stephan H, Frahm HD (1985b) Volume comparisons in the cerebellar complex of Primates. II. Cerebellar nuclei. Folia primatol. 44: 182-203

Mayr E (1976) Evolution and the Diversity of Life. Belknap Press, Cambridge (Mass.)-London

McDowell SB (1958) The Greater Antillean insectivores. Bull. Amer. Mus. Nat. Hist. 115: 113-214

McHaffie JG, Stein BE (1982) Eye movements evoked by electrical stimulation in the superior colliculus of rats and hamsters. Brain Res. 247: 243-253

Medway Lord (1967) A bat-eating bat, *Megaderma lyra* Geoffroy. Malay. Nat. J. 20: 107-110

Medway Lord (1969) The wild mammals of Malaya and offshore islands including Singapore. Oxford University Press, Kuala Lumpur-London

Mehler WR (1971) Idea of a new anatomy of the thalamus. J. Psychiat. Res. 8: 203-217

Mehler WR, Rubertone JA (1985) Anatomy of the vestibular nucleus complex. In: Paxinos G (ed) The Rat Nervous System. Vol 2, Hindbrain and spinal cord. Academic Press, Sydney, pp 185-219

Meibach RC, Siegel A (1975) The origin of fornix fibers which project to the mammillary bodies in the rat: a horseradish peroxidase study. Brain Res. 88: 508-512

Mellanby K (1966) Mole activity in woodlands, fens and other habitats. J. Zool., London 149: 35-41

Menétrey D, Roudier F, Besson JM (1983) Spinal neurons reaching the lateral reticular nucleus as studied in the rat by retrograde transport of horseradish peroxidase. J. comp. Neurol. 220: 439-452

Menner E (1929) Untersuchungen über die Retina mit besonderer Berücksichtigung der äußeren Körnerschicht. Z. vergl. Physiol. 8: 761-826

Merritt JF (1986) Winter survival adaptations of the short-tailed shrew (*Blarina brevicauda*) in an appalachian montane forest. J. Mammal. 67: 450- 464

Mezhzherin VA (1964) Dehnel's phenomenon and its possible explanation. Acta theriol. 8: 95-114

Mezhzherin VA, Melnikova GL (1966) Adaptive importance of seasonal changes in some morphophysiological indices in shrews. Acta theriol. 11: 503-521

Michielsen NC (1966) Intraspecific and interspeccific competition in the shrews *Sorex araneus* L. and *S. minutus* L. Arch. neerl. Zool. 17: 73-174

Middleton AD (1931) A contribution to the biology of the common shrew, *Sorex araneus* Linnaeus. Proc. Zool. Soc., London: 133-143

Miller GS (1942) Zoological results of the George Vanderbilt Sumatran Expedition, 1936-1939. Part V.-Mammals collected by Frederick A. Ulmer, Jr. on Sumatra and Nias. Proc. Acad. Sci. Phil. 94: 107-165

MilneEdwards A, Grandidier A (1872) Description d'un nouveau mammifère insectivore de Madagascar (*Geogale aurita*). Ann. Sci. Nat. Zool. 15: Article 19

Misek I (1988) Developmental anatomy of mole *Talpa europaea* (Insectivora, Talpidae). VIII. Ontogeny of visual organ. Folia Zool. 37: 333-342

Misek I (1989) Developmental anatomy of the mole *Talpa europaea* (Insectivora, Talpidae). IX. Prenatal development of the neocortex. Folia Zool. 38: 157-166

Misek I (1990a) Morphogenesis of the neocortex in *Sorex araneus*. Folia Zool. 39: 85-96

Misek I (1990b) Morphogenesis of the mammalian neocortex. II. Prenatal development of the neocortex in selected mammals. Acta Sc. Nat. Brno 24: 1-40

Mishkin M, Aggleton J (1981) Multiple functional contributions of the amygdala in the monkey. In: Ben-Ari Y (ed) The Amygdaloid Complex. Elsevier/North Holland, Amsterdam-New York, pp 409-420

Möhres FP, Kulzer E (1956a) Über die Orientierung der Flughunde (Chiroptera-Pteropodidae). Z. vergl. Physiol. 38: 1-29

Möhres FP, Kulzer E (1956b) Untersuchungen über die Ultraschallorientierung von vier afrikanischen Fledermausfamilien. Verh. Dtsch. Zool. Ges. Erlangen 1955, Zool. Anz. (Suppl.) 19: 59-65

Mohanakumar KP, Sood PP (1980) Histological and histoenzymological studies on the medulla oblongata and pons of hedgehog (*Paraechinus micropus*). Acta morph. neerl.-scand. 18: 291-304

Mohanty N, Kar S (1982) A biochemical study of the musk shrew, *Suncus murinus*. Pranikee 3: 70-77

Mohr E (1936a) Die äußere Nase bei Igel und Maulwurf. Zool. Anz. (Leipzig) 113: 93-95

Mohr E (1936b) Biologische Beobachtungen an *Solenodon paradoxus* Brandt in Gefangenschaft. I. Zool. Anz. (Leipzig) 113: 177-188

Mohr E (1936c) Biologische Beobachtungen an *Solenodon paradoxus* Brandt in Gefangenschaft. II. Zool. Anz. (Leipzig) 116: 65-76

Mohr E (1936d) Biologische Beobachtungen an *Solenodon paradoxus* Brandt in Gefangenschaft. III. Zool. Anz. (Leipzig) 117: 233-241

Mohr E (1938) Biologische Beobachtungen an *Solenodon paradoxus* Brandt in Gefangenschaft. IV. Zool. Anz. (Leipzig) 122: 132-143

Moore JK, Karapas F, Moore RY (1977) Projections of the inferior colliculus in insectivores and primates. Brain Behav. Evol. 14: 301-327

Moore RY (1973) Retinohypothalamic projection in mammals: a comparative study. Brain Res. 49: 403-409

Morrison PR, Pearson OP (1946) The metabolism of a very small mammal. Science 104: 287-289

Morrison PR, Ryser FA, Dawe AR (1959) Studies on the physiology of the masked shrew *Sorex cinereus*. Physiol. Zool.(Chic.) 32: 256-271

Mott FW (1907) The progressive evolution of the structure and functions of the visual cortex in mammalia. Arch. Neurol. 3: 1-48

Müller F, O'Rahilly R (1980) The early development of the nervous system in staged insectivore and primate embryos. J. comp. Neurol. 193: 741-751

Myrcha A (1967) Comparative studies on the morphology of the stomach in the Insectivora. Acta theriol. 12: 223-244

Naik DR, Dominic CJ (1971) Lead haematoxylin-positive cells in the hypophysis of the musk shrew, *Suncus murinus* L. (Insectivora). J. Reprod. Fertil. 27: 298

Naik DR, Dominic CJ (1972) The pituitary gland of the musk shrew, *Suncus murinus* L. (Insectivora), with special reference to the cytology of the adenohypophysis. Am. J. Anat. 134: 145-165

Naik DR, Dominic CJ (1978) Functional significance of the cells in the pars anterior of the pituitary gland of the musk shrew (*Suncus murinus* L.). J. Endocrin. 79: 91-102

Nakamura T (1930a) Der Rollersche Kern. Eine vergleichend anatomische Untersuchung. Arb. Neurol. Inst. Wien 32: 67-94

Nakamura T (1930b) Vergleichend-anatomische Untersuchungen über den sogenannten akzessorischen Abduzenskern. Arb. Neurol. Inst. Wien 32: 262-282

Nauta HJW, Cole M (1978) Efferent projections of the subthalamic nucleus: An autoradiographic study in monkey and cat. J. comp. Neurol. 180: 1-17

Negus V (1958) The comparative anatomy and physiology of the nose and paranasal sinuses. Livingstone, Edinburgh-London

Negus V (1965) The Biology of Respiration. Livingstone, Edinburgh-London

Nevo E (1979) Adaptive convergence and divergence of subterranean mammals. Ann. Rev. Ecol. Syst. 10: 269-308

Newman JR (1976) Population dynamics of the wandering shrew *Sorex vagrans*. Wasman J. Biol. 34: 235-250

Newman JR, Winans SS (1980) An experimental study of the ventral striatum of the golden hamster. II. Neuronal connections of the olfactory tubercle. J. comp. Neurol. 191: 193-212

Niethammer G (1970) Beobachtungen am Pyrenäen-Desman, *Galemys pyrenaica*. Bonner zool. Beitr. 21: 157-182

Niethammer J (1956) Das Gewicht der Waldspitzmaus, *Sorex araneus* Linné, 1758, im Jahreslauf. Säugetierkdl. Mitt. 4: 160-165

Niethammer J (1960) Über die Säugetiere der Niederen Tauern. Mitt. Zool. Mus. Berlin 36: 408-443

Niethammer J (1973) Zur Kenntnis der Igel (Erinaceidae) Afgahanistans. Z. Säugetierkd. 38: 271-276

Niethammer J (1977) Ein syntopes Vorkommen der Wasserspitzmäuse *Neomys fodiens* und *N. anomalus*. Z. Säugetierkd. 42: 1-6

Niethammer J (1978) Weitere Beobachtungen über syntope Wasserspitzmäuse der Arten *Neomys fodiens* und *N. anomalus*. Z. Säugetierkd. 43: 313-321

Nieuwenhuys R, Voogd J, VanHuijzen C (1978) The Human Central Nervous System. A Synopsis and Atlas. Springer, New York

Nieuwenhuys R, Voogd J, VanHuijzen C (1980) Das Zentralnervensystem des Menschen. Springer, Berlin-Heidelberg

Oakley DA (1979) Cerebral cortex and adaptive behaviour. In: Oakley DA, Plotkin HC (eds) Psychology in Progress: Brain, Behaviour and Evolution. Methuen & Co., London, pp 154-188

Ognev SI (1928) Mammals of Eastern Europe and Northern Asia. Vol 1, Insectivora and Chiroptera. Translation: 1962 Israel Program for Scientific Translations, Sivan Press, Jerusalem

Olszewski J (1950) On the anatomical and functional organization of the spinal trigeminal nucleus. J. comp. Neurol. 92: 401-413

Papadopoulos GC, Karamanlidis AN, Michaloudi H, Dinopoulos A, Antonopoulos J, Parnavelas JG (1985) The coexistence of oxytocin and cortico- tropin-releasing factor in the hypothalamus: an immunocytochemical study in the rat, sheep and hedgehog. Neuroscience Letters 62: 213-218

Papadopoulos GC, Karamanlidis AN, Dinopoulos A, Antonopoulos J (1986a) Somatostatinlike immunoreactive neurons in the hedgehog (*Erinaceus europaeus*) and the sheep (*Ovis aries*) central nervous system. J. comp. Neurol. 244: 174-192

Papadopoulos GC, Karamanlidis AN, Antonopoulos J, Dinopoulos A (1986b) Neurotensinlike immunoreactive neurons in the hedgehog (*Erinaceus europeus*) and the sheep (*Ovis aries*) central nervous system. J. comp. Neurol. 244: 193-203

Papez JW (1927) Subdivisions of the facial nucleus. J. comp. Neurol. 43: 159-191

Passingham RE (1975) The brain and intelligence. Brain, Behav. Evol. 11: 1-15

Patterson B (1957) Mammalian phylogeny. In: First Symposium on host specificity among parasites of Vertebrates. Institut de Zoologie, Université Neuchatel, pp 15-49

Pekelsky A (1922) Über die Kerne der Raphe und der benachbarten Anteile der retikulierten Substanz. Arb. Neurol. Inst. Wien 23: 21-73

Pernetta JC (1977) Anatomical and behavioural specialisations of shrews in relation to their diet. Can. J. Zool. 55: 1442-1453

Petrovicky P (1966) Formatio reticularis of the hedgehog (*Erinaceus europaeus*). Acta Univ. Carolinae Medica 12: 293-307

Petrovicky P (1971) Structure and incidence of Gudden's tegmental nuclei in some mammals. Acta anat. 80: 273-286

Petrovicky P, Kolesarova D (1989) Parabrachial nuclear complex. A comparative study of its cytoarchitectonics in birds and some mammals including man. J. Hirnforsch. 30: 539-550

Pevet P, Balemans MGM, DeReuver GF (1981) The pineal gland of the mole (*Talpa europaea* L.). VII. Activity of hydroxyindole-0-methyltransferase (HIOMT) in the formation of 5-methoxytryptophan, 5-methoxytryptamine, 5- methoxyindole-3-acetic acid, 5-methoxytryptophol and melatonin in the eyes and the pineal. J. Neural Transmission 51: 271-282

Peyre A, Herlant M (1961) Les modifications cytologiques de l'antéhypophyse du Desman (*Galemys pyrenaicus* G.). C. R. Acad. Sci. (Paris) 252: 463-465

Pieters A, Herlant M (1972) Modifications saisonnières des cellules à prolactine dans l'antéhypophyse de la Taupe male. C. R. Acad. Sci. (Paris) 274: 3002-3006

Pines JL (1927) Über ein bisher unbeachtetes Gebilde im Gehirn einiger Säugetiere: Das subfornicale Organ des III. Ventrikels. J. Psychol. Neurol. 34: 186-193

Pirlot P, Pottier J (1977) Encephalization and quantitative brain composition in bats in relation to their life-habits. Rev. Can. Biol. 36: 321-336

Pirlot P, Stephan H (1970) Encephalization in Chiroptera. Can. J. Zool. 48: 433-444

Pirogov AA (1977) Heterosensory interaction in neocortical neurons of the hedgehog *Erinaceus europaeus*. J. evol. Biochem. Physiol. 13: 334-340

Pirogov AA, Malyukova IV (1975) An investigation of projections of somatic afferentiation in neocortex of the hedgehog *Erinaceus europaeus*. J. evol. Biochem. Physiol. 11: 254-261

Pirogov AA, Malyukova IV (1976a) Heterosensory interaction in the neocortex of the hedgehog *Erinaceus europaeus* (russ.). In: Kreps EM (ed) Comparative Neurophysiology and Neurochemistry. Nauka, Leningrad, pp 136-143

Pirogov AA, Malyukova IV (1976b) Studies on the projections of teleceptive inputs in the neocortex of the hedgehog *Erinaceus europaeus* (russ.). In: Kreps EM (ed) Comparative Neurophysiology and Neurochemistry. Nauka, Leningrad, pp 144-151

Platt WJ, Blakley NR (1973) Short-term effects of shrew predation upon invertebrate prey. Proc. Iowa Acad. Sci. 80: 60-66

Platzer W, Firbas W (1966) Die Cochlea der Soricidae. (Ihre Windungszahl und absolute Größe). Anat. Anz. 118: 101-113

Poduschka W (1968) Über die Wahrnehmung von Ultraschall beim Igel, *Erinaceus europaeus roumanicus*. Z. vergl. Physiol. 61: 420-426

Poduschka W (1974) Das Paarungsverhalten des Großen Igel-Tenrek (*Setifer setosus*, Froriep 1806) und die Frage des phylogenetischen Alters einiger Paarungseinzelheiten. Z. Tierpsychol. 34: 345-358

Poduschka W (1977a) Insectivore communication. In: Sebeok TA (ed) How Animals Communicate. Indiana University Press, Bloomington-London, pp 600- 633

Poduschka W (1977b) Das Paarungsvorspiel des Osteuropäischen Igels (*Erinaceus e. roumanicus*) und theoretische Überlegungen zum Problem männlicher Sexualpheromone. Zool. Anz. (Jena) 199: 187-208

Poduschka W, Firbas W (1968) Das Selbstbespeicheln des Igels, *Erinaceus europaeus* Linné, 1758, steht in Beziehung zur Funktion des Jacobsonschen Organes. Z. Säugetierkd. 33: 160-172

Poduschka W, Poduschka C (1985) Zur Frage des Gattungsnamens von *"Geogale" aletris* Butler and Hopwood, 1957 (Mammalia: Insectivora) aus dem Miozän Ostafrikas. Z. Säugetierkd. 50: 129-140

Pokorny F (1926) Zur vergleichenden Anatomie der Hypophyse. Z. Anat. Entwickl.-Gesch. 78: 308-331

Poliakov GI Development and complication of the cortical part of the coupling mechanism in the evolution of verterbrates. J. Hirnforsch. 7: 253-273

Polyak S (1957) The vertebrate visual system. Univ. Chicago Press. : 1-1390

Pompeiano O, Brodal A (1957) Spino-vestibular fibers in the cat. An experimental study. J. comp. Neurol. 108: 353-381

Portmann A (1947) Études sur la cérébralisation chez les oiseaux. II. Les indices intra-cérébraux. Alauda, Rev. int. Ornithologie 15: 1-15

Portmann A (1962) Cerebralisation und Ontogenese. In: Bauer KF (ed) Medizinische Grundlagenforschung, Vol 4. Thieme, Stuttgart, pp 3-62

Pospisilova E, Malinsky J (1985) The accessory olfactory bulb of the hedgehog. Folia morphol. (Prague) 33: 34-37

Pospisilova E, Malinsky J (1986) Myelinated neurons in olfactory glomerules of the hedgehog. Verh. Anat. Ges. 80: 811-812

Powell TPS, Cowan WM (1962) An experimental study of the projection of the cochlea. J. Anat. 96: 269-284

Probst M (1900) Zur Anatomie und Physiologie experimenteller Zwischenhirnverletzungen. Dtsch. Z. Nervenheilk. 17: 141-168

Probst M (1901) Über den Hirnmechanismus der Motilität. Jb. Psychiat. Neurol. 20: 181-291

Pruitt WO (1954) Aging in the masked shrew, *Sorex cinereus cinereus* Kerr. J. Mammal. 35: 35-39

Pruitt WO (1957) A survey of the mammalian family Soricidae (Shrews). Säugetierkdl. Mitt. 5: 18-27

Pucek M (1965) Water contents and seasonal changes of the brain-weight in shrews. Acta theriol. 10: 353-367

Pucek Z (1955) Untersuchungen über die Veränderlichkeit des Schädels im Lebenszyklus von *Sorex araneus araneus* L. Ann. Univ. M. Curie-Sklodowska (Lublin). Sect. C. 9: 163-211

Pucek Z (1957) Histomorphologische Untersuchungen über die Winterdepression des Schädels bei *Sorex* L. und *Neomys* Kaup. Ann. Univ. M. Curie-Sklodowska (Lublin). Sect. C. 10: 399-425

Pucek Z (1963) Seasonal changes in the braincase of some representatives of the genus *Sorex* from the Palearctic. J. Mammal. 44: 523-536

Pucek Z (1964) Morphological changes in shrews kept in captivity. Acta theriol. 8: 137-166

Pucek Z (1965) Seasonal and age changes in the weight of internal organs of shrews. Acta theriol. 10: 369-438

Pucek Z (1970) Seasonal and age change in shrews as an adaptive process. Symp. Zool. Soc. Lond. 26: 189-207

Pucek Z, Markov G (1964) Seasonal changes in the skull of the common shrew from Bulgaria. Acta theriol. 9: 363-366

Purves HD, Bassett EG (1963) The staining reactions of pars intermedia cells and their differentiation from pars anterior cells. In: Benoit J, DaLage C (eds) Cytologie de l'adénohypophysis, Coll. Intern. CNRS Paris, pp 231-241

Putkonen P, Sarajas HSS, Suomalainen P (1964) Electrical activity of the olfactory bulb and cerebral cortex in hedgehogs arousing from hibernation. Ann. Acad. Sci. Fenn. A V, Medica 106/7: 1-11

Puttick GM, Jarvis JUM (1977) The functional anatomy of the neck and forelimbs of the cape golden mole, *Chrysochloris asiatica* (Lipotyphla: Chrysochloridae). Zool. Afr. 12: 445-458

Qayyum MA, Vidyasagar PSPV, Haleem Khan A (1975) Anatomical and neurohistological studies on the tongue of the Indian hedgehog, *Hemiechinus auratus*. Indian J. Zool. 3: 17-24

Quilliam TA (1964) Special features of the eye of the mole (*Talpa europaea*). Anat. Rec. 148: 396

Quilliam TA (1966a) The mole's sensory apparatus. J. Zool., London 149: 76-88

Quilliam TA (1966b) The problem of vision in the ecology of *Talpa europaea*. Exp. Eye Res. 5: 63-78

Quilliam TA, Armstrong J (1963) Mechanorezeptoren. Endeavour 22: 55-60

Rahm U (1960) Note sur les spécimens actuellement connus de *Micropotamogale (Mesopotamogale) ruwenzorii* et leur répartition. Mammalia 24: 511-515

Rahm U (1961) Beobachtungen an der ersten in Gefangenschaft gehaltenen *Mesopotamogale ruwenzorii* (Mammalia-Insectivora). Rev. Suisse Zool. 68: 73-90

Rami A, Bréhier A, Thomasset M, Rabié A (1987) The comparative immunocytochemical distribution of 28 kDA cholecalcin (CaBP) in the hippocampus of rat, guinea pig and hedgehog. Brain Res. 422: 149-153

Ravizza RJ, Diamond IT (1972) Projections from the auditory thalamus to neocortex in the hedgehog (*Paraechinus hypomelas*). Anat. Rec. 172: 390-391

Ravizza RJ, Heffner HE, Masterton B (1969) Hearing in primitive mammals, II: Hedgehog (*Hemiechinus auritus*). J. Aud. Res. 9: 8-11

Redlich E (1903) Zur vergleichenden Anatomie der Assoziationssysteme des Gehirns der Säugetiere. I. Das Cingulum. Arb. Neurol. Inst. Wien 10: 104-184

Reed CA (1951) Locomotion and appendicular anatomy in three soricoid insectivores. Am. Midl. Nat. 45: 513-671

Reep R (1984) Relationship between prefrontal and limbic cortex: A comparative anatomical review. Brain Behav. Evol. 25: 5-80

Regidor J, Divac I (1990) A bilateral thalamocortical projection in the hedgehog. Evolutionary implications. Personal communication

Rehkämper G (1981) Vergleichende Architektonik des Neocortex der Insectivora. Z. zool. Syst. Evol.-forsch. 19: 233-263

Rehkämper G, Stephan H, Poduschka W (1986) The brain of *Geogale aurita* Milne-Edwards and Grandidier 1872 (Tenrecidae, Insectivora). J. Hirnforsch. 27: 391-399

Reich Z (1909) Vom Aufbau der Mittelzone des Rückenmarks. Arb. Neurol. Inst. Wien 17: 314-357

Rempe U (1970) Morphometrische Untersuchungen an Iltisschädeln zur Klärung der Verwandtschaft von Steppeniltis, Waldiltis und Frettchen. Analyse eines "Grenzfalles" zwischen Unterart und Art. Z. wiss. Zool. 180: 185-367

Rempe U, Weber EE (1972) An illustration of the principal ideas of MANOVA. Biometrics 28: 235-238

Repenning CA (1967) Subfamilies and genera of the Soricidae. Classification, historical zoogeography, and temporal correlation of the shrews. Geological Survey Professional Paper 565. United States Government, Washington

Retzius G (1898) Zur äusseren Morphologie des Riechhirns der Säugethiere und des Menschen. Biologische Untersuchungen N.F. 8. Fischer, Jena

Ricardo JA (1980) Efferent connections of the subthalamic region in the rat. I. The subthalamic nucleus of Luys. Brain Res. 202: 257-271

Richard PB (1985) The sensorial world of the Pyrenean desman, *Galemys pyrenaicus*. Acta Zool. Fennica 173: 255-258

Richard PB, Viallard AV (1969) Le desman des Pyrénées (*Galemys pyrenaicus*): Premières notes sur sa biologie. Terre et Vie 3: 225-245

Ring G, Ganchrow D (1983) Projections of nucleus caudalis and spinal cord to brainstem and diencephalon in the hedgehog (*Erinaceus europaeus* and *Paraechinus aethiopicus*): a degeneration study. J. comp. Neurol. 216: 132-151

Rinvik E (1975) Demonstration of nigrothalamic connections in the cat by retrograde axonal transport of horseradish peroxidase. Brain Res. 90: 313-318

Ritter C (1898) Die Linse des Maulwurfes. Arch. mikr. Anat. 53: 397-403

Rochon-Duvigneaud A (1943) Les yeux et la vision des Vertébrés. Masson, Paris

Rochon-Duvigneaud A (1972) L'oeil et la vision. In: Grassé PP (ed) Traité de Zoologie. Vol 16. Masson, Paris, pp 607-703

Rose M (1912) Histologische Lokalisation der Großhirnrinde bei kleinen Säugetieren (Rodentia, Insectivora, Chiroptera). J. Psychol. Neurol. 19: 391-479

Rose M (1927a) Der Allocortex bei Tier und Mensch. I. Teil. J. Psychol. Neurol. 34: 1-111

Rose M (1927b) Die sog. Riechrinde beim Menschen und beim Affen. II. Teil des "Allocortex bei Tier und Mensch". J. Psychol. Neurol. 34: 261-401

Rose M (1928) Gyrus limbicus anterior and Regio retrosplenialis (Cortex holoprotoptychos quinquestratificatus). Vergleichende Architektonik bei Tier und Mensch. J. Psychol. Neurol. 35: 65-173

Rose M (1929) Die Inselrinde des Menschen und der Tiere. J. Psychol. Neurol. 37: 467-624

Rose S (1927) Vergleichende Messungen im Allocortex bei Tier und Mensch. J. Psychol. Neurol. 34: 250-255

Rosene DL, VanHoesen GW (1977) Hippocampal efferents reach widespread areas of cerebral cortex and amygdala in the rhesus monkey. Science 198: 315-317

Rubertone JA, Haines DE (1982) The vestibular complex in a prosimian primate (*Galago senegalensis*): Morphology and spinovestibular connections. Brain Behav. Evol. 20: 129-155

Rubertone JA, Mehler WR (1980) Afferents to the vestibular complex in rat. A horseradish peroxidase study. Soc. Neurosci. Abstr. 6: 225

Rudd RL (1965) Weight and growth in malaysian rain forest mammals. J. Mammal. 46: 588-594

Rudge MR (1968) The food of the common shrew *Sorex araneus* L. (Insectivora, Soricidae) in Britain. J. Anim. Ecol. 37: 565-581

Rütten A, Hewing M, Wittkowski W (1988) Seasonal ultrastructural changes of the hypophyseal pars tuberalis in the hedgehog (*Erinaceus europaeus* L.). Acta anat. 133: 217-223

Ryzen M, Campbell B (1955) Organization of the cerebral cortex. III. Cortex of *Sorex pacificus*. J. comp. Neurol. 102: 365-424

Saarikoski J (1967) Respiratory responses of cerebral cortex slices from a hibernating and a normal homeothermic mammal to electrical stimulation at 10-36°C. Ann. Acad. Sci. Fenn. A IV, Biologica 117: 3-22

Sabbath W (1909) Zur Histologie der vorderen Wurzeln des Rückenmarks der Säuger. Arb. Neurol. Inst. Wien 17: 175-189

Sachs E (1909) Eine vergleichende anatomische Studie des Thalamus opticus der Säugetiere. Arb. Neurol. Inst. Wien 17: 280-306

Sales G, Pye D (1974) Ultrasonic communication by animals. Chapman and Hall, London

Sando I (1965) The anatomical interrelationships of the cochlear nerve fibers. Acta otolaryng. 59: 417-436

Sanides D, Sanides F (1974) A comparative Golgi study of the neocortex in insectivores and rodents. Z. mikr.-anat. Forsch. 88: 957-977

Sanides F (1969) Comparative architectonics of the neocortex of mammals and their evolutionary interpretation. Ann. N.Y. Acad. Sci. 167: 404-423

Sanides F, Sanides D (1972) The "extraverted neurons" of the mammalian cerebral cortex. Z. Anat. Entwickl.-Gesch. 136: 272-293

Sano T (1909) Vergleichend-anatomische und physiologische Untersuchungen über die Substantia gelatinosa des Hinterhorns. Arb. Neurol. Inst. Wien 17: 1-71

Sas E, Sanides F (1970) A comparative Golgi study of Cajal foetal cells. Z. mikr.-anat. Forsch. 82: 385-396

Sato Y (1977) Comparative morphology of the visual system of some japanese species of Soricoidea (superfamily) in relation to life habits. J. Hirnforsch. 18: 531-546

Scalia F, Winans SS (1975) The differential projections of the olfactory bulb and accessory olfactory bulb in mammals. J. comp. Neurol. 161: 31-55

Schilder P (1910) Vergleichend histologische Untersuchungen über den Nucleus sacralis Stillingi. Arb. Neurol. Inst. Wien 18: 195- 206

Schmidt H, Lierse W (1968) Die Duraverhältnisse, Hirngefäße und Kapillardichte in den Gehirnen des europäischen Igels (*Erinaceus europaeus*) und des Wüstenigels (*Erinaceus algirus levaudeni*). Acta anat. 70: 465-492

Schmidt U (1979) Die Lokalisation vergrabenen Futters bei der Hausspitzmaus, *Crocidura russula* Hermann. Z. Säugetierkd. 44: 59-60

Schmidt U, Nadolski A (1979) Die Verteilung von olfaktorischem und respiratorischem Epithel in der Nasenhöhle der Hausspitzmaus, *Crocidura russula* (Soricidae). Z. Säugetierkd. 44: 18-25

Schneider GE (1969) Two Visual Systems. Brain mechanisms for localization and discrimination are dissociated by tectal and cortical lesions. Science 163: 895-902

Schroeder DM, Jane JA (1976) The intercollicular area of the inferior colliculus. Brain Behav. Evol. 13: 125-141

Schroeder DM, Jane JA, Nauta HJW (1970) Thalamic projections from the dorsal column and cerebellar nuclei and tectum in the hedgehog, *Erinaceus europaeus*. Anat. Rec. 166: 374

Schubarth H (1958) Zur Variabilität von *Sorex araneus araneus* L. Acta theriol. 2: 175-202

Schwarz S (1935/36) Über das Mausauge, seine Akkommodation, und über das Spitzmausauge. Jena. Z. Med. Naturw. 70: 113-158

Schwerdtfeger WK (1984) Structure and fiber connections of the hippocampus. A comparative study. Advances in Anatomy Embryology and Cell Biology 83: 1-74

Schwerdtfeger WK, Oelschläger HA, Stephan H (1984) Quantitative neuroanatomy of the brain of the La Plata dolphin, *Pontoporia blainvillei*. Anat. Embryol. 170: 11-19

Schwerdtfeger WK, Danscher G, Geiger H (1985) Entorhinal and prepiriform cortices of the european hedgehog. A histochemical and densitometric study based on a comparison between Timm's sulphide silver method and the selenium method. Brain Res. 348: 69-76

Sefton AJ, Dreher B (1985) Visual system. In: Paxinos G (ed) The Rat Nervous System. Vol 1, Forebrain and midbrain. Academic Press, Sydney, pp 169-221

Sharma DR (1958) Studies on the anatomy of the Indian Insectivore, *Suncus murinus*. J. Morph. 102: 427-553

Sheehan D (1933) Vergleichend anatomische Untersuchungen über die Kerne am Aquaeductus Sylvii. Arb. Neurol. Inst. Wien 35: 1-28

Shibkov AA (1979) The role of sensory systems in the proximal orientation of shrews of the genera *Sorex* and *Neomys* (russ.). Russkia Zool. Journ. 58: 76-81)

Shigenaga Y, Takabatake M, Sugimoto T, Sakai A (1979) Neurons in marginal layer of trigeminal nucleus caudalis projecting to ventrobasal complex (VB) and posterior nuclear group (PO) demonstrated by retrograde labeling with horseradish peroxidase. Brain Res. 166: 391-396

Shima R (1909) Zur vergleichenden Anatomie des dorsalen Vaguskerns. Arb. Neurol. Inst. Wien 17: 190-216

Shtark MB (1970/72) The Brain of Hibernating Animals. Nauka, Novosibirsk (russ.), Nasa TTF-619 (engl.)

Siegmund R, Kapischke HJ (1983) Untersuchungen zur Erfassung der motorischen und lokomotorischen Aktivität der Waldspitzmaus (*Sorex araneus* L.). Zool. Anz. (Jena) 210: 282-288

Siemen D (1976) Elektronenmikroskopische Untersuchungen zur Reduktion des Auges bei unterirdisch lebenden oder nachtaktiven Säugetieren: *Talpa europaea* (Linne, 1758), *Erinaceus europaeus* (Linne, 1758), *Echinops telfairi* (Martin, 1838). Dissertation, Kiel

Sigmund L (1985) Anatomy, morphometry and function of sense organs in shrews (Soricidae, Insectivora, Mammalia). In: Duncker HR, Fleischer G (eds) Functional Morphology in Vertebrates. Fortschritte der Zoologie 30. Fischer, Stuttgart, pp 661-665

Sigmund L, Claussen CP (1986) Zur Ultrastruktur der Retina von Waldspitzmaus (*Sorex araneus*) und Gartenspitzmaus (*Crocidura suaveolens*). Verh. Anat. Ges. 80: 817-818

Sigmund L, Sedlacek F (1985) Morphometry of the olfactory organ and olfactory thresholds of some fatty acids in *Sorex araneus*. Acta Zool. Fennica 173: 249-251

Sigmund L, Druga R, Siegmund R (1984) Retinal projections in *Crocidura suaveolens* (Soricidae, Insectivora, Mammalia). The primary optic pathway. Vest. cs. Spolec. zool. 48: 296-301

Sigmund L, Siegmund R, Claussen CP (1987) Bau und Funktion der optischen Sinnesorgane bei der Waldspitzmaus (*Sorex araneus*) und der Gartenspitzmaus (*Crocidura suaveolens*) und ihre Beziehung zum lokomotorischen Verhalten. Zool. Jb. (Physiol.) 91: 63-78

Siivonen L (1954) Über die Größenvariation der Säugetiere und die *Sorex macropygmaeus* Mill.-Frage in Fennoskandien. Ann. Acad. Sci. Fenn. A IV, Biologica 21: 1-24

Simpson GG (1945) The principles of classification and a classification of mammals. Bull. Amer. Mus. Nat. Hist. 85

Skeen LC, Hall WC (1977) Efferent projections of the main and the accessory olfactory bulb in the tree shrew (*Tupaia glis*). J. comp. Neurol. 172: 1-35

Skeen LC, Masterton RB (1982) Origins of anthropoid intelligence. IV. Role of prefrontal system in delayed alternation and spatial reversal learning in a conservative eutherian (*Paraechinus hypomelas*). Brain Behav. Evol. 21: 185-198

Skrzypiec Z, Jastrzebski M (1980) Nerve centres of the cerebellum in the mole and common shrew. Acta Theriol. 25: 25-30

Slifer HF (1924) Relative brain weights in animals. Med. J. Rec. 119: 100

Slonaker JR (1902) The eye of the common mole, *Scalops aquaticus machrinus*. J. comp. Neurol. 12: 335-366

Slonaker JR (1920) Some morphological changes for adaption in the mole. J. Morph. 34: 335-363

Smit-Vis JH (1962) Some aspects of the hibernation in the European hedgehog *Erinaceus europaeus* L. Arch. neerl. Zool. 14: 513-597

Smith GE (1897) The relation of the fornix to the margin of the cerebral cortex. J. Anat. 32: 23-58

Smith GE (1902a) Notes on the brain of Macroscelides and other Insectivora. J. Linn. Soc. (Zool.) 28: 443-448

Smith GE (1902b) Descriptive and illustrated catalogue of the physiological series of comparative anatomy in the Museum of the Royal College of Surgeons of England. Vol 2, Mammalia. London

Smith GE (1903) On the morphology of the brain in the mammalia, with special reference to that of the lemurs, recent and extinct. Trans. Linn. Soc., London, ser. 2 (Zool.) 8: 319-432

Smith RL (1973) The ascending fiber projections from the principal sensory trigeminal nucleus in the rat. J. comp. Neurol. 148: 423-445

Smythies JR (1967) Brain mechanisms and behaviour. Brain 90: 697- 706

Snell O (1892) Die Abhängigkeit des Hirngewichts von dem Körpergewicht und den geistigen Fähigkeiten. Arch. Psychiat. Nervenkrankh. 23: 436-446

Söllner B, Kraft R (1980) Anatomie und Histologie der Nasenhöhle der Europäischen Wasserspitzmaus, *Neomys fodiens* (Pennant 1771), und anderer mitteleuropäischer Soriciden. Spixiana 3: 251-272

Sokolova ZA (1964) Reduction of the eye in moles as a result of function limitation. Folia biol. (Krakow) 12: 183-201

Sokolova ZA (1965) Reduction of eye in *Talpa europaea* L. and *Sorex araneus* L. (comparative study). Arkh. Anat. Gistol. Embriol. 48: 13-17 (russian)

Sood PP, Mohanakumar KP (1980) On the distribution of acetylcholinesterase in the medulla oblongata of hedgehog (*Paraechinus micropus*). Acta morph. neerl.-scand. 18: 281-290

Sood PP, Mohanakumar KP, Chauhan AK (1982) Glycosidases and lipid metabolism in the central nervous system of the hedgehog (*Paraechinus micropus*). Acta anat. 114: 339-346

Spatz H, Stephan H (1961) Adaptive Konvergenz von Schädel und Gehirn bei "Kopfwühlern". Zool. Anz. (Leipzig) 166: 402-423

Spiegel EA (1919) Die Kerne im Vorderhirn der Säuger. Arb. Neurol. Inst. Wien 22: 418-497

Spiegel EA, Zweig H (1919) Zur Cytoarchitektonik des Tuber cinereum. Arb. Neurol. Inst. Wien 22: 278-295

Spitzenberger F (1966) Die Alpenspitzmaus (*Sorex alpinus* Schinz, 1837) in Österreich. Ann. Naturhist. Mus. Wien 69: 313-321

Spitzenberger F (1978) Die Alpenspitzmaus (*Sorex alpinus* Schinz)- Mammalia austriaca 1 (Mamm., Insectivora, Soricidae). Mitt. Abt. Zool. Landesmus. Joanneum Graz 7: 145-162

Spitzenberger F (1980) Sumpf- und Wasserspitzmaus (*Neomys anomalus* Cabrera 1907 und *Neomys fodiens* Pennant 1771) in Österreich.(Mammalia austriaca 3). Mitt. Abt. Zool. Landesmus. Joanneum Graz 9: 1-39

Starck D (1961) Bemerkungen über Grabanpassungen bei Säugetieren und über die äußere Form des Gehirns von *Notoryctes typhlops* Stirling 1889 (Marsupialia, Notoryctidae). Zool. Anz. (Leipzig) 166: 423-432

Starck D (1974) Die Säugetiere Madagaskars, ihre Lebensräume und ihre Geschichte. Sitzungsber. Wiss. Ges. Univ. Frankfurt. 11: 1-62

Starck D (1978) Vergleichende Anatomie der Wirbeltiere auf evolutionsbiologischer Grundlage. Bd. 1: Theoretische Grundlagen. Stammesgeschichte und Systematik unter Berücksichtigung der niederen Chordata. Springer, Berlin-New York

Stein BE (1981) Organization of the rodent superior colliculus: some comparisons with other mammals. Behav. Brain Res. 3: 175-188

Stein BM, Carpenter MB (1967) Central projections of portions of the vestibular ganglia innervating specific parts of the labyrinth in the rhesus monkey. Am. J. Anat. 120: 281-317

Stengaard-Pedersen K, Fredens K, Larsson LI (1983) Comparative localization of enkephalin and cholecystokinin immunoreactivities and heavy metals in the hippocampus. Brain Res. 273: 81-96

Stengel E (1924) Vergleichend-anatomische Studien über die Kerne an der hinteren Kommissur und im Ursprungsgebiete des hinteren Längsbündels. Arb. Neurol. Inst. Wien 26: 419-454

Stephan H (1956a) Vergleichend-anatomische Untersuchungen an Insektivorengehirnen. I. Hirnform, palaeo-neocortikale Grenze und relative Zusammensetzung der Cortexoberfläche. Morph. Jb. 97: 77-122

Stephan H (1956b) Vergleichend-anatomische Untersuchungen an Insektivorengehirnen. II. Oberflächenmessungen am Allocortex im Hinblick auf funktionelle und phylogenetische Probleme. Morph. Jb. 97: 123-142

Stephan H (1959) Vergleichend-anatomische Untersuchungen an Insektivorengehirnen. III. Hirn-Körpergewichtsbeziehungen. Morph. Jb. 99: 853-880

Stephan H (1960a) Die quantitative Zusammensetzung der Oberflächen des Allocortex bei Insektivoren und Primaten. In: Tower DB, Schadé JP (eds) Structure and Function of the Cerebral Cortex. Proc. 2nd Intern. Meeting of Neurobiologists. Elsevier, Amsterdam, pp 51-58

Stephan H (1960b) Methodische Studien über den quantitativen Vergleich architektonischer Struktureinheiten des Gehirns. Z. wiss. Zool. 164: 143-172

Stephan H (1961) Vergleichend-anatomische Untersuchungen an Insektivorengehirnen. V. Die quantitative Zusammensetzung der Oberflächen des Allocortex. Acta anat. 44: 12-59

Stephan H (1963) Vergleichend-anatomische Untersuchungen am Uncus bei Insectivoren und Primaten. In: Bargmann W, Schadé JP (eds) The Rhinencephalon and Related Structures. Progr. Brain Res., Vol 3. Elsevier, Amsterdam, pp 111-121

Stephan H (1965) Der Bulbus olfactorius accessorius bei Insektivoren und Primaten. Acta anat. 62: 215-253

Stephan H (1966) Größenänderungen im olfaktorischen und limbischen System während der phylogenetischen Entwicklung der Primaten. In: Hassler R, Stephan H (eds) Evolution of the Forebrain. Thieme, Stuttgart, pp 377-388

Stephan H (1967a) Quantitative Vergleiche zur phylogenetischen Entwicklung des Gehirns der Primaten mit Hilfe von Progressionsindices. Mitt. Max-Planck-Ges. 1967: 63-86

Stephan H (1967b) Zur Entwicklungshöhe der Insektivoren nach Merkmalen des Gehirns und die Definition der "Basalen Insektivoren". Zool. Anz. (Leipzig) 179: 177-199

Stephan H (1969) Vergleichende metrische und morphologische Untersuchungen an den circumventriculären Organen bei Insektivoren und Primaten. In: Sterba G (ed) Zirkumventrikuläre Organe und Liquor. Fischer, Jena, pp 139-145

Stephan H (1972) Evolution of primate brains: A comparative anatomical investigation. In: Tuttle R (ed) The Functional and Evolutionary Biology of Primates. Aldine/Atherton Inc., Chicago, pp 155-174

Stephan H (1975) Allocortex. In: Bargmann W (ed) Handbuch der mikroskopischen Anatomie des Menschen, Vol 4, Part 9. Springer, Berlin-Heidelberg

Stephan H (1976) Vergleichende Anatomie des Allocortex. Verh. Anat. Ges. 70: 217-251

Stephan H (1977) Encephalisationsgrad südamerikanischer Fledermäuse und Makromorphologie ihrer Gehirne. Morph. Jb. 123: 151-179

Stephan H (1979) Comparative volumetric studies on striatum in Insectivores and Primates. Evolutionary aspects. Appl. Neurophysiol. 42: 78-80

Stephan H (1983) Evolutionary trends in limbic structures. Neuroscience & Biobehavioral Reviews 7: 367-374

Stephan H, Andy OJ (1962) The septum. A comparative study on its size in insectivores and primates. J. Hirnforsch. 5: 229-244

Stephan H, Andy OJ (1964) Quantitative comparisons of brain structures from insectivores to primates. Am. Zool. 4: 59-74

Stephan H, Andy OJ (1969) Quantitative comparative neuroanatomy of Primates: An attempt at a phylogenetic interpretation. In: Comparative and evolutionary aspects of the vertebrate central nervous system. Ann. N.Y. Acad. Sci. 167: 370-387

Stephan H, Andy OJ (1970) The allocortex in primates. In: Noback CR, Montagna W (eds) The Primate Brain. Advances in Primatology, Vol 1. Appleton-Century-Crofts, New York, pp 109-135

Stephan H, Andy OJ (1977) Quantitative comparison of the amygdala in insectivores and primates. Acta anat. 98: 130-153

Stephan H, Andy OJ (1982) General brain characteristics and septal areas of the insectivores. In: Crosby EC, Schnitzlein HN (eds) Comparative correlative neuroanatomy of the vertebrate telencephalon. MacMillan, New York, pp 525-564

Stephan H, Bauchot R (1959) Le cerveau de *Galemys pyrenaicus* Geoffroy, 1811 (Insectivora Talpidae) et ses modifications dans l'adaptation à la vie aquatique. Mammalia 23: 1-18

Stephan H, Bauchot R (1960) Les cerveaux de *Chlorotalpa stuhlmanni* (Matschie) 1894 et de *Chrysochloris asiatica* (Linné) 1758 (Insectivora, Chrysochloridae). Mammalia 24: 495-510

Stephan H, Bauchot R (1965) Hirn-Körpergewichtsbeziehungen bei den Halbaffen (Prosimii). Acta Zool. 46: 209-231

Stephan H, Bauchot R (1968a) Vergleichende Volumenuntersuchungen an Gehirnen europäischer Maulwürfe (Talpidae). J. Hirnforsch. 10: 247-258

Stephan H, Bauchot R (1968b) Gehirn und Endocranialausguß von *Desmana moschata* (Insectivora Talpidae). Morph. Jb. 112: 213-225

Stephan H, Kuhn HJ (1982) The brain of *Micropotamogale lamottei* Heim de Balsac, 1954. Z. Säugetierkd. 47: 129-142

Stephan H, Manolescu J (1980) Comparative investigations on hippocampus in Insectivores and Primates. Z. mikr.-anat. Forsch. 94: 1025-1050

Stephan H, Nelson JE (1981) Brains of Australian Chiroptera. I. Encephalisation and macromorphology. Aust. J. Zool. 29: 653-670

Stephan H, Pirlot P (1970) Volumetric comparisons of brain structures in bats. Z. zool. Syst. Evol.-forsch. 8: 200-236

Stephan H, Spatz H (1962) Vergleichend-anatomische Untersuchungen an Insektivorengehirnen. IV. Gehirne afrikanischer Insektivoren. Versuch einer Zuordnung von Hirnbau und Lebensweise. Morph. Jb. 103: 108-174

Stephan et al. 1970 - 1988 (in cronological order):

Stephan H, Bauchot R, Andy OJ (1970) Data on size of the brain and of various brain parts in insectivores and primates. In: Noback CR, Montagna W (eds) The Primate Brain. Advances in Primatology, Vol 1. Appleton-Century-Crofts, New York, pp 289-297

Stephan H, Pirlot P, Schneider R (1974) Volumetric analysis of pteropid brains. Acta anat. 87: 161-192

Stephan H, Frahm H, Bauchot R (1977) Vergleichende Untersuchungen an den Gehirnen madagassischer Halbaffen. I. Encephalisation und Makromorphologie. J. Hirnforsch. 18: 115-147

Stephan H, Baron G, Schwerdtfeger WK (1980) The Brain of the Common Marmoset (*Callithrix jacchus*). A Stereotaxic Atlas. Springer, Berlin-Heidelberg

Stephan H, Frahm H, Baron G (1981a) New and revised data on volumes of brain structures in insectivores and primates. Folia primatol. 35: 1-29

Stephan H, Nelson JE, Frahm HD (1981b) Brain size comparison in Chiroptera. Z. zool. Syst. Evol.-forsch. 19: 195-222

Stephan H, Baron G, Frahm HD (1982) Comparison of brain structure volumes in Insectivora and Primates. II. Accessory olfactory bulb (AOB). J. Hirnforsch. 23: 575-591

Stephan H, Baron G, Fons R (1984a) Brains of Soricidae. II. Volume comparison of brain components. Z. zool. Syst. Evol.-forsch. 22: 328-342

Stephan H, Frahm HD, Baron G (1984b) Comparison of brain structure volumes in Insectivora and Primates. IV. Non-cortical visual structures. J. Hirnforsch. 25: 385-403

Stephan H, Baron G, Frahm HD, Stephan M (1986a) Größenvergleiche an Gehirnen und Hirnstrukturen von Säugern. Z. mikr.-anat. Forsch. 100: 189-212

Stephan H, Kabongo K Mubalamata, Stephan M (1986b) The brain of *Micropotamogale ruwenzorii* (DeWitte and Frechkop, 1955). Z. Säugetierkd. 51: 193-204

Stephan H, Frahm HD, Baron G (1987a) Comparison of brain structure volumes in Insectivora and Primates. VII. Amygdaloid components. J. Hirnforsch. 28: 571-584

Stephan H, Frahm HD, Stephan M, Baron G (1987b) Brains of Vespertilionids. II. Vespertilioninae with special reference to *Tylonycteris*. Z. zool. Syst. Evol.-forsch. 25: 147-157

Stephan H, Frahm HD, Baron G (1987c) Brains of Vespertilionids. III. Comparative cytoarchitectonics of the hippocampus. Z. zool. Syst. Evol.-forsch. 25: 205-211

Stephan H, Baron G, Frahm HD (1988) Comparative size of brains and brain components. In: Steklis HD, Erwin J (eds) Comparative Primate Biology. Vol 4, Neurosciences. Liss, New York, pp 1-38

Stewart WA, King RB (1963) Fiber projections from the nucleus caudalis of the spinal trigeminal nucleus. J. comp. Neurol. 121: 271-286

Stieda L (1870) Studien über das centrale Nervensystem der Wirbelthiere. Z. wiss. Zool. 20: 273-456

Stolzenburg JU, Reichenbach A, Neumann M (1989) Size and density of glial and neuronal cells within the cerebral neocortex of various insectivorian species. Glia 2: 78-84

Stone RD (1986) The social ecology of the European mole (*Talpa europaea* L.) and the Pyrenean desman (*Galemys pyrenaicus* G.). Thesis, Aberdeen

Stone RD (1987) The activity patterns of the Pyrenean desman (*Galemys pyrenaicus*) (Insectivora: Talpidae), as determined under natural conditions. J. Zool., London 213: 95-106

Strick PL (1976) Anatomical analysis of ventrolateral thalamic input to primate motor cortex. J. Neurophysiol. 39: 1020-1031

Sugiura Y, Kitoh J (1984) The median and lateral substantia gelatinosa in the cervical cord of the musk shrew (*Suncus murinus*) and its synaptic composition. Anat. Embryol. 170: 21-28

Suzuki H, Kurosumi K (1972) Fine structure of the cutaneous nerve endings in the mole snout. Arch. histol. Jap. 34: 35-50

Swanson LW, Cowan WM (1975) Hippocampo-hypothalamic connections: Origin in subicular cortex, not ammon's horn. Science 189: 303-304

Swanson LW, Cowan WM (1977) An autoradiographic study of the organization of the efferent connections of the hippocampal formation in the rat. J. comp. Neurol. 172: 49-84

Sweet G (1909) The eyes of *Chrysochloris hottentota* and *C. asiatica*. Quart. J. micr. Sci. 53: 327-338

Swenson RS, Castro AJ (1983a) The afferent connections of the inferior olivary complex in rats: a study using the retrograde transport of horseradish peroxidase. Am. J. Anat. 166: 329-341

Swenson RS, Castro AJ (1983b) The afferent connections of the inferior olivary complex in rats. An anterograde study using autoradiographic and axonal degeneration techniques. Neuroscience 8: 259-275

Switzer RC, Johnson JI, Kirsch JAW (1980) Phylogeny through brain traits. Relation of lateral olfactory tract fibers to the accessory olfactory formation as a palimpsest of mammalian descent. Brain Behav. Evol. 17: 339-363

Switzer R, de Olmos J, Heimer L (1985) Olfactory system. In: Paxinos G (ed) The Rat Nervous System. Vol 1, Forebrain and midbrain. Academic Press, Sydney, pp 1-36

Szteyn S, Gawronska B, Szatkowski E (1987) Topography and structure of corpus striatum in Insectivora. Acta theriol. 32: 95-104

Taber E (1961) The cytoarchitecture of the brain stem of the cat. I. Brain stem nuclei of cat. J. comp. Neurol. 116: 27-69

Takagi J (1925) Studien zur vergleichenden Anatomie des Nucleus vestibularis triangularis. II. Vergleichende anatomische Untersuchungen über den Nucleus intercalatus und Nucleus praepositus, das dorsale Längsbündel von Schütz und das Triangularis-Inter-calatus-Bündel von Fuse. Arb. Neurol. Inst. Wien 27: 235-282

Takahashi D (1913) Zur vergleichenden Anatomie des Seitenhorns im Rückenmark der Vertebraten. Arb. Neurol. Inst. Wien 20: 62-83

Tartuferi F (1878) Le eminenze bigemine anteriori e il tratto ottico della *Talpa europea*. Riv. sper. Freniatria Medicina Legale 4: 52-68 and 281- 306

Tauber ES, Michel F, Roffwarg HP (1968) Preliminary note on the sleep and waking cycle in the desert hedgehog (*Paraechinus hypomelas*). Psychophysiol. 5: 201

Thenius E (1969) Stammesgeschichte der Säugetiere (einschließlich der Hominiden). In: Helmcke JG, Starck D, Wermuth H (eds) Handbuch der Zoologie. Eine Naturgeschichte der Stämme des Tierreiches, Vol 8, Parts 47 and 48. Gruyter & Co., Berlin, pp 1-722

Thenius E (1972) Grundzüge der Verbreitungsgeschichte der Säugetiere. Fischer, Stuttgart

Thenius E (1979) Stammesgeschichte der Insektenesser. In: Grzimek B (ed) Grzimeks Tierleben. Vol 10, Säugetiere 1. Deutscher Taschenbuch Verlag dtv, München, p 170

Thenius E (1980) Grundzüge der Faunen- und Verbreitungsgeschichte der Säugetiere. Fischer, Stuttgart

Thenius E, Hofer H (1960) Stammesgeschichte der Säugetiere. Springer, Berlin-Göttingen-Heidelberg

Thiessen DD (1977) Thermoenergetics and the evolution of pheromone communication. In: Sprague JM, Epstein AN (eds) Progress in Psychobiology and Physiological Psychology. Academic Press, New York-San Francisco-London, pp 91-191

Tigges J, Tigges M (1969) The accessory optic system in *Erinaceus* (Insectivora) and *Galago* (Primates). J. comp. Neurol. 137: 59-69

Tolchenova GA, Demyanenko GP (1975) Neuronal organization of the "association" area of the sea urchin cortex. Dokl. Akad. Nauk USSR, Biol. 224: 463-465

Torvik A (1957) The ascending fibers from the main trigeminal sensory nucleus. Am. J. Anat. 100: 1-15

Tracey DJ (1985) Somatosensory system. In: Paxinos G (ed) The Rat Nervous System. Vol 2, Hindbrain and spinal cord. Academic Press, Sydney, pp 129-152

Travers JB (1985) Organization and projections of the orofacial motor nuclei. In: Paxinos G (ed) The Rat Nervous System. Vol 2, Hindbrain and spinal cord. Academic Press, Sydney, pp 111-128

Travers JB, Norgren R (1983) Afferent projections to the oral motor nuclei in the rat. J. comp. Neurol. 220: 280-298

Turner W (1891) The convolution of the brain: A study in comparative anatomy. J. Anat. Physiol. 25: 105-153

Uuspää VJ (1963a) The 5-hydroxytryptamine content of the brain and some other organs of the hedgehog (*Erinaceus europaeus*) during activity and hibernation. Experientia 19: 156-158

Uuspää VJ (1963b) The catecholamine content of the brain and heart of the hedgehog (*Erinaceus europaeus*) during hibernation and in an active state. Ann. Med. Exp. Biol. Fenn. 41: 340-348

Valeton MT (1908) Beitrag zur vergleichenden Anatomie des hinteren Vierhügels des Menschen und einiger Säugetiere. Arb. Neurol. Inst. Wien 14: 29-75

Valverde F (1983) A comparative approach to neocortical organization based on the study of the brain of the hedgehog (*Erinaceus europaeus*). In: Grisolia, Samson, Norton, Reinoso-Suarez (eds) Ramon y Cajal's Contribution to the Neurosciences. Elsevier Science Publishers, Amsterdam, pp 149-170

Valverde F (1986) Intrinsic neocortical organization: some comparative aspects. Neuroscience 18: 1-23

Valverde F, Facal-Valverde MV (1986) Neocortical layers I and II of the hedgehog (*Erinaceus europaeus*). I. Intrinsic organization. Anat. Embryol. 173: 413-430

Valverde F, Lopez-Mascaraque L (1981) Neocortical endeavor: basic neuronal organization in the cortex of hedgehog. In: 11. Intern. Congr. Anat.: Glia and Neuronal Cell Biology. Liss, New York, pp 281-290

Valverde F, DeCarlos JA, Lopez-Mascaraque L, Donate-Oliver F (1986) Neocortical layers I and II of the hedgehog (*Erinaceus europaeus*). II. Thalamo-cortical connections. Anat. Embryol. 175: 167-179

Valverde F, Lopez-Mascaraque L, DeCarlos JA (1989) Structure of the nucleus olfactorius anterior of the hedgehog (*Erinaceus europaeus*). J. comp. Neurol. 279: 581-600

VanderVloet V (1906) Ueber den Verlauf der Pyramidenbahn bei niederen Säugetieren. Anat. Anz. 29: 113-132

VanValen L (1967) New paleocene Insectivores and Insectivore classification. Bull. Amer. Mus. Nat. Hist. 135: 217-284

VanValkenburg CT (1911a) Zur Kenntnis der Radix spinalis nervi trigemini. Mschr. Psychiat. Neurol. 29: 407-437

VanValkenburg CT (1911b) Zur vergleichenden Anatomie des mesencephalen Trigeminusanteils. Folia neuro-biol. 5: 360-418

VanVeen T, Vigh-Teichmann I, Vigh B, Hartwig HG (1986) Light and electron microscopy of S-antigen- and opsin-immunoreactive photoreceptors in the retina of turtle, chicken, and hedgehog. Exp. Biol. 45: 1-14

Vaughan TA (1985) Mammalogy (third edition). CBS College Publishing, Philadelphia

Verhaart WJC (1970) Comparative anatomical aspects of the mammalian brain stem and the cord. Vol 1, Text. Vol 2, Illustrations and Tables. Van Gorcum, Assen

Verheyen WN (1961a) Recherches anatomiques sur *Micropotamogale ruwenzorii*. 1. La morphologie externe, les viscères et l'organe génital mâle. 2. La myologie de la tête, du cou et de la patte antérieure. Bull. Soc. Roy. Zool. Anvers 21

Verheyen WN (1961b) Recherches anatomiques sur *Micropotamogale ruwenzorii*. 3. La myologie de la patte postérieure, de la région sacrale et de la queue. 4. Observations ostéologiques et considérations générales. Bull. Soc. Roy. Zool. Anvers 22

Verma K (1962) On the structure of the skulls of Indian hedgehogs. Mammalia 26: 362-401

Verrier ML (1935) Les variations morphologiques de la rétine et leurs conséquences physiologiques. A propos de la rétine d'une musaraigne (*Crocidura mimula* Miller). Annales des Sciences Naturelles, Zoologie 18: 205-216

Victorov IV (1966) Neuronal structure of anterior corpora bigemina in insectivora and rodents. Arkh. Anat. Gistol. Embriol. 51: 82-89

Vigh B, Vigh-Teichmann I (1981) Light- and electron-microscopic demonstration of immunoreactive opsin in the pinealocytes of various vertebrates. Cell Tissue Res. 221: 451-463

Vigh-Teichmann I, Vigh B, Gery I, VanVeen T (1986) Different types of pinealocytes as revealed by immunoelectron microscopy of anti-S- antigen and antiopsin binding sites in the pineal organ of toad, frog, hedgehog and bat. Exp. Biol. 45: 27-43

Viret J (1955) Ordre des Insectivores. Insectivores fossiles et évolution de l'ordre. In: Grassé PP (ed) Traité de Zoologie. Vol 17, Anatomie-Systématique-Biologie. Masson, Paris, pp 1698-1704

Vlasak P (1972) The biology of reproduction and post-natal development of *Crocidura suaveolens* Pallas, 1811 under laboratory conditions. Acta Universitas Carolinae - Biologica 1970: 207-292

Völsch M (1906) Zur vergleichenden Anatomie des Mandelkerns und seiner Nachbargebilde. I. Teil. Arch. mikr. Anat. 68: 573-683

Völsch M (1911) Zur vergleichenden Anatomie des Mandelkerns und seiner Nachbargebilde. II. Teil. Arch. mikr. Anat. 76: 373-52

Vogel P (1974) Note sur le comportement arboricole de *Sylvisorex megalura* (Soricidae, Insectivora). Mammalia 38: 171-176

Vogel P (1976) Energy consumption of European and African shrews. Acta theriol. 21: 195-206

Vogel P (1980) Metabolic levels and biological strategies in shrews. In: Schmidt-Nielsen K, Bolis L, Taylor CR (eds) Comparative Physiology: Primitive Mammals. Cambridge University Press, Cambridge, pp 170-180

Vogel P (1983) Contribution à l'écologie et à la zoogéographie de *Micropotamogale lamottei* (Mammalia, Tenrecidae). Terre et Vie 38: 37-49

Vogel P, Genoud M, Frey H (1981) Rythme journalièr d'activité chez quelques Crocidurinae Africains et Européens (Soricidae, Insectivora). Terre et Vie 35: 97-108

Vogt C, Vogt O (1907) Zur Kenntnis der elektrisch erregbaren Hirnrindengebiete bei den Säugetieren. J. Psychol. Neurol. (suppl) 8: 277-456

Voneida T, Nauta HJW, Jane JA (1969) Interhemispheric connections in the European hedgehog (*Erinaceus europaeus*). Anat. Rec. 163: 280

Vysinskaja GA (1961) Personal information given to Blinkov and Glezer (1968)

Walberg F (1952) The lateral reticular nucleus of the medulla oblongata in mammals. J. comp. Neurol. 96: 283-343

Walberg F, Bowsher D, Brodal A (1958) The termination of primary vestibular fibers in the vestibular nuclei in the cat. An experimental study with silver methods. J. comp. Neurol. 110: 391-419

Walberg F, Pompeiano O, Brodal A, Jansen J (1962) The fastigiovestibular projection in the cat. An experimental study with silver impregnation methods. J. comp. Neurol. 118: 49-75

Walker EP (1975) Mammals of the World (third edition). Johns Hopkins University Press, Baltimore-London

Wall PD, Taub A (1962) Four aspects of trigeminal nucleus and a paradox. J. Neurophysiol. 25: 110-126

Walter G (1861) Über den feineren Bau des Bulbus olfactorius. Virchows Arch. path. Anat. 22: 241-259

Walther Y, Bruns V (1987) Das Innenohr von *Sorex araneus* und *Crocidura russula* (Soricidae). 61. Vers. Dtsch. Ges. Säugetierkd. Berlin, p. 51

Warkany J (1924) Vergleichende anatomische Untersuchungen über die Beziehungen des Globus pallidus zur Substantia nigra. Arb. Neurol. Inst. Wien 25: 195-206

Warncke P (1908) Mitteilung neuer Gehirn- und Körpergewichtsbestimmungen bei Säugern, nebst Zusammenstellung der gesamten bisher beobachteten absoluten und relativen Gehirngewichte bei den verschiedenen Spezies. J. Psychol. Neurol. 13: 355-403

Waterlot G (1912) Déterminations de poids encéphaliques et de grandeurs oculaires chez quelques vertébrés du Dahomey. Bull. Mus. Hist. Nat., Paris 18: 491-494

Watson CRR, Switzer RC (1978) Trigeminal projections to cerebellar tactile areas in the rat - origin mainly from N. interpolaris and N. principalis. Neuroscience Letters 10: 77-82

Watson CRR, Sakai S, Armstrong W (1982) Organization of the facial nucleus in the rat. Brain Behav. Evol. 20: 19-28

Watson GA (1907) The mammalian cerebral cortex, with special reference to its comparative histology. I. - Order Insectivora. Arch. Neurol. (Mott) 3: 49-117

Weber E (1906) Über Beziehungen der Großhirnrinde zur unwillkürlichen Bewegung der Stacheln des Igels und Schwanzhaare von Katze, Eichhorn und Marder. Zentralblatt Physiol. 20: 353-358

Weber M (1896) Vorstudien über das Hirngewicht der Säugethiere. Festschrift f. Gegenbaur 3: 105-123

Weber M (1928) Die Säugetiere. Einführung in die Anatomie und Systematik der recenten und fossilen Mammalia. Vol 2, Systematischer Teil. Fischer, Jena

Webster WR (1985) Auditory system. In: Paxinos G (ed) The Rat Nervous System. Vol 2, Hindbrain and spinal cord. Academic Press, Sydney, pp 153-184

Weidenreich F (1899) Zur Anatomie der centralen Kleinhirnkerne der Säuger. Z. Morphol. Anthropol. 1: 259-312

Weiss E (1916) Zur vergleichenden Anatomie des Facialiskerns. Arb. Neurol. Inst. Wien 21: 51-78

Welcker H, Brandt A (1903) Gewichtswerthe der Körperorgane bei dem Menschen und den Thieren. Arch. Anthropol. 28: 1-89

Werkman H (1913) L'évolution ontogénique de la paroi antérieure du cerveau intermédiaire et des commissures du cerveau antérieur chez les mammifères inférieurs. Arch. neerl. sci. exact nat., III b 2: 1-90

West MJ (1990) A quantitative comparison of the hippocampal subdivisions of diverse species including hedgehogs, laboratory rodents, wild mice and men. In preparation

West MJ, Gaarskjaer FB, Danscher G (1984) The Timm-stained hippocampus of the european hedgehog: A basal mammalian form. J. comp. Neurol. 226: 477-488

West MJ, Schwerdtfeger WK (1985) An allometric study of hippocampal components. A comparative study of the brains of the European hedgehog (*Erinaceus europaeus*), the tree shrew (*Tupaia glis*), and the marmoset monkey (*Callithrix jacchus*). Brain Behav. Evol. 27: 93-105

Wiedemeyer GL (1974) Vergleichende stereologische Bestimmung der Massen und der Oberflächen der verschiedenen Großhirnanteile sowie der Zelldichte bei niedrigen Säugern (Opossum, Igel, Maulwurf). Z. mikr.-anat. Forsch. 88: 299-324

Williams EM (1909) Vergleichend anatomische Studien über den Bau und die Bedeutung der Oliva inferior der Säugetiere und Vögel. Arb. Neurol. Inst. Wien 17: 118-149

Winkler C, Potter A (1914) An anatomical guide to experimental researches on the cat's brain. Versluys, Amsterdam

Wirz K (1950) Studien über die Cerebraliation zur quantitativen Bestimmung der Rangordnung bei Säugetieren. Acta anat. 9: 134-196

Wöhrmann-Repenning A (1975) Zur vergleichenden makro- und mikroskopischen Anatomie der Nasenhöhle europäischer Insektivoren. Morph. Jb. 121: 698-756

Wöhrmann-Repenning A, Meinel W (1977) A comparative study on the nasal fossae of *Tupaia glis* and four Insectivores. Anat. Anz. 142: 331-345

Yalden DW (1966) The anatomy of mole locomotion. J. Zool., London 149: 55-64

Yaskin VA (1984) Seasonal changes in brain morphology in small mammals. In: Merritt JF (ed) Winter Ecology of Small Mammals. Carnegie Mus.Nat. Hist., pp 183-191

Yates TL (1984) Insectivores, Elephant Shrews, Tree Shrews, and Dermopterans. In: Anderson S, Jones JK (eds) Orders and Families of Recent Mammals of the world. John Wiley & Sons, New York, pp 117-144

Yellin AM, Jerison HJ (1980) Photically evoked potentials and afterpotentials recorded from the visual cortex of the unanesthetized hedgehog. Brain Res. 182: 79-84

Ziehen T (1897a) Das Centralnervensystem der Monotremen und Marsupialier. I. Teil. Makroskopische Anatomie. In: Semon R (ed) Zoologische Forschungsreisen. Fischer, Jena,

Ziehen T (1897b) Ueber die motorische Rindenregion von *Didelphys virginiana*. Zentralblatt Physiol. 11: 457-461

Ziehen T (1899a) Centralnervensystem. In: Bardeleben K (ed) Handbuch der Anatomie des Menschen. IV/1. Fischer, Jena, pp 1-576

Ziehen T (1899b) Zur vergleichenden Anatomie der Pyramidenbahn. Anat. Anz. 16: 446-452

Ziehen T (1899c) Ein Beitrag zur Lehre von den Beziehungen zwischen Lage und Funktion im Bereich der motorischen Region der Grosshirnrinde mit specieller Rücksicht auf das Rindenfeld des Orbicularis oculi. Arch. Physiol : 158-173

Ziehen T (1901) Über vergleichend-anatomische Gehirnwägungen. Mschr. Psychiat. Neurol. 9: 316-318

Zilles K, Rehkämper G, Schleicher A (1979) Autoradiographische und morphometrische Untersuchungen des visuellen Cortex bei einem basalen Insectivoren (*Echinops telfairi*). Verh. Anat. Ges. 73: 1085-1088

Zuckerkandl E (1887) Über das Riechzentrum. Eine vergleichend-anatomische Studie. Enke, Stuttgart

Zuckerkandl E (1907) Zur Anatomie und Entwicklungsgeschichte des Indusium griseum corporis callosi. Arb. Neurol. Inst. Wien 15: 17- 51

10 Tables

Table 1. List of abbreviations used for brain structures and key to tables containing data on these structures.
Size is given in mm^3 if not indicated otherwise.
N = number of species.

Abbr.	structure name	N	Size	Index	Aver. Index	%
			Table numbers			
LIN	linear brain measures (mm)	46	2	3	4	
BrW	brain weight (mg)	53	5,6	6	7	6,7
NET	net brain volume	50	8	6	7	8,9
VENT	ventricles	50	8	-	-	8,9
REST	rests	50	8	-	-	8,9
OBL	medulla oblongata	50	10	11	12	
OBL in	species with OBL subdivision	30	16	17	18	
TR	compl. sensorius trigeminalis	30	13	14	15	
TSO	ncl. tractus solitarii	30	13	14	15	
FUN	funicular nuclei	30	13	14	15	
FGR	ncl. fascicularis gracilis	30	13	14	15	
FCM	ncl. fasc. cuneatus medialis	30	13	14	15	
FCE	ncl. fasc. cuneatus externus	30	13	14	15	
RLN	relay nuclei	30				
REL	ncl. reticularis lateralis	30	16	17	18	
INO	complexus olivaris inferior	30	16	17	18	
PRP	ncl. prepositus hypoglossi	30	16	17	18	
COE	ncl. locus coeruleus	30	16	17	18	
MOT	motor nuclei	30	19	20	21	
V	ncl. mot. nervi trigemini	30	19	20	21	
VII	ncl. nervi facialis	30	19	20	21	
X	ncl. dorsalis mot. nervi vagi	30	19	20	21	
XII	ncl. nervi hypoglossi	30	19	20	21	
VC	vestibular complex	30	22	23	24	
VM	ncl. vestibularis medialis	30	22	23	24	
VI	ncl. vestibularis inferior	30	22	23	24	
VL	ncl. vestibularis lateralis	30	22	23	24	
VS	ncl. vestibularis superior	30	22	23	24	
AUD	auditory (acoustic) nuclei	30	25	26	27	
CON	cochlear nuclei	30	25	26	27	
DCO	ncl. cochlearis dorsalis	30	25	26	27	
VCO	ncl. cochlearis ventralis	30	25	26	27	
OLS	complexus olivaris superior	30	25	26	27	

Table 1. (cont.)

Abbr.	structure name	N	Table numbers Vol.	Index	Aver. Index
MES	mesencephalon	50	10	11	12
MES in species with MES subdivision		10	28	29	30
MTG	tegmentum mesencephali	10	28	29	30
NLL	ncl. lemnisci lateralis	10	28	29	30
MTC	tectum mesencephali	10	28	29	30
SUC	colliculi superiores	10	28	29	30
INC	colliculi inferiores	10	28	29	30
SCB	corpus subcommissuralis	23	40	41	42
CER	cerebellum	50	10	11	12
CER in species with CER ncl. measurements		10	31	32	33
VPO	ventral pons	2	31	32	33
TCN	total cerebellar nuclei	10	31	32	33
MCN	ncl. medialis cerebelli	10	31	32	33
ICN	ncl. interpositus cerebelli	10	31	32	33
LCN	ncl. lateralis cerebelli	10	31	32	33
DIE	diencephalon	50	10	11	12
DIE in species with DIE subdivision		10	34	35	36
ETH	epithalamus	10	34	35	36
HAM	ncl. habenularis medialis	38	40	41	42
EPI	epiphysis	38	40	41	42
THA	thalamus	10	34	35	36
CGM	corpus geniculatum mediale	10	37	38	39
CGL	corpus geniculatum laterale	10	37	38	39
GLD	dorsal part of CGL	10	37	38	39
GLV	ventral part of CGL	10	37	38	39
STH	subthalamus	10	34	35	36
PALL	pallidum	10	37	38	39
LUY	corpus luysi	10	37	38	39
HTH	hypothalamus	10	34	35	36
CAI	capsula interna	10	34	35	36
TRO	tractus opticus	10	34	35	36
TEL	telencephalon	50	10	11	12
BOL	olfactory bulbs	50	43	44	45
PAL	paleocortex	50	43	44	45
STR	striatum	50	43	44	45
SEP	septum	50	43	44	45
AMY	amygdala	50	43	44	45
HIP	hippocampus	50	43	44	45
SCH	schizocortex	50	43	44	45
NEO	neocortex	50	43	44	45

Table 1. (cont.)

Abbr.	structure name	N	Vol.	Index	Aver. Index
			Table numbers		
MOB	main olfactory bulb	50	46	47	48
L 1+2	ext. fiber + glomerular layers	43	46	47	48
L 3	external plexiform layer	43	46	47	48
L 4-6	mitral, int. plex., and gran. layers	43	46	47	48
PvZ	periventricular zone	43	46	47	48
AOB	accessory olfactory bulb	50	49	50	51
L1+2	vomeronasal fiber and glom. layers	41	49	50	51
L3-5	ext. plex., mitral and int. plex. layers	41	49	50	51
L6	granular layer	41	49	50	51
PAL	paleocortex	50	43	44	45
RB	regio retrobulbaris	50	52	53	54
PRPI	regio prepiriformis	50	52	53	54
TOL	tuberculum olfactorium	50	52	53	54
OLC	olfactory cortices	50	52	53	54
TRL	tractus olfactorius lateralis	50	52	53	54
COA	commissura anterior	50	52	53	54
SIN	substantia innominata	50	52	53	54
SEP	septum	50	43	44	45
SET	ncl. triangularis septi	23	40	41	42
SFB	corpus subfornicale	23	40	41	42
AMY	amygdala, amygdaloid complex	50	43	44	45
MAM	medial amygdaloid division	50	55	56	57
NTO	ncl. tract. olf. lateralis	38	55	56	57
LAM	lateral amygdaloid division	50	55	56	57
MCB	basal ncl., magnocell. part	38	55	56	57
HIP	hippocampus	50	43	44	45
HIP in species with HIP subdivision		10	58	59	60
HR	hipp. retrocommissuralis	10	58	59	60
HA	hipp. anterior (HS + HP)	10	58	59	60
HS	hipp. supracommissuralis	10	58	59	60
HP	hipp. precommissuralis	10	58	59	60
NEO	neocortex	50	43	44	45
NW	neocortical white matter	50	61	62	63
NG	neocortical grey matter	50	61	62	63
NG 1	neoc. grey, molecular layer	50	61	62	63
NG 2-6	neoc. grey, cell dense layers	50	61	62	63
ASG	area striata grey	10	64	64	64

Table 1. (cont.)

Abbr.	structure name	N	Vol.	Index	Aver. Index
			Table numbers		
Bauchot's data of diencephalic components (converted)					
ETH	epithalamus	20	65	66	67
THA	thalamus	20	65	66	67
STH	subthalamus	20	65	66	67
HTH	hypothalamus	20	65	66	67
RTH	ncl. reticularis	20	65	66	67
PTE	regio pretectalis	20	65	66	67
Bauchot's data of thalamic structures (converted)					
ANT	ncl. anterior thalami	20	68	69	70
MLT	ncl. mediales thalami	20	68	69	70
MNT	ncl. mediani thalami	20	68	69	70
DOT	ncl. dorsales thalami	20	68	69	70
VET	ncl. ventrales thalami	20	68	69	70
CGM	corpus geniculatum mediale	20	68	69	70
CGL	corpus geniculatum laterale	20	68	69	70
COM	area measures of commissures (mm^2)	43	71	72	73
Summarized size characteristics of all structures					74

Table 2. Linear measures of brains in mm

For proportion indices in species see Table 3
For average proportion indices in supraspecific taxonomic units see Table 4

	n	BL	HemL	BW	BH	n	CCL
Tenrecidae							
Tenrec ecaudatus	3	26.6	11.7	16.2	10.6	1	2.01
Setifer setosus	6	20.6	9.4	16.2	9.3	1	1.82
Hemicentetes semispin.	5	14.7	7.8	12.7	8.0	1	1.64
Echinops telfairi	6	16.4	7.9	11.7	7.1	1	1.24
Geogale aurita	1	8.5	4.4	7.0	4.0	1	0.33
Oryzorictes talpoides	1	12.6	7.3	11.6	6.9	1	2.68
Microgale dobsoni	2	14.6	7.7	11.0	7.0	1	3.13
M. talazaci	7	16.3	8.8	11.4	7.9	1	3.93
Limnogale mergulus	2	16.7	9.5	14.4	7.9	1	4.29
Potamogale velox	8	28.7	17.6	19.2	13.5	1	9.60
Micropotam. ruwenzorii	4	18.0	11.1	13.8	8.6	1	4.03
Chrysochloridae							
Chrysochloris asiatica	1	12.1	7.8	13.5	9.0	1	-.-
C. stuhlmanni	2	12.5	8.6	14.7	9.0	1	2.86
Solenodontidae							
Solenodon paradoxus	2	34.5	16.9	21.0	12.8	1	4.30
Erinaceidae							
Echinosorex gymnurus	3	35.2	19.9	23.7	15.5	1	7.53
Hylomys suillus	1	18.6	10.4	13.1	8.5	1	3.73
Atelerix algirus	4	29.9	16.4	19.4	11.8	1	3.49
Erinaceus europaeus	2	26.4	14.3	20.5	11.3	1	2.54
Hemiechinus auritus	7	23.9	12.8	17.5	10.7	1	3.61
Talpidae							
Desmana moschata	3	24.9	19.3	23.2	11.5	1	9.93
Galemys pyrenaicus	2	17.7	14.1	15.2	10.4	1	7.48
Talpa europaea	2	17.6	11.2	15.3	7.3	1	5.31
T. micrura	1	16.5	10.8	13.4	6.5	1	-.-
Parascalops breweri	6	17.2	10.7	13.8	7.4	1	4.19
Scalopus aquaticus	1	20.0	12.1	18.2	7.4	1	5.04

Table 2 (cont.)

	n	BL	HemL	BW	BH	n	CCL
Soricidae							
Sorex alpinus	-	-.-	-.-	-.-	-.-	1	2.48
S. araneus	2	9.7	6.7	9.2	4.8	1	2.39
S. cinereus	29	8.8	5.9	7.8	4.7	1	2.15
S. fumeus	2	10.3	6.9	8.7	5.4	1	2.86
S. minutus	-	-.-	-.-	-.-	-.-	1	1.85
Microsorex hoyi	7	7.8	4.9	6.2	3.8	1	1.91
Neomys anomalus	1	9.8	7.0	9.6	6.0	1	2.39
N. fodiens	4	11.1	7.4	10.3	6.0	1	2.59
Blarina brevicauda	40	11.8	7.5	11.6	6.3	1	3.18
Cryptotis parva	6	9.9	6.0	9.8	5.2	1	2.26
Anourosorex squamipes	7	11.2	6.8	11.9	6.3	1	1.87
Crocidura attenuata	1	10.0	5.6	8.8	4.4	1	-.-
C. flavescens	26	13.8	7.3	11.1	5.6	4	1.89
C. hildegardeae	4	10.5	5.4	8.3	4.5	1	-.-
C. russula	21	9.9	5.5	8.8	4.5	1	1.44
C. suaveolens	1	9.4	5.3	8.5	4.3	1	-.-
Suncus etruscus	6	6.7	3.8	5.8	2.8	1	0.97
S. murinus	13	14.4	7.1	10.8	5.1	1	2.24
Scutisorex somereni	4	14.8	8.1	11.9	7.0	1	2.16
Sylvisorex granti	6	9.2	5.2	7.9	4.5	1	1.45
S. megalura	11	10.2	5.3	7.9	4.6	1	1.68
Ruwenzorisorex suncoides	1	12.2	7.0	10.6	5.4	1	2.68
Myosorex babaulti	1	11.6	7.1	10.5	5.8	1	2.75

Number of individuals = n

Abbreviations:

BH	brain height
BL	brain length
BW	brain width
CCL	corpus callosum length
HemL	hemisphere length

Table 3. Proportion indices (= linear measures of brains and corpus callosum given in Table 2 related to the cube roots of the standard brain volumes)

For linear measures in species see Table 2
For average proportion indices in supraspecific taxonomic units see Table 4

	BrV	CR	BL/CR	HemL/CR	BW/CR	BH/CR	CCL/CR
Tenrecidae							
Tenrec ecaudatus	2498	13.57	1.96	0.86	1.19	0.78	0.15
Setifer setosus	1463	11.35	1.81	0.83	1.43	0.82	0.16
Hemicentetes semispin.	810	9.32	1.58	0.84	1.36	0.86	0.18
Echinops telfairi	601	8.44	1.94	0.94	1.39	0.84	0.15
Geogale aurita	125	5.00	1.70	0.88	1.40	0.80	0.07
Oryzorictes talpoides	560	8.24	1.53	0.89	1.41	0.84	0.33
Microgale dobsoni	538	8.13	1.80	0.95	1.35	0.86	0.38
M. talazaci	739	9.04	1.80	0.97	1.26	0.87	0.43
Limnogale mergulus	1110	10.35	1.61	0.92	1.39	0.76	0.41
Potamogale velox	4008	15.88	1.81	1.11	1.21	0.85	0.60
Micropotam. ruwenzorii	1095	10.31	1.74	1.08	1.34	0.83	0.39
Chrysochloridae							
Chrysochloris asiatica	676	8.77	1.38	0.89	1.54	1.03	-.-
C. stuhlmanni	710	8.92	1.40	0.96	1.65	1.01	0.32
Solenodontidae							
Solenodon paradoxus	4559	16.58	2.08	1.02	1.27	0.77	0.26
Erinaceidae							
Echinosorex gymnurus	5873	18.04	1.95	1.10	1.31	0.86	0.42
Hylomys suillus	1158	10.50	1.77	0.99	1.25	0.81	0.36
Atelerix algirus	3151	14.66	2.04	1.12	1.32	0.80	0.24
Erinaceus europaeus	3250	14.81	1.78	0.97	1.38	0.76	0.17
Hemiechinus auritus	1815	12.20	1.96	1.05	1.43	0.88	0.30
Talpidae							
Desmana moschata	3861	15.69	1.59	1.23	1.48	0.73	0.63
Galemys pyrenaicus	1283	10.87	1.63	1.30	1.40	0.96	0.69
Talpa europaea	988	9.96	1.77	1.12	1.54	0.73	0.53
T. micrura	788	9.24	1.79	1.17	1.44	0.70	-.-
Parascalops breweri	849	9.47	1.82	1.13	1.45	0.78	0.44
Scalopus aquaticus	1264	10.81	1.85	1.12	1.68	0.68	0.47

Table 3 (cont.)

	BrV	CR	BL/CR	HemL/CR	BW/CR	BH/CR	CCL/CR
Soricidae							
Sorex alpinus	253	6.32	-.-	-.-	-.-	-.-	0.39
S. araneus	208	5.93	1.63	1.13	1.55	0.81	0.40
S. cinereus	162	5.45	1.62	1.07	1.43	0.87	0.39
S. fumeus	233	6.15	1.67	1.11	1.41	0.87	0.47
S. minutus	111	4.81	-.-	-.-	-.-	-.-	0.38
Microsorex hoyi	93	4.53	1.73	1.09	1.38	0.84	0.42
Neomys anomalus	272	6.48	1.51	1.08	1.48	0.93	0.37
N. fodiens	317	6.82	1.62	1.08	1.51	0.87	0.38
Blarina brevicauda	379	7.24	1.62	1.03	1.60	0.87	0.44
Cryptotis parva	236	6.18	1.61	0.98	1.58	0.84	0.37
Anourosorex squamipes	375	7.21	1.55	0.95	1.65	0.87	0.26
Crocidura attenuata	202	5.86	1.71	0.96	1.50	0.75	-.-
C. flavescens	407	7.41	1.86	0.99	1.49	0.75	0.26
C. hildegardeae	206	5.90	1.78	0.92	1.41	0.76	-.-
C. russula	190	5.75	1.72	0.96	1.52	0.78	0.25
C. suaveolens	183	5.68	1.65	0.94	1.49	0.76	-.-
Suncus etruscus	60	3.91	1.71	0.97	1.49	0.71	0.25
S. murinus	372	7.19	2.00	0.99	1.50	0.71	0.31
Scutisorex somereni	618	8.52	1.74	0.95	1.40	0.82	0.25
Sylvisorex granti	159	5.42	1.70	0.96	1.45	0.82	0.37
S. megalura	181	5.66	1.80	0.93	1.39	0.82	0.30
Ruwenzorisorex suncoides	346	7.02	1.74	1.00	1.51	0.77	0.38
Myosorex babaulti	347	7.03	1.65	1.01	1.49	0.83	0.39

Abbreviations:

BH	brain height
BL	brain length
BrV	standard brain volume = $\frac{\text{BrW}}{1.036}$
BW	brain width
CCL	corpus callosum length
CR	cube root of standard BrV
HemL	hemisphere length

Table 4. Average proportion indices (= linear measures of brains and corpus callosum related to the cube roots of the standard brain volumes)

For linear measures in species see Table 2
For proportion indices in species see Table 3

	N	BL/CR	HemL/CR	BW/CR	BH/CR	CCL/CR
Insectivora (AvIS)	46	1.73	1.01	1.44	0.82	0.35
Tenrecidae	10	1.76	0.94	1.33	0.83	0.30
Tenrecinae*	4	1.82	0.87	1.34	0.83	0.16
Geogalinae*	1	1.70	0.88	1.40	0.80	0.07
Oryzorictinae*	4	1.69	0.93	1.35	0.83	0.39
Microgale	2	1.80	0.96	1.31	0.87	0.41
terr.Oryzorictinae	3	1.71	0.94	1.34	0.86	0.38
Limnogale	1	1.61	0.92	1.39	0.76	0.41
Potamogalinae*	2	1.78	1.10	1.28	0.84	0.50
Chrysochloridae*	2	1.39	0.93	1.60	1.02	0.32
Solenodontidae*	1	2.08	1.02	1.27	0.77	0.26
Erinaceidae	5	1.90	1.05	1.34	0.82	0.30
Echinosoricinae*	2	1.86	1.05	1.28	0.84	0.39
Erinaceinae*	3	1.93	1.05	1.38	0.81	0.24
Talpidae	6	1.74	1.18	1.50	0.76	0.55
Desmaninae*	2	1.61	1.27	1.44	0.85	0.66
Talpinae*	4	1.81	1.14	1.53	0.72	0.48
Talpa	2	1.78	1.15	1.49	0.72	0.53
Soricidae	21	1.70	1.00	1.49	0.81	0.35
Soricinae*	9	1.62	1.06	1.51	0.86	0.39
Sorex	3	1.64	1.10	1.46	0.85	0.41
terr. Soricinae	7	1.63	1.05	1.51	0.85	0.39
Neomys	2	1.57	1.08	1.50	0.90	0.37
Crocidurinae*	12	1.76	0.97	1.47	0.77	0.30
Crocidura	5	1.74	0.95	1.48	0.76	0.25
Suncus	2	1.86	0.98	1.50	0.71	0.28
Sylvisorex	2	1.75	0.95	1.42	0.82	0.28
AvIF of the 12 families and subfamilies marked by *		1.75	1.02	1.40	0.83	0.34

Number of species = N
Abbreviations:

BH	brain height	CCL	corpus callosum length
BL	brain length	CR	cube root of standard BrV
BW	brain width	HemL	hemisphere length

Table 5. Average body weights (BoW) and brain weights (BrW) of 53 species of Insectivora. n = number of individuals; CV = coefficient of variation.

	BoW in g	CV in %	BrW in mg	CV in %	n BoW	n BrW	Source of data
Tenrecidae							
Tenrecinae							
Tenrec ecaudatus	845	14.6	2633	13.3	9	9	8,40
	885	1.6	2490	10.4	2	4	8,10a,40
	852		2588		11	13	
Setifer setosus	248	27.7	1516	11.6	16	16	40
	223	28.4			12	0	40
	237		1516		28	16	
Hemicentetes semispin.	117	12.7	839	3.9	7	7	40
	113	24.4			2	0	40
	116		839		9	7	
Echinops telfairi	87.5	25.9	623	10.2	6	6	S
Geogalinae							
Geogale aurita	10		130				**b)**
Oryzorictinae							
Oryzorictes talpoides	44.2		580		1	1	40
Microgale dobsoni	31.9	9.6	557	1.0	3	3	S
M. talazaci	48.2	16.0	783	8.0	13	13	S
	48.2	21.8	693	3.0	4	3	S
	48.2		766		17	16	
Limnogale mergulus	92		1150				40
Potamogalinae							
Potamogale velox	618	12.9	4152	6.9	13	13	S
Micropotamog. lamottei	64.2		800				57
M. ruwenzorii	98.5	14.9	1134	9.9	4	4	S
	95.0	28.8			4	0	30
	96.8		1134		8	4	
Chrysochloridae							
Chrysochloris asiatica	49		700		1	1	S
C. stuhlmanni	39.9	15.5	741	7.7	9	9	S
	45.3	15.2			5	0	S
	41.8		741		14	9	
mm+ff median	40.2		736				**a)**
mm	46.0		780		9	5	
ff	34.4		692		5	4	

Table 5 (cont.)

	BoW in g	CV in %	BrW in mg	CV in %	n BoW	n BrW	Source of data
Solenodontidae							
Solenodon paradoxus	674	3.7	4795	5.0	4	4	42,S
	663		4580	0.3	1	2	S
	672		4723		5	6	
Erinaceidae							
Echinosoricinae							
Echinosorex gymnurus	823	12.5	6208	5.7	2	2	S
			5960		0	2	S
	823		6084		2	4	
Hylomys suillus	57		1200				c)
Erinaceinae							
Atelerix algirus	736	14.4	3264	4.8	7	7	S
Erinaceus europaeus	839	15.5	3378	6.4	21	21	2,3,4,8,13, 23,32,59,S
	860		3300		16	3	5,7,8,15,S
	849		3367		35	22	
Hemiechinus auritus	235	26.2	1880	8.3	7	7	S
Talpidae							
Desmaninae							
Desmana moschata	443	10.9	4000	2.5	3	3	40
Galemys pyrenaicus	61.1	4.6	1328	3.8	3	3	26,40
	59.6		1330		44	2	26,40
	59.7		1329		47	5	
Talpinae							
Talpa europaea	82.5	11.4	1016	6.9	11	11	5,6,11,
	78.0		1055		1	3	32,S
	82.1		1024		12	14	
T. micrura	41.4		816		1	1	S
Parascalops breweri	53.8	5.7	880	4.1	6	6	S
Scalopus aquaticus	115		1310		1	1	S
Soricidae							
Soricinae							
Sorex alpinus	10.3	24.7	259	5.2	4	4	S
	11.1		265	4.1	41	4	50,S
	11.0		262		45	8	
S. araneus	8.5	18.4	205	7.7	40	40	5,6,9,40,S
	10.3		218		717	249	12,15,17,19,
	10.2		216		757	289	20,22,24,27, 29,31,35,37, 38,39,47,S

Table 5 (cont.)

	BoW in g	CV in %	BrW in mg	CV in %	n BoW	n BrW	Source of data
Soricinae (cont.)							
Sorex cinereus	4.20	25.0	174	7.5	29	29	S,B
	3.9				28	0	16,18,25,34,B
	4.05		174		57	29	
	5.16	22.8	168	10.3	11	11	S,B (definite adults)
S. fumeus	8.4	9.9	241	3.5	5	2	S,B
S. minutus	4.11	22.8	120	13.9	11	11	S
	4.42		115		338	113	1,12,17,21,28,
	4.41		115		349	124	29,31,40,41, 43,44,47
S. palustris	14.6		278		1	1	B
Microsorex hoyi	2.63	7.5	96	12.0	7	7	S,B
Neomys anomalus	9.42	2.4	282	3.5	5	5	S
	11.8				76	0	17,47,48,53
	11.6		282		81	5	
N. fodiens	15.0	13.2	327	5.1	17	17	S
	15.4		333		254	4	15,17,29,31,
	15.3		328		271	21	36,41,43,47, 48,49,53,S
Blarina brevicauda	18.2	16.4	397	4.8	42	42	S,B
	17.4		348		68	68	14 (preserved)
	19.3				16	0	7,16,34,B
	17.9		367		126	110	
	19.7	18.9	393	2.9	7	7	S,B (definite adults)
Cryptotis parva	9.5	8.3	244	1.8	5	5	S
	10.3		248		4	1	S
	9.9		245		9	6	
Anourosorex squamipes	20.0	9.4	389	5.7	7	7	S
	20.5	17.2			2	0	S
	20.1		389		9	7	
Crocidurinae							
Crocidura attenuata	9.3		209		1	1	S
C. flavescens	31.6	23.5	426	9.0	52	52	S
	34.3		394		58	7	51,56,S
	33.0		422		110	59	
mm + ff median	29.3		414				**a)**
mm	32.7		432		63	43	
ff	25.8		396		22	16	
C. giffardi	82.0		545		1	1	10

Table 5 (cont.)

	BoW in g	CV in %	BrW in mg	CV in %	n BoW	n BrW	Source of data
Crocidurinae (cont.)							
C. hildegardeae	9.1	17.5	213	10.5	9	9	S
	10.5				41	0	40,51,S
	10.2		213		50	9	
C. jacksoni	10.3		250		1	1	S
	13.25				4	0	S
	12.7		250		5	1	
C. leucodon	13.5		190		1	1	S
C. russula	11.4	12.5	195	6.8	24	24	58,S
	11.0		201		61	12	22,54,55,56,
	11.1		197		85	36	58,F,S
C. suaveolens	10.3		190		1	1	F,S
Suncus etruscus	2.07	10.1	62	3.7	6	6	F,S
	1.85				194	0	45,46,52,56
	1.86		62		200	6	
S. murinus	34.1	31.9	383	8.7	26	26	33,S
	37.6	26.4	449		6	1	33,S
	34.8		385		32	27	
mm + ff median	33.8		383				**a)**
mm	41.9		400		18	15	
ff	25.7		366		14	12	
Scutisorex somereni	58.9	16.3	640	2.5	4	4	S
	63.7				70	0	51
	63.4		640		74	4	
Sylvisorex granti	3.88	5.5	165	3.1	6	6	S
	3.90				25	0	51,H
	3.90		165		31	6	
S. megalura	5.55	9.9	188	7.1	13	13	S
	5.47				40	0	40,51,56,S
	5.49		188		53	13	
Ruwenzorisorex suncoides	18.2		370		1	1	S
	18.5		358		1	1	S **d)**
Myosorex babaulti	17.0		360		1	1	S
	17.0				12	0	51
	17.0		360		13	1	

Notes to Table 5:

In general there are 3 lines for each species. The first line includes all the pairs of body and brain weights we are aware of, the second line single values (either body weights or brain weights), and the third line the combined data from the two preceding lines. If there are more than three lines, they include data of preserved material (as e.g. in *Blarina*), or data from definite adults (if appropriate).

Table 5 (cont.)

a) Median values between males and females are given in the case of statistically significant sex differences (99% for BoW, P < .01; 95% for BrW, P < .05).
b) From *Geogale aurita* only secondary inferred data, taken from Rehkämper et al. (1986), are available (see text).
c) From *Hylomys suillus* we have only one brain of a juvenile animal (BoW 31.4 g, BrW 957 mg). Its brain was used to measure the brain composition. Based on BoW data of Miller (1942), Rudd (1965), Harrison (1966), Lim and Heyneman (1968) and Medway (1969) adults have a BoW of about 57 g. Based on cranial capacities (n = 7) the expected brain weight is 1200 mg.
d) In the second line of *Ruwenzorisorex suncoides* a specimen of uncertain determination is listed , which in Fons et al. (1984) was designated as *Sylvisorex* (*lunaris*?). According to its BrW it may have been a *Ruwenzorisorex suncoides*.

Key to the source of body and brain weight data:
1: Snell (1892); 2: Weber (1896); 3: Dubois (1897); 4: Flatau and Jacobsohn (1899); 5: Ziehen (1899a); 6: Welcker und Brandt (1903); 7: Hrdlicka (1905); 8: Warncke (1908); 9: Lapicque (1912); 10: Waterlot (1912); 10a: Brodmann (1913); 11: Slifer (1924); 12: Middleton (1931); 13: Bonin (1937); 14: Crile and Quiring (1940); 15: Cantuel (1943); 16: Morrison and Pearson (1946); 17: Borowski and Dehnel (1953); 18: Pruitt (1954); 19: Siivonen (1954); 20: Pucek (1955); 21: Cabon (1956); 22: Niethammer (1956); 23: Haug (1958); 24: Schubarth (1958); 25: Morrison et al. (1959); 26: Stephan and Bauchot (1959); 27: Bielak and Pucek (1960); 28: Hawkins et al. (1960); 29: Niethammer (1960); 30: Rahm (1960, 1961); 31: Hawkins and Jewell (1962); 32: Portmann (1962); 33: Bauchot and Stephan (1964); 34: Buckner (1964); 35: Pucek, Z. (1964); 36: Gebczynska and Gebczynski (1965); 37: Gebczynski (1965); 38: Pucek, M. (1965); 39: Pucek, Z. (1965); 40: Bauchot and Stephan (1966); 41: Mangold-Wirz (1966); 42: Boller (1969); 43: Gebczynski (1971a); 44: Gebczynski (1971b); 45: Fons (1974); 46: Fons et Saint Girons (1975); 47: Gebczynski (1977); 48: Niethammer (1977); 49: Niethammer (1978); 50: Spitzenberger (1978); 51: Dieterlen und Heim de Balsac (1979); 52: Fons (1979); 53: Spitzenberger (1980); 54: Genoud (1981); 55: Genoud and Vogel (1981); 56: Vogel et al. (1981); 57: Stephan and Kuhn (1982); 58: Schwerdtfeger (1984); 59: Hofer (pers. comm.).

F: collection Roger Fons; H: collection Rainer Hutterer; S: collection Heinz Stephan in cooperation with Georg Baron (B), Kunwar Bhatnagar, Roland Bauchot, Petra Friesleber, Kabongo Ka Mubalamata, Linjanja Lulyo, Otmar Lütt, José Regidor, Gerd Rehkämper, Wolfbernhard Spatz, Michael and Roland Stephan, Gottfried Vauk, Songsakdi Yenbutra, and others.

Table 6. Standards of body and brain weights, percentages of brain weight relative to body weight, encephalization indices (EIs) derived from the standards and size indices (SIs) of NET brain volumes. EIs and SIs are related to the average of the four Tenrecinae (AvTn = 100)

For average encephalization indices in supraspecific taxonomic units see Table 7
For NET brain volumes see Table 8

	Standards		BrW	BrW	NET
	BoW	BrW	BoW	EI	SI
slope				.66	.66
y-intercept				1.544	1.506
Tenrecidae					
Tenrec ecaudatus	852	2588	0.30	86	85
Setifer setosus	237	1516	0.64	117	119
Hemicentetes semispinosus	116	839	0.72	104	104
Echinops telfairi	87.5	623	0.71	93	93
Geogale aurita	10.0	130	1.30	81	83
Oryzorictes talpoides	44.2	580	1.31	136	138
Microgale dobsoni	31.9	557	1.75	162	162
M. talazaci	48.2	766	1.59	170	174
Limnogale mergulus	92.0	1150	1.25	166	165
Potamogale velox	618	4152	0.67	171	174
Micropotamogale lamottei	64.2	800	1.25	147	149
M. ruwenzorii	96.8	1134	1.17	158	162
Chrysochloridae					
Chrysochloris asiatica	49.0	700	1.43	153	157
C. stuhlmanni	40.2	736	1.83	184	188
Solenodontidae					
Solenodon paradoxus	672	4723	0.70	184	184
Erinaceidae					
Echinosorex gymnurus	823	6084	0.74	207	209
Hylomys suillus	57.0	1200	2.11	238	241
Atelerix algirus	736	3264	0.44	120	117
Erinaceus europaeus	849	3367	0.40	112	112
Hemiechinus auritus	235	1880	0.80	146	144
Talpidae					
Desmana moschata	443	4000	0.90	205	202
Galemys pyrenaicus	59.7	1329	2.23	255	258
Talpa europaea	82.1	1024	1.25	160	163
T. micrura	41.4	816	1.97	200	205
Parascalops breweri	53.8	880	1.64	181	188
Scalopus aquaticus	115	1310	1.14	163	170

Table 6 (cont.)

	Standards BoW	Standards BrW	BrW/BoW	BrW EI	NET SI
Soricidae					
Sorex alpinus	11.0	262	2.38	154	160
S. araneus	10.2	216	2.12	133	137
S. cinereus	5.16	168	3.26	163	169
S. fumeus	8.40	241	2.87	169	175
S. minutus	4.41	115	2.61	123	128
S. palustris	14.6	278	1.90	135	-
Microsorex hoyi	2.63	96	3.65	145	150
Neomys anomalus	11.6	282	2.43	160	166
N. fodiens	15.3	328	2.14	155	159
Blarina brevicauda	19.7	393	1.99	157	162
Cryptotis parva	9.90	245	2.47	154	160
Anourosorex squamipes	20.1	389	1.94	153	158
Crocidura attenuata	9.30	209	2.25	137	142
C. flavescens	29.3	414	1.41	127	130
C. giffardi	82.0	545	0.66	85	-
C. hildegardeae	10.2	213	2.09	131	136
C. jacksoni	12.7	250	1.97	133	136
C. leucodon	13.5	190	1.41	97	-
C. russula	11.1	197	1.77	115	118
C. suaveolens	10.3	190	1.84	116	120
Suncus etruscus	1.86	62	3.33	118	121
S. murinus	33.8	383	1.13	107	110
Scutisorex somereni	63.4	640	1.01	118	122
Sylvisorex granti	3.90	165	4.23	192	199
S. megalura	5.49	188	3.42	175	181
Ruwenzorisorex suncoides	18.2	370	2.03	156	161
Myosorex babaulti	17.0	360	2.12	159	164

Abbreviations and notes:

BoW	body weight in g
BrW	brain weight in mg
EI BrW	encephalization indices
SI NET	size indices of the NET brain volume

Table 7. Average percentages of brain weight related to body weight, and averages of the encephalization indices (EIs) of BrW and size indices (SIs) of the NET brain volumes in supraspecific taxonomic units

For encephalization indices in species see Table 6

	N BrW	N NET	BrW / BoW	BrW EI	NET SI
slope				.66	.66
y-intercept				1.544	1.506
Insectivora (AvIS)	53	50	1.71	148	154
Tenrecidae	12	12	1.06	133	134
Tenrecinae*	4	4	0.59	**100**	**100**
Geogalinae*	1	1	1.30	81	83
Oryzorictinae*	4	4	1.47	158	160
Microgale	2	2	1.67	166	168
terr.Oryzorictinae	3	3	1.55	156	158
Limnogale	1	1	1.25	166	165
Potamogalinae*	3	3	1.03	159	162
Micropotamogale	2	2	1.21	153	155
Chrysochloridae*	2	2	1.63	169	172
Solenodontidae*	1	1	0.70	184	184
Erinaceidae	5	5	0.90	165	165
Echinosoricinae*	2	2	1.42	222	225
Erinaceinae*	3	3	0.55	126	124
Talpidae	6	6	1.52	194	198
Desmaninae*	2	2	1.56	230	230
Talpinae*	4	4	1.50	176	181
Talpa	2	2	1.61	180	184
Soricidae	27	24	2.24	140	148
Soricinae*	12	11	2.48	151	157
Sorex	6	5	2.52	146	154
terr. Soricinae	10	9	2.52	149	155
Neomys	2	2	2.29	157	162
Crocidurinae*	15	13	2.05	137	142
Crocidura	8	6	1.68	118	130
Suncus	2	2	2.23	112	115
Sylvisorex	2	2	3.83	183	190
AvIF of the 12 families and subfamilies marked by *			1.36	158	160

Number of species = N
Abbreviations as in Table 6

Table 8. Volumes of net brain (NET), ventricles (VENT) and rests (REST) and their percentages relative to standard brain volumes (= 100%)

For standard brain volumes (BrV) see Table 3
For average percentages in supraspecific taxonomic units see Table 9

	n	NET mm^3	NET %	VENT mm^3	VENT %	REST mm^3	REST %
Tenrecidae							
Tenrec ecaudatus	3	2331	93.3	65.7	2.6	101.6	4.1
Setifer setosus	3	1404	96.0	22.0	1.5	36.8	2.5
Hemicentetes semispin.	2	770	95.0	16.9	2.1	23.3	2.9
Echinops telfairi	2	569	94.6	8.4	1.4	24.0	4.0
Geogale aurita	2	122	97.4	1.2	0.9	2.1	1.7
Oryzorictes talpoides	1	538	96.0	4.8	0.9	17.3	3.1
Microgale dobsoni	1	511	95.0	5.3	1.0	21.8	4.0
M. talazaci	2	718	97.2	5.6	0.7	15.5	2.1
Limnogale mergulus	1	1046	94.3	12.7	1.1	51.0	4.6
Potamogale velox	5	3874	96.7	18.4	0.4	115.4	2.9
Micropotamogale lam.	3	743	96.2	11.9	1.6	17.2	2.2
M. ruwenzorii	4	1064	97.2	8.4	0.8	22.7	2.0
Chrysochloridae							
Chrysochloris asiatica	1	657	97.2	8.1	1.2	10.7	1.6
C. stuhlmanni	2	690	97.1	7.0	1.0	13.7	1.9
Solenodontidae							
Solenodon paradoxus	2	4331	95.0	57.1	1.3	171.1	3.7
Erinaceidae							
Echinosorex gymnurus	3	5638	96.0	121.4	2.1	112.9	1.9
Hylomys suillus	1	1116	96.3	23.2	2.0	19.4	1.7
Atelerix algirus	3	2916	92.6	123.4	3.9	111.1	3.5
Erinaceus europaeus	5	3065	94.3	105.8	3.3	79.2	2.4
Hemiechinus auritus	3	1692	93.3	81.7	4.5	40.7	2.2
Talpidae							
Desmana moschata	1	3620	93.8	137.6	3.5	103.7	2.7
Galemys pyrenaicus	2	1229	95.8	27.4	2.2	25.9	2.0
Talpa europaea	2	957	96.8	10.9	1.1	20.7	2.1
T. micrura	2	768	97.5	9.8	1.3	9.7	1.2
Parascalops breweri	3	835	98.3	10.3	1.2	4.5	0.5
Scalopus aquaticus	1	1246	98.5	8.6	0.7	10.0	0.8

Table 8 (cont.)

	n	NET mm³	NET %	VENT mm³	VENT %	REST mm³	REST %
Soricidae							
Sorex alpinus	3	250	98.7	1.2	0.5	2.2	0.8
S. araneus	5	203	97.5	1.3	0.6	4.0	1.9
S. cinereus	3	160	98.5	0.8	0.5	1.7	1.0
S. fumeus	2	229	98.3	1.8	0.8	2.1	0.9
S. minutus	4	109	98.2	0.4	0.4	1.5	1.4
Microsorex hoyi	3	91	98.4	0.5	0.6	1.0	1.0
Neomys anomalus	3	268	98.5	1.5	0.5	2.6	1.0
N. fodiens	2	308	97.4	1.7	0.5	6.7	2.1
Blarina brevicauda	3	372	98.1	2.7	0.7	4.7	1.2
Cryptotis parva	3	233	98.4	1.5	0.6	2.3	1.0
Anourosorex squamipes	4	368	98.0	2.9	0.8	4.5	1.2
Crocidura attenuata	1	198	98.3	1.4	0.7	2.1	1.0
C. flavescens	3	387	96.9	2.7	0.7	9.6	2.4
C. hildegardeae	4	202	98.2	1.1	0.5	2.6	1.3
C. jacksoni	1	234	97.0	1.2	0.5	6.0	2.5
C. russula	5	186	97.7	1.5	0.8	2.9	1.5
C. suaveolens	3	180	98.0	1.6	0.9	2.2	1.1
Suncus etruscus	3	58	97.8	0.5	0.8	0.9	1.4
S. murinus	3	359	97.0	3.1	0.9	7.8	2.1
Scutisorex somereni	3	604	97.8	4.6	0.8	8.9	1.4
Sylvisorex granti	3	157	98.4	0.9	0.5	1.7	1.1
S. megalura	3	178	98.3	1.1	0.6	2.0	1.1
Ruwenzorisorex sunc.	1	351	98.3	2.1	0.6	3.9	1.1
Myosorex babaulti	1	342	98.3	1.5	0.5	4.2	1.2

Number of individuals = n

Abbreviations:

NET net brain volume

REST rests including meninges, nerves, hypophysis, ect.

VENT ventricle volumes

Table 9. Average percentages of NET brain, ventricles and rests relative to standard brain volumes (BrV = 100%)

For volumes and percentages in species see Table 8

	N	NET	VENT	REST
Insectivora (AvIS)	50	96.9	1.2	1.9
Tenrecidae	12	95.7	1.3	3.0
Tenrecinae*	4	94.7	1.9	3.4
Geogalinae*	1	97.4	0.9	1.7
Oryzorictinae*	4	95.6	0.9	3.5
Microgale	2	96.1	0.9	3.0
terr.Oryzorictinae	3	96.1	0.8	3.1
Limnogale	1	94.3	1.1	4.6
Potamogalinae*	3	96.7	0.9	2.4
Micropotamogale	2	96.7	1.2	2.1
Chrysochloridae*	2	97.1	1.1	1.8
Solenodontidae*	1	95.0	1.3	3.7
Erinaceidae	5	94.5	3.1	2.4
Echinosoricinae*	2	96.2	2.0	1.8
Erinaceinae*	3	93.4	3.9	2.7
Talpidae	6	96.8	1.6	1.6
Desmaninae*	2	94.8	2.8	2.4
Talpinae*	4	97.8	1.1	1.1
Talpa	2	97.1	1.2	1.7
Soricidae	24	98.0	0.6	1.4
Soricinae*	11	98.2	0.6	1.2
Sorex	5	98.2	0.6	1.2
terr. Soricinae	9	98.2	0.6	1.2
Neomys	2	98.0	0.5	1.5
Crocidurinae*	13	97.8	0.7	1.5
Crocidura	6	97.7	0.7	1.6
Suncus	2	97.4	0.8	1.8
Sylvisorex	2	98.3	0.6	1.1
AvIF of the 12 families and subfamilies marked by *		96.2	1.5	2.3

Number of species = N
Abbreviations as in Table 8

Table 10. Volumes of the five fundamental brain parts

For size indices in species see Table 11
For average size indices in supraspecific taxonomic units see Table 12

	n	OBL	MES	CER	DIE	TEL
Tenrecidae						
Tenrec ecaudatus	3	315.5	133.3	311.5	153.7	1417
Setifer setosus	3	180.6	76.6	146.8	93.1	907
Hemicentetes semispinosus	2	117.0	45.9	120.5	47.7	439
Echinops telfairi	2	90.3	44.0	58.2	40.3	336
Geogale aurita	2	34.0	12.7	12.5	7.9	55
Oryzorictes talpoides	1	61.0	26.6	81.3	36.9	332
Microgale dobsoni	1	58.7	37.6	57.0	32.6	325
M. talazaci	2	84.0	45.5	94.9	45.4	449
Limnogale mergulus	1	203.8	67.4	176.0	81.0	518
Potamogale velox	5	722.8	259.9	622.7	300.3	1968
Micropotamogale lamottei	3	134.9	58.6	82.7	62.8	404
M. ruwenzorii	4	198.0	72.5	149.3	88.8	555
Chrysochloridae						
Chrysochloris asiatica	1	58.5	43.2	72.6	52.8	430
C. stuhlmanni	2	68.0	41.7	74.3	51.6	454
Solenodontidae						
Solenodon paradoxus	2	471.6	208.7	761.2	272.3	2617
Erinaceidae						
Echinosorex gymnurus	3	622.4	360.4	801.0	395.1	3459
Hylomys suillus	1	125.4	83.9	157.2	85.9	663
Atelerix algirus	3	343.0	156.4	364.7	251.0	1801
Erinaceus europaeus	5	335.0	151.3	395.8	242.4	1941
Hemiechinus auritus	3	185.2	109.0	196.1	140.2	1062
Talpidae						
Desmana moschata	1	402.4	218.7	514.5	343.8	2140
Galemys pyrenaicus	2	147.1	62.1	183.6	104.5	732
Talpa europaea	2	95.7	50.0	174.2	80.7	556
T. micrura	2	77.1	45.0	116.0	60.8	469
Parascalops breweri	3	74.7	35.6	121.0	66.1	537
Scalopus aquaticus	1	118.5	57.7	197.7	91.7	780

Table 10 (cont.)

	n	OBL	MES	CER	DIE	TEL
Soricidae						
Sorex alpinus	3	31.2	12.5	31.5	17.0	157
S. araneus	5	29.1	11.5	22.0	14.8	126
S. cinereus	3	17.2	8.3	16.5	10.9	107
S. fumeus	2	22.6	11.1	23.4	15.8	156
S. minutus	4	15.0	6.4	12.1	7.7	68
Microsorex hoyi	3	10.9	4.8	8.7	6.9	60
Neomys anomalus	3	34.2	16.0	32.4	21.5	164
N. fodiens	2	45.3	19.2	41.0	24.1	179
Blarina brevicauda	3	43.0	18.8	42.3	26.3	242
Cryptotis parva	3	30.1	12.7	24.2	17.1	149
Anourosorex squamipes	4	44.1	18.8	39.7	25.4	240
Crocidura attenuata	1	32.4	15.0	24.3	13.0	114
C. flavescens	3	59.9	26.7	51.9	25.1	224
C. hildegardeae	4	31.4	14.9	23.0	14.0	118
C. jacksoni	1	42.3	17.9	28.5	16.3	129
C. russula	5	28.7	13.7	21.0	13.9	109
C. suaveolens	3	27.0	13.7	21.9	12.7	105
Suncus etruscus	3	11.1	4.7	5.9	4.4	32
S. murinus	3	54.4	24.6	42.8	24.1	213
Scutisorex somereni	3	82.9	36.6	83.9	40.7	360
Sylvisorex granti	3	21.5	11.4	18.7	10.1	95
S. megalura	3	26.8	14.2	24.9	12.3	100
Ruwenzorisorex suncoides	1	54.9	25.0	52.7	26.2	192
Myosorex babaulti	1	46.2	19.0	42.4	23.6	211

Number of individuals = n

Abbreviations and notes:

CER cerebellum; for volumes of components see Table 31
DIE diencephalon; for volumes of components see Tables 34, 37 and 40
MES mesencephalon; for volumes of components see Table 28
OBL medulla oblongata; for volumes of components see Tables 13, 16, 19, 22, and 25
TEL telencephalon; for volumes of components see Table 43

Table 11. Size indices of the five fundamental brain parts related to the average of the four Tenrecinae (AvTn = 100)

For volumes in species see Table 10
For average size indices in supraspecific taxonomic units see Table 12

	n	OBL	MES	CER	DIE	TEL
slope		.61	.58	.68	.65	.68
y-intercept		.776	.481	.552	.350	1.242
Tenrecidae						
Tenrec ecaudatus	3	86	88	89	85	83
Setifer setosus	3	108	106	100	119	126
Hemicentetes semispin.	2	108	96	133	97	99
Echinops telfairi	2	99	109	78	98	92
Geogale aurita	2	140	110	73	79	66
Oryzorictes talpoides	1	101	98	173	141	145
Microgale dobsoni	1	119	167	152	153	177
M. talazaci	2	132	159	191	163	184
Limnogale mergulus	1	216	162	228	192	137
Potamogale velox	5	240	207	221	206	143
Micropotamogale lam.	3	178	173	137	188	137
M. ruwenzorii	4	204	169	187	203	142
Chrysochloridae						
Chrysochloris asiatica	1	91	149	144	188	175
C. stuhlmanni	2	120	162	169	209	211
Solenodontidae						
Solenodon paradoxus	2	149	158	255	177	179
Erinaceidae						
Echinosorex gymnurus	3	174	243	234	225	206
Hylomys suillus	1	178	266	282	277	243
Atelerix algirus	3	102	112	115	154	116
Erinaceus europaeus	5	92	100	113	135	113
Hemiechinus auritus	3	111	152	134	180	148
Talpidae						
Desmana moschata	1	164	211	229	292	195
Galemys pyrenaicus	2	203	192	319	327	260
Talpa europaea	2	109	128	244	205	159
T. micrura	2	133	171	259	242	214
Parascalops breweri	3	110	117	226	221	205
Scalopus aquaticus	1	110	122	220	188	177

Table 11 (cont.)

	n	OBL	MES	CER	DIE	TEL
Soricidae						
Sorex alpinus	3	121	103	173	160	176
S. araneus	5	118	99	127	146	149
S. cinereus	3	106	106	151	167	200
S. fumeus	2	104	107	154	177	210
S. minutus	4	102	90	124	131	142
Microsorex hoyi	3	101	91	127	165	178
Neomys anomalus	3	128	128	171	195	177
N. fodiens	2	144	131	180	183	160
Blarina brevicauda	3	117	110	156	169	182
Cryptotis parva	3	124	111	143	172	179
Anourosorex squamipes	4	119	109	145	161	179
Crocidura attenuata	1	139	136	150	136	143
C. flavescens	3	128	125	146	125	129
C. hildegardeae	4	128	128	133	139	140
C. jacksoni	1	150	135	142	139	131
C. russula	5	111	112	114	129	121
C. suaveolens	3	109	117	126	124	123
Suncus etruscus	3	127	109	109	132	121
S. murinus	3	106	105	110	109	111
Scutisorex somereni	3	110	109	140	122	123
Sylvisorex granti	3	157	171	208	186	216
S. megalura	3	159	174	220	182	180
Ruwenzorisorex suncoides	1	157	154	205	177	153
Myosorex babaulti	1	137	121	173	167	176

Number of individuals = n

Abbreviations and notes:

CER cerebellum; for size indices of components see Table 32
DIE diencephalon; for size indices of components see Tables 35, 38 and 41
MES mesencephalon; for size indices of components see Table 29
OBL medulla oblongata; for size indices of components see Tables 14, 17, 20, 23, and 26
TEL telencephalon; for size indices of components see Table 44

Table 12. Average size indices of the five fundamental brain parts related to the average of the four Tenrecinae (AvTn = 100)

	N	OBL	MES	CER	DIE	TEL
slope		.61	.58	.68	.65	.68
y-intercept		.776	.481	.552	.350	1.242
Insectivora (AvIS)	50	132	136	167	169	159
Tenrecidae	12	144	137	147	144	128
Tenrecinae*	4	**100**	**100**	**100**	**100**	**100**
Geogalinae*	1	140	110	73	79	66
Oryzorictinae*	4	142	146	186	162	161
Microgale	2	126	163	171	158	180
terr. Oryzorictinae	3	117	141	172	152	169
Limnogale	1	216	162	228	192	137
Potamogalinae*	3	207	183	182	199	140
Micropotamogale	2	191	171	162	195	139
Chrysochloridae*	2	105	156	157	198	193
Solenodontidae*	1	149	158	255	177	179
Erinaceidae	5	131	175	176	194	165
Echinosoricinae*	2	176	254	258	251	225
Erinaceinae*	3	102	121	121	156	126
Talpidae	6	138	157	249	246	202
Desmaninae*	2	184	201	274	310	227
Talpinae*	4	116	134	237	214	189
Talpa	2	121	150	251	223	186
Soricidae	24	125	120	151	154	158
Soricinae*	11	117	108	150	166	176
Sorex	5	110	101	146	156	175
terr. Soricinae	9	112	103	144	161	177
Neomys	2	136	129	176	189	169
Crocidurinae*	13	132	130	152	144	144
Crocidura	6	127	125	135	132	131
Suncus	2	117	107	109	120	116
Sylvisorex	2	158	173	214	184	198
AvIF of the 12 families and subfamilies marked by *		139	150	179	180	161

Number of species = N Abbreviations and notes:

CER cerebellum; for av. size indices of components see Table 33
DIE diencephalon; for av. size indices of components see Tables 36, 39 and 42
MES mesencephalon; for av. size indices of components see Table 30
OBL medulla oblongata; for av. size indices of components see Tables 15, 18, 21, 24 and 27
TEL telencephalon; for av. size indices of components see Table 45

Table 13. Volumes of the sensory trigeminal complex and of the tractus solitarii and funicular nuclei of the medulla oblongata

For size indices in species see Table 14
For average size indices in supraspecific taxonomic units see Table 15

	n	TR	TSO	FUN	FGR	FCM	FCE
Tenrecidae							
Tenrec ecaudatus	1	49.3	5.37	5.56	1.351	2.325	1.879
Setifer setosus	1	27.5	2.37	3.02	0.770	1.511	0.737
Hemicentetes semispin.	1	17.6	2.20	2.40	0.599	0.990	0.809
Echinops telfairi	1	12.7	1.71	2.58	0.707	0.986	0.883
Microgale talazaci	1	13.8	1.73	1.70	0.444	0.752	0.506
Potamogale velox	1	151.4	9.70	8.25	2.813	2.820	2.616
Micropotamogale ruwenz.	1	47.6	3.63	2.70	0.989	1.293	0.422
Chrysochloridae							
Chrysochloris stuhlmanni	1	14.4	1.21	1.59	0.466	0.705	0.414
Solenodontidae							
Solenodon paradoxus	1	69.2	9.89	11.03	3.579	5.373	2.073
Erinaceidae							
Echinosorex gymnurus	1	90.5	9.53	10.66	3.417	3.094	4.153
Hylomys suillus	1	14.5	2.18	2.27	0.484	0.633	1.149
Atelerix algirus	1	39.9	6.18	9.23	2.079	3.073	4.075
Erinaceus europaeus	1	44.8	7.18	10.49	1.837	3.385	5.266
Hemiechinus auritus	1	23.9	3.63	4.46	0.941	1.485	2.029
Talpidae							
Desmana moschata	1	97.0	9.00	9.12	2.559	3.932	2.632
Galemys pyrenaicus	1	32.3	2.97	3.23	0.680	1.738	0.812
Talpa europaea	1	13.3	1.88	4.07	0.626	2.468	0.973
Soricidae							
Sorex araneus	1	6.0	0.70	0.31	0.095	0.132	0.087
S. cinereus	1	4.2	0.42	0.22	0.067	0.102	0.051
Microsorex hoyi	1	2.2	0.30	0.18	0.059	0.078	0.044
Neomys fodiens	1	9.7	1.24	0.66	0.212	0.254	0.195
Blarina brevicauda	1	8.1	1.26	0.70	0.216	0.260	0.227
Crocidura flavescens	1	10.8	1.07	0.89	0.384	0.408	0.094
C. hildegardeae	1	6.8	0.70	0.40	0.154	0.183	0.066
C. russula	1	5.3	0.53	0.32	0.151	0.126	0.039
Suncus etruscus	1	2.3	0.22	0.17	0.059	0.070	0.037
S. murinus	1	9.8	1.10	0.84	0.247	0.302	0.287
Scutisorex somereni	1	15.5	1.53	1.54	0.452	0.642	0.448
Sylvisorex megalura	1	4.9	0.61	0.51	0.168	0.173	0.165
Ruwenzorisorex sunc.	1	12.0	1.12	0.82	0.267	0.350	0.205

Number of individuals = n
Abbreviations as in Table 15

Table 14. Size indices of the sensory trigeminal complex and of the tractus solitarii and funicular nuclei of the medulla oblongata (AvTn = 100)

For volumes in species see Table 13
For average size indices in supraspecific taxonomic units see Table 15

	n	TR	TSO	FUN	FGR	FCM	FCE
slope		.59	.58	.62	.62	.62	.61
y-intercept		-.001	-.926	-.927	-1.515	-1.300	-1.400
Tenrecidae							
Tenrec ecaudatus	1	92	90	72	67	71	77
Setifer setosus	1	110	84	86	85	102	66
Hemicentetes semispinosus	1	107	118	106	103	104	112
Echinops telfairi	1	91	108	136	145	123	145
Microgale talazaci	1	140	154	130	131	136	120
Potamogale velox	1	342	197	130	171	105	130
Micropotamogale ruwenzorii	1	322	216	134	190	151	65
Chrysochloridae							
Chrysochloris stuhlmanni	1	163	120	136	154	142	109
Solenodontidae							
Solenodon paradoxus	1	149	191	165	207	189	98
Erinaceidae							
Echisosorex gymnurus	1	173	164	140	174	96	174
Hylomys suillus	1	133	176	156	129	103	245
Atelerix algirus	1	81	113	130	114	102	183
Erinaceus europaeus	1	84	121	135	92	103	216
Hemiechinus auritus	1	96	129	128	104	100	182
Talpidae							
Desmana moschata	1	267	221	176	192	179	161
Galemys pyrenaicus	1	290	234	216	176	275	168
Talpa europaea	1	99	123	224	133	320	166
Soricidae							
Sorex araneus	1	153	153	63	74	63	53
S. cinereus	1	161	138	67	79	74	47
Microsorex hoyi	1	122	145	84	106	86	61
Neomys fodiens	1	194	215	103	128	93	93
Blarina brevicauda	1	140	189	94	111	82	93
Crocidura flavescens	1	148	127	92	155	100	30
C. hildegardeae	1	175	155	82	121	88	40
C. russula	1	128	111	60	111	57	23
Suncus etruscus	1	162	127	96	132	95	64
S. murinus	1	123	121	80	91	68	84
Scutisorex somereni	1	134	116	99	113	98	90
Sylvisorex megalura	1	179	191	149	191	120	147
Ruwenzorisorex suncoides	1	217	175	115	145	116	88

Number of individuals = n Abbreviations as in Table 15

Table 15. Average size indices of the sensory trigeminal complex and of the tractus solitarii and funicular nuclei related to the average of the four Tenrecinae (AvTn = 100)

For volumes in species see Table 13
For size indices in species see Table 14

	N	TR	TSO	FUN	FGR	FCM	FCE
slope		.59	.58	.62	.62	.62	.61
y-intercept		-.001	-.926	-.927	-1.515	-1.300	-1.400
Insectivora (AvIS)	30	159	151	119	131	118	111
Tenrecidae	7	172	138	113	127	113	102
Tenrecinae*	4	**100**	**100**	**100**	**100**	**100**	**100**
Oryzorictinae*	1	140	154	130	131	136	120
Potamogalinae*	2	332	206	132	181	128	98
Chrysochloridae*	1	163	120	136	154	142	109
Solenodontidae*	1	149	191	165	207	189	98
Erinaceidae	5	113	141	138	123	101	200
Echinosoricinae*	2	153	170	148	152	100	209
Erinaceinae*	3	87	121	131	103	102	194
Talpidae	3	219	193	205	167	258	165
Desmaninae*	2	279	228	196	184	227	165
Talpinae*	1	99	123	224	133	320	166
Soricidae	13	157	151	91	120	88	70
Soricinae*	5	154	168	82	100	79	69
Sorex	2	157	145	65	76	68	50
terr. Soricinae	4	144	156	77	93	76	63
Neomys	1	194	215	103	128	93	93
Crocidurinae*	8	158	140	97	132	93	71
Crocidura	3	150	131	78	129	82	31
Suncus	2	143	124	88	112	82	74
AvIF of the 11 families and subfamilies marked by *		165	156	140	143	147	127

Number of species = N

Abbreviations:

FCE	ncl. fasc. cuneatus externus
FCM	ncl. fasc. cuneatus medialis
FGR	nucleus fascicularis gracilis
FUN	funicular nuclei
TR	compl. sensorius trigeminalis
TSO	nucleus tractus solitarii

Table 16. Volumes of some relay nuclei of the medulla oblongata

For size indices in species see Table 17
For average size indices in supraspecific taxonomic units see Table 18

	OBL	n	REL	INO	PRP	COE
Tenrecidae						
Tenrec ecaudatus	315.5	1	2.083	2.989	1.664	0.317
Setifer setosus	180.6	1	0.717	1.019	0.802	0.184
Hemicentetes semispinosus	117.0	1	0.799	1.009	0.826	0.172
Echinops telfairi	90.3	1	0.596	0.640	0.599	0.153
Microgale talazaci	84.0	1	0.616	0.894	0.603	0.105
Potamogale velox	722.8	1	4.799	5.567	3.089	0.651
Micropotamogale ruwenz.	198.0	1	1.327	1.377	1.088	0.299
Chrysochloridae						
Chrysochloris stuhlmanni	68.0	1	0.433	1.007	0.523	0.074
Solenodontidae						
Solenodon paradoxus	471.6	1	4.181	4.933	2.894	0.566
Erinaceidae						
Echinosorex gymnurus	622.4	1	4.947	7.576	3.669	0.594
Hylomys suillus	125.4	1	1.381	1.452	1.007	0.172
Atelerix algirus	343.0	1	3.590	3.401	1.923	-.-
Erinaceus europaeus	335.0	1	3.130	5.299	2.175	0.621
Hemiechinus auritus	185.2	1	1.813	1.959	0.987	0.324
Talpidae						
Desmana moschata	402.4	1	2.916	4.894	2.837	0.436
Galemys pyrenaicus	147.1	1	0.889	1.807	0.775	0.103
Talpa europaea	95.7	1	0.994	1.714	0.570	0.128
Soricidae						
Sorex araneus	29.1	1	0.240	0.342	0.171	0.035
S. cinereus	17.2	1	0.218	0.224	0.169	0.014
Microsorex hoyi	10.9	1	0.084	0.120	0.095	0.019
Neomys fodiens	45.3	1	0.242	0.486	0.351	0.039
Blarina brevicauda	43.0	1	0.307	0.558	0.379	0.036
Crocidura flavescens	59.9	1	0.554	0.458	0.422	0.041
C. hildegardeae	31.4	1	0.291	0.303	0.239	0.051
C. russula	28.7	1	0.165	0.285	0.213	0.032
Suncus etruscus	11.1	1	0.092	0.131	0.090	0.008
S. murinus	54.4	1	0.414	0.612	0.330	0.043
Scutisorex somereni	82.9	1	0.630	0.980	0.659	0.085
Sylvisorex megalura	26.8	1	0.268	0.268	0.207	0.020
Ruwenzorisorex suncoides	54.9	1	0.454	0.467	0.384	0.064

Number of individuals = n

Abbreviations:

COE nucleus locus coeruleus
INO complexus olivaris inferior
OBL medulla oblongata
PRP nucleus prepositus hypoglossi
REL nucleus reticularis lateralis

Table 17. Size indices of some relay nuclei of the medulla oblongata related to the average of the four Tenrecinae (AvTn = 100)

For volumes in species see Table 16
For average size indices in supraspecific taxonomic units see Table 18

	OBL	n	REL	INO	PRP	COE
slope	.61		.58	.61	.52	.64
y-intercept	.776		-1.379	-1.340	-1.250	-2.176
Tenrecidae						
Tenrec ecaudatus	86	1	100	107	89	63
Setifer setosus	108	1	72	79	83	83
Hemicentetes semispinosus	108	1	121	121	124	123
Echinops telfairi	99	1	107	92	104	131
Microgale talazaci	132	1	156	184	143	132
Potamogale velox	240	1	276	242	194	160
Micropotamogale ruwenz.	204	1	224	185	179	240
Chrysochloridae						
Chrysochloris stuhlmanni	120	1	122	231	136	104
Solenodontidae						
Solenodon paradoxus	149	1	229	203	174	132
Erinaceidae						
Echinosorex gymnurus	174	1	241	276	199	121
Hylomys suillus	178	1	317	270	219	194
Atelerix algirus	102	1	187	133	110	--
Erinaceus europaeus	92	1	150	189	116	124
Hemiechinus auritus	111	1	183	153	103	148
Talpidae						
Desmana moschata	164	1	204	260	212	132
Galemys pyrenaicus	203	1	199	326	164	113
Talpa europaea	109	1	185	255	102	114
Soricidae						
Sorex araneus	118	1	149	181	91	119
S. cinereus	106	1	201	180	128	72
Microsorex hoyi	101	1	115	146	102	154
Neomys fodiens	144	1	119	201	151	102
Blarina brevicauda	117	1	130	198	143	80
Crocidura flavescens	128	1	187	128	130	71
C. hildegardeae	128	1	183	163	128	177
C. russula	111	1	98	144	108	104
Suncus etruscus	127	1	154	196	116	79
S. murinus	106	1	129	156	94	67
Scutisorex somereni	110	1	136	171	135	89
Sylvisorex megalura	159	1	239	207	152	101
Ruwenzorisorex suncoides	157	1	202	174	151	150

Number of individuals = n Abbreviations as in Tables 16 and 18

Table 18. Average size indices of some relay nuclei of the medulla oblongata related to the average of the four Tenrecinae (AvTn = 100)

For volumes in species see Table 16
Foe size indices in species see Table 17

	N	OBL	REL	INO	PRP	COE
slope		.61	.58	.61	.52	.64
y-intercept		.776	-1.379	-1.340	-1.250	-2.176
Insectivora (AvIS)	30+)	133	170	185	136	120
Tenrecidae	7	140	151	144	131	133
Tenrecinae*	4	**100**	**100**	**100**	**100**	**100**
Oryzorictinae*	1	132	156	184	143	132
Potamogalinae*	2	222	250	213	187	200
Chrysochloridae*	1	120	122	231	136	104
Solenodontidae*	1	149	229	203	174	132
Erinaceidae	5 +)	131	216	204	149	147
Echinosoricinae*	2	176	279	273	209	158
Erinaceinae*	3 +)	102	173	158	110	136
Talpidae	3	159	196	280	159	120
Desmaninae*	2	184	201	293	188	123
Talpinae*	1	109	185	255	102	114
Soricidae	13	124	157	173	125	105
Soricinae*	5	117	143	181	123	105
Sorex	2	112	175	181	109	96
terr. Soricinae	4	111	149	176	116	106
Neomys	1	144	119	201	151	102
Crocidurinae*	8	128	166	167	127	105
Crocidura	3	122	156	145	122	117
Suncus	2	117	141	176	105	73
AvIF of the 11 families and subfamilies marked by *		140	182	205	145	128

Number of species = N +) for COE = N – 1

Abbreviations:

COE nucleus locus coeruleus
INO complexus olivaris inferior
OBL medulla oblongata
PRP nucleus prepositus hypoglossi
REL nucleus reticularis lateralis

Table 19. Volumes of motor nuclei of the medulla oblongata

For size indices in species see Table 20
For average size indices in supraspecific taxonomic units see Table 21

	n	MOT	V	VII	X	XII
Tenrecidae						
Tenrec ecaudatus	1	7.88	1.890	3.585	0.829	1.572
Setifer setosus	1	4.14	0.832	2.099	0.372	0.841
Hemicentetes semispinosus	1	2.57	0.427	1.342	0.278	0.521
Echinops telfairi	1	2.66	0.385	1.444	0.281	0.549
Microgale talazaci	1	2.00	0.391	1.109	0.208	0.295
Potamogale velox	1	15.81	2.255	10.839	1.117	1.598
Micropotamogale ruwenzorii	1	3.77	0.783	2.154	0.355	0.482
Chrysochloridae						
Chrysochloris stuhlmanni	1	1.53	0.328	0.691	0.190	0.322
Solenodontidae						
Solenodon paradoxus	1	9.48	2.378	4.325	0.990	1.785
Erinaceidae						
Echinosorex gymnurus	1	16.68	3.378	8.810	1.460	3.036
Hylomys suillus	1	3.41	0.601	1.814	0.391	0.601
Atelerix algirus	1	9.65	1.553	4.560	0.920	2.612
Erinaceus europaeus	1	11.14	1.693	5.366	1.293	2.785
Hemiechinus auritus	1	5.30	0.818	2.468	0.486	1.530
Talpidae						
Desmana moschata	1	8.45	1.295	5.003	0.799	1.355
Galemys pyrenaicus	1	2.86	0.528	1.496	0.271	0.564
Talpa europaea	1	2.01	0.385	1.023	0.284	0.323
Soricidae						
Sorex araneus	1	0.76	0.143	0.344	0.098	0.171
S. cinereus	1	0.35	0.049	0.175	0.039	0.087
Microsorex hoyi	1	0.28	0.038	0.124	0.044	0.071
Neomys fodiens	1	1.03	0.181	0.500	0.116	0.234
Blarina brevicauda	1	1.04	0.169	0.418	0.216	0.236
Crocidura flavescens	1	1.11	0.186	0.549	0.160	0.216
C. hildegardeae	1	0.61	0.108	0.303	0.089	0.108
C. russula	1	0.58	0.070	0.329	0.075	0.107
Suncus etruscus	1	0.24	0.030	0.130	0.028	0.048
S. murinus	1	1.49	0.217	0.798	0.152	0.322
Scutisorex somereni	1	1.71	0.241	0.875	0.262	0.334
Sylvisorex megalura	1	0.54	0.068	0.298	0.075	0.103
Ruwenzorisorex suncoides	1	1.12	0.136	0.614	0.157	0.216

Number of individuals = n

Abbreviations:

MOT sum of measured motor nuclei
V ncl. motorius nervi trigemini
VII ncl. nervi facialis
X ncl. dorsalis motorius nervi vagi
XII ncl. nervi hypoglossi

Table 20. Size indices of motor nuclei of the medulla oblongata related to the average of the four Tenrecinae (AvTn = 100)

For volumes in species see Table 19
For average size indices in supraspecific taxonomic units see Table 21

	n	MOT	V	VII	X	XII
slope		.62	.67	.62	.62	.59
y-intercept		-.853	-1.705	-1.146	-1.844	-1.476
Tenrecidae						
Tenrec ecaudatus	1	86	104	76	88	88
Setifer setosus	1	100	108	99	88	100
Hemicentetes semispinosus	1	96	90	99	102	94
Echinops telfairi	1	118	98	126	123	117
Microgale talazaci	1	129	148	140	131	90
Potamogale velox	1	210	154	282	145	108
Micropotamogale ruwenzorii	1	158	185	177	146	97
Chrysochloridae						
Chrysochloris stuhlmanni	1	110	140	98	134	109
Solenodontidae						
Solenodon paradoxus	1	119	154	107	122	115
Erinaceidae						
Echinosorex gymnurus	1	185	191	192	159	173
Hylomys suillus	1	198	203	207	223	166
Atelerix algirus	1	115	94	107	107	159
Erinaceus europaeus	1	121	94	115	138	156
Hemiechinus auritus	1	128	107	117	115	183
Talpidae						
Desmana moschata	1	138	111	160	128	111
Galemys pyrenaicus	1	161	173	166	150	151
Talpa europaea	1	93	102	93	129	72
Soricidae						
Sorex araneus	1	128	153	114	162	130
S. cinereus	1	90	83	88	99	98
Microsorex hoyi	1	108	101	95	168	119
Neomys fodiens	1	135	148	129	149	140
Blarina brevicauda	1	117	116	92	238	122
Crocidura flavescens	1	98	98	95	138	88
C. hildegardeae	1	104	117	102	148	83
C. russula	1	93	70	104	117	77
Suncus etruscus	1	115	99	124	134	100
S. murinus	1	120	104	126	120	121
Scutisorex somereni	1	93	76	93	140	86
Sylvisorex megalura	1	135	109	145	182	113
Ruwenzorisorex suncoides	1	132	99	142	181	117

Number of individuals = n Abbreviations as in Tables 19 and 21

Table 21. Average size indices of motor nuclei of the medulla oblongata related to the average of the four Tenrecinae (AvTn = 100)

For volumes in species see Table 19
For size indices in species see Table 20

	N	MOT	V	VII	X	XII
slope		.62	.67	.62	.62	.59
y-intercept		-.853	-1.705	-1.146	-1.844	-1.476
Insectivora (AvIS)	30	124	121	127	140	116
Tenrecidae	7	128	127	143	118	99
Tenrecinae*	4	**100**	**100**	**100**	**100**	**100**
Oryzorictinae*	1	129	148	140	131	90
Potamogalinae*	2	184	170	230	145	103
Chrysochloridae*	1	110	140	98	134	109
Solenodontidae*	1	119	154	107	122	115
Erinaceidae	5	149	138	148	148	167
Echinosoricinae*	2	192	197	200	191	169
Erinaceinae*	3	121	98	113	120	166
Talpidae	3	131	129	140	136	111
Desmaninae*	2	150	142	163	139	131
Talpinae*	1	93	102	93	129	72
Soricidae	13	113	106	111	152	107
Soricinae*	5	116	120	104	163	122
Sorex	2	109	118	101	131	114
terr. Soricinae	4	111	113	97	167	117
Neomys	1	135	148	129	149	140
Crocidurinae*	8	111	97	116	145	98
Crocidura	3	98	95	100	134	83
Suncus	2	117	102	125	127	110
AvIF of the 11 families and subfamilies marked by *		130	133	133	138	116

Number of species = N
Abbreviations:

MOT	sum of measured motor nuclei
V	ncl. motorius nervi trigemini
VII	ncl. nervi facialis
X	ncl. dorsalis motorius nervi vagi
XII	ncl. nervi hypoglossi

Table 22. Volumes of the vestibular nuclei of the medulla oblongata

For size indices in species see Table 23
For average size indices in supraspecific taxonomic units see Table 24

	n	VC	VM	VI	VL	VS
Tenrecidae						
Tenrec ecaudatus	1	14.44	5.152	4.270	2.885	2.132
Setifer setosus	1	7.18	2.705	2.009	1.090	1.379
Hemicentetes semispinosus	1	5.69	2.125	1.823	0.995	0.748
Echinops telfairi	1	3.85	1.482	1.111	0.625	0.634
Microgale talazaci	1	3.44	1.519	0.883	0.681	0.361
Potamogale velox	1	25.00	8.039	9.248	3.451	4.266
Micropotamogale ruwenzorii	1	8.58	3.259	2.626	1.277	1.421
Chrysochloridae						
Chrysochloris stuhlmanni	1	3.04	1.286	0.796	0.504	0.456
Solenodontidae						
Solenodon paradoxus	1	22.05	8.800	7.285	2.936	3.030
Erinaceidae						
Echinosorex gymnurus	1	33.49	14.390	8.991	7.053	3.055
Hylomys suillus	1	8.52	3.603	2.332	1.462	1.122
Atelerix algirus	1	15.56	6.417	4.774	2.399	1.972
Erinaceus europaeus	1	15.90	6.438	3.979	3.256	2.231
Hemiechinus auritus	1	8.59	3.422	2.273	1.716	1.180
Talpidae						
Desmana moschata	1	19.49	8.125	6.485	2.541	2.341
Galemys pyrenaicus	1	6.16	2.612	2.085	0.793	0.666
Talpa europaea	1	4.14	1.671	1.418	0.460	0.590
Soricidae						
Sorex araneus	1	1.29	0.539	0.266	0.264	0.219
S. cinereus	1	0.88	0.367	0.262	0.148	0.106
Microsorex hoyi	1	0.55	0.206	0.156	0.108	0.076
Neomys fodiens	1	2.32	0.964	0.728	0.356	0.270
Blarina brevicauda	1	2.22	0.997	0.668	0.224	0.326
Crocidura flavescens	1	2.15	0.924	0.554	0.422	0.246
C. hildegardeae	1	1.57	0.674	0.470	0.237	0.185
C. russula	1	1.19	0.544	0.346	0.171	0.131
Suncus etruscus	1	0.52	0.265	0.133	0.059	0.061
S. murinus	1	2.14	0.932	0.595	0.385	0.230
Scutisorex somereni	1	3.55	1.416	1.069	0.613	0.448
Sylvisorex megalura	1	1.40	0.644	0.325	0.273	0.158
Ruwenzorisorex suncoides	1	2.51	1.129	0.886	0.235	0.259

Number of individuals = n

Abbreviations:

VC vestibular complex
VI nucleus vestibularis inferior
VL nucleus vestibularis lateralis
VM nucleus vestibularis medialis
VS nucleus vestibularis superior

Table 23. Size indices of the vestibular nuclei of the medulla oblongata related to the average of the four Tenrecinae (AvTn = 100)

For volumes in species see Table 22
For average size indices in supraspecific taxonomic units see Table 24

	n	VC	VM	VI	VL	VS
slope		.53	.53	.55	.54	.51
y-intercept		-.393	-.822	-.966	-1.179	-1.149
Tenrecidae						
Tenrec ecaudatus	1	100	96	97	114	96
Setifer setosus	1	98	99	92	86	120
Hemicentetes semispinosus	1	113	114	123	115	93
Echinops telfairi	1	89	92	88	84	91
Microgale talazaci	1	109	129	97	127	70
Potamogale velox	1	205	177	249	162	227
Micropotamogale ruwenzorii	1	188	192	196	163	194
Chrysochloridae						
Chrysochloris stuhlmanni	1	106	121	97	104	98
Solenodontidae						
Solenodon paradoxus	1	173	185	188	132	154
Erinaceidae						
Echinosorex gymnurus	1	236	272	207	284	140
Hylomys suillus	1	247	281	233	249	201
Atelerix algirus	1	116	129	117	103	96
Erinaceus europaeus	1	110	120	90	129	101
Hemiechinus auritus	1	118	126	104	136	103
Talpidae						
Desmana moschata	1	191	213	210	143	147
Galemys pyrenaicus	1	174	198	203	132	117
Talpa europaea	1	99	107	116	64	88
Soricidae						
Sorex araneus	1	93	104	69	114	94
S. cinereus	1	91	102	98	92	65
Microsorex hoyi	1	81	82	85	97	66
Neomys fodiens	1	135	151	150	123	95
Blarina brevicauda	1	113	136	120	68	100
Crocidura flavescens	1	89	102	80	103	62
C. hildegardeae	1	114	132	122	103	81
C. russula	1	82	101	85	70	54
Suncus etruscus	1	92	127	87	64	62
S. murinus	1	82	96	79	87	54
Scutisorex somereni	1	97	104	101	98	76
Sylvisorex megalura	1	140	173	118	164	93
Ruwenzorisorex suncoides	1	133	161	166	74	83

Number of individuals = n Abbreviations as in Tables 22 and 24

Table 24. Average size indices of vestibular nuclei of the medulla oblongata related to the average of the four Tenrecinae (AvTn = 100)

For volumes in species see Table 22
For size indices in species see Table 23

	N	VC	VM	VI	VL	VS
slope		.53	.53	.55	.54	.51
y-intercept		-.393	-.822	-.966	-1.179	-1.149
Insectivora (AvIS)	30	127	141	129	119	104
Tenrecidae	7	129	128	135	122	127
Tenrecinae*	4	**100**	**100**	**100**	**100**	**100**
Oryzorictinae*	1	109	129	97	127	70
Potamogalinae*	2	197	184	223	163	211
Chrysochloridae*	1	106	121	97	104	98
Solenodontidae*	1	173	185	188	132	154
Erinaceidae	5	165	186	150	180	128
Echinosoricinae*	2	241	276	220	266	171
Erinaceinae*	3	115	125	104	122	100
Talpidae	3	155	173	176	113	117
Desmaninae*	2	182	206	207	137	132
Talpinae*	1	99	107	116	64	88
Soricidae	13	103	121	105	97	76
Soricinae*	5	103	115	104	99	84
Sorex	2	92	103	83	103	80
terr. Soricinae	4	95	106	93	93	81
Neomys	1	135	151	150	123	95
Crocidurinae*	8	104	125	105	96	71
Crocidura	3	95	112	96	92	66
Suncus	2	87	111	83	75	58
AvIF of the 11 families and subfamilies marked by *		139	152	142	128	116

Number of species = N
Abbreviations:

VC	vestibular complex
VI	nucleus vestibularis inferior
VL	nucleus vestibularis lateralis
VM	nucleus vestibularis medialis
VS	nucleus vestibularis superior

Table 25. Volumes of the auditory nuclei of the medulla oblongata

For size indices in species see Table 26
For average size indices in supraspecific taxonomic units see Table 27

	n	AUD	CON	DCO	VCO	OLS
Tenrecidae						
Tenrec ecaudatus	1	14.14	8.43	4.23	4.20	5.71
Setifer setosus	1	9.59	5.54	2.00	3.54	4.05
Hemicentetes semispinosus	1	4.78	2.87	1.01	1.86	1.91
Echinops telfairi	1	6.81	4.39	1.63	2.76	2.42
Microgale talazaci	1	4.83	3.13	1.31	1.82	1.70
Potamogale velox	1	19.72	10.48	4.06	6.42	9.24
Micropotamogale ruwenzorii	1	5.36	2.94	1.33	1.61	2.42
Chrysochloridae						
Chrysochloris stuhlmanni	1	3.09	1.71	0.77	0.94	1.38
Solenodontidae						
Solenodon paradoxus	1	19.32	11.70	4.65	7.05	7.62
Erinaceidae						
Echinosorex gymnurus	1	35.20	22.16	11.82	10.34	13.04
Hylomys suillus	1	8.27	4.69	1.91	2.78	3.58
Atelerix algirus	1	14.93	8.56	3.55	5.01	6.37
Erinaceus europaeus	1	10.95	6.84	2.15	4.69	4.11
Hemiechinus auritus	1	11.01	5.95	2.07	3.88	5.06
Talpidae						
Desmana moschata	1	11.45	5.31	2.47	2.84	6.14
Galemys pyrenaicus	1	3.25	1.88	0.85	1.03	1.37
Talpa europaea	1	3.53	2.04	0.99	1.05	1.49
Soricidae						
Sorex araneus	1	0.90	0.61	0.30	0.31	0.29
S. cinereus	1	0.62	0.38	0.18	0.20	0.24
Microsorex hoyi	1	0.34	0.22	0.10	0.12	0.12
Neomys fodiens	1	1.22	0.72	0.33	0.39	0.50
Blarina brevicauda	1	1.42	0.82	0.37	0.45	0.60
Crocidura flavescens	1	3.11	1.92	1.01	0.91	1.19
C. hildegardeae	1	1.89	1.12	0.50	0.62	0.77
C. russula	1	1.47	0.90	0.45	0.45	0.57
Suncus etruscus	1	0.50	0.30	0.14	0.16	0.20
S. murinus	1	2.87	1.86	0.83	1.03	1.01
Scutisorex somereni	1	5.45	3.59	1.36	2.23	1.86
Sylvisorex megalura	1	1.74	1.07	0.52	0.55	0.67
Ruwenzorisorex suncoides	1	2.53	1.59	0.77	0.82	0.94

Number of individuals = n

Abbreviations:

AUD	auditory (acoustic) nuclei		
CON	cochlear nuclei	OLS	complexus olivaris superior
DCO	nucleus cochlearis dorsalis	VCO	nucleus cochlearis ventralis

Table 26. Size indices of the auditory nuclei related to the average of the four Tenrecinae (AvTn = 100)

For volumes in species see Table 25
For average size indices in supraspecific taxonomic units see Table 27

	n	AUD	CON	DCO	VCO	OLS
slope		.51	.50	.50	.51	.54
y-intercept		-.267	-.461	-.867	-.700	-.742
Tenrecidae						
Tenrec ecaudatus	1	84	83	107	67	82
Setifer setosus	1	109	104	96	109	117
Hemicentetes semispinosus	1	78	77	69	82	81
Echinops telfairi	1	129	136	128	141	119
Microgale talazaci	1	124	130	139	126	116
Potamogale velox	1	138	122	120	121	159
Micropotamogale ruwenzorii	1	96	86	99	78	113
Chrysochloridae						
Chrysochloris stuhlmanni	1	87	78	90	71	104
Solenodontidae						
Solenodon paradoxus	1	129	130	132	128	125
Erinaceidae						
Echinosorex gymnurus	1	212	223	303	169	192
Hylomys suillus	1	195	180	187	177	223
Atelerix algirus	1	95	91	96	87	99
Erinaceus europaeus	1	65	68	54	75	59
Hemiechinus auritus	1	126	112	100	120	147
Talpidae						
Desmana moschata	1	95	73	86	64	126
Galemys pyrenaicus	1	75	70	81	64	83
Talpa europaea	1	69	65	81	56	76
Soricidae						
Sorex araneus	1	51	55	70	47	46
S. cinereus	1	50	49	57	45	54
Microsorex hoyi	1	39	39	43	38	40
Neomys fodiens	1	56	53	62	49	64
Blarina brevicauda	1	57	53	61	50	66
Crocidura flavescens	1	103	103	137	82	106
C. hildegardeae	1	108	102	116	96	122
C. russula	1	80	78	100	66	86
Suncus etruscus	1	67	64	78	57	77
S. murinus	1	88	93	105	86	83
Scutisorex somereni	1	122	130	126	134	109
Sylvisorex megalura	1	135	132	162	116	148
Ruwenzorisorex suncoides	1	107	108	132	94	109

Number of individuals = n   Abbreviations as in Tables 25 and 27

Table 27. Average size indices of the auditory nuclei related to the average of the four Tenrecinae (AvTn = 100)

For volumes in species see Table 25
For size indices in species see Table 26

	N	AUD	CON	DCO	VCO	OLS
slope		.51	.50	.50	.51	.54
y-intercept		-.267	-.461	-.867	-.700	-.742
Insectivora (AvIS)	30	99	96	107	90	104
Tenrecidae	7	108	105	108	103	112
Tenrecinae*	4	**100**	**100**	**100**	**100**	**100**
Oryzorictinae*	1	124	130	139	126	116
Potamogalinae*	2	117	104	110	100	136
Chrysochloridae*	1	87	78	90	71	104
Solenodontidae*	1	129	130	132	128	125
Erinaceidae	5	139	135	148	126	144
Echinosoricinae*	2	203	201	245	173	207
Erinaceinae*	3	95	90	83	94	102
Talpidae	3	80	69	83	61	95
Desmaninae*	2	85	72	84	64	105
Talpinae*	1	69	65	81	56	76
Soricidae	13	82	81	96	74	85
Soricinae*	5	51	50	59	46	54
Sorex	2	50	52	63	46	50
terr. Soricinae	4	49	49	58	45	51
Neomys	1	56	53	62	49	64
Crocidurinae*	8	101	101	120	91	105
Crocidura	3	97	94	118	81	105
Suncus	2	78	78	92	72	80
AvIF of the 11 families and subfamilies marked by *		106	102	113	95	112

Number of species = N
Abbreviations:

AUD auditory (acoustic) nuclei
CON cochlear nuclei
DCO nucleus cochlearis dorsalis
OLS complexus olivaris superior
VCO nucleus cochlearis ventralis

Table 28. Volumes of mesencephalic components

For size indices in species see Table 29
For average size indices in supraspecific taxonomic units see Table 30
For volumes of subcommissural body (SCB) see Table 40

	n	MES	n	MTG	NLL	MTC	SUC	INC
Tenrecidae								
Tenrec ecaudatus	3	133.3	1	72.82	4.48	60.49	24.73	35.76
Setifer setosus	3	76.6	1	44.94	3.24	31.68	9.69	21.99
Hemicentetes semisp.	2	45.9	1	27.17	2.40	18.69	8.33	10.36
Echinops telfairi	2	44.0	1	19.72	2.23	24.29	8.50	15.79
Erinaceidae								
Erinaceus europaeus	5	151.3	1	92.55	3.72	58.74	29.57	29.17
Soricidae								
Sorex araneus	5	11.5	1	7.15	0.62	4.34	2.69	1.65
S. minutus	4	6.4	1	3.90	0.19	2.53	1.62	0.91
Crocidura flavescens	3	26.7	1	14.58	1.33	12.15	5.54	6.61
C. russula	5	13.7	1	7.78	0.65	5.89	3.04	2.85
Suncus murinus	3	24.6	1	13.98	1.01	10.61	5.09	5.52

Number of individuals = n
Abbreviations:

INC colliculi inferiores
MES mesencephalon
MTC tectum mesencephali (colliculi superiores + inferiores)
MTG tegmentum mesencephali
NLL nucleus lemnisci lateralis
SUC colliculi superiores

Table 29. Size indices of mesencephalic components related to the average of the four Tenrecinae (AvTn = 100)

For volumes in species see Table 28
For average size indices in supraspecific taxonomic units see Table 30
For size indices of subcommissural body (SCB) see Table 41

	n	MES	n	MTG	NLL	MTC	SUC	INC
slope		.58		.58	.58	.58	.58	.58
y-intercept		.481		.213	-.865	.141	-.287	-.061
Tenrecidae								
Tenrec ecaudatus	3	88	1	89	66	87	96	82
Setifer setosus	3	106	1	115	100	96	79	106
Hemicentetes semispin.	2	96	1	106	112	86	102	76
Echinops telfairi	2	109	1	90	122	131	123	136
Erinaceidae								
Erinaceus europaeus	5	100	1	113	54	85	115	67
Soricidae								
Sorex araneus	5	99	1	114	117	82	135	49
S. minutus	4	90	1	101	60	77	132	44
Crocidura flavescens	3	125	1	126	137	124	151	107
C. russula	5	112	1	118	118	105	146	81
Suncus murinus	3	105	1	111	96	100	128	82

Number of individuals = n
Abbreviations:

INC colliculi inferiores
MES mesencephalon
MTC tectum mesencephali (colliculi superiores + inferiores)
MTG tegmentum mesencephali
NLL nucleus lemnisci lateralis
SUC colliculi superiores

Table 30. Average size indices of mesencephalic components related to the average of the four Tenrecinae (AvTn = 100)

For volumes in species see Table 28
For size indices in species see Table 29
For average size indices of subcommissural body (SCB) see Table 42

	N	MES	MTG	NLL	MTC	SUC	INC
slope		.58	.58	.58	.58	.58	.58
y-intercept		.481	.213	-.865	.141	-.287	-.061
Insectivora (AvIS)	10	103	108	98	97	121	83
Tenrecidae							
Tenrecinae*	4	**100**	**100**	**100**	**100**	**100**	**100**
Erinaceidae							
Erinaceinae*	1	100	113	54	85	115	67
Soricidae	5	106	114	106	98	138	73
Soricinae*	2	94	107	88	79	134	47
Crocidurinae*	3	114	118	117	110	142	90
Crocidura	2	118	122	128	115	148	94
AvIF of the 4 subfamilies marked by *		102	110	90	94	123	76

Number of species = N
Abbreviations:

INC	colliculi inferiores
MES	mesencephalon
MTC	tectum mesencephali (colliculi superiores + inferiores)
MTG	tegmentum mesencephali
NLL	nucleus lemnisci lateralis
SUC	colliculi superiores

Table 31. Volumes of cerebellar nuclei and ventral pons

For size indices in species see Table 32
For average size indices in supraspecific taxonomic units see Table 33

	n	CER	n	VPO	TCN	MCN	ICN	LCN
Tenrecidae								
Tenrec ecaudatus	3	311.5	1	2.524	9.890	2.074	4.465	3.351
Setifer setosus	3	146.8	1	-.-	4.331	1.239	1.845	1.247
Hemicentetes semisp.	2	120.5	1	-.-	3.332	0.563	1.572	1.197
Echinops telfairi	2	58.2	1	0.494	1.720	0.390	0.870	0.460
Erinaceidae								
Erinaceus europaeus	5	395.8	1	-.-	12.716	3.315	5.556	3.845
Soricidae								
Sorex araneus	5	22.0	1	-.-	0.579	0.179	0.185	0.215
S. minutus	4	12.1	1	-.-	0.359	0.095	0.128	0.136
Crocidura flavescens	3	51.9	1	-.-	1.145	0.268	0.415	0.462
C. russula	5	21.0	1	-.-	0.564	0.114	0.225	0.225
Suncus murinus	3	42.8	1	-.-	1.077	0.238	0.389	0.450

Number of individuals = n
Abbreviations:

CER	cerebellum
ICN	nucleus interpositus cerebelli
LCN	nucleus lateralis cerebelli
MCN	nucleus medialis cerebelli
TCN	total cerebellar nuclei
VPO	ventral pons

Table 32. Size indices of cerebellar nuclei and ventral pons related to the average of the four Tenrecinae (AvTn = 100)

For volumes in species see Table 31
For average size indices in supraspecific taxonomic units see Table 33

	n	CER	n	VPO	TCN	MCN	ICN	LCN
slope		.68		.68	.68	.68	.68	.68
y-intercept		.552		-1.608	-.979	-1.638	-1.314	-1.476
Tenrecidae								
Tenrec ecaudatus	3	89	1	104	96	92	94	102
Setifer setosus	3	100	1	--	100	131	92	91
Hemicentetes semispin.	2	133	1	--	125	97	128	141
Echinops telfairi	2	78	1	96	78	81	86	66
Erinaceidae								
Erinaceus europaeus	5	113	1	--	124	147	117	117
Soricidae								
Sorex araneus	5	127	1	--	114	160	79	133
S. minutus	4	124	1	--	125	150	96	148
Crocidura flavescens	3	146	1	--	110	117	86	139
C. russula	5	114	1	--	105	96	90	131
Suncus murinus	3	110	1	--	94	94	73	123

Number of individuals = n
Abbreviations:

CER	cerebellum
ICN	nucleus interpositus cerebelli
LCN	nucleus lateralis cerebelli
MCN	nucleus medialis cerebelli
TCN	total cerebellar nuclei
VPO	ventral pons

Table 33. Average size indices of cerebellar nuclei and ventral pons related to the average of the four Tenrecinae (AvTn = 100)

For volumes in species see Table 31
For size indices in species see Table 32

	N	CER	TCN	MCN	ICN	LCN
slope		.68	.68	.68	.68	.68
y-intercept		.552	-.979	-1.638	-1.314	-1.476
Insectivora (AvIS)	10	113	107	117	94	119
Tenrecidae						
Tenrecinae*	4	**100**	**100**	**100**	**100**	**100**
Erinaceidae						
Erinaceinae*	1	113	124	147	117	117
Soricidae	5	124	110	123	85	135
Soricinae*	2	126	119	155	87	140
Crocidurinae*	3	123	103	102	83	131
Crocidura	2	130	107	107	88	135
AvIF of the 4 subfamilies marked by *		115	112	126	97	122

Number of species = N
Abbreviations:

CER cerebellum
ICN nucleus interpositus cerebelli
LCN nucleus lateralis cerebelli
MCN nucleus medialis cerebelli
TCN total cerebellar nuclei

Table 34. Volumes of diencephalic components

For size indices in species see Table 35
For average size indices in supraspecific taxonomic units see Table 36
For volumes of subthalamic components and geniculate bodies see Table 37
For volumes of medial nuclei of habenula (HAM) see Table 40

	n	DIE	ETH	THA	STH	HTH	CAI	TRO
Tenrecidae								
Tenrec ecaudatus	2	153.7	2.890	66.67	12.29	54.94	14.03	2.903
Setifer setosus	1	93.1	1.557	39.89	5.58	36.82	8.24	1.029
Hemicentetes semispin.	1	47.7	1.261	20.66	3.60	18.59	3.05	0.551
Echinops telfairi	1	40.3	0.965	17.67	2.65	16.19	2.20	0.597
Erinaceidae								
Erinaceus europaeus	2	242.4	3.789	119.50	14.49	81.51	19.09	4.041
Soricidae								
Sorex araneus	2	14.8	0.269	7.38	1.11	5.26	0.62	0.146
S. minutus	1	7.7	0.150	3.65	0.66	2.86	0.27	0.099
Crocidura flavescens	2	25.1	0.564	12.50	1.52	8.40	1.88	0.217
C. russula	1	13.9	0.360	7.31	0.85	4.51	0.71	0.113
Suncus murinus	1	24.1	0.498	11.56	1.59	9.08	1.11	0.218

Number of individuals = n
Abbreviations:

CAI	capsula interna
DIE	diencephalon
ETH	epithalamus
HTH	hypothalamus
STH	subthalamus
THA	thalamus
TRO	tractus opticus

Table 35. Size indices of diencephalic components related to the average of the four Tenrecinae (AvTn = 100)

For volumes in species see Table 34
For average size indices in supraspecific taxonomic units see Table 36
For size indices of subthalamic components and geniculate bodies see Table 38
For size indices of medial nuclei of the habenula (HAM) see Table 41

	n	DIE	ETH	THA	STH	HTH	CAI	TRO
slope		.65	.63	.64	.71	.62	.78	.61
y-intercept		.350	-1.276	.010	-.948	.007	-1.083	-1.416
Tenrecidae								
Tenrec ecaudatus	2	85	78	87	91	82	88	123
Setifer setosus	1	119	94	118	102	122	140	95
Hemicentetes semispin.	1	97	119	96	109	96	91	79
Echinops telfairi	1	98	109	99	98	100	81	102
Erinaceidae								
Erinaceus europaeus	2	135	102	156	107	123	120	172
Soricidae								
Sorex araneus	2	146	117	163	189	123	123	92
S. minutus	1	131	111	138	204	112	102	104
Crocidura flavescens	2	125	127	141	123	102	163	72
C. russula	1	129	149	153	137	100	132	68
Suncus murinus	1	109	102	119	116	101	86	66

Number of individuals = n
Abbreviations:

CAI capsula interna
DIE diencephalon
ETH epithalamus
HTH hypothalamus
STH subthalamus
THA thalamus
TRO tractus opticus

Table 36. Average size indices of diencephalic components related to the average of the four Tenrecidae (AvTn = 100)

For volumes in species see Table 34
For size indices in species see Table 35
For av. size indices of subthalamic components and geniculate bodies see Table 39
For average size indices of medial nuclei of the habenula (HAM) see Table 42

	N	DIE	ETH	THA	STH	HTH	CAI	TRO
slope		.65	.63	.64	.71	.62	.78	.61
y-intercept		.350	-1.276	.010	-.948	.007	-1.083	-1.416
Insectivora (AvIS)	10	117	111	127	127	106	113	97
Tenrecidae								
Tenrecinae*	4	**100**	**100**	**100**	**100**	**100**	**100**	**100**
Erinaceidae								
Erinaceinae*	1	135	102	156	107	123	120	172
Soricidae	5	128	121	143	154	108	121	80
Soricinae*	2	138	114	151	196	117	112	98
Crocidurinae*	3	121	126	138	125	101	127	69
Crocidura	2	127	138	147	130	101	147	70
AvIF of the 4 subfamilies marked by *		124	111	136	132	110	115	110

Number of species = N
Abbreviations:

CAI	capsula interna
DIE	diencephalon
ETH	epithalamus
HTH	hypothalamus
STH	subthalamus
THA	thalamus
TRO	tractus opticus

Table 37. Volumes of subthalamic components and geniculate bodies

For size indices in species see Table 38
For average size indices in supraspecific taxonomic units see Table 39
For volumes of other diencephalic components see Tables 65 and 68

	n	STH	PALL	LUY	CGM	CGL	GLD	GLV
Tenrecidae								
Tenrec ecaudatus	2	12.29	11.78	0.513	10.406	4.419	3.134	1.285
Setifer setosus	1	5.58	5.28	0.304	5.288	2.916	2.022	0.894
Hemicentetes semisp.	1	3.60	3.39	0.205	2.624	1.194	0.798	0.396
Echinops telfairi	1	2.65	2.41	0.239	2.124	1.126	0.736	0.390
Erinaceidae								
Erinaceus europaeus	2	14.49	13.56	0.927	13.984	10.752	7.956	2.796
Soricidae								
Sorex araneus	2	1.11	1.00	0.107	0.670	0.264	0.150	0.114
S. minutus	1	0.66	0.61	0.053	0.306	0.159	0.110	0.049
Crocidura flavescens	2	1.52	1.40	0.122	1.204	0.670	0.478	0.192
C. russula	1	0.85	0.76	0.087	0.656	0.239	0.150	0.089
Suncus murinus	1	1.59	1.48	0.109	1.458	0.492	0.336	0.156

Number of individuals = n
Abbreviations:

- CGL corpus geniculatum laterale
- CGM corpus geniculatum mediale
- GLD dorsal part of CGL
- GLV ventral part of CGL
- LUY nucleus subthalamicus Luysi
- PALL globus pallidus, pallidum
- STH subthalamus

Table 38. Size indices of subthalamic components and geniculate bodies related to the average of the four Tenrecinae (AvTn = 100)

For volumes in species see Table 37
For average size indices in supraspecific taxonomic units see Table 39
For size indices of other diencephalic components see Tables 66 and 69

	n	STH	PALL	LUY	CGM	CGL	GLD	GLV
slope		.71	.71	.71	.72	.60	.60	.60
y-intercept		-.948	-.975	-2.162	-1.052	-1.080	-1.246	-1.579
Tenrecidae								
Tenrec ecaudatus	2	91	92	62	91	93	96	85
Setifer setosus	1	102	103	91	116	132	134	127
Hemicentetes semispin.	1	109	110	102	97	83	81	87
Echinops telfairi	1	98	95	145	96	93	89	101
Erinaceidae								
Erinaceus europaeus	2	107	107	112	123	226	245	185
Soricidae								
Sorex araneus	2	189	181	297	142	79	66	107
S. minutus	1	204	199	268	119	78	80	77
Crocidura flavescens	2	123	120	161	119	106	111	96
C. russula	1	137	131	229	131	68	62	80
Suncus murinus	1	116	115	130	130	72	72	72

Number of individuals = n
Abbreviations:

CGL	corpus geniculatum laterale
CGM	corpus geniculatum mediale
GLD	dorsal part of CGL
GLV	ventral part of CGL
LUY	nucleus subthalamicus Luysi
PALL	globus pallidus, pallidum
STH	subthalamus

Table 39. Average size indices of subthalamic components and geniculate bodies related to the average of the four Tenrecidae (AvTn = 100)

For volumes in species see Table 37
For size indices in species see Table 38
For average size indices of other diencephalic components see Tables 67 and 70

	N	STH	PALL	LUY	CGM	CGL	GLD	GLV
slope		.71	.71	.71	.72	.60	.60	.60
y-intercept		-.948	-.975	-2.162	-1.052	-1.080	-1.246	-1.579
Insectivora (AvIS)	10	127	125	160	116	103	104	102
Tenrecidae								
Tenrecinae*	4	**100**	**100**	**100**	**100**	**100**	**100**	**100**
Erinaceidae								
Erinaceinae*	1	107	107	112	123	226	245	185
Soricidae	5	154	149	217	128	81	78	86
Soricinae*	2	196	190	283	130	79	73	92
Crocidurinae*	3	125	122	173	127	82	82	83
Crocidura	2	130	125	195	125	87	87	88
AvIF of the 4 subfamilies marked by *		132	130	167	120	122	125	115

Number of species = N
Abbreviations:

CGL corpus geniculatum laterale
CGM corpus geniculatum mediale
GLD dorsal part of CGL
GLV ventral part of CGL
LUY nucleus subthalamicus Luysi
PALL globus pallidus, pallidum
STH subthalamus

Table 40. Volumes of some periventricular and septal structures

For size indices in species see Table 41
For average size indices in supraspecific taxonomic units see Table 42'

	n	SET	n	SFB	n	HAM	n	EPI	n	SCB
Tenrecidae										
Tenrec ecaudatus	1	0.702	1	0.124	1	0.986	1	0.479	1	0.0112
Setifer setosus	1	0.329	1	0.060	1	0.426	1	0.445	1	0.0031
Hemicentetes semis.	1	0.344	1	0.033	1	0.432	1	0.184	1	0.0039
Echinops telfairi	1	0.338	1	0.029	1	0.365	1	0.243	1	0.0021
Oryzorictes talpoides	1	0.331	1	0.018	1	0.257	1	0.051	1	0.0007
Microgale dobsoni	1	0.278	1	0.008	1	0.241	1	0.035	1	0.0010
M. talazaci	1	0.367	1	0.019	1	0.370	1	0.087	1	0.0015
Limnogale mergulus	1	0.602	1	0.019	1	0.469	1	0.122	1	0.0020
Potamogale velox	1	1.108	1	0.090	1	1.462	1	0.003	1	0.0017
Micropotamogale ruwenz.	1	0.367	1	0.070	1	0.592	1	0.013	-	-.-
Chrysochloridae										
Chrysochloris asiatica	1	0.485	1	0.015	1	0.449	1	0.025	1	0.0081
C. stuhlmanni	6	0.394	6	0.019	6	0.379	6	0.096	6	(0.0)
Solenodontidae										
Solenodon paradoxus	1	0.884	1	0.084	1	2.109	1	1.289	1	0.0442
Erinaceidae										
Echinosorex gymnurus	1	1.664	1	0.110	1	2.400	1	0	-	-.-
Hylomys suillus	1	0.480	1	0.028	1	0.591	1	0.197	-	-.-
Atelerix algirus	1	1.199	1	0.063	1	0.948	1	0.303	1	0.0226
Erinaceus europaeus	2	0.885	2	0.059	2	1.177	2	0.424	2	0.0313
Hemiechinus auritus	1	0.419	1	0.080	1	0.747	1	0.121	-	-.-
Talpidae										
Desmana moschata	1	1.058	1	0.035	1	1.051	1	0.459	1	0.0278
Galemys pyrenaicus	1	0.446	5	0.038	1	0.424	1	0.071	1	0.0055
Talpa europaea	1	0.329	5	0.009	1	0.365	1	0.050	1	0.0102
T. micrura	1	0.386	1	0.018	1	0.470	1	0.058	-	-.-
Parascalops breweri	1	0.530	1	0.018	1	0.458	1	0.068	-	-.-
Scalopus aquaticus	1	0.504	1	0.019	1	0.603	1	0.095	1	0.0260

Table 40 (cont.)

	n	SET	n	SFB	n	HAM	n	EPI	n	SCB
Soricidae										
Sorex araneus	2	0.124	10	0.004	1	0.081	10	0.011	1	0.0003
S. minutus	1	0.074	1	0.003	1	0.062	1	0.004	1	0.0003
Microsorex hoyi	1	0.071	1	0.003	1	0.065	1	0.003	-	-.-
Neomys fodiens	1	0.269	1	0.007	1	0.210	1	0.028	1	0.0004
Anourosorex squamipes	1	0.215	1	0.019	1	0.181	1	0.025	-	-.-
Crocidura flavescens	1	0.261	1	0.018	1	0.191	1	0.037	1	0.0006
C. russula	2	0.126	2	0.010	1	0.141	5	0.029	1	0.0008
Suncus etruscus	1	0.052	1	0.004	1	0.070	1	0.007	-	-.-
S. murinus	1	0.203	1	0.015	1	0.170	1	0.064	1	0.0022
Scutisorex somereni	1	0.370	1	0.026	1	0.283	1	0.035	-	-.-
Sylvisorex granti	1	0.081	1	0.006	1	0.088	1	0.017	-	-.-
S. megalura	1	0.084	1	0.007	1	0.117	1	0.016	-	-.-
Ruwenzorisorex sunc.	1	0.204	1	0.011	1	0.266	1	0.020	-	-.-
Myosorex babaulti	1	0.148	1	0.009	1	0.157	1	0.003	-	-.-

Number of individuals = n

Abbreviations:

EPI epiphysis cerebri, corpus pineale
HAM nucleus medialis habenulae
SCB corpus subcommissuralis
SET nucleus triangularis septi
SFB corpus subfornicale

Table 41. Size indices of some periventricular and septal structures related to the average of the four Tenrecinae (AvTn = 100)

For volumes in species see Table 40
For average size indices in supraspecific taxonomic units see Table 42

	n	SET	n	SFB	n	HAM	n	EPI	n	SCB
slope		.51		.44		.58		.58		.55
y-intercept		-1.570		-2.304		-1.637		-1.838		-3.652
Tenrecidae										
Tenrec ecaudatus	1	84	1	128	1	85	1	66	1	123
Setifer setosus	1	75	1	109	1	77	1	129	1	69
Hemicentetes semispin.	1	113	1	81	1	119	1	80	1	127
Echinops telfairi	1	128	1	82	1	118	1	125	1	81
Oryzorictes talpoides	1	178	1	68	1	124	1	39	1	39
Microgale dobsoni	1	177	1	33	1	140	1	33	1	66
M. talazaci	1	189	1	68	1	169	1	63	1	78
Limnogale mergulus	1	223	1	53	1	148	1	61	1	75
Potamogale velox	1	155	1	107	1	153	1	0.5	1	23
Micropotamogale ruwenz.	1	132	1	189	1	181	1	6	-	-.-
Chrysochloridae										
Chrysochloris asiatica	1	248	1	53	1	204	1	18	1	427
C. stuhlmanni	6	223	6	75	6	193	6	77	6	(0)
Solenodontidae										
Solenodon paradoxus	1	119	1	97	1	209	1	203	1	553
Erinaceidae										
Echinosorex gymnurus	1	202	1	115	1	212	1	0	-	-.-
Hylomys suillus	1	227	1	95	1	245	1	130	-	-.-
Atelerix algirus	1	154	1	70	1	89	1	45	1	269
Erinaceus europaeus	2	105	2	61	2	102	2	58	2	344
Hemiechinus auritus	1	96	1	146	1	136	1	35	-	-.-
Talpidae										
Desmana moschata	1	176	1	49	1	133	1	92	1	437
Galemys pyrenaicus	1	206	5	126	1	171	1	46	1	261
Talpa europaea	1	129	5	25	1	123	1	26	1	404
T. micrura	1	215	1	69	1	235	1	46	-	-.-
Parascalops breweri	1	258	1	64	1	197	1	46	-	-.-
Scalopus aquaticus	1	166	1	48	1	167	1	42	1	857

Table 41 (cont.)

	n	SET	n	SFB	n	HAM	n	EPI	n	SCB
Soricidae										
Sorex araneus	2	141	10	26	1	91	10	20	1	40
S. minutus	1	129	1	31	1	114	1	11	1	63
Microsorex hoyi	1	161	1	45	1	160	1	13	-	-.-
Neomys fodiens	1	249	1	43	1	187	1	39	1	41
Anourosorex squamipes	1	173	1	104	1	137	1	30	-	-.-
Crocidura flavescens	1	173	1	81	1	117	1	36	1	40
C. russula	2	137	2	72	1	151	5	50	1	100
Suncus etruscus	1	141	1	60	1	212	1	35	-	-.-
S. murinus	1	125	1	64	1	96	1	57	1	144
Scutisorex somereni	1	166	1	85	1	111	1	21	-	-.-
Sylvisorex granti	1	150	1	65	1	172	1	54	-	-.-
S. megalura	1	130	1	70	1	190	1	41	-	-.-
Ruwenzorisorex sunc.	1	172	1	63	1	214	1	26	-	-.-
Myosorex babaulti	1	129	1	51	1	131	1	4	-	-.-

Number of individuals = n

Abbreviations:

EPI	epiphysis cerebri, corpus pineale
HAM	nucleus medialis habenulae
SCB	corpus subcommissuralis
SET	nucleus triangularis septi
SFB	corpus subfornicale

Table 42. Average size indices of some periventricular and septal structures related to the average of the four Tenrecinae (AvTn = 100)

For volumes in species see Table 40
For size indices in species see Table 41

	N	SET	SFB	HAM	EPI	N	SCB
slope		.51	.44	.58	.58		.55
y-intercept		-1.570	-2.304	-1.637	-1.838		-3.652
Insectivora (AvIS)	38	162	75	153	50	23	203
Tenrecidae	10	145	92	131	60	9	75
Tenrecinae*	4	**100**	**100**	**100**	**100**	4	**100**
Oryzorictinae*	4	192	56	145	49	4	64
Microgale	2	183	50	155	48	2	72
terr. Oryzorictinae	3	181	56	144	45	3	61
Limnogale	1	223	53	148	61	1	75
Potamogalinae*	2	144	148	167	3	1	23
Chrysochloridae*	2	235	64	198	48	1	427
Solenodontidae*	1	119	97	209	203	1	553
Erinaceidae	5	157	97	157	54	-	--
Echinosoricinae*	2	214	105	229	65	-	--
Erinaceinae*	3	118	92	109	46	2	306
Talpidae	6	192	63	171	50	4	490
Desmaninae*	2	191	87	152	69	2	349
Talpinae*	4	192	51	180	40	2	630
Talpa	2	172	47	179	36	1	404
Soricidae	14	155	61	149	31	6	72
Soricinae*	5	170	50	138	23	3	48
Sorex	2	135	28	103	15	2	52
Crocidurinae*	9	147	68	155	36	3	95
Crocidura	2	155	76	134	43	2	70
Suncus	2	133	62	154	46	1	144
Sylvisorex	2	140	67	181	48	-	--
AvIF of the families and subfamilies marked by *	(11)	166	83	162	62	(10)	260

Number of species = N
Abbreviations:

EPI	epiphysis cerebri, corpus pineale
HAM	nucleus medialis habenulae
SCB	corpus subcommissuralis
SET	nucleus triangularis septi
SFB	corpus subfornicale

Table 43. Volumes of telencephalic components

For size indices in species see Table 44
For average size indices in supraspecific taxonomic units see Table 45

	BOL	PAL	STR	SEP	AMY	HIP	SCH	NEO
Tenrecidae								
Tenrec ecaudatus	316.7	378.3	98.81	37.15	88.80	147.1	77.07	272.8
Setifer setosus	213.3	262.9	57.65	24.76	67.62	96.6	47.98	136.6
Hemicentetes semisp.	93.2	114.6	27.78	13.33	20.56	63.1	29.14	76.9
Echinops telfairi	71.8	94.5	24.20	11.54	22.51	46.7	13.18	51.7
Geogale aurita	11.7	13.3	3.81	2.04	4.12	10.9	2.02	7.2
Oryzorictes talpoides	47.0	67.2	32.53	10.05	20.75	64.3	23.47	66.5
Microgale dobsoni	53.5	76.2	22.95	9.52	16.89	66.7	16.26	62.6
M. talazaci	68.4	107.1	32.30	11.37	24.44	93.5	22.61	88.9
Limnogale mergulus	43.2	60.6	57.90	19.54	26.82	78.2	36.41	195.6
Potamogale velox	83.9	128.4	148.49	49.30	62.16	322.4	123.63	1050.0
Micropotamogale lam.	34.1	48.2	32.64	15.06	19.05	77.7	32.92	144.4
M. ruwenzorii	33.7	59.5	59.47	17.25	22.10	97.8	37.22	227.8
Chrysochloridae								
Chrysochloris asiatica	58.6	94.9	44.32	12.12	20.05	66.3	20.93	112.6
C. stuhlmanni	60.5	107.3	43.24	11.29	23.17	58.8	21.32	128.6
Solenodontidae								
Solenodon paradox.	469.8	561.6	257.51	66.17	135.36	299.7	158.14	668.6
Erinaceidae								
Echinosorex gymnur.	427.4	682.8	209.44	76.75	170.14	609.1	198.79	1085.0
Hylomys suillus	85.8	128.0	38.38	17.50	37.43	127.0	51.47	177.8
Atelerix algirus	292.4	365.0	143.04	43.73	103.30	231.8	111.50	510.3
Erinaceus europaeus	335.6	398.6	147.95	51.97	108.67	237.9	134.80	525.1
Hemiechinus auritus	174.6	205.3	80.25	28.07	71.06	128.3	67.94	306.2
Talpidae								
Desmana moschata	141.6	229.5	221.87	42.62	90.80	267.3	163.27	983.4
Galemys pyrenaicus	39.2	70.0	74.38	20.11	36.43	108.6	44.82	338.7
Talpa europaea	60.2	85.9	60.44	15.07	26.41	87.1	39.76	181.2
T. micrura	53.2	78.6	48.08	13.30	22.69	67.4	28.93	157.2
Parascalops breweri	58.9	89.5	60.43	13.42	22.58	83.0	33.83	175.6
Scalopus aquaticus	86.6	139.0	76.18	17.34	30.87	109.9	67.52	252.9

Table 43 (cont.)

	BOL	PAL	STR	SEP	AMY	HIP	SCH	NEO
Soricidae								
Sorex alpinus	16.6	27.9	16.63	4.02	8.05	28.9	9.73	45.5
S. araneus	17.2	27.9	11.50	3.39	7.70	22.8	7.26	28.2
S. cinereus	11.7	22.2	11.77	2.63	4.89	19.5	5.45	28.8
S. fumeus	15.1	29.3	14.42	3.74	7.43	27.8	10.01	48.0
S. minutus	8.9	14.0	7.05	1.82	3.46	12.4	4.24	15.9
Microsorex hoyi	6.4	11.5	6.28	1.69	3.28	11.7	3.76	15.2
Neomys anomalus	13.1	23.5	18.73	4.29	8.18	25.2	12.00	59.0
N. fodiens	16.4	26.0	20.00	5.71	8.59	34.4	10.93	56.6
Blarina brevicauda	28.9	44.4	22.28	6.79	10.93	44.7	14.89	68.7
Cryptotis parva	20.5	30.0	12.88	4.11	9.10	26.2	9.64	36.2
Anourosorex squamip.	32.3	50.4	20.13	6.11	15.35	34.0	14.32	67.6
Crocidura attenuata	19.7	25.5	7.58	3.06	6.66	19.3	6.59	25.3
C. flavescens	35.4	48.2	16.93	6.01	12.63	38.3	13.25	53.1
C. hildegardeae	19.5	30.4	8.87	3.31	6.89	17.8	5.65	26.0
C. jacksoni	19.4	30.9	9.74	3.42	7.62	22.4	7.11	28.6
C. russula	17.1	24.0	8.64	3.22	6.65	17.9	5.85	25.3
C. suaveolens	16.3	23.3	8.06	2.89	5.94	17.2	6.24	24.6
Suncus etruscus	4.7	7.1	2.71	1.04	1.73	6.2	1.89	7.1
S. murinus	36.9	47.1	15.16	5.98	11.09	33.4	12.77	50.5
Scutisorex somereni	58.8	82.8	26.08	10.86	17.85	57.4	19.19	87.2
Sylvisorex granti	15.4	23.4	7.62	2.68	5.06	15.8	4.33	20.6
S. megalura	17.0	25.3	9.55	2.93	5.58	12.0	3.88	23.9
Ruwenzorisorex sunc.	17.6	29.0	17.43	6.02	10.25	39.5	13.02	59.5
Myosorex babaulti	34.6	52.3	18.62	5.58	12.08	33.8	8.27	45.3

Number of individuals as in Table 10

Abbreviations and notes:

AMY amygdala; for volumes of components see Table 55
BOL bulbi olfactorii (main + accessory); for volumes of components see Tables 46 and 49
HIP hippocampus; for volumes of components see Table 58
NEO neocortex; for volumes of components see Table 61
PAL paleocortex; for volumes of components see Table 52
SCH schizocortex
SEP septum
STR striatum

Table 44. Size indices of telencephalic components related to the average of the four Tenrecinae (AvTn = 100)

For volumes in species see Table 43
For average size indices in supraspecific taxonomic units see Table 45

	BOL	PAL	STR	SEP	AMY	HIP	SCH	NEO
slope	.64	.63	.67	.61	.60	.57	.64	.73
y-intercept	.682	.799	.089	-.129	.232	.581	.062	.346
Tenrecidae								
Tenrec ecaudatus	88	86	88	81	91	82	89	89
Setifer setosus	134	133	120	118	149	112	126	114
Hemicentetes semispin.	93	91	94	99	70	110	121	108
Echinops telfairi	85	90	99	101	90	96	65	89
Geogale aurita	56	50	66	67	61	77	40	61
Oryzorictes talpoides	86	98	209	134	125	195	180	189
Microgale dobsoni	121	137	184	155	124	243	154	225
M. talazaci	119	148	196	144	140	270	164	237
Limnogale mergulus	50	56	228	167	104	156	175	325
Potamogale velox	29	36	163	131	77	217	175	434
Micropotamogale lam.	49	56	164	160	92	190	199	312
M. ruwenzorii	38	53	226	143	83	189	173	365
Chrysochloridae								
Chrysochloris asiatica	101	130	266	152	114	189	150	296
C. stuhlmanni	118	166	297	159	148	188	174	391
Solenodontidae								
Solenodon paradoxus	151	148	268	168	160	192	213	260
Erinaceidae								
Echinosorex gymnurus	121	158	190	172	178	348	235	364
Hylomys suillus	134	159	208	200	194	333	336	419
Atelerix algirus	89	91	140	105	115	141	141	186
Erinaceus europaeus	93	90	131	114	111	134	156	172
Hemiechinus auritus	110	105	169	135	157	150	179	257
Talpidae								
Desmana moschata	60	78	305	139	137	218	287	519
Galemys pyrenaicus	60	85	391	223	184	277	284	772
Talpa europaea	75	85	257	138	110	185	205	327
T. micrura	102	120	323	185	142	212	231	468
Parascalops breweri	96	115	341	159	121	225	229	432
Scalopus aquaticus	86	111	258	129	105	193	281	357

Table 44 (cont.)

	BOL	PAL	STR	SEP	AMY	HIP	SCH	NEO
Soricidae								
Sorex alpinus	74	98	272	125	112	193	182	356
S. araneus	81	102	198	111	112	159	142	233
S. cinereus	85	125	319	130	107	201	165	391
S. fumeus	80	122	282	137	121	217	222	458
S. minutus	72	87	213	99	83	140	142	243
Microsorex hoyi	72	99	268	126	108	176	175	339
Neomys anomalus	57	80	295	129	110	164	217	445
N. fodiens	60	74	262	145	98	191	165	348
Blarina brevicauda	89	108	246	148	107	214	192	351
Cryptotis parva	98	112	226	137	135	186	193	306
Anourosorex squam.	98	121	220	132	149	161	182	341
Crocidura attenuata	98	99	139	106	102	142	137	224
C. flavescens	85	91	144	103	98	147	132	203
C. hildegardeae	92	112	153	108	100	124	111	215
C. jacksoni	79	99	145	97	97	138	121	202
C. russula	76	84	140	100	92	119	109	197
C. suaveolens	76	85	138	94	86	120	122	202
Suncus etruscus	65	76	146	96	70	113	110	203
S. murinus	81	81	117	94	79	118	116	174
Scutisorex somereni	86	96	132	116	87	142	117	190
Sylvisorex granti	134	158	249	157	131	191	157	345
S. megalura	119	138	249	139	118	119	113	311
Ruwenzorisorex sunc.	57	74	203	138	105	198	176	323
Myosorex babaulti	117	139	227	133	129	176	117	258

Number of individuals as in Table 10

Abbreviations and notes:

AMY amygdala; for size indices of components see Table 56
BOL bulbi olfactorii (main + accessory); for size indices of components see Tables 47 and 50
HIP hippocampus; for size indices of components see Table 59
NEO neocortex; for size indices of components see Table 62
PAL paleocortex; for size indices of components see Table 53
SCH schizocortex
SEP septum
STR striatum

Table 45. Average size indices of telencephalic components related to the average of the four Tenrecinae (AvTn = 100)

	BOL	PAL	STR	SEP	AMY	HIP	SCH	NEO
slope	.64	.63	.67	.61	.60	.57	.64	.73
y-intercept	.682	.799	.089	-.129	.232	.581	.062	.346
Insectivora (AvIS)	88	103	207	132	114	175	168	292
Tenrecidae	79	86	153	125	101	161	138	212
Tenrecinae*	**100**	**100**	**100**	**100**	**100**	**100**	**100**	**100**
Geogalinae*	56	50	66	67	61	77	40	61
Oryzorictinae*	94	110	204	150	123	216	168	244
Microgale	120	142	190	149	132	257	159	231
terr.Oryzorictinae	109	128	196	144	130	236	166	217
Limnogale	50	56	228	167	104	156	175	325
Potamogalinae*	39	48	184	145	84	199	182	370
Micropotamogale	44	54	195	151	88	190	186	338
Chrysochloridae*	110	148	281	156	131	189	162	344
Solenodontidae*	151	148	268	168	160	192	213	260
Erinaceidae	109	121	168	145	151	221	209	280
Echinosoricinae*	128	159	199	186	186	340	285	391
Erinaceinae*	97	95	147	118	128	142	159	205
Talpidae	80	99	313	162	133	218	253	479
Desmaninae*	60	82	348	181	161	247	285	645
Talpinae*	90	108	295	152	120	204	237	396
Talpa	88	102	290	161	126	199	218	397
Soricidae	85	103	208	121	106	160	151	286
Soricinae*	79	103	255	129	113	182	180	347
Sorex	78	107	257	120	107	182	171	336
terr. Soricinae	83	108	249	127	115	183	177	335
Neomys	58	77	279	137	104	177	191	397
Crocidurinae*	90	103	168	114	100	142	126	234
Crocidura	84	95	143	101	96	132	122	207
Suncus	73	79	131	95	74	116	113	188
Sylvisorex	126	148	249	148	124	155	135	328
AvIF of the 12 families and subfamilies marked by *	91	104	210	139	122	186	178	300

Number of species as in Table 12

Abbreviations and notes:

- AMY amygdala; for av. size indices of components see Table 57
- BOL bulbi olfactorii (main + accessory); for av. size indices of components see Tables 48 and 51
- HIP hippocampus; for av. size indices of components see Table 60
- NEO neocortex; for av. size indices of components see Table 63
- PAL paleocortex; for av. size indices of components see Table 54
- SEP septum SCH schizocortex STR striatum

Table 46. Volumes of main olfactory bulbs and some of its layers

For size indices in species see Table 47
For average size indices in supraspecific taxonomic units see Table 48

	n	MOB	n	L1+2	L3	L4-6	PvZ
Tenrecidae							
Tenrec ecaudatus	2	315.9	1	130.15	57.36	102.49	25.93
Setifer setosus	2	212.8	1	73.13	51.41	81.99	6.23
Hemicentetes semis.	1	93.0	1	33.39	22.24	34.13	3.24
Echinops telfairi	2	71.6	1	27.06	17.14	26.12	1.26
Geogale aurita	1	11.6	1	3.90	2.93	4.62	0.13
Oryzorictes talpoides	2	46.9	1	11.58	13.83	20.65	0.80
Microgale dobsoni	1	53.4	1	17.88	14.69	20.20	0.59
M. talazaci	1	68.0	1	19.20	21.82	26.16	0.84
Limnogale mergulus	1	42.9	1	16.20	11.40	14.70	0.57
Potamogale velox	2	81.7	1	27.30	20.31	31.34	2.76
Micropotamogale lam.	1	34.1	1	11.45	9.53	12.79	0.29
M. ruwenzorii	4	33.7	1	10.22	8.87	14.44	0.21
Chrysochloridae							
Chrysochloris asiatica	1	58.5	1	19.96	13.63	23.26	1.62
C. stuhlmanni	2	60.3	1	19.42	14.70	24.36	1.80
Solenodontidae							
Solenodon paradoxus	2	469.6	1	224.95	84.14	133.09	27.37
Erinaceidae							
Echinosorex gymnurus	1	424.2	1	148.10	83.88	160.71	31.53
Hylomys suillus	1	84.4	1	32.82	19.15	29.59	2.85
Atelerix algirus	2	290.2	1	121.53	49.28	97.03	22.39
Erinaceus europaeus	2	333.7	1	138.08	60.47	109.72	25.38
Hemiechinus auritus	3	173.3	1	62.93	37.34	60.60	12.42
Talpidae							
Desmana moschata	1	141.4	1	47.52	27.42	60.25	6.24
Galemys pyrenaicus	2	39.0	1	12.57	10.52	15.07	0.86
Talpa europaea	2	59.9	1	21.02	14.33	22.75	1.77
T. micrura	1	53.0	1	19.46	12.33	20.54	0.70
Parascalops breweri	1	58.8	1	20.32	14.03	23.24	1.25
Scalopus aquaticus	1	86.5	1	34.05	20.63	30.63	1.14

Table 46 (cont.)

	n	MOB	n	L1+2	L3	L4-6	PvZ
Soricidae							
Sorex alpinus	1	16.5	--	-.-	-.-	-.-	-.-
S. araneus	2	17.1	1	4.85	4.77	7.33	0.15
S. cinereus	1	11.6	--	-.-	-.-	-.-	-.-
S. fumeus	1	15.0	--	-.-	-.-	-.-	-.-
S. minutus	1	8.9	1	2.72	2.72	3.41	0.07
Microsorex hoyi	1	6.4	1	2.01	1.85	2.53	0.02
Neomys anomalus	1	13.0	1	3.92	3.40	5.40	0.25
N. fodiens	2	16.3	1	5.51	5.04	5.60	0.15
Blarina brevicauda	1	28.9	1	9.77	7.93	10.86	0.32
Cryptotis parva	1	20.5	1	5.90	5.90	8.56	0.14
Anourosorex squamipes	1	32.2	1	9.10	8.98	13.47	0.68
Crocidura attenuata	2	19.6	--	-.-	-.-	-.-	-.-
C. flavescens	1	35.2	1	11.72	9.85	13.01	0.60
C. hildegardeae	1	19.4	--	-.-	-.-	-.-	-.-
C. jacksoni	2	19.3	--	-.-	-.-	-.-	-.-
C. russula	1	17.0	1	5.89	4.84	6.16	0.14
C. suaveolens	1	16.2	--	-.-	-.-	-.-	-.-
Suncus etruscus	1	4.7	1	1.58	1.36	1.69	0.02
S. murinus	1	36.7	1	11.88	10.16	13.20	1.44
Scutisorex somereni	1	58.6	1	20.00	14.50	22.90	1.21
Sylvisorex granti	1	15.4	1	4.80	4.43	6.10	0.08
S. megalura	1	16.9	1	5.44	4.62	6.71	0.14
Ruwenzorisorex sunc.	1	17.5	1	5.77	4.45	7.06	0.26
Myosorex babaulti	1	34.5	1	10.63	9.26	14.45	0.16

Number of individuals = n

Abbreviations:

- L1+2 external fiber and glomerular layers
- L3 external plexiform layer
- L4-6 mitral, internal plexiform, and granular layers
- MOB main olfactory bulb
- PvZ periventricular zone

Table 47. Size indices of the main olfactory bulbs and of some of the MOB layers related to the average of the four Tenrecinae (AvTn = 100)

For volumes in species see Table 46
Foe average size indices in supraspecific taxonomic units see Table 48

	n	MOB	n	L1+2	L3	L4-6	PvZ
slope		.64		.64	.64	.64	.64
y-intercept		.681		.248	.037	.241	-.721
Tenrecidae							
Tenrec ecaudatus	2	88	1	98	70	78	182
Setifer setosus	2	134	1	125	143	142	99
Hemicentetes semispin.	1	93	1	90	97	94	81
Echinops telfairi	2	85	1	87	90	86	38
Geogale aurita	1	55	1	50	62	61	16
Oryzorictes talpoides	2	86	1	58	112	105	37
Microgale dobsoni	1	121	1	110	147	126	34
M. talazaci	1	119	1	91	168	126	37
Limnogale mergulus	1	49	1	51	58	47	17
Potamogale velox	2	28	1	25	31	29	24
Micropotamogale lam.	1	49	1	45	61	51	10
M. ruwenzorii	4	38	1	31	44	44	6
Chrysochloridae							
Chrysochloris asiatica	1	101	1	93	104	111	70
C. stuhlmanni	2	118	1	103	127	132	89
Solenodontidae							
Solenodon paradoxus	2	152	1	197	120	118	223
Erinaceidae							
Echinosorex gymnurus	1	120	1	114	105	126	226
Hylomys suillus	1	132	1	139	132	128	113
Atelerix algirus	2	88	1	100	66	81	172
Erinaceus europaeus	2	93	1	104	74	84	178
Hemiechinus auritus	3	110	1	108	104	106	198
Talpidae							
Desmana moschata	1	60	1	54	51	70	66
Galemys pyrenaicus	2	59	1	52	71	63	33
Talpa europaea	2	74	1	71	78	78	55
T. micrura	1	102	1	101	104	109	34
Parascalops breweri	1	96	1	90	101	104	51
Scalopus aquaticus	1	86	1	92	91	84	29

Table 47 (cont.)

	n	MOB	n	L1+2	L3	L4-6	PvZ
Soricidae							
Sorex alpinus	1	74	--	--	--	--	--
S. araneus	2	81	1	62	99	95	18
S. cinereus	1	85	--	--	--	--	--
S. fumeus	1	80	--	--	--	--	--
S. minutus	1	72	1	60	97	76	13
Microsorex hoyi	1	72	1	61	91	78	6
Neomys anomalus	1	56	1	46	65	65	28
N. fodiens	2	59	1	54	81	56	14
Blarina brevicauda	1	89	1	82	108	93	25
Cryptotis parva	1	99	1	77	125	113	17
Anourosorex squamipes	1	98	1	75	121	113	52
Crocidura attenuata	2	98	--	--	--	--	--
C. flavescens	1	84	1	76	104	86	36
C. hildegardeae	1	91	--	--	--	--	--
C. jacksoni	2	79	--	--	--	--	--
C. russula	1	76	1	71	95	76	16
C. suaveolens	1	76	--	--	--	--	--
Suncus etruscus	1	65	1	60	84	65	8
S. murinus	1	80	1	70	98	80	79
Scutisorex somereni	1	86	1	79	94	92	45
Sylvisorex granti	1	134	1	113	170	147	18
S. megalura	1	118	1	103	143	129	25
Ruwenzorisorex sunc.	1	57	1	51	64	63	21
Myosorex babaulti	1	117	1	98	139	135	14

Number of individuals = n

Abbreviations:

L1+2	external fiber and glomerular layers
L3	external plexiform layer
L4-6	mitral, internal plexiform, and granular layers
MOB	main olfactory bulb
PvZ	periventricular zone

Table 48. Average size indices of the main olfactory bulbs and of some of the MOB layers related to the average of the four Tenrecinae (AvTn = 100)

	N	MOB	N	L1+2	L3	L4-6	PvZ
slope		.64		.64	.64	.64	.64
y-intercept		.681		.248	.037	.241	-.721
Insectivora (AvIS)	50	87	43	82	97	92	59
Tenrecidae	12	79	12	72	90	82	48
Tenrecinae*	4	**100**	4	**100**	**100**	**100**	**100**
Geogalinae*	1	55	1	50	62	61	16
Oryzorictinae*	4	94	4	77	121	101	31
Microgale	2	120	2	100	157	126	35
terr.Oryzorictinae	3	109	3	86	142	119	36
Limnogale	1	49	1	51	58	47	17
Potamogalinae*	3	38	3	34	45	42	13
Micropotamogale	2	44	2	38	52	48	8
Chrysochloridae*	2	110	2	98	115	121	80
Solenodontidae*	1	152	1	197	120	118	223
Erinaceidae	5	109	5	113	96	105	178
Echinosoricinae*	2	126	2	127	119	127	169
Erinaceinae*	3	97	3	104	81	90	183
Talpidae	6	80	6	77	83	85	45
Desmaninae*	2	60	2	53	61	67	50
Talpinae*	4	90	4	89	94	94	42
Talpa	2	88	2	86	91	93	45
Soricidae	24	85	17	73	105	92	26
Soricinae*	11	79	8	65	98	86	22
Sorex	5	78	2	61	98	85	16
terr. Soricinae	9	83	6	69	107	95	22
Neomys	2	58	2	50	73	60	21
Crocidurinae*	13	89	9	80	110	97	29
Crocidura	6	84	2	74	100	81	26
Suncus	2	73	2	65	91	72	44
Sylvisorex	2	126	2	108	156	138	22
AvIF of the 12 families and subfamilies marked by *		91		90	94	92	80

Number of species = N

Abbreviations:

L1+2 external fiber and glomerular layers
L3 external plexiform layer
L4-6 mitral, internal plexiform, and granular layers
MOB main olfactory bulbs
PvZ periventricular zone

Table 49. Volumes of accessory olfactory bulbs and some of its layers

For size indices in species see Table 50
For average size indices in supraspecific taxonomic units see Table 51

	n	AOB	n	L1+2	L3-5	L6
Tenrecidae						
Tenrec ecaudatus	2	0.730	1	0.393	0.218	0.119
Setifer setosus	2	0.544	1	0.276	0.184	0.084
Hemicentetes semispin.	1	0.194	1	0.086	0.097	0.011
Echinops telfairi	2	0.255	1	0.114	0.107	0.034
Geogale aurita	1	0.086	1	0.023	0.045	0.018
Oryzorictes talpoides	2	0.120	1	0.037	0.076	0.007
Microgale dobsoni	1	0.177	1	0.059	0.099	0.019
M. talazaci	1	0.338	1	0.093	0.168	0.077
Limnogale mergulus	1	0.323	1	0.080	0.165	0.078
Potamogale velox	2	2.202	1	1.194	0.702	0.306
Micropotamogale lam.	1	0.067	1	0.026	0.035	0.006
M. ruwenzorii	4	0.015	--	-.-	-.-	-.-
Chrysochloridae						
Chrysochloris asiatica	1	0.132	1	0.037	0.078	0.017
C. stuhlmanni	2	0.202	1	0.066	0.108	0.028
Solenodontidae						
Solenodon paradoxus	2	0.222	1	0.052	0.149	0.021
Erinaceidae						
Echinosorex gymnurus	1	3.208	1	1.413	0.952	0.843
Hylomys suillus	1	1.347	1	0.511	0.490	0.346
Atelerix algirus	2	2.163	1	0.756	0.839	0.568
Erinaceus europaeus	2	1.922	1	0.627	0.922	0.373
Hemiechinus auritus	3	1.337	1	0.426	0.614	0.297
Talpidae						
Desmana moschata	1	0.218	1	0.096	0.092	0.030
Galemys pyrenaicus	2	0.206	1	0.050	0.128	0.028
Talpa europaea	2	0.300	1	0.090	0.141	0.069
T. micrura	1	0.119	1	0.045	0.046	0.028
Parascalops breweri	1	0.075	1	0.028	0.033	0.014
Scalopus aquaticus	1	0.108	1	0.041	0.050	0.017

Table 49 (cont.)

	n	AOB	n	L1+2	L3-5	L6
Soricidae						
Sorex alpinus	1	0.050	--	-.-	-.-	-.-
S. araneus	2	0.087	1	0.025	0.040	0.022
S. cinereus	1	0.022	--	-.-	-.-	-.-
S. fumeus	1	0.046	1	0.016	0.025	0.005
S. minutus	1	0.021	1	0.006	0.011	0.004
Microsorex hoyi	1	0.021	--	-.-	-.-	-.-
Neomys anomalus	1	0.079	1	0.031	0.038	0.010
N. fodiens	2	0.103	1	0.035	0.050	0.018
Blarina brevicauda	1	0.060	--	-.-	-.-	-.-
Cryptotis parva	1	0.019	1	0.006	0.009	0.004
Anourosorex squamipes	1	0.031	--	-.-	-.-	-.-
Crocidura attenuata	2	0.103	1	0.038	0.051	0.014
C. flavescens	1	0.208	1	0.080	0.089	0.039
C. hildegardeae	1	0.085	--	-.-	-.-	-.-
C. jacksoni	2	0.082	--	-.-	-.-	-.-
C. russula	1	0.045	1	0.013	0.025	0.007
C. suaveolens	1	0.078	--	-.-	-.-	-.-
Suncus etruscus	1	0.024	1	0.008	0.013	0.003
S. murinus	1	0.195	1	0.082	0.077	0.036
Scutisorex somereni	1	0.219	1	0.071	0.109	0.039
Sylvisorex granti	1	0.032	1	0.009	0.020	0.003
S. megalura	1	0.050	1	0.017	0.029	0.004
Ruwenzorisorex sunc.	1	0.079	1	0.030	0.044	0.005
Myosorex babaulti	1	0.088	1	0.031	0.047	0.010

Number of individuals = n

Abbreviations:

AOB accessory olfactory bulb
L1+2 vomeronasal fiber and glomerular layers
L3-5 external plexiform, mitral, and internal plexiform layers
L6 internal granular layer

Table 50. Size indices of the accessory olfactory bulbs, and some of the AOB layers, related to the average of the four Tenrecinae (AvTn = 100)

For volumes in species see Table 49
For average size indices in supraspecific taxonomic units see Table 51

	n	AOB	n	L1+2	L3-5	L6
slope		.64		.64	.64	.64
y-intercept		-1.903		-2.219	-2.318	-2.785
Tenrecidae						
Tenrec ecaudatus	2	78	1	87	60	97
Setifer setosus	2	131	1	138	116	154
Hemicentetes semispinosus	1	74	1	68	96	32
Echinops telfairi	2	116	1	107	128	117
Geogale aurita	1	157	1	85	215	250
Oryzorictes talpoides	2	85	1	54	139	38
Microgale dobsoni	1	154	1	106	225	126
M. talazaci	1	226	1	129	291	392
Limnogale mergulus	1	143	1	73	190	263
Potamogale velox	2	288	1	323	239	306
Micropotamogale lamottei	1	37	1	30	51	23
M. ruwenzorii	4	6	--	--	--	--
Chrysochloridae						
Chrysochloris asiatica	1	87	1	51	134	86
C. stuhlmanni	2	152	1	103	211	159
Solenodontidae						
Solenodon paradoxus	2	28	1	13	48	20
Erinaceidae						
Echinosorex gymnurus	1	349	1	319	270	700
Hylomys suillus	1	810	1	636	766	1587
Atelerix algirus	2	253	1	183	255	507
Erinaceus europaeus	2	205	1	139	256	303
Hemiechinus auritus	3	325	1	214	388	550
Talpidae						
Desmana moschata	1	35	1	32	39	37
Galemys pyrenaicus	2	120	1	60	194	124
Talpa europaea	2	143	1	88	175	251
T. micrura	1	88	1	69	89	155
Parascalops breweri	1	47	1	37	53	64
Scalopus aquaticus	1	41	1	32	50	49

Table 50 (cont.)

	n	AOB	n	L1+2	L3-5	L6
Soricidae						
Sorex alpinus	1	86	--	--	--	--
S. araneus	2	157	1	92	189	307
S. cinereus	1	62	--	--	--	--
S. fumeus	1	93	1	68	132	73
S. minutus	1	65	1	40	84	99
Microsorex hoyi	1	90	--	--	--	--
Neomys anomalus	1	132	1	106	165	132
N. fodiens	2	144	1	101	182	196
Blarina brevicauda	1	71	--	--	--	--
Cryptotis parva	1	35	1	23	45	49
Anourosorex squamipes	1	36	--	--	--	--
Crocidura attenuata	2	197	1	149	254	209
C. flavescens	1	191	1	152	214	270
C. hildegardeae	1	154	--	--	--	--
C. jacksoni	2	129	--	--	--	--
C. russula	1	77	1	45	112	97
C. suaveolens	1	140	--	--	--	--
Suncus etruscus	1	129	1	87	178	139
S. murinus	1	164	1	142	168	233
Scutisorex somereni	1	123	1	82	159	166
Sylvisorex granti	1	108	1	66	177	66
S. megalura	1	133	1	92	203	84
Ruwenzorisorex suncoides	1	98	1	77	142	49
Myosorex babaulti	1	115	1	84	158	103

Number of individuals = n

Abbreviations:

AOB accessory olfactory bulb
L1+2 vomeronasal fiber and glomerular layers
L3-5 external plexiform, mitral, and internal plexiform layers
L6 granular layer

Table 51. Average size indices of the accessory olfactory bulbs and of some of the AOB layers related to the average of the four Tenrecinae (AvTn = 100)

	N	AOB	N	AOB	L1+2	L3-5	L6
slope		.64		.64	.64	.64	.64
y-intercept		-1.903		-1.903	-2.219	-2.318	-2.785
Insectivora (AvIS)	50	138	41	150	112	177	211
Tenrecidae	12	125	11	135	109	159	163
Tenrecinae*	4	**100**	4	**100**	**100**	**100**	**100**
Geogalinae*	1	157	1	157	85	215	250
Oryzorictinae*	4	152	4	152	91	211	205
Microgale	2	190	2	190	117	258	259
terr.Oryzorictinae	3	155	3	155	96	219	185
Limnogale	1	143	1	143	73	190	263
Potamogalinae*	3	111	2	163	177	145	164
Micropotamogale	2	22	1	37	30	51	23
Chrysochloridae*	2	120	2	120	77	173	123
Solenodontidae*	1	28	1	28	13	48	20
Erinaceidae	5	388	5	388	298	387	729
Echinosoricinae*	2	580	2	580	478	518	1143
Erinaceinae*	3	261	3	261	179	300	453
Talpidae	6	79	6	79	53	100	113
Desmaninae*	2	78	2	78	46	116	81
Talpinae*	4	80	4	80	57	92	130
Talpa	2	115	2	115	79	132	203
Soricidae	24	114	16	123	88	160	142
Soricinae*	11	88	6	104	72	133	143
Sorex	5	93	3	105	67	135	160
terr. Soricinae	9	77	4	88	56	113	132
Neomys	2	138	2	138	103	174	164
Crocidurinae*	13	135	10	134	98	176	142
Crocidura	6	148	3	155	115	193	192
Suncus	2	146	2	146	114	173	186
Sylvisorex	2	121	2	121	79	190	75
AvIF of the 12 families and subfamilies marked by *		158		163	123	186	246

Number of species = N

Abbreviations:

- AOB accessory olfactory bulb
- L1+2 vomeronasal fiber and glomerular layers
- L3-5 external plexiform, mitral, and internal plexiform layers
- L6 granular layer

Table 52. Volumes of paleocortical components

For volumes of total paleocortex (PAL) see Table 43
For size indices in species see Table 53
For average size indices in supraspecific taxonomic units see Table 54

	n	RB	PRPI	TOL	OLC	TRL	COA	SIN
Tenrecidae								
Tenrec ecaudatus	1	50.44	227.1	51.37	328.9	15.248	24.242	9.881
Setifer setosus	1	39.47	146.5	45.78	231.7	12.157	10.908	8.080
Hemicentetes semis.	1	14.70	68.5	19.84	103.0	4.109	3.090	4.353
Echinops telfairi	1	12.38	58.3	15.20	85.9	2.870	2.443	3.264
Geogale aurita	2	1.85	7.9	2.14	11.8	0.568	0.404	0.487
Oryzorictes talpoides	1	9.77	37.2	14.18	61.2	2.094	1.257	2.657
Microgale dobsoni	1	10.12	47.8	12.96	70.8	2.387	1.364	1.570
M. talazaci	1	13.70	67.7	17.50	98.9	3.692	2.630	1.818
Limnogale mergulus	1	7.91	33.4	13.18	54.5	3.092	1.411	1.558
Potamogale velox	1	18.19	71.6	17.57	107.4	5.033	6.181	9.800
Micropotamogale lam.	1	7.65	27.1	8.69	43.5	2.428	0.922	1.407
M. ruwenzorii	1	7.46	33.6	11.62	52.7	1.725	1.847	3.207
Chrysochloridae								
Chrysochloris asiatica	1	14.15	52.3	20.71	87.1	3.047	2.387	2.269
C. stuhlmanni	1	14.88	59.7	24.45	99.0	3.332	2.539	2.356
Solenodontidae								
Solenodon paradoxus	1	72.65	316.2	116.90	505.8	19.529	21.167	15.183
Erinaceidae								
Echinosorex gymnurus	3	95.41	384.0	114.40	593.8	30.474	39.034	19.447
Hylomys suillus	1	18.82	73.9	22.99	115.7	4.232	4.961	3.114
Atelerix algirus	1	53.86	220.8	54.38	329.0	13.017	13.859	9.133
Erinaceus europaeus	1	61.93	228.7	56.46	347.1	14.777	23.427	13.274
Hemiechinus auritus	1	21.53	127.4	35.41	184.4	6.812	7.718	6.394
Talpidae								
Desmana moschata	1	34.28	135.8	39.97	210.0	6.662	4.396	8.407
Galemys pyrenaicus	1	7.95	36.5	18.74	63.2	1.841	0.928	4.052
Talpa europaea	1	11.71	45.6	19.17	76.4	2.510	4.055	2.895
T. micrura	2	10.41	44.5	16.98	71.9	2.208	2.211	2.331
Parascalops breweri	3	12.78	49.5	19.80	82.0	2.571	1.753	3.125
Scalopus aquaticus	1	18.57	82.5	25.86	127.0	3.716	3.440	4.926

Table 52 (cont.)

	n	RB	PRPI	TOL	OLC	TRL	COA	SIN
Soricidae								
Sorex alpinus	3	4.37	15.4	6.28	26.1	0.464	0.648	0.715
S. araneus	1	3.96	15.9	5.98	25.8	0.676	0.530	0.826
S. cinereus	1	3.21	12.5	4.83	20.5	0.482	0.630	0.530
S. fumeus	1	4.36	17.0	6.03	27.4	0.523	0.735	0.660
S. minutus	1	1.75	7.2	3.75	12.7	0.342	0.363	0.578
Microsorex hoyi	1	1.87	5.8	2.86	10.5	0.293	0.245	0.461
Neomys anomalus	3	3.34	13.1	5.23	21.7	0.533	0.399	0.853
N. fodiens	1	4.35	12.4	6.04	22.8	0.712	0.842	1.663
Blarina brevicauda	1	6.88	23.5	9.61	40.0	1.442	1.492	1.470
Cryptotis parva	3	4.77	16.2	6.64	27.6	0.752	0.701	0.943
Anourosorex squam.	4	7.62	28.7	10.12	46.4	1.194	1.191	1.641
Crocidura attenuata	1	3.56	15.2	4.91	23.7	0.491	0.662	0.630
C. flavescens	1	8.14	27.6	7.21	42.9	1.682	1.906	1.653
C. hildegardeae	1	4.46	17.9	6.09	28.5	0.713	0.657	0.615
C. jacksoni	1	4.47	18.6	5.32	28.4	1.037	0.947	0.550
C. russula	1	3.64	13.7	4.67	22.0	0.678	0.597	0.747
C. suaveolens	3	3.55	13.7	4.41	21.6	0.531	0.465	0.655
Suncus etruscus	1	1.17	4.0	1.45	6.6	0.204	0.141	0.145
S. murinus	1	7.30	29.1	6.75	43.1	1.316	1.000	1.631
Scutisorex somereni	1	13.11	49.1	13.90	76.1	2.533	2.225	1.868
Sylvisorex granti	1	3.33	14.1	4.35	21.8	0.565	0.581	0.487
S. megalura	1	3.27	15.5	4.91	23.7	0.557	0.555	0.512
Ruwenzorisorex sunc.	1	4.12	17.8	5.05	27.0	0.709	0.625	0.632
Myosorex babaulti	1	7.30	30.0	11.00	48.3	1.435	1.368	1.213

Number of individuals = n

Abbreviations:

COA complex of internal fibers passing through the anterior commissure
OLC olfactory cortices (RB + PRPI + TOL)
PRPI regio prepiriformis
RB regio retrobulbaris (=nucleus olfactorius anterior)
SIN substantia innominata
TOL tuberculum olfactorium
TRL tractus olfactorius lateralis

Table 53. Size indices of paleocortical components related to the average of the four Tenrecinae (AvTn = 100)

For size indices of total paleocortex (PAL) see Table 44
For volumes in species see Table 52
For average size indices in supraspecific taxonomic units see Table 54

	n	RB	PRPI	TOL	OLC	TRL	COA	SIN
slope		.63	.59	.53	.59	.67	.76	.64
y-intercept		-.064	.662	.242	.841	-.704	-.921	-.716
Tenrecidae								
Tenrec ecaudatus	1	83	92	82	89	84	120	68
Setifer setosus	1	146	127	145	133	158	143	127
Hemicentetes semispin.	1	85	90	92	90	86	69	108
Echinops telfairi	1	86	91	81	89	73	68	97
Geogale aurita	2	50	44	36	44	61	58	58
Oryzorictes talpoides	1	104	87	109	94	84	59	122
Microgale dobsoni	1	132	135	118	132	119	82	89
M. talazaci	1	138	150	128	145	139	115	79
Limnogale mergulus	1	53	51	69	55	76	38	45
Potamogale velox	1	37	35	33	35	34	39	83
Micropotamogale lam.	1	64	51	55	54	76	33	51
M. ruwenzorii	1	48	49	59	51	41	48	89
Chrysochloridae								
Chrysochloris asiatica	1	141	115	151	126	114	103	98
C. stuhlmanni	1	168	147	198	162	142	128	115
Solenodontidae								
Solenodon paradoxus	1	139	148	212	157	126	125	122
Erinaceidae								
Echinosorex gymnurus	3	161	159	187	163	172	198	138
Hylomys suillus	1	171	148	155	154	143	191	122
Atelerix algirus	1	98	98	94	97	79	77	69
Erinaceus europaeus	1	102	93	91	94	82	116	92
Hemiechinus auritus	1	80	111	112	106	89	102	101
Talpidae								
Desmana moschata	1	85	81	91	83	57	36	88
Galemys pyrenaicus	1	70	71	123	82	60	35	154
Talpa europaea	1	84	74	106	82	66	119	90
T. micrura	2	116	108	135	115	92	109	112
Parascalops breweri	3	120	103	137	113	90	71	127
Scalopus aquaticus	1	108	109	120	111	78	78	123

Table 53 (cont.)

	n	RB	PRPI	TOL	OLC	TRL	COA	SIN
Soricidae								
Sorex alpinus	3	112	82	101	91	47	87	80
S. araneus	1	106	88	100	95	72	76	97
S. cinereus	1	132	103	116	112	81	151	96
S. fumeus	1	132	106	112	113	64	122	88
S. minutus	1	80	65	98	76	64	98	116
Microsorex hoyi	1	118	71	98	86	78	98	129
Neomys anomalus	3	83	67	82	74	52	52	92
N. fodiens	1	90	54	82	66	58	88	151
Blarina brevicauda	1	122	88	113	99	99	129	113
Cryptotis parva	3	130	91	113	103	82	102	113
Anourosorex squam.	4	133	106	118	114	81	101	125
Crocidura attenuata	1	101	89	86	92	56	101	79
C. flavescens	1	112	82	69	84	89	122	99
C. hildegardeae	1	120	99	102	104	76	94	72
C. jacksoni	1	105	90	79	91	96	114	56
C. russula	1	92	72	75	77	68	80	83
C. suaveolens	3	95	75	73	79	56	66	77
Suncus etruscus	1	92	60	60	66	68	73	51
S. murinus	1	92	79	60	78	63	57	89
Scutisorex somereni	1	111	93	88	95	79	79	68
Sylvisorex granti	1	164	138	121	141	115	172	106
S. megalura	1	129	124	114	125	90	127	90
Ruwenzorisorex sunc.	1	77	70	62	70	51	57	51
Myosorex babaulti	1	142	123	140	131	109	132	103

Number of individuals = n

Abbreviations:

COA	complex of internal fibers passing through the anterior commissure
OLC	olfactory cortices (RB + PRPI + TOL)
PRPI	regio prepiriformis
RB	regio retrobulbaris (=nucleus olfactorius anterior)
SIN	substantia innominata
TOL	tuberculum olfactorium
TRL	tractus olfactorius lateralis

Table 54. Average size indices of paleocortical components related to the average of the four Tenrecinae (AvTn = 100)

	N	RB	PRPI	TOL	OLC	TRL	COA	SIN
slope		.63	.59	.53	.59	.67	.76	.64
y-intercept		-.064	.662	.242	.841	-.704	-.921	-.716
Insectivora (AvIS)	50	107	94	104	98	84	95	96
Tenrecidae	12	86	83	84	84	86	73	85
Tenrecinae*	4	**100**	**100**	**100**	**100**	**100**	**100**	**100**
Geogalinae*	1	50	44	36	44	61	58	58
Oryzorictinae*	4	107	106	106	107	104	73	84
Microgale	2	135	142	123	139	129	99	84
terr.Oryzorictinae	3	125	124	119	124	114	85	97
Limnogale	1	53	51	69	55	76	38	45
Potamogalinae*	3	50	45	49	47	50	40	75
Micropotamogale	2	56	50	57	52	58	40	70
Chrysochloridae*	2	155	131	174	144	128	116	106
Solenodontidae*	1	139	148	212	157	126	125	122
Erinaceidae	5	122	122	128	123	113	137	104
Echinosoricinae*	2	166	154	171	158	157	195	130
Erinaceinae*	3	93	101	99	99	83	98	88
Talpidae	6	97	91	119	98	74	74	116
Desmaninae*	2	78	76	107	82	58	35	121
Talpinae*	4	107	98	125	105	82	94	113
Talpa	2	100	91	121	99	79	114	101
Soricidae	24	111	88	94	94	75	99	93
Soricinae*	11	113	84	103	93	71	100	109
Sorex	5	112	89	105	97	66	107	96
terr. Soricinae	9	118	89	108	99	74	107	107
Neomys	2	87	61	82	70	55	70	122
Crocidurinae*	13	110	92	87	95	78	98	79
Crocidura	6	104	85	81	88	73	96	78
Suncus	2	82	70	60	72	66	65	70
Sylvisorex	2	147	131	118	133	102	150	98
AvIF of the 12 families and subfamilies marked by *		106	98	114	103	92	94	99

Number of species = N

Abbreviations:

COA	complex of anterior commissure		
OLC	olfactory cortices (RB + PRPI + TOL)		
PRPI	regio prepiriformis	TOL	tuberculum olfactorium
RB	regio retrobulbaris (ncl. olf. ant.)		
SIN	substantia innominata	TRL	tractus olfactorius lateralis

Table 55. Volumes of amygdala and some amygdaloid components

For size indices in species see Table 56
For average size indices in supraspecific taxonomic units see Table 57

	AMY	n	MAM	LAM	n	NTO	MCB
Tenrecidae							
Tenrec ecaudatus	88.80	2	39.33	49.47	1	3.129	9.151
Setifer setosus	67.62	3	31.63	35.99	1	2.826	7.960
Hemicentetes semis.	20.56	2	9.52	11.04	1	1.185	2.221
Echinops telfairi	22.51	2	11.17	11.34	1	0.705	3.203
Geogale aurita	4.12	1	2.04	2.08	--	-.-	-.-
Oryzorictes talpoides	20.75	1	9.66	11.09	1	0.639	1.210
Microgale dobsoni	16.89	1	6.73	10.16	1	0.749	1.709
M. talazaci	24.44	1	11.72	12.72	1	0.724	1.647
Limnogale mergulus	26.82	2	11.04	15.78	1	0.310	2.530
Potamogale velox	62.16	1	28.65	33.51	1	0.455	5.323
Micropotamogale lam.	19.05	1	8.74	10.31	1	0.268	1.388
M. ruwenzorii	22.10	1	9.25	12.85	1	0.466	1.639
Chrysochloridae							
Chrysochloris asiatica	20.05	1	9.08	10.97	1	0.831	1.628
C. stuhlmanni	23.17	1	12.15	11.02	1	0.956	2.113
Solenodontidae							
Solenodon paradoxus	135.36	1	51.31	84.05	1	2.522	14.260
Erinaceidae							
Echinosorex gymnurus	170.14	3	69.57	100.57	--	-.-	-.-
Hylomys suillus	37.43	1	17.21	20.22	--	-.-	-.-
Atelerix algirus	103.30	1	46.82	56.48	1	3.134	7.650
Erinaceus europaeus	108.67	3	45.00	63.67	1	3.748	11.518
Hemiechinus auritus	71.06	1	29.55	41.51	1	2.767	4.878
Talpidae							
Desmana moschata	90.80	1	40.87	49.93	1	0.094	12.178
Galemys pyrenaicus	36.43	1	14.95	21.48	1	0.231	5.669
Talpa europaea	26.41	1	11.37	15.04	1	0.565	3.497
T. micrura	22.69	2	11.03	11.66	--	-.-	-.-
Parascalops breweri	22.58	3	11.14	11.44	--	-.-	-.-
Scalopus aquaticus	30.87	1	14.49	16.38	--	-.-	-.-

Table 55 (cont.)

	AMY	n	MAM	LAM	n	NTO	MCB
Soricidae							
Sorex alpinus	8.05	3	3.70	4.35	--	-.-	-.-
S. araneus	7.70	3	3.74	3.96	2	0.275	0.758
S. cinereus	4.89	1	2.38	2.51	1	0.244	0.541
S. fumeus	7.43	1	3.67	3.76	1	0.330	0.642
S. minutus	3.46	2	1.83	1.63	1	0.179	0.572
Microsorex hoyi	3.28	1	1.35	1.93	1	0.106	0.232
Neomys anomalus	8.18	3	3.94	4.24	--	-.-	-.-
N. fodiens	8.59	1	4.13	4.46	1	0.239	1.161
Blarina brevicauda	10.93	1	5.02	5.91	1	0.454	0.894
Cryptotis parva	9.10	3	4.55	4.55	--	-.-	-.-
Anourosorex squamipes	15.35	4	7.40	7.95	--	-.-	-.-
Crocidura attenuata	6.66	1	3.47	3.19	--	-.-	-.-
C. flavescens	12.63	2	6.20	6.43	1	0.415	1.203
C. hildegardeae	6.89	1	3.53	3.36	1	0.336	0.592
C. jacksoni	7.62	1	3.50	4.12	1	0.237	0.539
C. russula	6.65	2	3.24	3.41	1	0.240	0.762
C. suaveolens	5.94	3	2.94	3.00	--	-.-	-.-
Suncus etruscus	1.73	1	0.92	0.81	1	0.084	0.118
S. murinus	11.09	2	5.43	5.66	1	0.442	0.793
Scutisorex somereni	17.85	1	8.02	9.83	1	0.790	1.664
Sylvisorex granti	5.06	1	2.42	2.64	1	0.320	0.541
S. megalura	5.58	1	2.90	2.68	1	0.268	0.454
Ruwenzorisorex sunc.	10.25	1	4.74	5.51	1	0.178	0.694
Myosorex babaulti	12.08	1	6.22	5.86	1	0.584	1.095

Number of individuals of the total amygdaloid complex (AMY) as in Table 10
Number of individuals = n
Abbreviations:

- AMY total amygdaloid complex, amygdala
- LAM lateral amygdaloid division, including basal, cortical, and lateral nuclei
- MAM medial amygdaloid division, including central and medial nuclei, and anterior amygdaloid area
- MCB magnocellular part of the basal nucleus
- NTO bed nucleus of the lateral olfactory tract

Table 56. Size indices of the amygdala and some amygdaloid components related to the average of the four Tenrecinae (AvTn = 100)

For volumes in species see Table 55
For average size indices in supraspecific taxonomic units see Table 57

	AMY	n	MAM	LAM	n	NTO	MCB
slope	.60		.60	.60		.60	.60
y-intercept	.232		-.098	-.043		-1.159	-.696
Tenrecidae							
Tenrec ecaudatus	91	2	86	95	1	79	79
Setifer setosus	149	3	149	149	1	153	149
Hemicentetes semispin.	70	2	69	70	1	99	64
Echinops telfairi	90	2	96	86	1	69	109
Geogale aurita	61	1	64	58	--	--	--
Oryzorictes talpoides	125	1	125	126	1	95	62
Microgale dobsoni	124	1	106	140	1	135	106
M. talazaci	140	1	144	137	1	102	80
Limnogale mergulus	104	2	92	116	1	30	83
Potamogale velox	77	1	76	78	1	14	56
Micropotamogale lam.	92	1	90	94	1	32	57
M. ruwenzorii	83	1	75	91	1	43	52
Chrysochloridae							
Chrysochloris asiatica	114	1	110	117	1	116	78
C. stuhlmanni	148	1	166	133	1	150	114
Solenodontidae							
Solenodon paradoxus	160	1	129	187	1	73	142
Erinaceidae							
Echinosorex gymnurus	178	3	155	198	--	--	--
Hylomys suillus	194	1	191	197	--	--	--
Atelerix algirus	115	1	112	119	1	86	72
Erinaceus europaeus	111	3	99	123	1	95	100
Hemiechinus auritus	157	1	140	173	1	151	92
Talpidae							
Desmana moschata	137	1	132	142	1	4	156
Galemys pyrenaicus	184	1	161	204	1	29	242
Talpa europaea	110	1	101	118	1	58	123
T. micrura	142	2	148	138	--	--	--
Parascalops breweri	121	3	128	116	--	--	--
Scalopus aquaticus	105	1	105	105	--	--	--

Table 56 (cont.)

	AMY	n	MAM	LAM	n	NTO	MCB
Soricidae							
Sorex alpinus	112	3	110	114	--	--	--
S. araneus	112	3	116	109	2	98	93
S. cinereus	107	1	111	103	1	131	100
S. fumeus	121	1	128	116	1	133	89
S. minutus	83	2	94	74	1	106	117
Microsorex hoyi	108	1	94	119	1	86	64
Neomys anomalus	110	3	114	108	--	--	--
N. fodiens	98	1	101	96	1	67	112
Blarina brevicauda	107	1	105	109	1	109	74
Cryptotis parva	135	3	144	127	--	--	--
Anourosorex squamipes	149	4	153	145	--	--	--
Crocidura attenuata	102	1	114	92	--	--	--
C. flavescens	98	2	102	94	1	79	79
C. hildegardeae	100	1	110	92	1	120	73
C. jacksoni	97	1	95	99	1	74	58
C. russula	92	2	96	89	1	82	89
C. suaveolens	86	3	91	82	--	--	--
Suncus etruscus	70	1	79	62	1	83	40
S. murinus	79	2	82	76	1	77	48
Scutisorex somereni	87	1	83	90	1	94	69
Sylvisorex granti	131	1	134	129	1	204	119
S. megalura	118	1	131	106	1	139	81
Ruwenzorisorex suncoides	105	1	104	107	1	45	60
Myosorex babaulti	129	1	142	118	1	154	99

Number of individuals of the total amygdaloid complex (AMY) as in Table 10
Number of individuals = n
Abbreviations:

AMY total amygdaloid complex, amygdala
LAM lateral amygdaloid division, including basal, cortical, and lateral nuclei
MAM medial amygdaloid division, including central and medial nuclei, and anterior amygdaloid area
MCB magnocellular part of the basal nucleus
NTO bed nucleus of the lateral olfactory tract

Table 57. Average size indices of the amygdala and some amygdaloid components related to the average of the four Tenrecinae (AvTn = 100)

	N	AMY	MAM	LAM	N	NTO	MCB
slope		.60	.60	.60		.60	.60
y-intercept		.232	-.098	-.043		-1.159	-.696
Insectivora (AvIS)	50	114	114	115	38	92	92
Tenrecidae	12	101	98	103	11	77	82
Tenrecinae*	4	**100**	**100**	**100**	4	**100**	**100**
Geogalinae*	1	61	64	58	--	--	--
Oryzorictinae*	4	123	116	130	4	90	83
Microgale	2	132	125	139	2	119	93
terr.Oryzorictinae	3	130	125	135	3	111	83
Limnogale	1	104	92	116	1	30	83
Potamogalinae*	3	84	80	88	3	30	55
Micropotamogale	2	88	82	92	2	38	55
Chrysochloridae*	2	131	138	125	2	133	96
Solenodontidae*	1	160	129	187	1	73	142
Erinaceidae	5	151	139	162	3	110	88
Echinosoricinae*	2	186	173	198	--	--	--
Erinaceinae*	3	128	117	138	3	110	88
Talpidae	6	133	129	137	3	30	174
Desmaninae*	2	161	147	173	2	16	199
Talpinae*	4	120	121	119	1	58	123
Talpa	2	126	125	128	1	58	123
Soricidae	24	106	110	102	18	105	81
Soricinae*	11	113	116	111	7	104	93
Sorex	5	107	112	103	4	117	100
terr. Soricinae	9	115	118	113	6	111	90
Neomys	2	104	107	102	1	67	112
Crocidurinae*	13	100	105	95	11	105	74
Crocidura	6	96	101	91	4	89	75
Suncus	2	74	81	69	2	80	44
Sylvisorex	2	124	132	117	2	172	100
AvTn of the families and subfamilies marked by *	(12)	122	117	127	(10)	82	105

Number of species = N

Abbreviations:

AMY	total amygdaloid complex		
LAM	lateral amygdaloid division	MCB	large-celled part of basal nucleus
MAM	medial amygdaloid division	NTO	bed nucleus of the lateral olfactory tract

Table 58. Volumes of hippocampal components

For size indices in species see Table 59
For average size indices in supraspecific taxonomic units see Table 60

	n	HIP	n	HR	HA	HS	HP
Tenrecidae							
Tenrec ecaudatus	3	147.1	1	136.3	10.82	2.086	8.73
Setifer setosus	3	96.6	1	89.9	6.69	1.041	5.65
Hemicentetes semispin.	2	63.1	1	58.3	4.84	0.644	4.20
Echinops telfairi	2	46.7	1	44.7	2.01	0.240	1.77
Erinaceidae							
Erinaceus europaeus	5	237.9	1	223.2	14.76	2.350	12.41
Soricidae							
Sorex araneus	5	22.8	1	21.2	1.51	0.244	1.27
S. minutus	4	12.4	1	11.7	0.68	0.120	0.56
Crocidura flavescens	3	38.3	1	35.8	2.58	0.515	2.06
C. russula	5	17.9	1	16.7	1.22	0.175	1.04
Suncus murinus	3	33.4	1	31.0	2.37	0.605	1.77

Number of individuals = n
Abbreviations:

HA	hippocampus anterior (hippocampus supra- + precommissuralis, HS + HP)
HIP	total hippocampus
HP	hippocampus precommissuralis
HR	hippocampus retrocommissuralis
HS	hippocampus supracommissuralis

Table 59. Size indices of hippocampal components related to the average of the four Tenrecinae (AvTn = 100)

For volumes in species see Table 58
For average size indices in supraspecific taxonomic units see Table 60

	n	HIP	n	HR	HA	HS	HP
slope		.57		.57	.57	.57	.57
y-intercept		.581		.552	-.599	-1.420	-.670
Tenrecidae							
Tenrec ecaudatus	3	82	1	82	92	117	87
Setifer setosus	3	112	1	112	118	121	117
Hemicentetes semispin.	2	110	1	109	128	113	131
Echinops telfairi	2	96	1	98	62	49	65
Erinaceidae							
Erinaceus europaeus	5	134	1	134	126	132	124
Soricidae							
Sorex araneus	5	159	1	159	160	171	158
S. minutus	4	140	1	141	116	135	112
Crocidura flavescens	3	147	1	146	149	198	141
C. russula	5	119	1	119	122	117	123
Suncus murinus	3	118	1	117	127	214	111

Number of individuals = n
Abbreviations:

HA	hippocampus anterior (hippocampus supra- + precommissuralis, HS + HP)
HIP	total hippocampus
HP	hippocampus precommissuralis
HR	hippocampus retrocommissuralis
HS	hippocampus supracommissuralis

Table 60. Average size indices of hippocampal components related to the average of the four Tenrecinae (AvTn = 100)

For volumes in species see Table 58
For size indices in species see Table 59

	N	HIP	HR	HA	HS	HP
slope		.57	.57	.57	.57	.57
y-intercept		.581	.552	-.599	-1.420	-.670
Insectivora (AvIS)	10	122	122	120	137	117
Tenrecidae						
Tenrecinae*	4	**100**	**100**	**100**	**100**	**100**
Erinaceidae						
Erinaceinae*	1	134	134	126	132	124
Soricidae	5	137	136	135	167	129
Soricinae*	2	150	150	138	153	135
Crocidurinae*	3	128	127	133	176	125
Crocidura	2	133	133	136	157	132
AvIF of the 4 subfamilies marked by *		128	128	124	140	121

Number of species = N
Abbreviations:

- HA hippocampus anterior (hippocampus supra- + precommissuralis, HS + HP)
- HIP total hippocampus
- HP hippocampus precommissuralis
- HR hippocampus retrocommissuralis
- HS hippocampus supracommissuralis

Table 61. Volumes of neocortical components

For size indices in species see Table 62
For average size indices in supraspecific taxonomic units see Table 63

	n	NEO	n	NW	NG	NG1	NG2-6
Tenrecidae							
Tenrec ecaudatus	3	272.8	1	35.76	237.0	73.32	163.7
Setifer setosus	3	136.6	1	17.73	118.9	38.18	80.7
Hemicentetes semis.	2	76.9	1	6.14	70.7	24.02	46.7
Echinops telfairi	2	51.7	1	5.25	46.5	15.33	31.1
Geogale aurita	2	7.2	1	0.53	6.7	1.81	4.9
Oryzorictes talpoides	1	66.5	1	5.91	60.6	18.63	41.9
Microgale dobsoni	1	62.6	1	5.59	57.0	16.79	40.2
M. talazaci	2	88.9	1	9.71	79.2	23.59	55.6
Limnogale mergulus	1	195.6	1	18.09	177.5	42.50	135.0
Potamogale velox	5	1050.0	1	140.89	909.1	141.57	767.6
Micropotamogale lamottei	3	144.4	1	14.36	130.0	30.25	99.8
M. ruwenzorii	4	227.8	1	23.33	204.4	41.12	163.3
Chrysochloridae							
Chrysochloris asiatica	1	112.6	1	5.80	106.8	29.05	77.8
C. stuhlmanni	2	128.3	1	7.61	120.7	31.25	89.4
Solenodontidae							
Solenodon paradoxus	2	668.6	1	68.55	600.1	162.66	437.4
Erinaceidae							
Echinosorex gymnurus	3	1085.0	1	224.98	860.0	194.52	665.5
Hylomys suillus	1	177.8	1	18.12	159.6	44.82	114.8
Atelerix algirus	3	510.3	1	64.88	445.4	119.32	326.1
Erinaceus europaeus	5	525.1	1	68.70	456.4	129.79	326.6
Hemiechinus auritus	3	306.2	1	30.47	275.8	74.24	201.5
Talpidae							
Desmana moschata	1	983.4	1	129.55	853.8	157.34	696.5
Galemys pyrenaicus	2	338.7	1	33.73	305.0	53.73	251.2
Talpa europaea	2	181.2	1	14.13	167.1	38.01	129.1
T. micrura	2	157.2	1	10.49	146.7	33.06	113.6
Parascalops breweri	3	175.6	1	10.85	164.8	34.98	129.8
Scalopus aquaticus	1	252.9	1	20.26	232.6	58.99	173.6

Table 61 (cont.)

	n	NEO	n	NW	NG	NG1	NG2-6
Soricidae							
Sorex alpinus	3	45.5	1	2.34	43.2	10.42	32.8
S. araneus	5	28.2	3	2.55	25.6	5.56	20.0
S. cinereus	3	28.8	3	1.97	26.8	6.77	20.0
S. fumeus	2	48.0	2	2.80	45.2	10.45	34.8
S. minutus	4	15.9	1	1.19	14.7	3.20	11.5
Microsorex hoyi	3	15.2	3	0.85	14.4	4.05	10.3
Neomys anomalus	3	59.0	1	3.96	55.1	12.32	42.8
N. fodiens	2	56.6	2	5.19	51.4	11.10	40.3
Blarina brevicauda	3	68.7	3	5.17	63.5	15.33	48.2
Cryptotis parva	3	36.2	1	2.67	33.5	8.18	25.3
Anourosorex squamipes	4	67.6	1	3.32	64.3	17.71	46.6
Crocidura attenuata	1	25.3	1	1.90	23.4	6.63	16.8
C. flavescens	3	53.1	3	4.18	48.9	14.81	34.1
C. hildegardeae	4	26.0	1	1.92	24.1	7.00	17.1
C. jacksoni	1	28.6	1	2.16	26.4	6.82	19.6
C. russula	5	25.3	5	1.77	23.6	6.89	16.7
C. suaveolens	3	24.6	1	1.68	22.9	5.91	17.0
Suncus etruscus	3	7.1	3	0.48	6.6	2.11	4.5
S. murinus	3	50.5	3	4.63	45.9	13.62	32.3
Scutisorex somereni	3	87.2	1	6.81	80.3	21.79	58.6
Sylvisorex granti	3	20.6	1	1.42	19.2	5.25	14.0
S. megalura	3	23.9	1	1.33	22.6	6.04	16.6
Ruwenzorisorex suncoides	1	59.5	1	3.96	55.6	15.23	40.4
Myosorex babaulti	1	45.3	1	3.68	41.6	10.63	31.0

Number of individuals = n

Abbreviations:

NEO total neocortex
NG neocortical grey matter
NG1 molecular layer of neocortical grey matter
NG2-6 cell layers 2-6 of neocortical grey matter
NW neocortical white matter

Table 62. Size indices of neocortical components related to the average of the four Tenrecinae (AvTn = 100)

For volumes in species see Table 61
For average size indices in supraspecific taxonomic units see Table 63

	n	NEO	n	NW	NG	NG1	NG2-6
slope		.73		.83	.72	.68	.73
y-intercept		.346		-.847	.318	-.077	.124
Tenrecidae							
Tenrec ecaudatus	3	89	1	93	88	89	89
Setifer setosus	3	114	1	133	112	111	112
Hemicentetes semispin.	2	108	1	84	111	113	109
Echinops telfairi	2	89	1	90	89	87	89
Geogale aurita	2	61	1	55	61	45	68
Oryzorictes talpoides	1	189	1	179	190	169	198
Microgale dobsoni	1	225	1	222	227	190	241
M. talazaci	2	237	1	274	234	202	247
Limnogale mergulus	1	325	1	298	329	234	374
Potamogale velox	5	434	1	478	428	214	529
Micropotamogale lam.	3	312	1	319	312	213	359
M. ruwenzorii	4	365	1	369	365	219	436
Chrysochloridae							
Chrysochloris asiatica	1	296	1	161	312	246	341
C. stuhlmanni	2	390	1	250	406	303	453
Solenodontidae							
Solenodon paradoxus	2	260	1	217	266	232	284
Erinaceidae							
Echinosorex gymnurus	3	364	1	602	329	242	372
Hylomys suillus	1	419	1	444	418	342	451
Atelerix algirus	3	186	1	190	185	160	198
Erinaceus europaeus	5	172	1	179	171	158	179
Hemiechinus auritus	3	257	1	231	260	216	281
Talpidae							
Desmana moschata	1	519	1	579	510	298	612
Galemys pyrenaicus	2	772	1	796	772	398	954
Talpa europaea	2	327	1	256	336	227	389
T. micrura	2	468	1	335	483	314	564
Parascalops breweri	3	432	1	279	450	278	532
Scalopus aquaticus	1	357	1	278	367	280	409

Table 62 (cont.)

	n	NEO	n	NW	NG	NG1	NG2-6
Soricidae							
Sorex alpinus	3	356	1	225	369	244	428
S. araneus	5	233	3	261	231	137	276
S. cinereus	3	391	3	354	395	265	454
S. fumeus	2	458	2	336	470	293	553
S. minutus	4	243	1	244	244	139	294
Microsorex hoyi	3	339	3	267	344	250	383
Neomys anomalus	3	445	1	364	454	278	537
N. fodiens	2	348	2	379	347	207	413
Blarina brevicauda	3	351	3	306	357	241	411
Cryptotis parva	3	306	1	280	309	206	357
Anourosorex squamipes	4	341	1	193	356	275	392
Crocidura attenuata	1	224	1	210	226	174	248
C. flavescens	3	203	3	178	206	178	218
C. hildegardeae	4	215	1	196	218	172	236
C. jacksoni	1	202	1	184	204	145	231
C. russula	5	197	5	169	200	160	216
C. suaveolens	3	202	1	170	205	144	233
Suncus etruscus	3	203	3	202	203	165	214
S. murinus	3	174	3	175	175	148	186
Scutisorex somereni	3	190	1	153	195	155	213
Sylvisorex granti	3	345	1	322	347	249	389
S. megalura	3	311	1	227	319	226	359
Ruwenzorisorex sunc.	1	323	1	250	331	253	365
Myosorex babaulti	1	258	1	246	260	185	294

Number of individuals = n

Abbreviations:

NEO	total neocortex
NG	neocortical grey matter
NG1	molecular layer of neocortical grey matter
NG2-6	cell layers 2-6 of neocortical grey matter
NW	neocortical white matter

Table 63. Average size indices of neocortical components related to the average of the four Tenrecinae (AvTn = 100)

	N	NEO	NW	NG	NG1	NG2-6
slope		.73	.83	.72	.68	.73
y-intercept		.346	-.847	.318	-.077	.124
Insectivora (AvIS)	50	292	266	296	209	335
Tenrecidae	12	212	216	212	157	238
Tenrecinae*	4	**100**	**100**	**100**	**100**	**100**
Geogalinae*	1	61	55	61	45	68
Oryzorictinae*	4	244	243	245	199	265
Microgale	2	231	248	230	196	244
terr. Oryzorictinae	3	217	225	217	187	229
Limnogale	1	325	298	329	234	374
Potamogalinae*	3	370	389	368	215	441
Micropotamogale	2	338	344	339	216	398
Chrysochloridae*	2	343	205	359	274	397
Solenodontidae*	1	260	217	266	232	284
Erinaceidae	5	279	329	273	224	296
Echinosoricinae*	2	391	523	373	292	412
Erinaceinae*	3	205	200	205	178	219
Talpidae	6	479	421	486	299	577
Desmaninae*	2	645	688	641	348	783
Talpinae*	4	396	287	409	274	473
Talpa	2	397	296	410	270	476
Soricidae	24	286	246	290	204	329
Soricinae*	11	347	292	352	230	409
Sorex	5	336	284	342	216	401
terr. Soricinae	9	335	274	342	228	394
Neomys	2	397	372	400	243	475
Crocidurinae*	13	234	207	238	181	262
Crocidura	6	207	185	210	162	230
Suncus	2	188	189	189	157	200
Sylvisorex	2	328	275	333	237	374
AvIF of the 12 families and subfamilies marked by *		300	284	301	214	343

Number of species = N

Abbreviations:

NEO	total neocortex
NG	neocortical grey matter
NG1	molecular layer of neocortical grey matter
NG2-6	cell layers 2-6 of neocortical grey matter
NW	neocortical white matter

Table 64. Volumes of neocortical grey (NG), area striata grey (ASG), percentage of ASG relative to NG, and size indices of ASG related to the average of the four Tenrecinae (AvTn = 100)

ASG was estimated from GLD volumes (taken from Table 37) according to the interrelationship found in Primates: log GLD = -0.8801 + 0.8476 × log ASG

	NG mm^3	ASG mm^3	ASG / NG	ASG Index
				.55 = slope
Tenrecidae				-.054 = y-intercept
Tenrecinae				
Tenrec ecaudatus	237.00	42.04	17.7	116
Setifer setosus	118.90	25.07	21.1	140
Hemicentetes semispin.	70.77	8.37	11.8	69
Echinops telfairi	46.46	7.61	16.4	74
Erinaceidae				
Erinaceinae				
Erinaceus europaeus	456.43	126.18	27.6	350
Soricidae				
Soricinae				
Sorex araneus	25.60	1.16	4.5	37
S. minutus	14.75	0.81	5.5	41
Crocidurinae				
Crocidura flavescens	48.87	4.57	9.4	81
C. russula	23.56	1.16	4.9	35
Suncus murinus	45.90	3.02	6.6	49
Average values	N			
Insectivora (AvIS)	10		12.6	99
Tenrecinae*	4		16.8	**100**
Erinaceinae*	1		27.6	350
Soricinae*	2		5.0	39
Crocidurinae*	3		7.0	55
AvIF of the 4 subfamilies marked by *			14.1	136

Number of species = N

Abbreviations and notes:

- ASG area striata grey matter
- GLD dorsal part of lateral geniculate body
- NG total neocortex grey matter

Table 65. Volumes of diencephalon components based on data given by Bauchot (1979b)

For size indices in species see Table 66
For average size indices in supraspecific taxonomic units see Table 67

	DIE	ETH	THA	STH	HTH	RTH	PTE
Tenrecidae							
Tenrec ecaudatus	153.7	3.175	60.86	38.27	43.52	6.20	1.704
Setifer setosus	93.1	2.256	37.55	20.40	28.45	3.52	0.933
Hemicentetes semispin.	47.7	1.043	20.01	10.44	13.63	2.07	0.516
Echinops telfairi	40.3	0.878	17.10	8.69	11.01	2.07	0.532
Microgale talazaci	45.4	1.023	19.98	10.91	10.91	2.05	0.517
Limnogale mergulus	81.0	1.150	34.76	20.87	19.52	4.06	0.672
Potamogale velox	300.3	3.036	143.70	69.98	64.19	16.71	2.703
Chrysochloridae							
Chrysochloris stuhlm.	51.6	1.164	20.55	16.88	10.58	2.02	0.373
Solenodontidae							
Solenodon paradoxus	272.3	5.630	111.16	85.99	56.97	10.30	2.225
Erinaceidae							
Atelerix algirus	251.0	4.236	98.41	67.15	69.53	9.09	2.544
Erinaceus europaeus	242.4	4.940	92.41	57.84	76.36	8.12	2.756
Talpidae							
Desmana moschata	343.8	4.126	125.89	126.91	70.90	13.03	2.914
Galemys pyrenaicus	104.5	1.532	41.67	35.14	21.46	3.93	0.732
Talpa europaea	80.7	1.366	29.58	27.83	18.51	2.87	0.534
Soricidae							
Sorex araneus	14.8	0.304	5.80	3.88	4.12	0.57	0.125
S. minutus	7.7	0.145	3.04	1.92	2.21	0.31	0.068
Neomys fodiens	24.1	0.444	9.67	7.51	5.27	0.98	0.197
Crocidura flavescens	25.1	0.514	10.92	6.02	6.15	1.20	0.285
C. russula	13.9	0.297	5.47	3.59	3.75	0.59	0.167
Suncus murinus	24.1	0.440	9.88	5.53	6.85	1.09	0.269

Abbreviations:

DIE	diencephalon
ETH	epithalamus
HTH	hypothalamus
PTE	regio pretectalis
RTH	nucleus reticularis
STH	subthalamus
THA	thalamus

Table 66. Size indices of diencephalon components related to the average of the four Tenrecinae (AvTn = 100) based on data given by Bauchot (1979b)

For volumes in species see Table 65
For average size indices in supraspecific taxonomic units see Table 67

	DIE	ETH	THA	STH	HTH	RTH	PTE
slope	.65	.63	.64	.71	.62	.63	.62
y-intercept	.350	-1.256	-.013	-.438	-.121	-.970	-1.531
Tenrecidae							
Tenrec ecaudatus	85	82	84	87	88	83	88
Setifer setosus	119	130	117	115	127	105	107
Hemicentetes semispin.	97	94	98	98	95	97	92
Echinops telfairi	98	95	101	100	91	115	113
Microgale talazaci	163	161	172	191	130	166	159
Limnogale mergulus	192	120	198	231	156	220	138
Potamogale velox	206	96	242	200	158	272	171
Chrysochloridae							
Chrysochloris stuhlm.	209	205	199	336	141	183	128
Solenodontidae							
Solenodon paradoxus	177	168	178	232	133	159	133
Erinaceidae							
Atelerix algirus	154	119	148	170	153	133	144
Erinaceus europaeus	135	127	127	132	154	108	143
Talpidae							
Desmana moschata	292	160	263	460	214	262	226
Galemys pyrenaicus	327	210	313	528	225	279	197
Talpa europaea	205	153	181	334	159	166	118
Soricidae							
Sorex araneus	146	127	135	205	129	122	101
S. minutus	131	103	121	184	116	112	92
Neomys fodiens	183	144	174	297	128	165	123
Crocidura flavescens	125	110	129	150	100	133	119
C. russula	129	118	121	178	111	120	128
Suncus murinus	109	86	107	124	102	111	103

Abbreviations:

DIE	diencephalon		
ETH	epithalamus	RTH	nucleus reticularis
HTH	hypothalamus	STH	subthalamus
PTE	regio pretectalis	THA	thalamus

Table 67. Average size indices of diencephalon components related to the average of the four Tenrecinae (AvTn = 100) based on data given by Bauchot (1979b)

For volumes in species see Table 65
For size indices in species see Table 66

	N	DIE	ETH	THA	STH	HTH	RTH	PTE
slope		.65	.63	.64	.71	.62	.63	.62
y-intercept		.350	-1.256	-.013	-.438	-.121	-.970	-1.531
Insectivora (AvIS)	20	164	130	160	218	136	156	131
Tenrecidae	7	137	111	145	146	121	151	124
Tenrecinae*	4	**100**	**100**	**100**	**100**	**100**	**100**	**100**
Oryzorictinae*	2	177	140	185	211	143	193	149
Potamogalinae*	1	206	96	242	200	158	272	171
Chrysochloridae*	1	209	205	199	336	141	183	128
Solenodontidae*	1	177	168	178	232	133	159	133
Erinaceidae								
Erinaceinae*	2	144	123	138	151	154	120	144
Talpidae	3	275	174	253	441	199	236	180
Desmaninae*	2	310	185	288	494	219	270	212
Talpinae*	1	205	153	181	334	159	166	118
Soricidae	6	137	115	131	190	115	127	111
Soricinae*	3	153	124	143	228	125	133	105
Sorex	2	138	115	128	194	123	117	96
Neomys	1	183	144	174	297	128	165	123
Crocidurinae*	3	121	105	119	151	104	121	117
Crocidura	2	127	114	125	164	106	127	123
Suncus	1	109	86	107	124	102	111	103
AvIF of the 10 families and sub-families marked by *		180	140	177	244	144	172	138

Number of species = N
Abbreviations:

- DIE diencephalon
- ETH epithalamus
- HTH hypothalamus
- PTE regio pretectalis
- RTH nucleus reticularis
- STH subthalamus
- THA thalamus

Table 68. Volumes of thalamus components based on data given by Bauchot (1979b)

For size indices in species see Table 69
For average size indices in supraspecific taxonomic units see Table 70

	ANT	MLT	MNT	DOT	VET	CGM	CGL
Tenrecidae							
Tenrec ecaudatus	2.48	6.02	6.12	4.36	27.28	10.03	4.58
Setifer setosus	2.14	4.29	4.35	2.69	15.71	5.77	2.60
Hemicentetes semispin.	1.31	2.01	2.18	1.38	8.56	3.15	1.42
Echinops telfairi	1.05	1.49	1.82	0.94	7.56	2.78	1.47
Microgale talazaci	2.31	2.27	2.59	2.06	5.77	3.77	1.22
Limnogale mergulus	1.61	3.89	4.26	3.70	7.62	12.47	1.21
Potamogale velox	6.68	17.78	10.71	21.67	31.30	51.26	4.32
Chrysochloridae							
Chrysochloris stuhlm.	1.36	2.60	3.56	2.72	4.86	4.95	0.50
Solenodontidae							
Solenodon paradoxus	7.00	13.03	19.58	12.50	39.54	14.54	4.98
Erinaceidae							
Atelerix algirus	6.48	15.29	11.40	11.99	34.61	12.72	5.93
Erinaceus europaeus	4.61	12.70	10.02	10.44	34.39	12.64	7.62
Talpidae							
Desmana moschata	5.87	17.75	17.37	19.80	23.03	37.72	4.34
Galemys pyrenaicus	2.66	6.40	6.49	7.59	6.61	10.82	1.10
Talpa europaea	1.92	3.92	4.76	3.68	7.22	7.36	0.72
Soricidae							
Sorex araneus	0.44	0.73	0.77	0.55	1.84	1.20	0.26
S. minutus	0.20	0.33	0.41	0.22	1.04	0.68	0.16
Neomys fodiens	0.61	1.01	1.25	0.97	2.08	3.40	0.35
Crocidura flavescens	0.80	1.30	1.25	1.13	3.50	2.28	0.65
C. russula	0.33	0.64	0.60	0.48	1.84	1.20	0.38
Suncus murinus	0.56	1.10	0.95	0.89	3.49	2.28	0.62

Abbreviations:

ANT nuclei anteriores thalami
CGL corpus geniculatum laterale
CGM corpus geniculatum mediale
DOT nuclei dorsales thalami
MLT nuclei mediales thalami
MNT nuclei mediani thalami
VET nuclei ventrales thalami

Table 69. Size indices of thalamus components related to the average of the four Tenrecinae (AvTn = 100) based on data given by Bauchot (1979b)

For volumes in species see Table 68
For average size indices in supraspecific taxonomic units see Table 70

	ANT	MLT	MNT	DOT	VET	CGM	CGL
slope	.58	.62	.64	.68	.64	.72	.60
y-intercept	-1.123	-.964	-.978	-1.281	-.377	-.996	-1.046
Tenrecidae							
Tenrec ecaudatus	66	84	77	85	87	77	89
Setifer setosus	119	133	125	125	113	112	109
Hemicentetes semispin.	110	97	99	104	97	102	91
Echinops telfairi	104	86	99	86	103	110	111
Microgale talazaci	324	189	206	282	115	229	133
Limnogale mergulus	155	217	224	326	100	476	89
Potamogale velox	213	304	167	524	122	497	102
Chrysochloridae							
Chrysochloris stuhlmanni	212	242	318	421	109	343	61
Solenodontidae							
Solenodon paradoxus	213	212	289	285	146	133	111
Erinaceidae							
Atelerix algirus	187	235	158	257	121	109	125
Erinaceus europaeus	122	179	127	203	109	98	148
Talpidae							
Desmana moschata	227	374	334	600	111	465	125
Galemys pyrenaicus	330	466	451	899	115	564	105
Talpa europaea	198	234	270	351	102	305	57
Soricidae							
Sorex araneus	153	160	166	218	99	224	71
S. minutus	112	120	149	153	96	231	74
Neomys fodiens	168	171	207	290	86	473	76
Crocidura flavescens	150	147	137	218	96	199	95
C. russula	107	132	122	179	94	211	100
Suncus murinus	97	114	95	155	87	179	83

Abbreviations:

ANT nuclei anteriores thalami
CGL corpus geniculatum laterale
CGM corpus geniculatum mediale
DOT nuclei dorsales thalami
MLT nuclei mediales thalami
MNT nuclei mediani thalami
VET nuclei ventrales thalami

Table 70. Average size indices of thalamus components related to the average of the four Tenrecinae (AvTn = 100) based on data given by Bauchot (1979b)

For volumes in species see Table 68
For size indices in species see Table 69

	N	ANT	MLT	MNT	DOT	VET	CGM	CGL
slope		.58	.62	.64	.68	.64	.72	.60
y-intercept		-1.123	-.964	-.978	-1.281	-.377	-.996	-1.046
Insectivora (AvIS)	20	168	195	191	288	105	257	98
Tenrecidae	7	156	159	142	219	105	229	103
Tenrecinae*	4	**100**	**100**	**100**	**100**	**100**	**100**	**100**
Oryzorictinae*	2	239	203	215	304	108	353	111
Potamogalinae*	1	213	304	167	524	122	497	102
Chrysochloridae*	1	212	242	318	421	109	343	61
Solenodontidae*	1	213	212	289	285	146	133	111
Erinaceidae								
Erinaceinae*	2	155	207	143	230	115	103	137
Talpidae	3	252	358	351	617	109	445	95
Desmaninae*	2	279	420	392	749	113	515	115
Talpinae*	1	198	234	270	351	102	305	57
Soricidae	6	131	141	146	202	93	253	83
Soricinae*	3	144	150	174	220	94	309	74
Sorex	2	132	140	157	185	97	228	73
Neomys	1	168	171	207	290	86	473	76
Crocidurinae*	3	118	131	118	184	92	196	93
Crocidura	2	129	140	130	198	95	205	98
Suncus	1	97	114	95	155	87	179	83
AvIF of the 10 families and subfamilies marked by *		187	220	219	337	110	285	96

Abbreviations:

ANT nuclei anteriores thalami
CGL corpus geniculatum laterale
CGM corpus geniculatum mediale
DOT nuclei dorsales thalami
MLT nuclei mediales thalami
MNT nuclei mediani thalami
VET nuclei ventrales thalami

Table 71. Area measurements of commissures and chiasma opticum in mm^2

For size indices in species see Table 72
For average size indices in supraspecific taxonomic units see Table 73

	n	CCA	COA	CHO
Tenrecidae				
Tenrec ecaudatus	1	0.859	1.025	0.209
Setifer setosus	1	0.691	0.734	0.092
Hemicentetes semispin.	1	0.542	0.412	0.046
Echinops telfairi	1	0.230	0.384	0.121
Geogale aurita	1	0.068	0.139	0.022
Oryzorictes talpoides	1	0.646	0.284	0.017
Microgale dobsoni	1	0.689	0.300	0.029
M. talazaci	1	0.890	0.358	0.043
Limnogale mergulus	1	1.649	0.238	0.051
Potamogale velox	1	6.241	0.392	0.129
Micropotamog. ruwenzorii	1	1.098	0.156	0.038
Chrysochloridae				
Chrysochloris stuhlmanni	1	0.754	0.362	0.011
Solenodontidae				
Solenodon paradoxus	1	2.730	1.175	0.094
Erinaceidae				
Echinosorex gymnurus	1	7.016	1.347	0.407
Hylomys suillus	1	1.441	0.278	0.308
Atelerix algirus	2	2.535	0.967	0.397
Erinaceus europaeus	1	1.480	0.554	0.235
Hemiechinus auritus	3	1.574	0.541	0.236
Talpidae				
Desmana moschata	1	5.059	0.466	0.071
Galemys pyrenaicus	1	2.055	0.183	0.042
Talpa europaea	1	1.134	0.377	0.010
Parascalops breweri	1	0.739	0.292	0.011
Scalopus aquaticus	1	1.859	0.424	-.-

Table 71 (cont.)

	n	CCA	COA	CHO
Soricidae				
Sorex alpinus	1	0.379	0.158	0.016
S. araneus	1	0.312	0.146	0.034
S. cinereus	1	0.312	0.118	0.014
S. fumeus	1	0.543	0.129	0.020
S. minutus	1	0.145	0.048	0.015
Microsorex hoyi	1	0.194	0.081	0.009
Neomys anomalus	1	0.338	0.083	0.019
N. fodiens	1	0.376	0.095	0.021
Blarina brevicauda	1	0.534	0.205	0.023
Cryptotis parva	1	0.336	0.139	-.-
Anourosorex squamipes	1	0.327	0.181	0.008
Crocidura flavescens	4	0.362	0.213	0.039
C. russula	1	0.242	0.158	0.018
Suncus etruscus	1	0.095	0.057	0.008
S. murinus	1	0.440	0.183	0.033
Scutisorex somereni	1	0.556	0.315	0.033
Sylvisorex granti	1	0.209	0.096	0.012
S. megalura	1	0.169	0.122	0.018
Ruwenzorisorex suncoides	1	0.494	0.170	0.020
Myosorex babaulti	1	0.622	0.294	0.030

Number of individuals = n

Abbreviations:

CCA corpus callosum area

CHO chiasma opticum

COA commissura anterior

Table 72. Size indices of commissural areas and chiasma opticum area related to the average of the four Tenrecinae (AvTn = 100)

For area measurements in species see Table 71
For average size indices in supraspecific taxonomic units see Table 73

	n	CCA	COA	CHO
slope		.52	.44	.37
y-intercept		-1.474	-1.253	-1.821
Tenrecidae				
Tenrec ecaudatus	1	77	94	114
Setifer setosus	1	120	119	81
Hemicentetes semispin.	1	136	91	52
Echinops telfairi	1	67	96	153
Geogale aurita	1	61	90	61
Oryzorictes talpoides	1	268	96	27
Microgale dobsoni	1	339	117	53
M. talazaci	1	353	116	67
Limnogale mergulus	1	468	58	63
Potamogale velox	1	658	42	79
Micropotamog. ruwenzorii	1	303	37	46
Chrysochloridae				
Chrysochloris stuhlmanni	1	329	128	19
Solenodontidae				
Solenodon paradoxus	1	275	120	56
Erinaceidae				
Echinosorex gymnurus	1	637	126	225
Hylomys suillus	1	524	84	457
Atelerix algirus	2	244	95	229
Erinaceus europaeus	1	132	51	129
Hemiechinus auritus	3	274	88	207
Talpidae				
Desmana moschata	1	634	57	49
Galemys pyrenaicus	1	730	54	61
Talpa europaea	1	341	97	13
Parascalops breweri	1	277	91	16
Scalopus aquaticus	1	470	94	--

Table 72 (cont.)

	n	CCA	COA	CHO
Soricidae				
Sorex alpinus	1	324	99	44
S. araneus	1	278	94	94
S. cinereus	1	396	102	51
S. fumeus	1	535	91	59
S. minutus	1	200	45	56
Microsorex hoyi	1	349	94	43
Neomys anomalus	1	281	50	49
N. fodiens	1	271	51	51
Blarina brevicauda	1	338	99	51
Cryptotis parva	1	304	91	--
Anourosorex squamipes	1	205	87	17
Crocidura flavescens	4	186	86	75
C. russula	1	206	98	49
Suncus etruscus	1	204	77	43
S. murinus	1	210	70	59
Scutisorex somereni	1	191	91	46
Sylvisorex granti	1	307	95	50
S. megalura	1	208	103	62
Ruwenzorisorex suncoides	1	325	85	46
Myosorex babaulti	1	425	151	68

Number of individuals = n

Abbreviations:

CCA corpus callosum area

CHO chiasma opticum

COA commissura anterior

Table 73. Average size indices of commissural and chiasma opticum areas related to the average of the four Tenrecinae (AvTn = 100)

For area measurements in species see Table 71
For size indices in species see Table 72

	N	CCA	COA	CHO
slope		.52	.44	.37
y-intercept		-1.474	-1.253	-1.821
Insectivora (AvIS)	43 §)	313	88	80
Tenrecidae	11	259	87	72
Tenrecinae*	4	**100**	**100**	**100**
Geogalinae*	1	61	90	61
Oryzorictinae*	4	357	97	53
Microgale	2	346	117	60
terr. Oryzorictinae	3	320	110	49
Limnogale	1	468	58	63
Potamogalinae*	2	480	39	63
Chrysochloridae*	1	329	128	19
Solenodontidae*	1	275	120	56
Erinaceidae	5	362	89	249
Echinosoricinae*	2	581	105	341
Erinaceinae*	3	217	78	188
Talpidae	5 +)	490	79	49
Desmaninae*	2	682	56	55
Talpinae*	3 +)	363	94	15
Soricidae	20 +)	287	88	53
Soricinae*	11 +)	316	82	52
Sorex	5	347	86	61
terr. Soricinae	9 +)	325	89	52
Neomys	2	276	51	50
Crocidurinae*	9	251	95	55
Crocidura	2	196	92	62
Suncus	2	207	74	51
Sylvisorex	2	257	99	56
AvIF of the 12 families and subfamilies marked by *		334	90	91

Number of species = N   +) for CHO = N – 1   §) for CHO = N – 2

Abbreviations:

CCA corpus callosum area
CHO chiasma opticum
COA commissura anterior

Table 74. Size characteristics of the various brain structures in the average Insectivora relative to the average of the four Tenrecinae (AvTn = 100)

	N	Min	Max	Max/Min	AvIS	CV	NN	AvIF	CV
NET	50	83	258	3.1	154	23.6	12	160	27.6
OBL	50	86	240	2.8	132	25.9	12	139	24.9
MES	50	88	266	3.0	136	29.1	12	150	29.8
CER	50	73	319	4.4	167	32.6	12	179	36.7
DIE	50	79	327	4.1	169	29.3	12	180	34.7
TEL	50	66	260	3.9	159	25.6	12	161	30.0
OBL	50	86	240	2.8	132	25.9	12	139	24.9
OBL limited	30	86	240	2.8	133	28.0	11	140	27.8
TR	30	81	342	4.2	159	42.7	11	165	45.7
TSO	30	84	234	2.8	151	27.3	11	156	25.8
FUN									
FGR	30	67	207	3.1	131	29.3	11	143	25.2
FCM	30	57	320	5.6	118	49.1	11	147	49.4
FCE	30	23	245	10.7	111	51.4	11	127	38.1
RLN									
REL	30	72	317	4.4	170	33.4	11	182	27.8
INO	30	79	326	4.1	185	30.9	11	205	27.3
PRP	30	83	219	2.6	136	27.5	11	145	26.2
COE	29	63	240	3.8	120	33.7	11	128	23.2
MOT									
V	30	70	203	2.9	121	29.1	11	133	24.9
VII	30	76	282	3.7	127	34.3	11	133	34.4
X	30	88	238	2.7	140	24.5	11	138	17.3
XII	30	72	183	2.5	116	25.4	11	116	25.9
VC									
VM	30	82	281	3.4	141	36.2	11	152	35.7
VI	30	69	249	3.6	129	40.3	11	142	38.4
VL	30	64	284	4.4	119	41.2	11	128	41.1
VS	30	54	227	4.2	104	41.6	11	116	39.0
AUD									
DCO	30	43	303	7.0	107	46.4	11	113	44.2
VCO	30	38	177	4.7	90	41.1	11	95	38.8
OLS	30	40	223	5.6	104	39.5	11	112	34.5

Table 74 (cont.)

	N	Min	Max	Max/Min	AvIS	CV	NN	AvIF	CV
MES	50	88	266	3.0	136	29.1	12	150	29.8
MES limited	10	88	125	1.4	103	10.6	4	102	8.3
MTG	10	89	126	1.4	108	11.0	4	110	7.1
NLL	10	54	137	2.5	98	29.4	4	90	29.7
MTC	10	77	131	1.7	97	18.6	4	94	15.1
SUC	10	79	151	1.9	121	19.0	4	123	15.4
INC	10	44	136	3.1	83	33.1	4	76	31.3
SCB	23	23	857	37.3	203	104.7	10	260	86.5
CER	50	73	319	4.4	167	32.6	12	179	36.7
CER limited	10	78	146	1.9	113	18.1	4	115	10.2
TCN	10	78	125	1.6	107	14.4	4	112	10.6
MCN	10	81	160	2.0	117	24.5	4	126	23.1
ICN	10	73	128	1.8	94	17.7	4	97	15.8
LCN	10	66	148	2.2	119	21.8	4	122	14.3
DIE	50	79	327	4.1	169	29.3	12	180	34.7
DIE limited	10	85	146	1.7	117	16.6	4	124	14.0
ETH	10	78	149	1.9	111	17.5	4	111	10.9
HAM	38	77	245	3.2	153	29.2	11	162	24.7
EPI	38	0	203	∞	50	82.5	11	62	85.6
THA	10	87	163	1.9	127	21.4	4	136	18.6
CGM	10	91	142	1.6	116	14.5	4	120	11.4
CGL	10	68	226	3.3	103	45.8	4	122	57.6
GLD	10	62	245	4.0	104	52.4	4	125	64.6
GLV	10	72	185	2.6	102	33.2	4	115	41.0
STH	10	91	204	2.2	127	30.3	4	132	33.3
PALL	10	92	199	2.2	125	29.0	4	130	31.8
LUY	10	62	297	4.8	160	49.6	4	167	50.1
HTH	10	82	123	1.5	106	12.7	4	110	10.5
CAI	10	81	163	2.0	113	24.3	4	115	10.1
TRO	10	66	172	2.6	97	32.8	4	110	40.0

Number of species = N Number of subfamilies and families = NN

Abbreviations and notes:

AvIF average of families and subfamilies AvIS average of species
CV coefficient of variation (standard deviation of the mean in percent of the mean)
Max maximum index in species Min minimum index in species
Max/Min quotient of maximum through minimum division

Table 74 (cont.)

	N	Min	Max	Max/Min	AvIS	CV	NN	AvIF	CV
TEL	50	66	260	3.9	159	25.6	12	161	30.0
TEL components									
BOL	50	29	151	5.2	88	30.1	12	91	33.9
PAL	50	36	166	4.6	103	29.7	12	104	33.6
STR	50	66	391	5.9	207	35.4	12	210	39.7
SEP	50	67	223	3.3	132	23.0	12	139	25.0
AMY	50	61	194	3.2	114	25.8	12	122	28.5
HIP	50	77	348	4.5	175	32.1	12	186	37.0
SCH	50	40	336	8.4	168	34.1	12	178	40.1
NEO	50	61	772	12.7	292	43.9	12	300	51.4
MOB	50	28	152	5.4	87	30.1	12	91	34.3
L1+2	43	25	197	7.9	82	38.5	12	90	47.9
L3	43	31	170	5.5	97	32.8	12	94	27.8
L4-6	43	29	147	5.1	92	31.5	12	92	27.5
PvZ	43	6	226	37.7	59	104.2	12	80	91.5
AOB	50	6	810	135	138	87.9	12	158	91.9
AOB limited	41	28	810	28.9	150	86.8	12	163	87.9
L1+2	41	13	636	48.9	112	96.1	12	123	99.1
L3-5	41	39	766	19.6	177	69.2	12	186	66.9
L6	41	20	1587	79.4	211	126.0	12	246	122.8
PAL	50	36	166	4.6	103	29.7	12	104	33.6
PAL components									
RB	50	37	171	4.6	107	29.5	12	106	34.3
PRPI	50	35	159	4.5	94	31.9	12	98	35.3
TOL	50	33	212	6.4	104	35.8	12	114	44.4
TRL	50	34	172	5.1	84	35.3	12	92	35.5
COA	50	33	198	6.0	95	40.8	12	94	45.1
SIN	50	45	154	3.4	96	27.1	12	99	22.3
SEP	50	67	223	3.3	132	23.0	12	139	25.0
SEP components									
SET	38	75	258	3.4	162	28.4	11	166	26.2
SFB	38	25	189	7.6	75	45.6	11	83	35.4

Abbreviations and notes on page 499

Table 74 (cont.)

	N	Min	Max	Max/Min	AvIS	CV	NN	AvIF	CV
TEL (cont.)									
AMY	50	61	194	3.2	114	25.8	12	122	28.5
AMY components									
MAM	50	64	191	3.0	114	24.3	12	117	24.7
NTO	38	4	204	51.0	92	47.4	10	82	45.5
LAM	50	58	204	3.5	115	29.2	12	127	33.0
MCB	38	40	242	6.1	92	40.9	10	105	38.8
HIP	50	77	348	4.5	175	32.1	12	186	37.0
HIP limited	10	82	159	1.9	122	19.2	4	128	16.3
HR	10	82	159	1.9	122	19.3	4	128	16.3
HS	10	49	214	4.4	137	34.5	4	140	23.0
HP	10	65	158	2.4	117	22.5	4	121	12.3
NEO	50	61	772	12.7	292	43.9	12	300	51.4
NEO components									
NW	50	55	796	14.5	266	50.9	12	284	62.2
NG	50	61	772	12.7	296	43.8	12	301	50.9
NG1	50	45	398	8.8	209	33.2	12	214	38.8
NG2-6	50	68	954	14.0	335	48.1	12	343	55.5
ASG	10	35	350	10.0	99	95.5	4	136	106.6

Bauchot's data of the diencephalic components (converted)

	N	Min	Max	Max/Min	AvIS	CV	NN	AvIF	CV
ETH	20	82	210	2.6	130	28.1	10	140	26.6
THA	20	84	313	3.7	160	37.4	10	177	32.1
STH	20	87	528	6.1	218	55.0	10	244	47.5
HTH	20	88	225	2.6	136	27.2	10	144	23.4
RTH	20	83	279	3.4	156	38.4	10	172	34.9
PTE	20	88	226	2.6	131	27.1	10	138	24.5

Bauchot's data of thalamic structures (converted)

	N	Min	Max	Max/Min	AvIS	CV	NN	AvIF	CV
ANT	20	66	330	5.0	168	42.3	10	187	30.1
MLT	20	84	466	5.5	195	50.2	10	220	41.6
MNT	20	77	451	5.9	191	51.1	10	219	43.7
DOT	20	85	899	10.6	288	69.3	10	337	56.0
VET	20	86	146	1.7	105	13.7	10	110	14.3
CGM	20	77	564	7.3	257	61.6	10	285	52.9
CGL	20	57	148	2.6	98	24.6	10	96	26.4

Abbreviations and notes on page 499

11 Atlas of the Brain of *Atelerix algirus*

The brains for this atlas were obtained through the cooperation with
Dr. José Regidor
from the Anatomical Institute of the University of Las Palmas, Gran Canaria, who made available three specimens of *Atelerix algirus* and all laboratory facilities of his Institute to allow for the stereotaxic procedures.

Zero coordinates

The zero planes, based on anatomical landmarks of the head, and intersecting each other at right angles, were defined as follows:

1. The horizontal plane (H 0) passes through the centers of the external auditory meatus and the lower margins of the upper jaws between I^1 and I^2.

2. The anterior-posterior plane (AP 0) lies perpendicular to the horizontal plane and also passes through the centers of the external auditory meatus.

3. The left-right zero plane (LR 0) is the median sagittal plane.

Reference coordinates

Under deep Nembutal anesthesia the hedgehogs (*Atelerix algirus*) were perfused intracardially with 250 ml of 0.9% NaCl followed by 500 ml of Bouin's fluid as fixative. Then the heads were fixed in a stereotaxic headholder for rats (David Kopf Co.). Five reference needles were inserted into the intracranial brain by aid of the stereotaxic instrument in order (1) to orientate the brain for sectioning in one of the three fundamental planes (transversal, sagittal, horizontal), (2) to identify the position of each section and (3) to estimate the shrinkage during the embedding, cutting and staining procedures. The needles have the following coordinates and functions:

1. A pair of needles was inserted perpendicular to the horizontal plane at A 5.0 in left-right positions of L 6.0 and R 6.0. The distance of these vertical needles from the median sagittal plane in the sections indicate the degree of shrinkage in the left-right direction.

2. A pair of needles was inserted parallel to the horizontal plane in a distance of H +10.0 at AP 0 and A 15.0. These needles serve to determine the average thickness of the transversal section when related to the fresh brain. In the atlas brain (4989) the number of sections between A 15.0 and AP 0 (that is in a distance of 15 000 μm) is 622 instead of the expected 750 20 μm sections. Thus one section corresponds to 24.12 μm in the fresh brain, a value which is used to calculate the position of each transversal section in the AP direction.

3. A horizontal needle was inserted in a left position of L 3.0 at H +7.5 parallel to sagittal and horizontal planes. This longitudinal needle is used to identify each point in the transversal section according to its H level. The distance from the median sagittal plane is an additional criterion for the shrinkage in the left-right direction.

Histological procedures

Following the stereotaxic procedures the brains were removed and kept for four days in the fixative. At the beginning of the washing procedure the brains were photographed. Washing (4-6 weeks) was performed in an ascending series of ethanols. After being passed through methylbenzoate and benzene the brains were embedded in paraffine, orientated for sectioning by aid of the needles and cut in the transversal plan (brains 4989, 4990), sagittal plane (left hemisphere of 4988) and horizontal plane (right hemisphere of brain 4988). The serial sections were stained for cells with gallocyanine. Sagittal and horizontal sections were additionally used to determine the structural borders given in the atlas (transversal sections only).

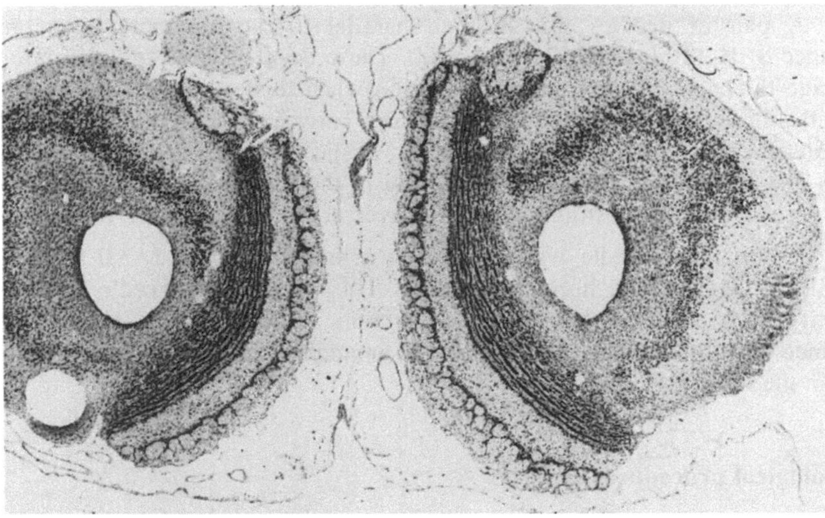

A 12.5

A 12.5

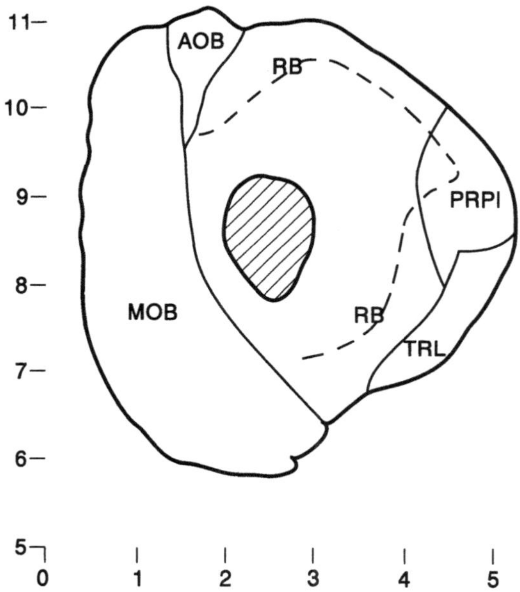

AOB	accessory olfactory bulb	RB	retrobulbar region
MOB	main olfactory bulb	TRL	lateral olfactory tract
PRPI	prepiriform region		

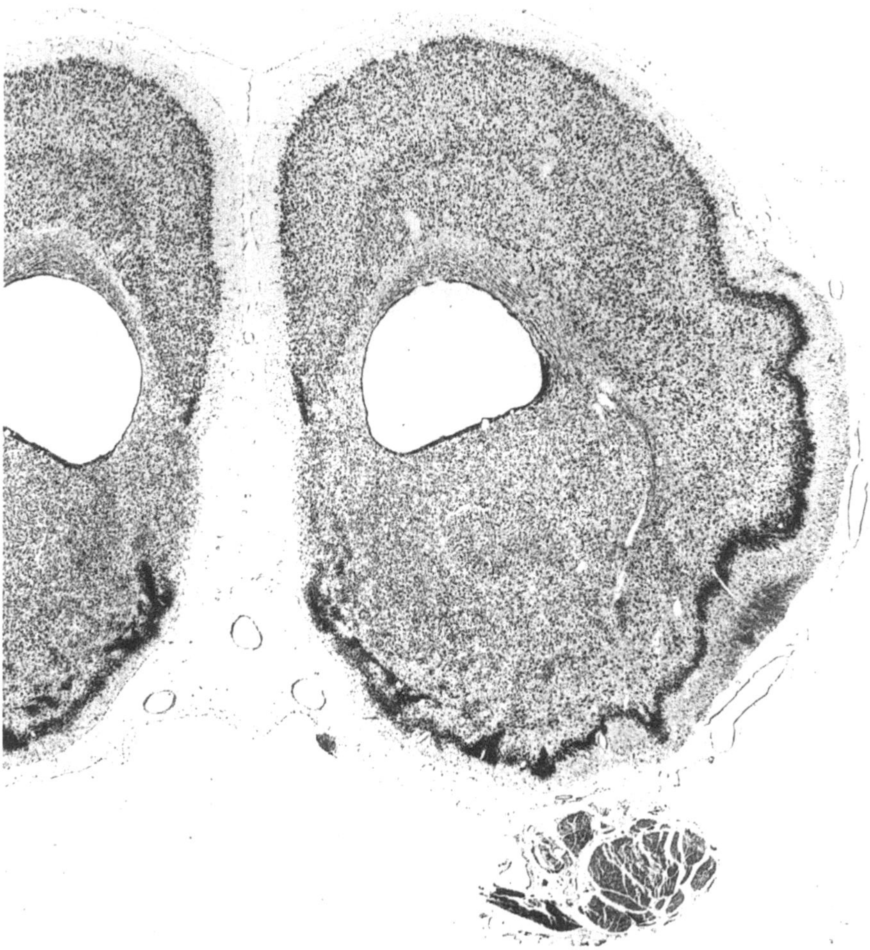

A 10

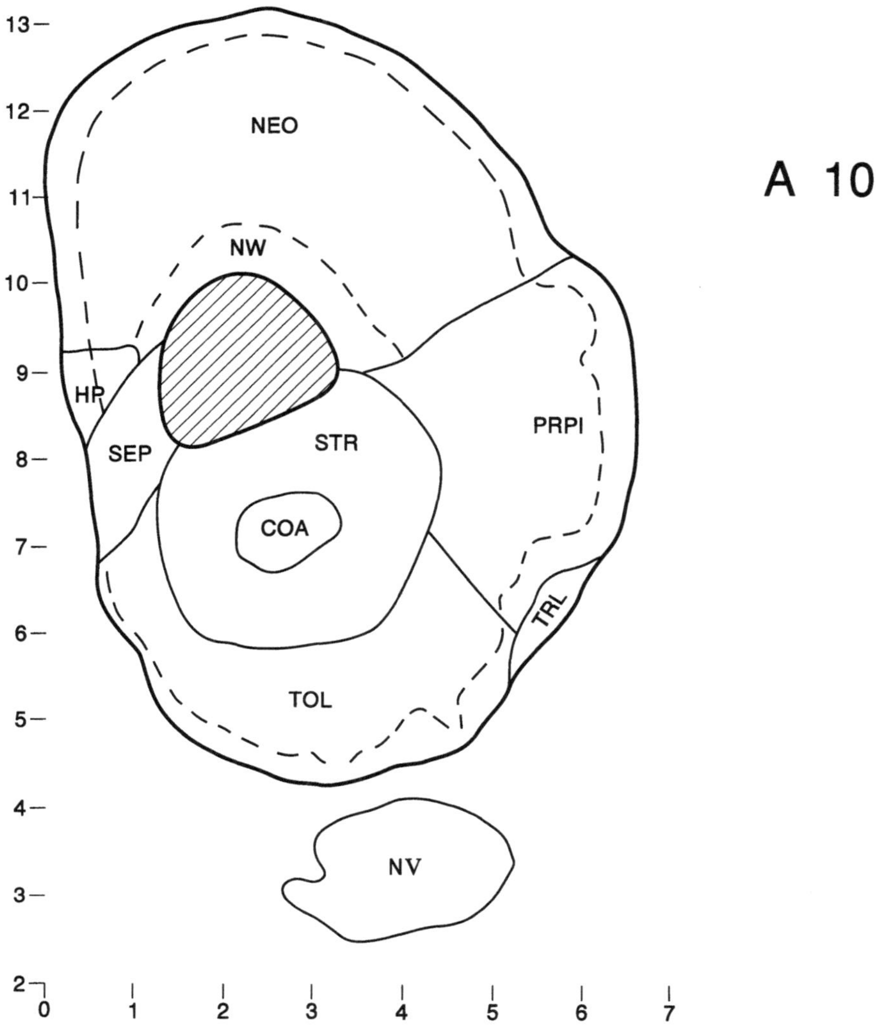

COA	anterior commissure	PRPI	prepiriform region
HP	hipp. precommissuralis	SEP	telencephalic septum
NEO	neocortex	STR	striatum
NW	neocortical white matter	TOL	olfactory tubercle
N V	trigeminal nerve	TRL	lateral olfactory tract

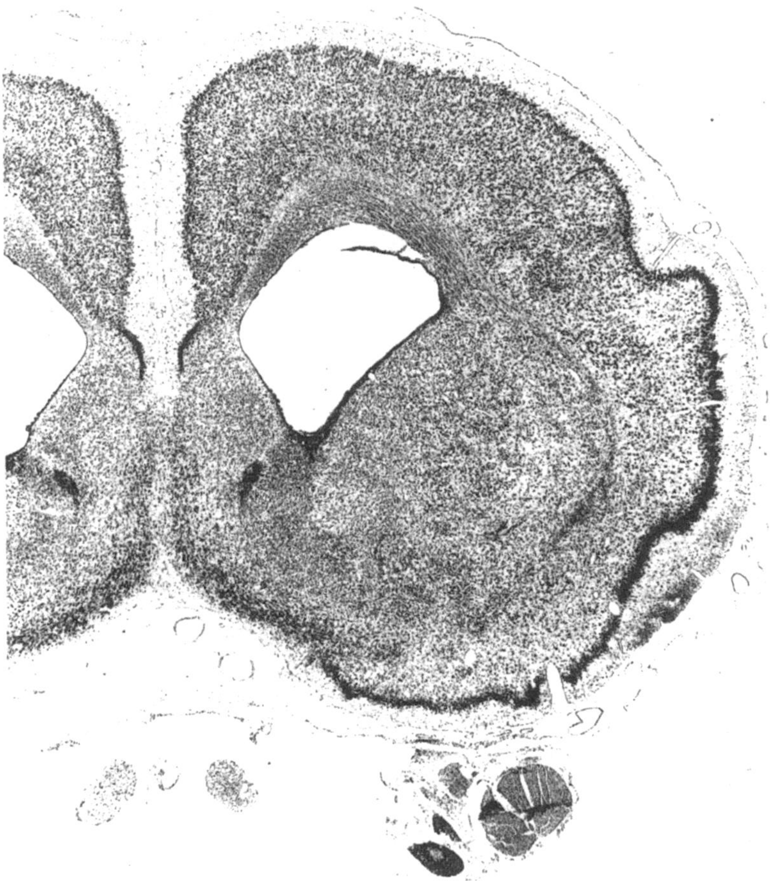

A 9

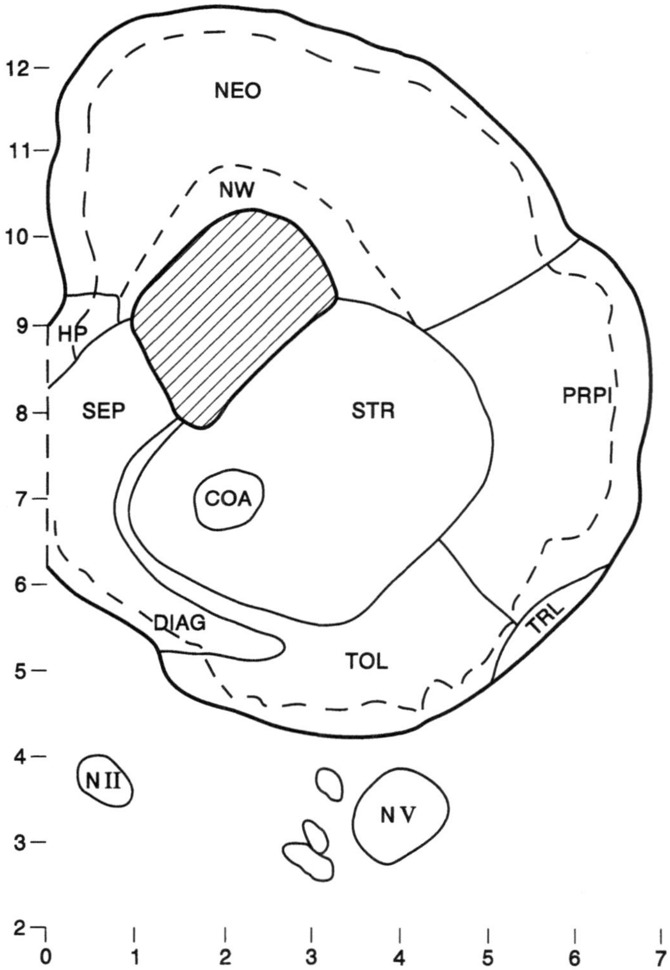

A 9

COA	anterior commissure	N V	trigeminal nerve
DIAG	diagonal band of Broca	PRPI	prepiriform region
HP	hipp. precommissuralis	SEP	telencephalic septum
NEO	neocortex	STR	striatum
NW	neocortical white matter	TOL	olfactory tubercle
N II	optic nerve	TRL	lateral olfactory tract

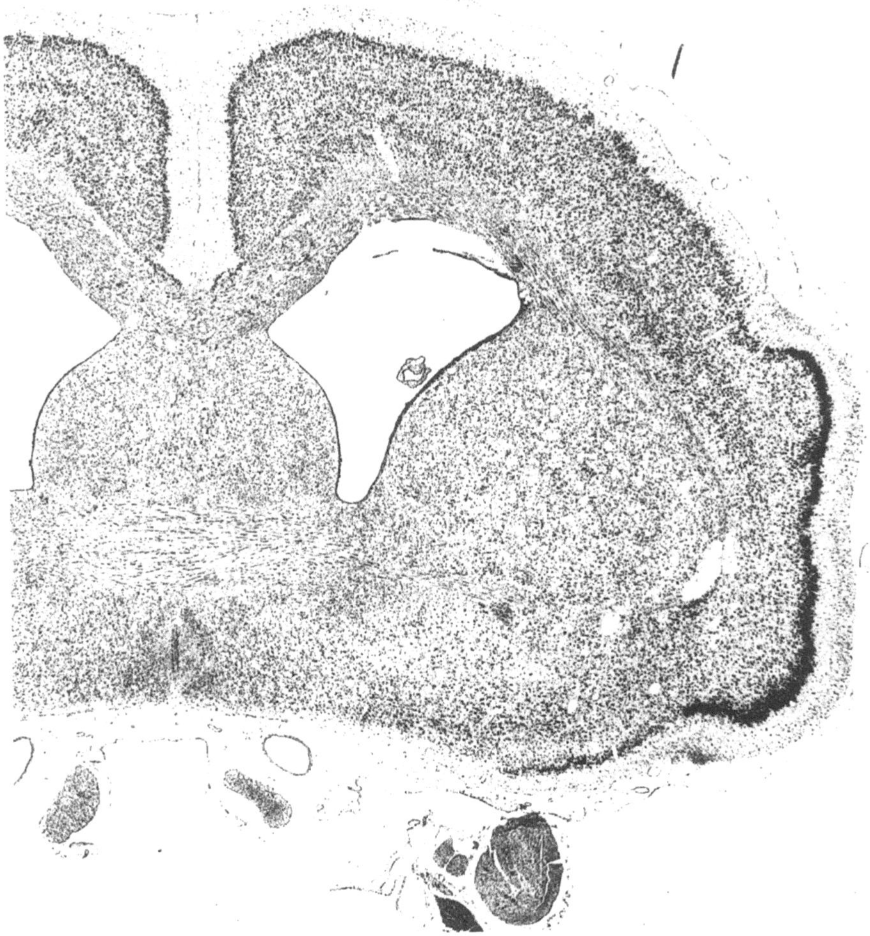

A 8

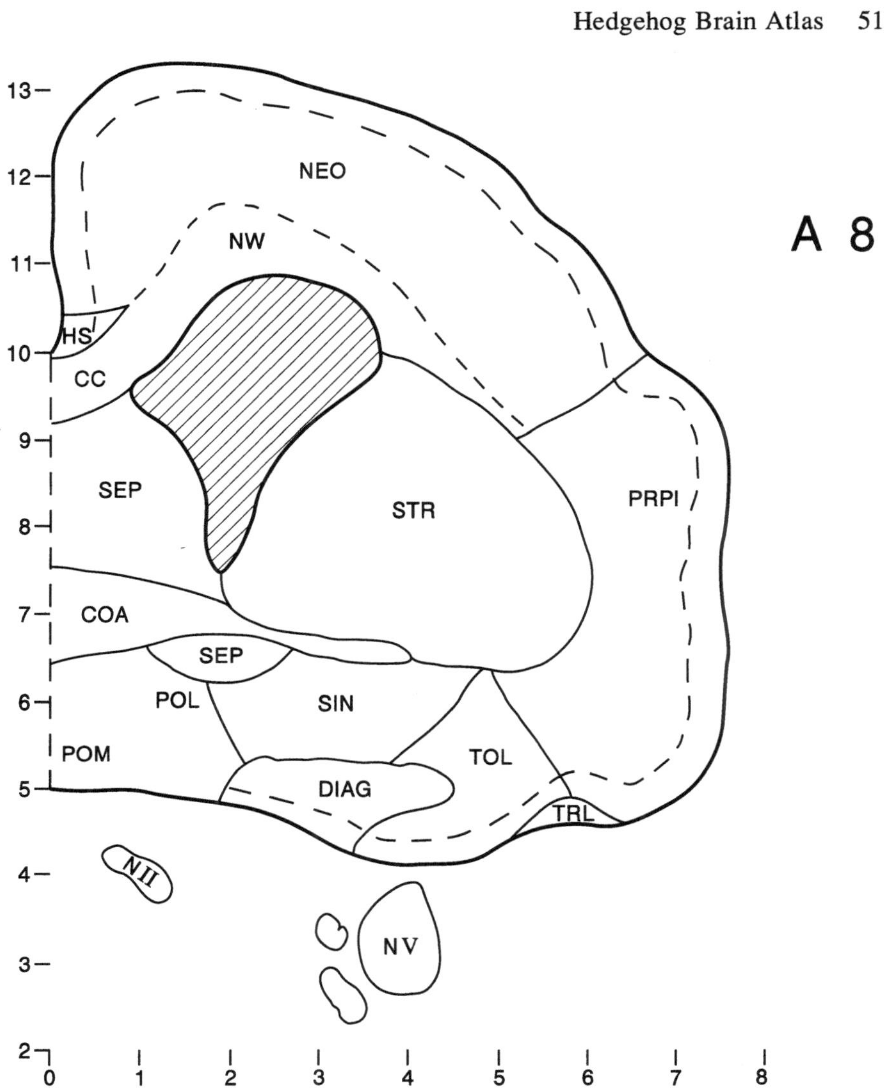

CC	corpus callosum	POL	lateral preoptic area
COA	anterior commissure	POM	medial preoptic area
DIAG	diagonal band of Broca	PRPI	prepiriform region
HS	hipp. supracommissuralis	SEP	telencephalic septum
NEO	neocortex	SIN	substantia innominata
NW	neocortical white matter	STR	striatum
N II	optic nerve	TOL	olfactory tubercle
N V	trigeminal nerve	TRL	lateral olfactory tract

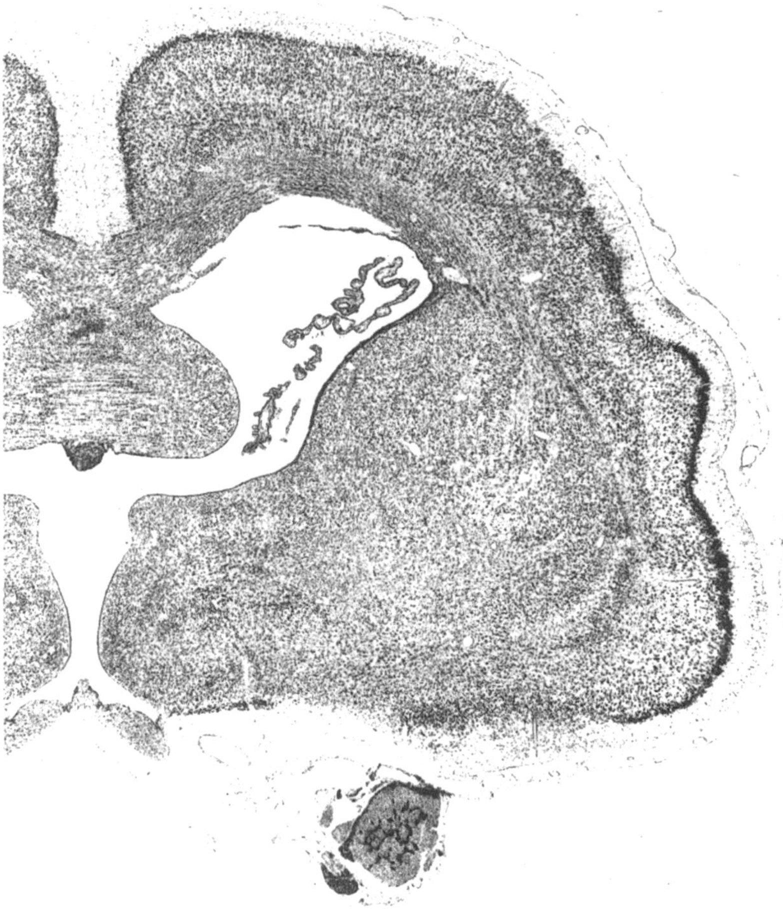

A 7

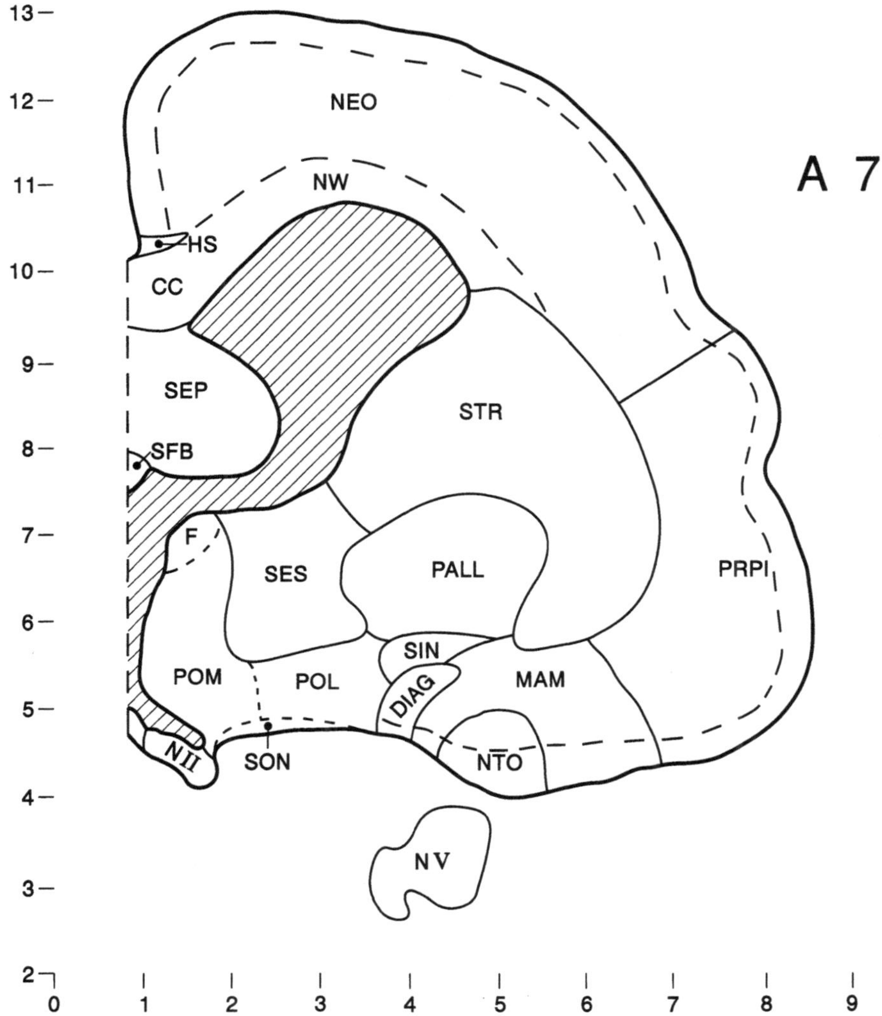

CC	corpus callosum	PALL	pallidum, globus pallidus
DIAG	diagonal band of Broca	POL	lateral preoptic area
F	fimbria-fornix complex	POM	medial preoptic area
HS	hipp. supracommissuralis	PRPI	prepiriform region
MAM	medial amygdaloid division	SEP	telencephalic septum
NEO	neocortex	SES	stria terminalis nuclei
NTO	nucleus of the lateral olfactory tract	SFB	subfornical body
NW	neocortical white matter	SIN	substantia innominata
N II	optic nerve	SON	supraoptic nucleus
N V	trigeminal nerve	STR	striatum

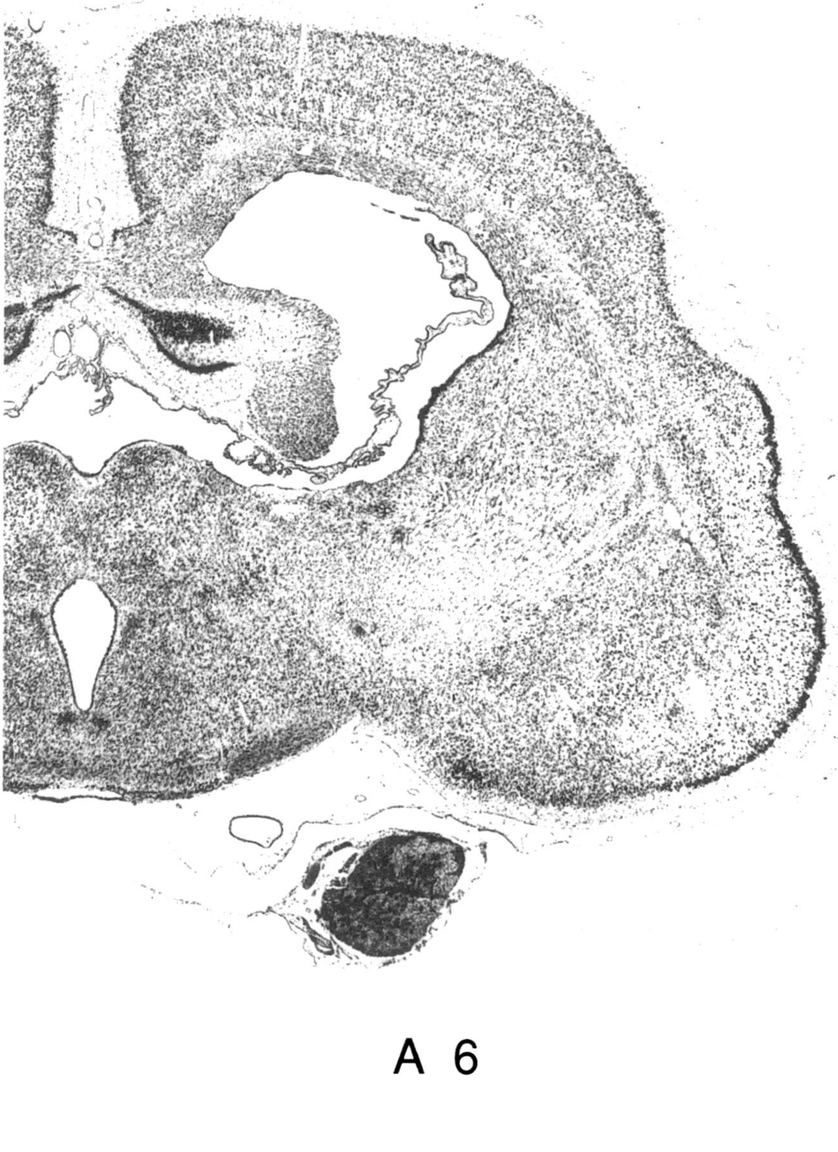

A 6

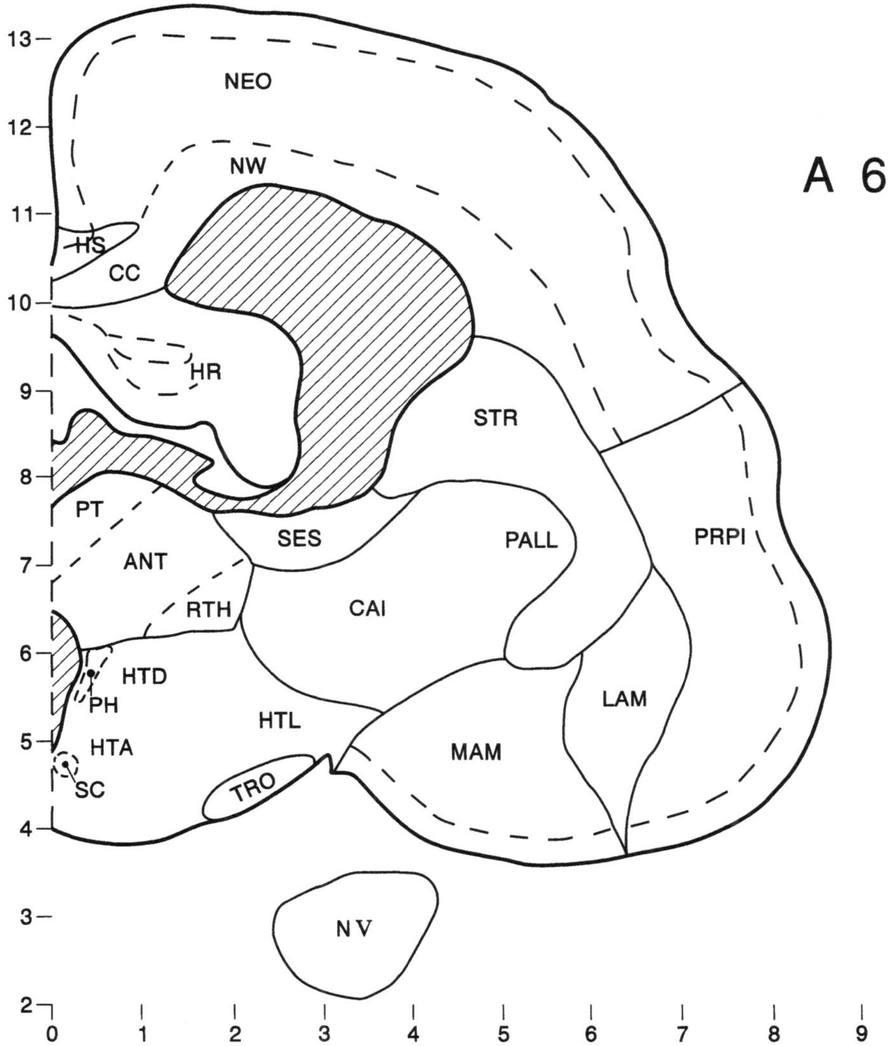

ANT	nucleus anterior thalami	NW	neocortical white matter
CAI	capsula interna	N V	trigeminal nerve
CC	corpus callosum	PALL	pallidum, globus pallidus
HR	hipp. retrocommissuralis	PH	ncl. parahypothalamicus
HS	hipp. supracommissuralis	PRPI	prepiriform region
HTA	anterior hypothalamic area	PT	ncl. paratenialis thalami
HTD	dorsal hypothalamic area	RTH	nuclei reticulares thalami
HTL	lateral hypothalamus	SC	ncl. suprachiasmaticus
LAM	lateral amygdaloid division	SES	stria terminalis nuclei
MAM	medial amygdaloid division	STR	striatum
NEO	neocortex	TRO	optic tract

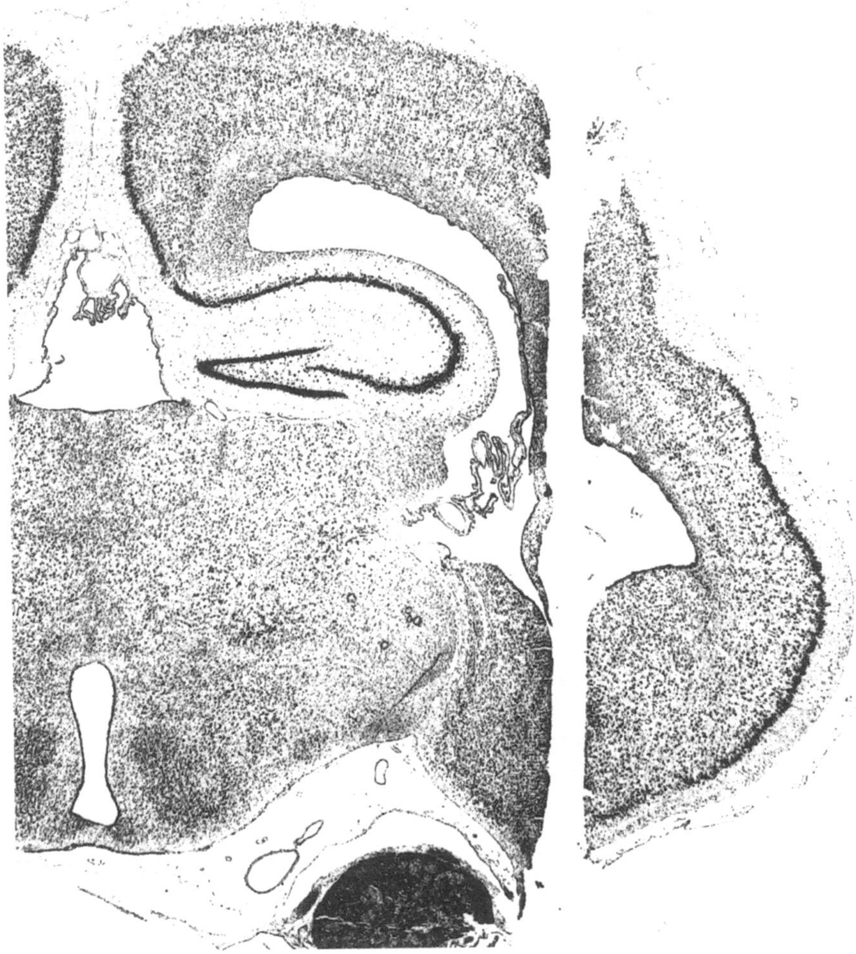

A 5

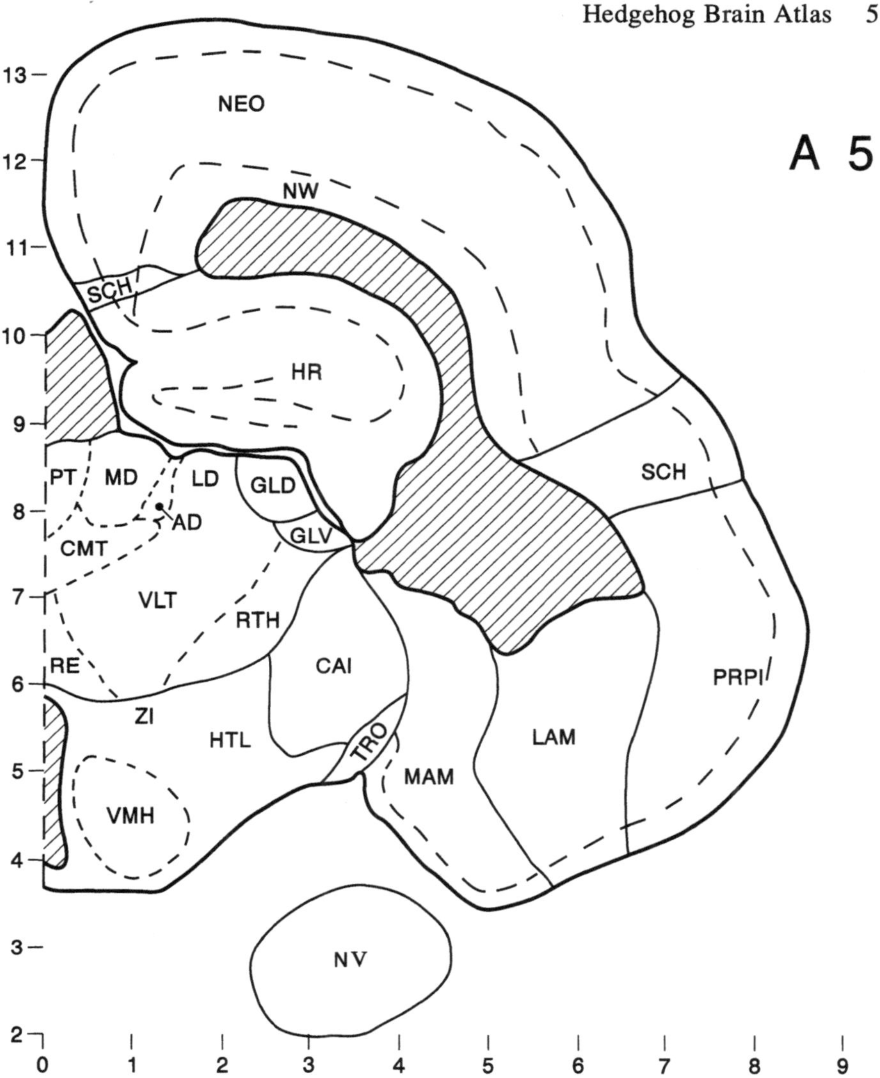

AD	ncl. anterodorsalis thalami
CAI	capsula interna
CMT	ncl. centromedialis thalami
GLD	dorsal nucleus of CGL
GLV	ventral nucleus of CGL
HR	hipp. retrocommissuralis
HTL	lateral hypothalamus
LAM	lateral amygdaloid division
LD	ncl. laterodorsalis thalami
MAM	medial amygdaloid division
MD	ncl. med. thalami dorsalis
NEO	neocortex
NW	neocortical white matter
N V	trigeminal nerve
PRPI	prepiriform region
PT	ncl. paratenialis thalami
RE	nucleus reuniens thalami
RTH	nuclei reticulares thalami
SCH	schizocortex
TRO	optic tract
VLT	ncl. ventrolateralis thalami
VMH	ncl. ventromed. hypothalami
ZI	zona incerta

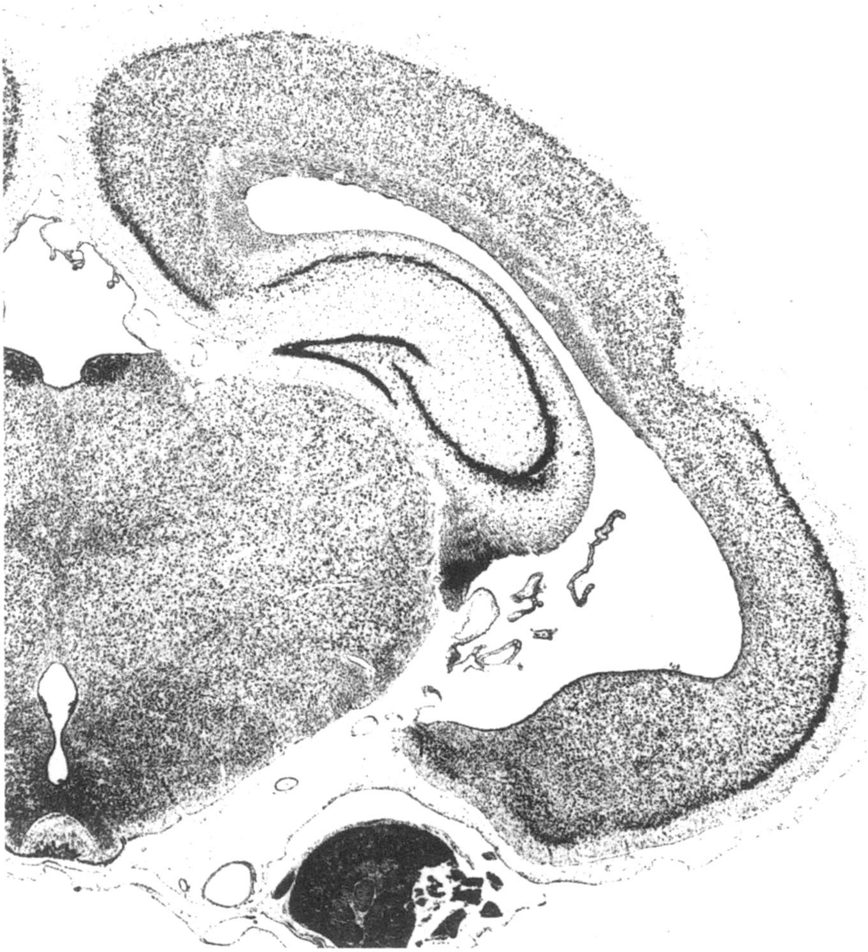

A 4

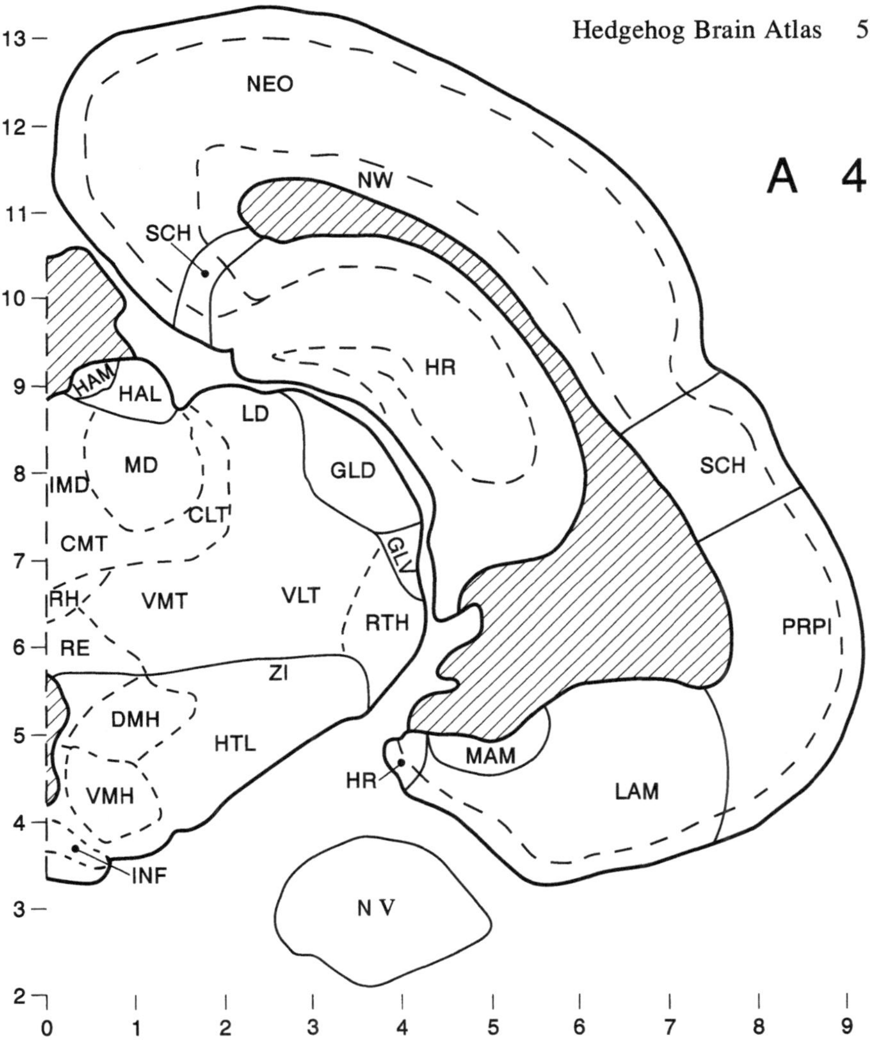

CLT	ncl. centrolateralis thalami	MD	ncl. med. thalami dorsalis
CMT	ncl. centromedialis thalami	NEO	neocortex
DMH	ncl. dorsomed. hypothalami	NW	neocortical white matter
GLD	dorsal nucleus of CGL	N V	trigeminal nerve
GLV	ventral nucleus of CGL	PRPI	prepiriform region
HAL	ncl. habenularis lateralis	RE	nucleus reuniens thalami
HAM	ncl. habenularis medialis	RH	ncl. rhomboides thalami
HR	hipp. retrocommissuralis	RTH	nuclei reticulares thalami
HTL	hypothalamus lateralis	SCH	schizocortex
IMD	ncl. intermediodors. thalami	VLT	ncl. ventrolateralis thalami
INF	nucleus infundibularis	VMH	ncl. ventromed. hypothalami
LAM	lateral amygdaloid division	VMT	ncl. ventromed. thalami
LD	ncl. laterodorsalis thalami	ZI	zona incerta
MAM	medial amygdaloid division		

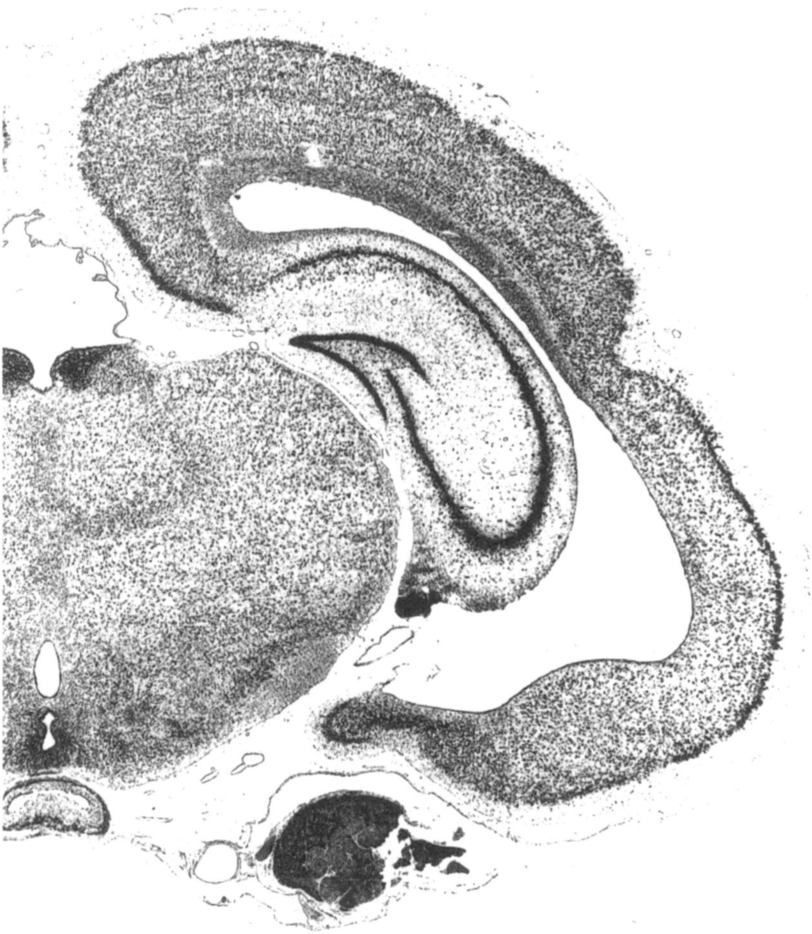

A 3.5

A 3.5

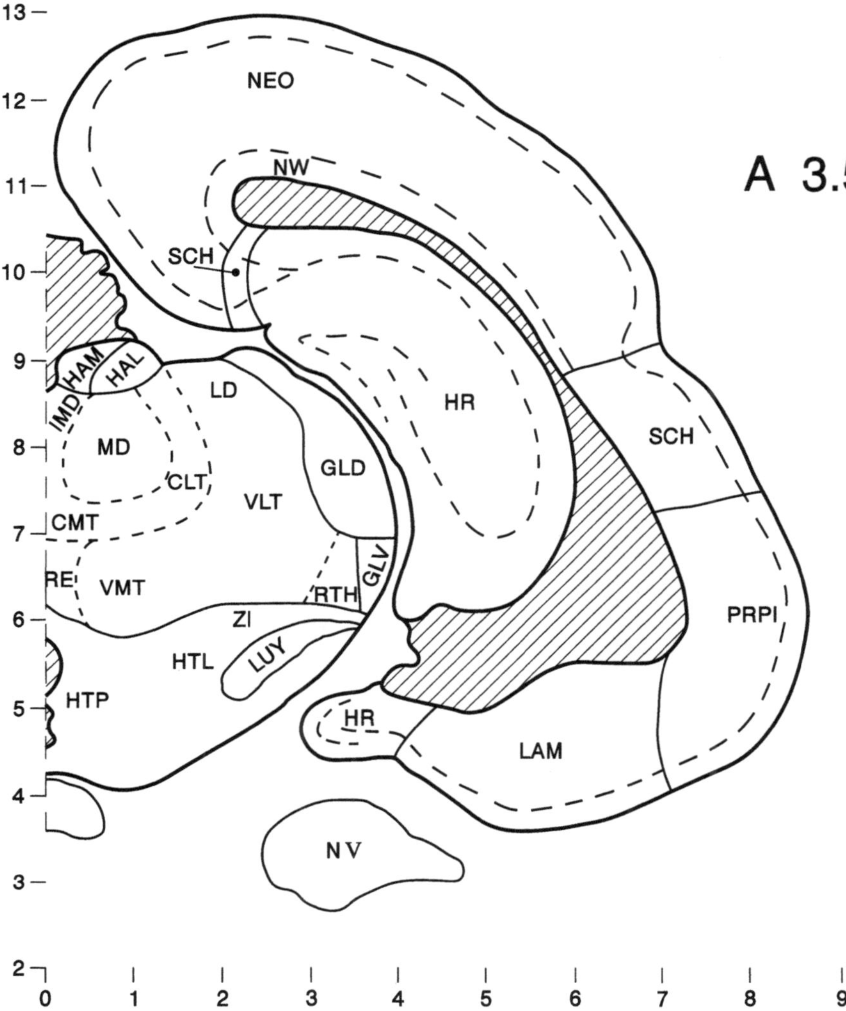

CLT	ncl. centrolateralis thalami
CMT	ncl. centromedialis thalami
GLD	dorsal nucleus of CGL
GLV	ventral nucleus of CGL
HAL	ncl. habenularis lateralis
HAM	ncl. habenularis medialis
HR	hipp. retrocommissuralis
HTL	hypothalamus lateralis
HTP	area hypothalamica posterior
IMD	ncl. intermediodors. thalami
LAM	lateral amygdaloid division
LD	ncl. laterodorsalis thalami
LUY	nucleus subthalamicus Luysi
MD	ncl. med. thalami dorsalis
NEO	neocortex
NW	neocortical white matter
N V	trigeminal nerve
PRPI	prepiriform region
RE	nucleus reuniens thalami
RTH	nuclei reticulares thalami
SCH	schizocortex
VLT	ncl. ventrolateralis thalami
VMT	ncl. ventromed. thalami
ZI	zona incerta

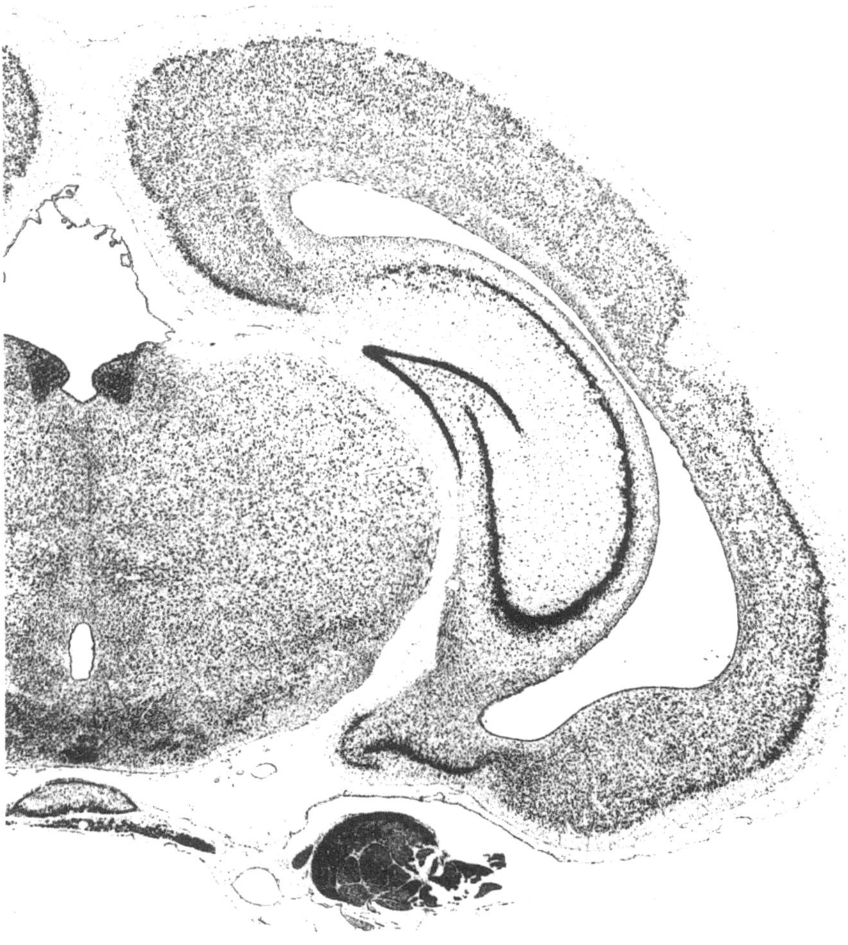

A 3

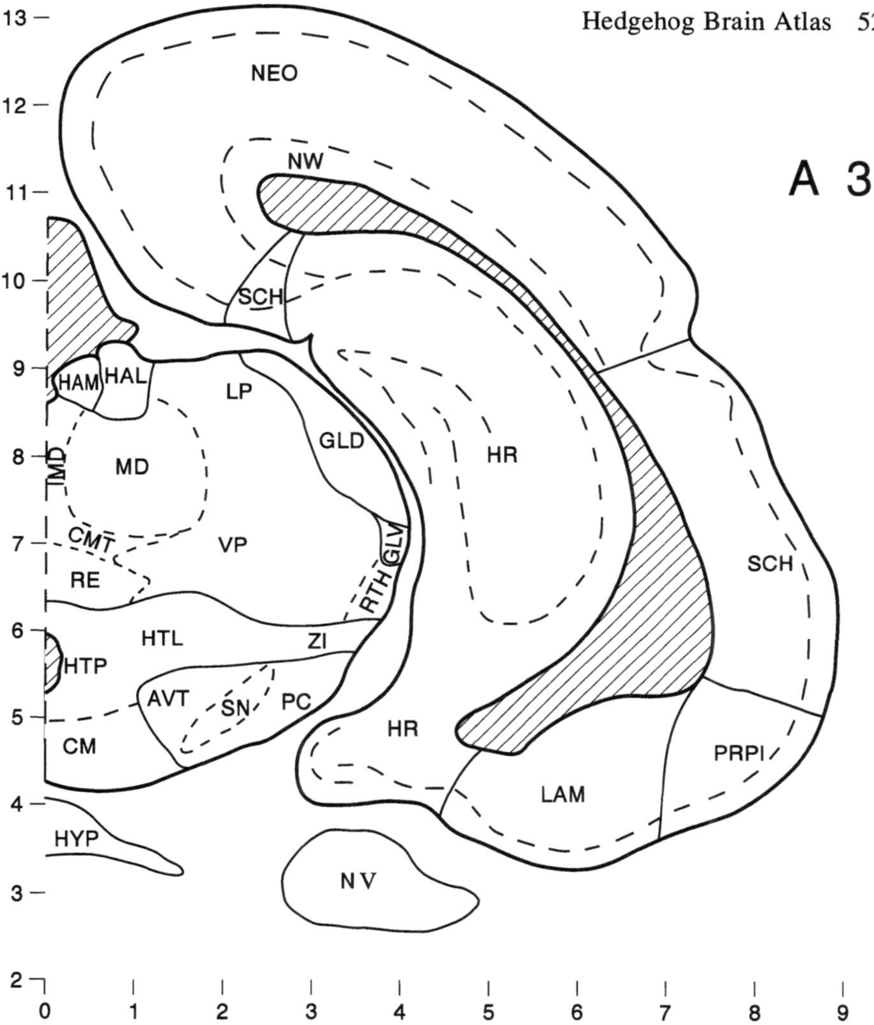

AVT	area ventralis tegmenti	LP	ncl. lateralis post. thalami
CM	corpus mamillare	MD	ncl. med. thalami dorsalis
CMT	ncl. centromedialis thalami	NEO	neocortex
GLD	dorsal nucleus of CGL	NW	neocortical white matter
GLV	ventral nucleus of CGL	N V	trigeminal nerve
HAL	ncl. habenularis lateralis	PC	pedunculus cerebri
HAM	ncl. habenularis medialis	PRPI	prepiriform region
HR	hipp. retrocommissuralis	RE	nucleus reuniens thalami
HTL	hypothalamus lateralis	RTH	nuclei reticulares thalami
HTP	area hypothalamica posterior	SCH	schizocortex
HYP	hypophysis	SN	substantia nigra
IMD	ncl. intermediodors. thalami	VP	thalamus ventroposterior
LAM	lateral amygdaloid division	ZI	zona incerta

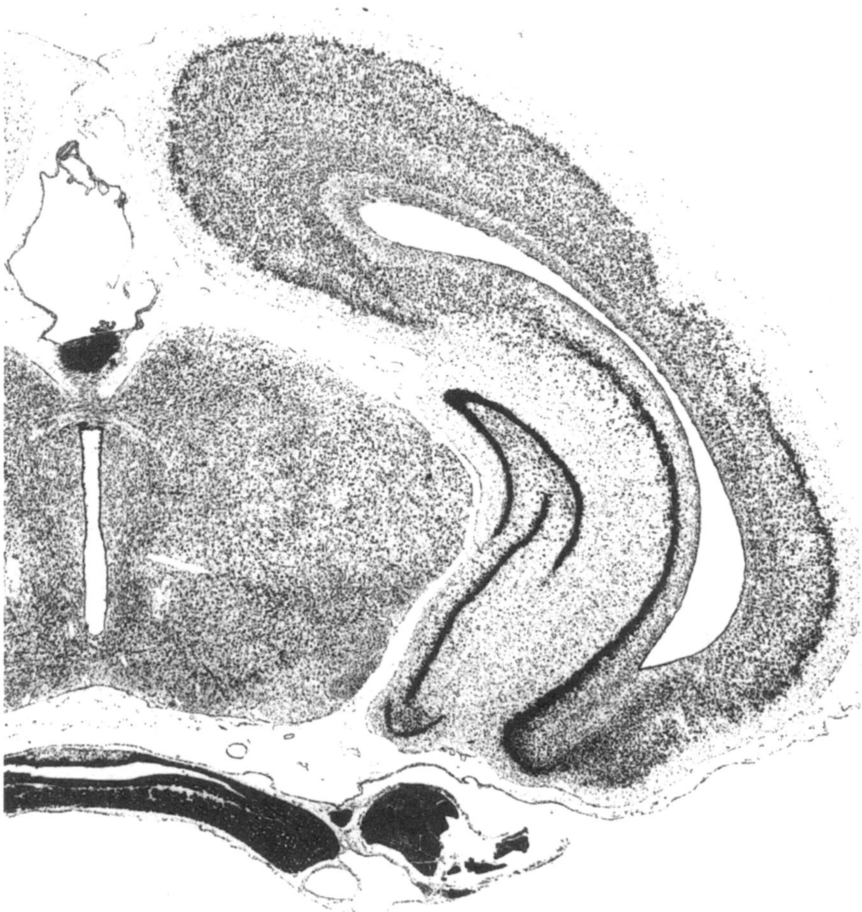

A 2

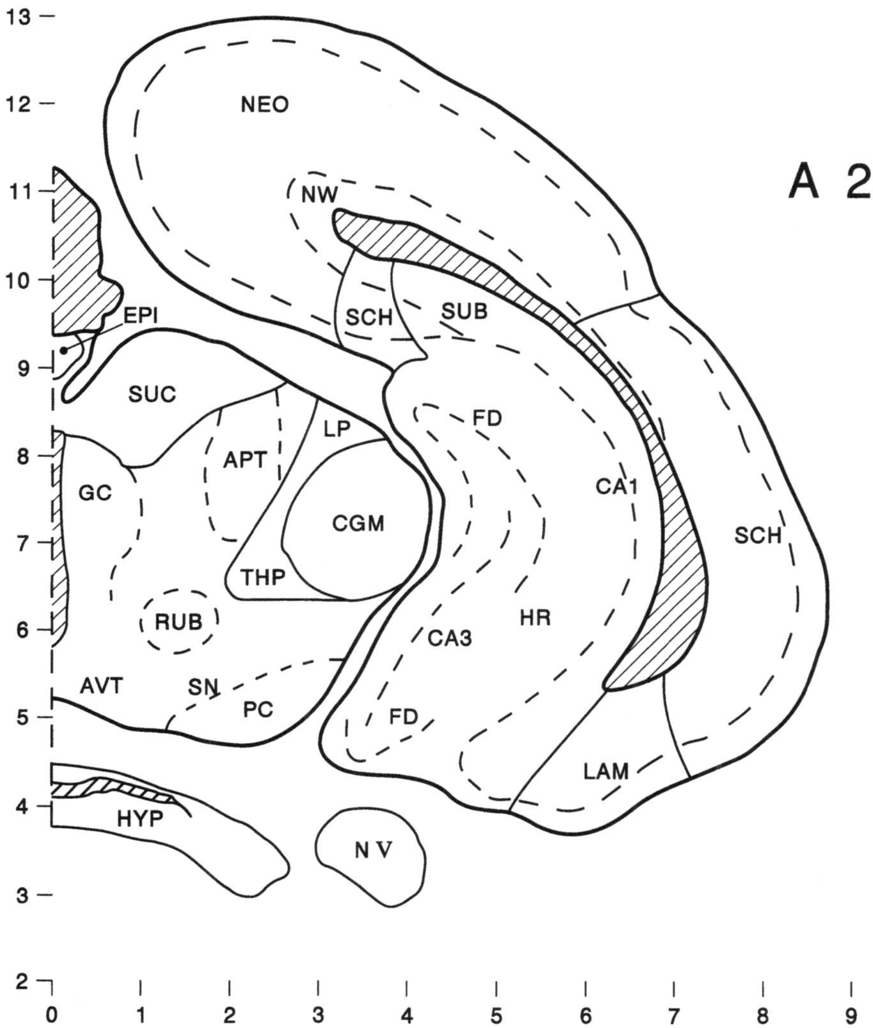

APT	area pretectalis	LP	ncl. lateralis post. thalami
AVT	area ventralis tegmenti	NEO	neocortex
CA 1	area 1 of hippocampus (HR)	NW	neocortical white matter
CA 3	area 3 of hippocampus (HR)	N V	trigeminal nerve
CGM	corpus geniculatum mediale	PC	pedunculus cerebri
EPI	epiphysis, pineal body	RUB	nucleus ruber
FD	fascia dentata of HR	SCH	schizocortex
GC	griseum centrale	SN	substantia nigra
HR	hipp. retrocommissuralis	SUB	subiculum (hippocampi)
HYP	hypophysis	SUC	colliculus superior
LAM	lateral amygdaloid division	THP	thalamus posterior

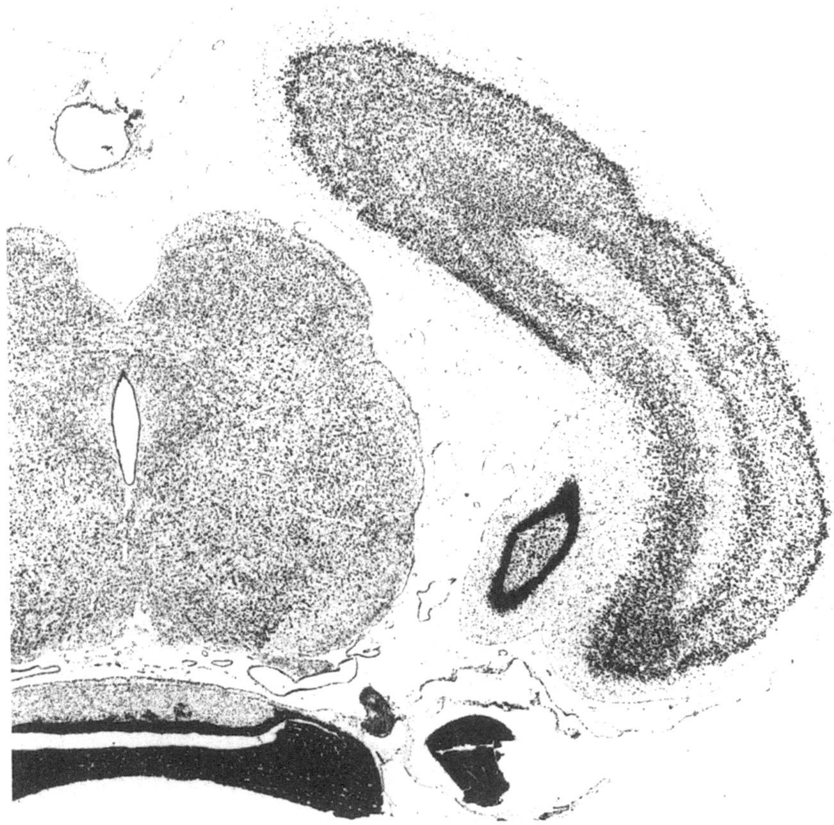

A 1

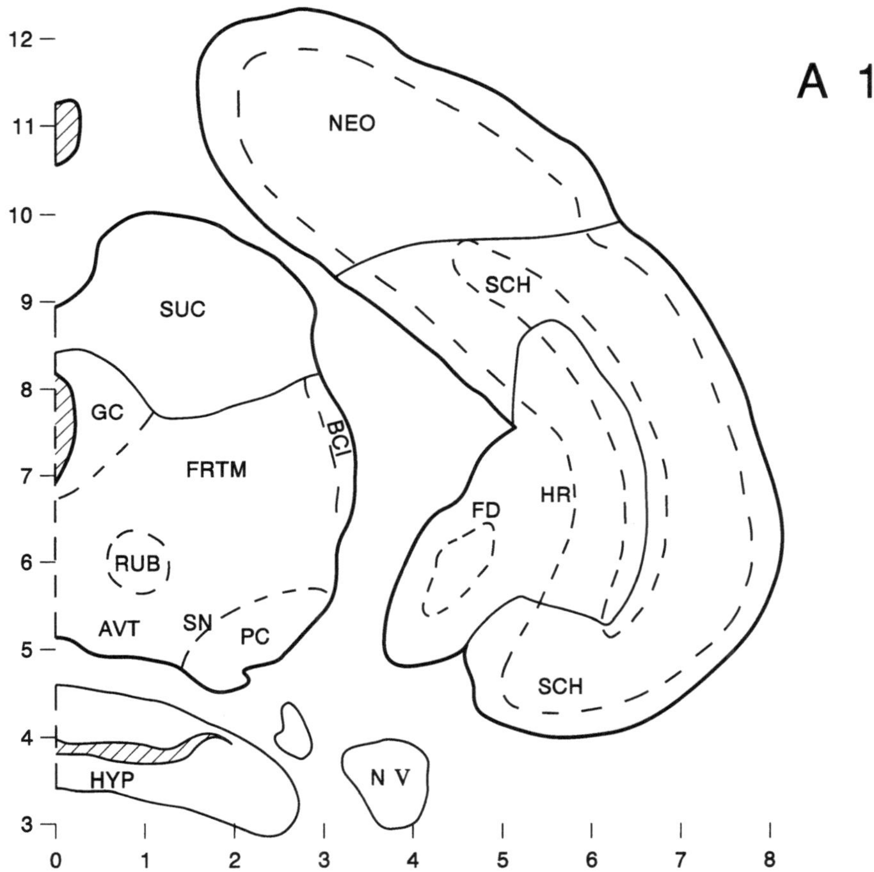

AVT	area ventralis tegmenti	NEO	neocortex
BCI	brachium colliculi inferioris	N V	trigeminal nerve
FD	fascia dentata of HR	PC	pedunculus cerebri
FRTM	formatio reticularis tegmenti mesencephali	RUB	nucleus ruber
GC	griseum centrale	SCH	schizocortex
HR	hipp. retrocommissuralis	SN	substantia nigra
HYP	hypophysis	SUC	colliculus superior

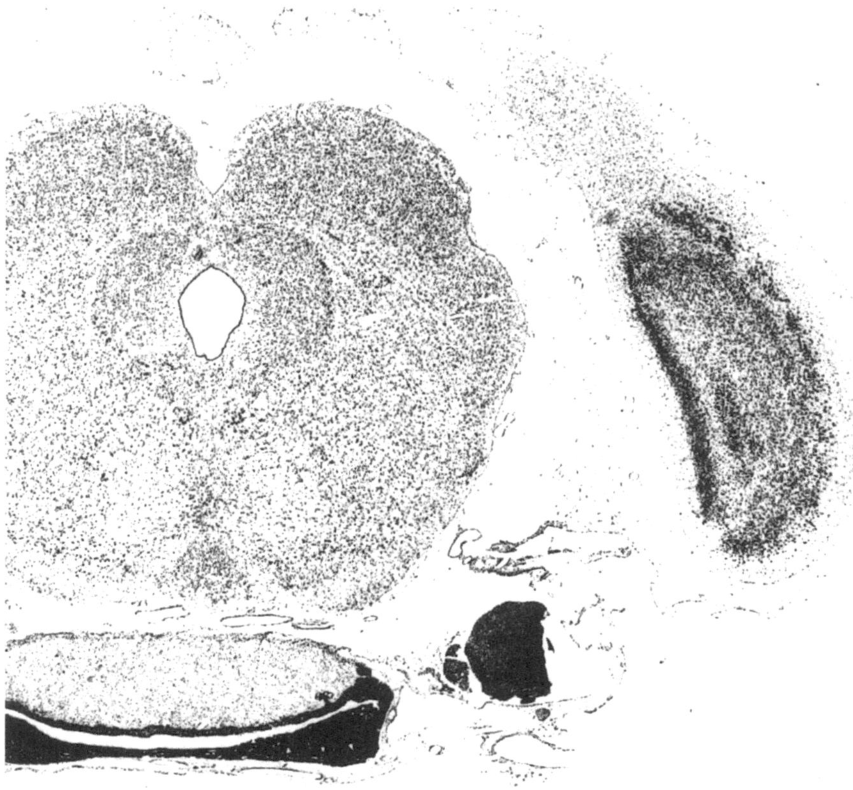

AP 0

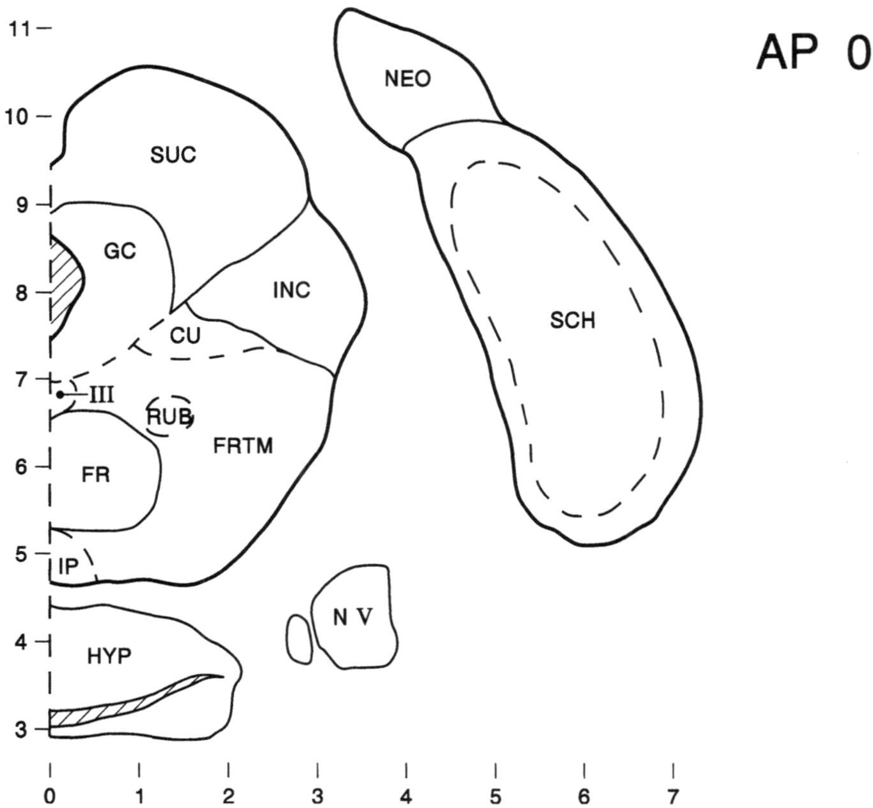

CU	nucleus cuneiformis	IP	nucleus interpeduncularis
FR	formatio reticularis	NEO	neocortex
FRTM	formatio reticularis tegmenti mesencephali	N V	trigeminal nerve
GC	griseum centrale	RUB	nucleus ruber
HYP	hypophysis	SCH	schizocortex
INC	colliculus inferior	SUC	colliculus superior
		III	nucleus oculomotorius

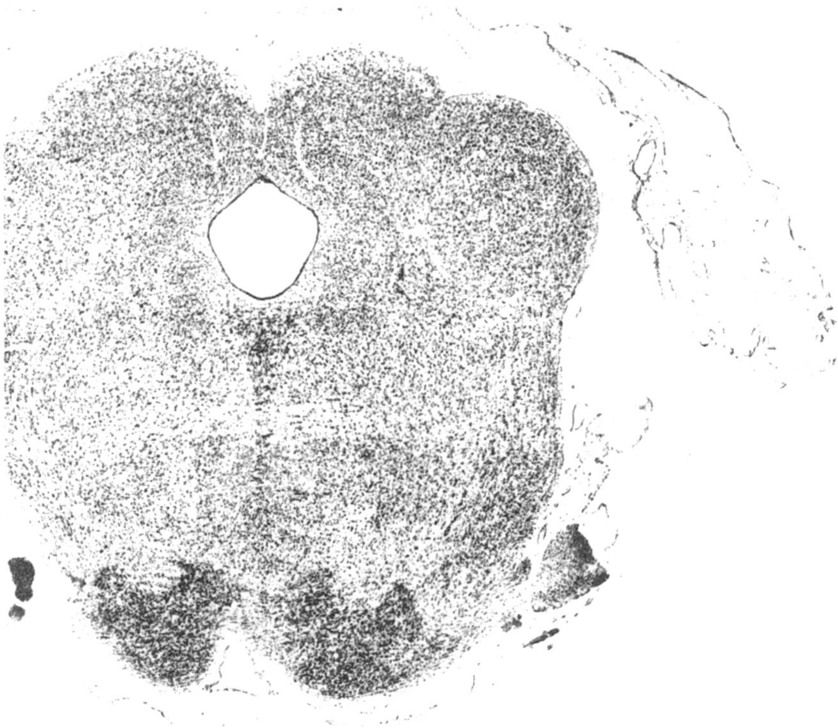

P 1

P 1

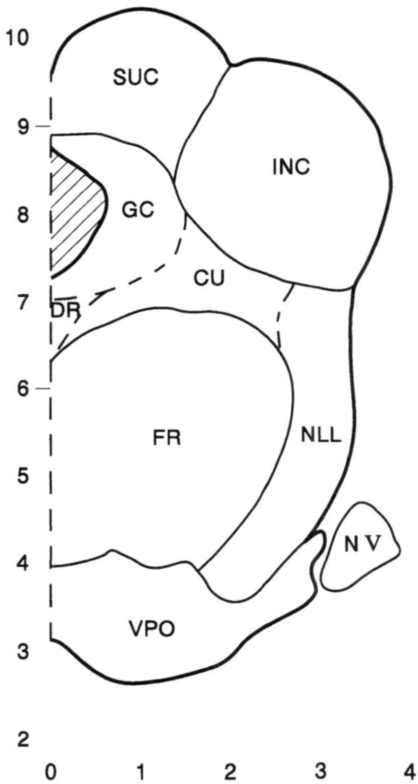

CU	nucleus cuneiformis	NLL	nuclei of lateral lemniscus
DR	nucleus dorsalis raphe	N V	trigeminal nerve
FR	formatio reticularis	SUC	colliculus superior
GC	griseum centrale	VPO	ventral pons
INC	colliculus inferior		

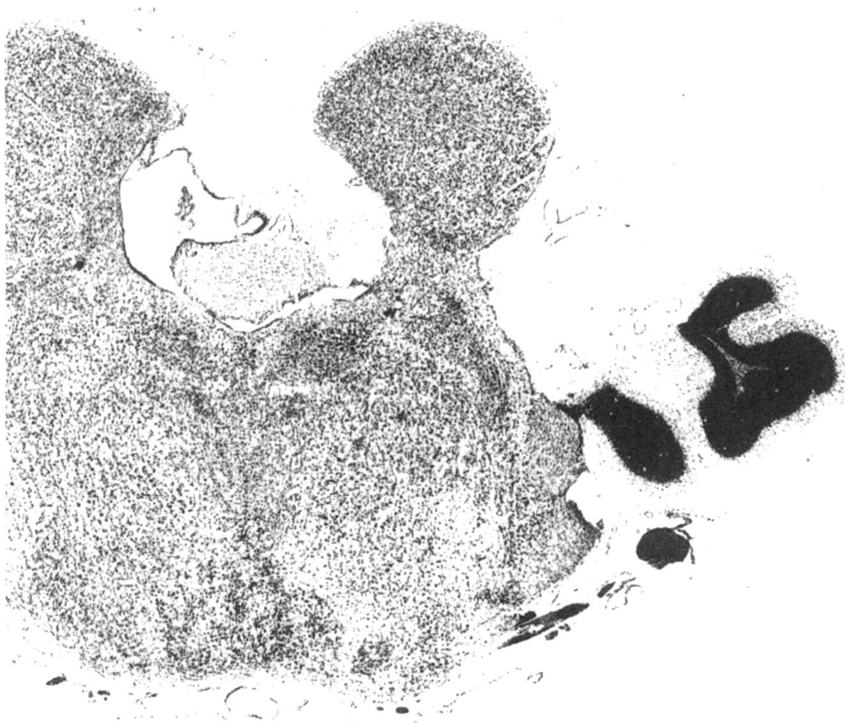

P 2

P 2

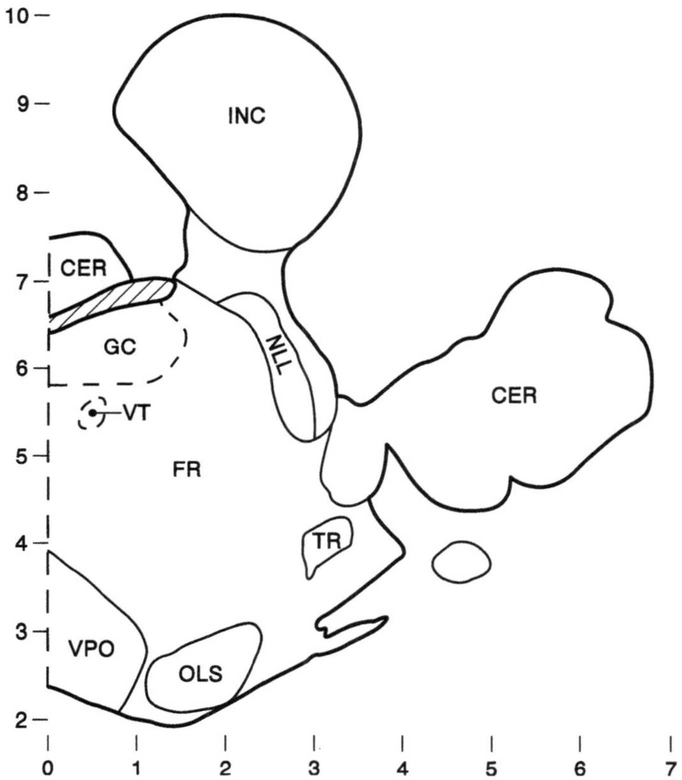

CER	cerebellum
FR	formatio reticularis
GC	griseum centrale
INC	colliculus inferior
NLL	nuclei of lateral lemniscus
OLS	superior olive
TR	complexus sensorius nervi trigemini
VPO	ventral pons
VT	ncl. ventralis tegmenti

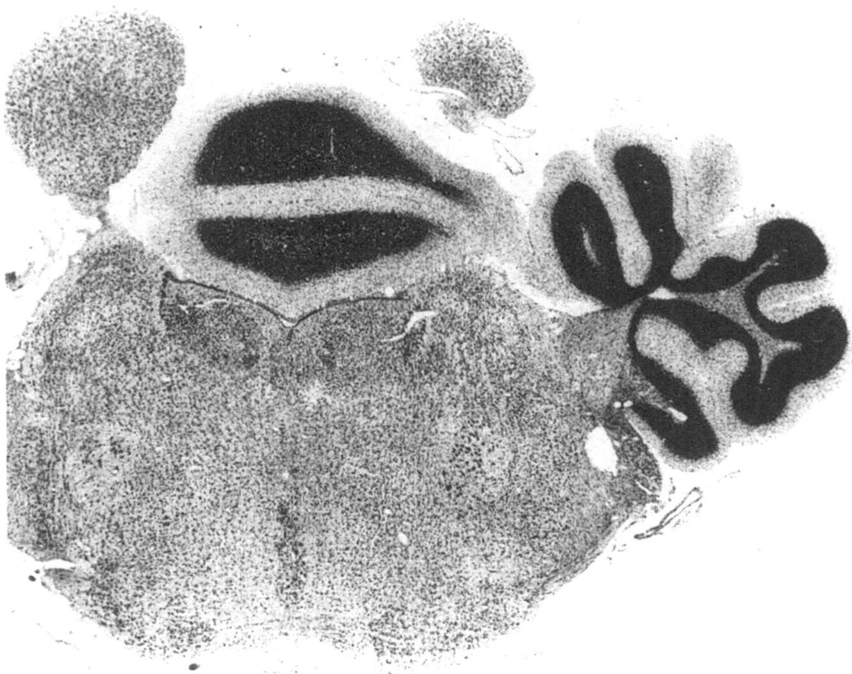

P 2.5

P 2.5

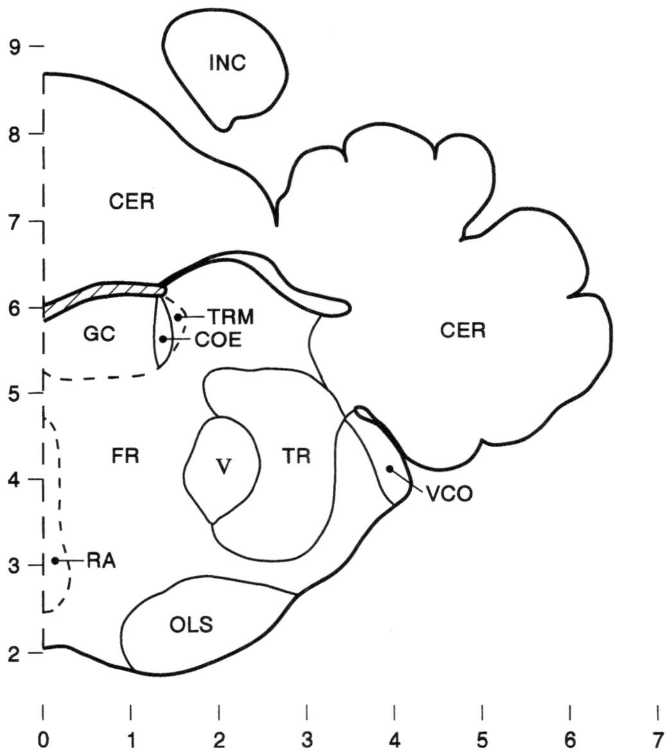

CER	cerebellum	TR	complexus sensorius nervi trigemini
COE	nucleus locus coeruleus	TRM	nucleus mesencephalicus nervi trigemini
FR	formatio reticularis	VCO	ncl. cochlearis ventralis
GC	griseum centrale	V	nucleus motorius nervi trigemini
INC	colliculus inferior		
OLS	superior olive		
RA	nucleus raphes		

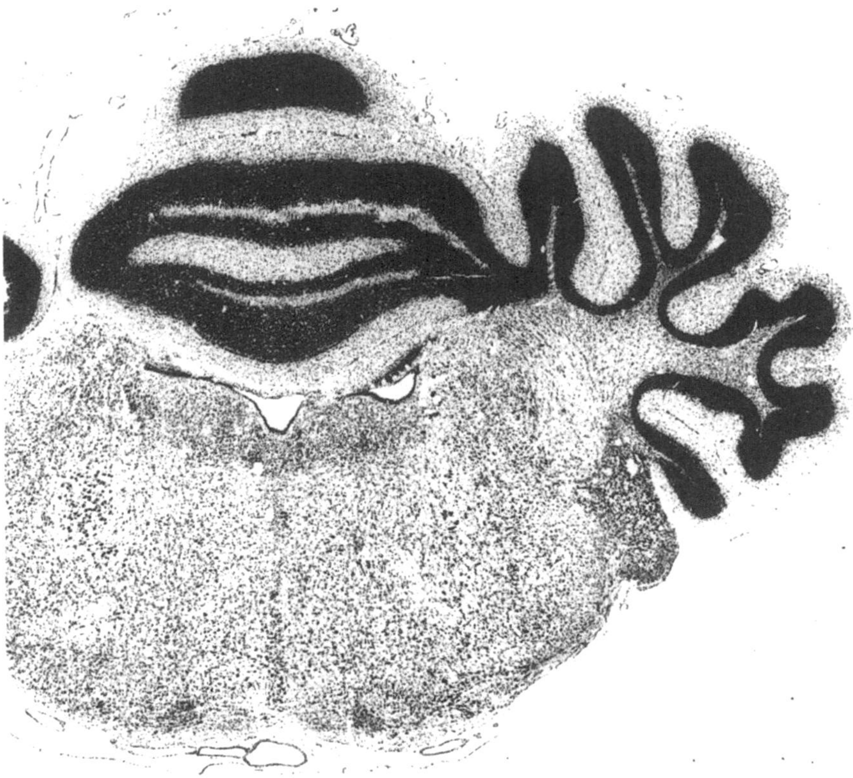

P 3

P 3

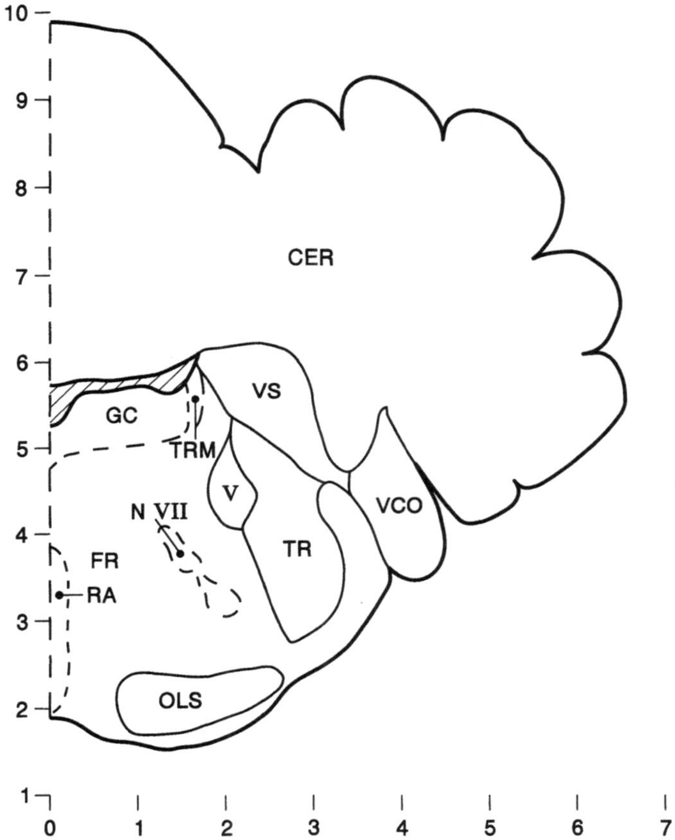

CER	cerebellum
FR	formatio reticularis
GC	griseum centrale
N VII	nervus facialis
OLS	superior olive
RA	nucleus raphes
TR	complexus sensorius nervi trigemini
TRM	nucleus mesencephalicus nervi trigemini
VCO	ncl. cochlearis ventralis
VS	ncl. vestibularis superior
V	nucleus motorius nervi trigemini

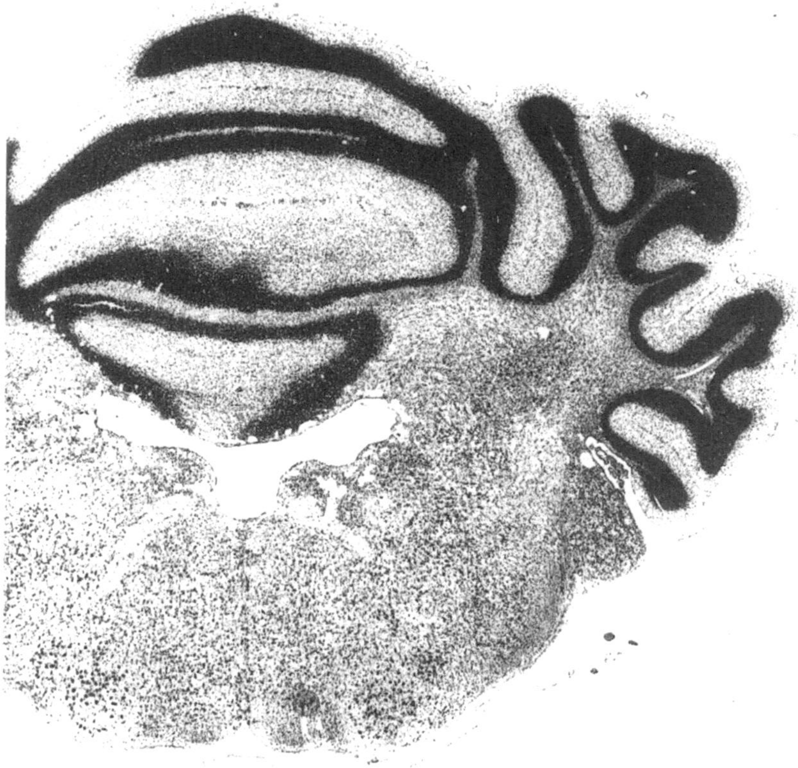

P 3.5

P 3.5

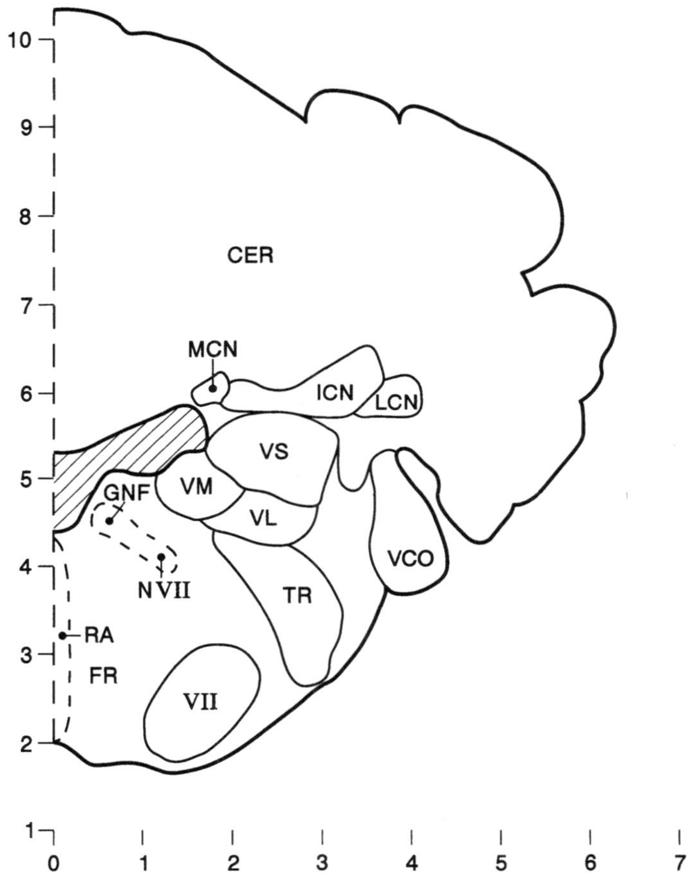

CER	cerebellum	RA	nucleus raphes
FR	formatio reticularis	TR	complexus sensorius nervi trigemini
GNF	genu nervi facialis	VCO	ncl. cochlearis ventralis
ICN	nucleus interpositus cerebelli	VL	ncl. vestibularis lateralis
LCN	nucleus lateralis cerebelli	VM	ncl. vestibularis medialis
MCN	nucleus medialis cerebelli	VS	ncl. vestibularis superior
N VII	nervus facialis		

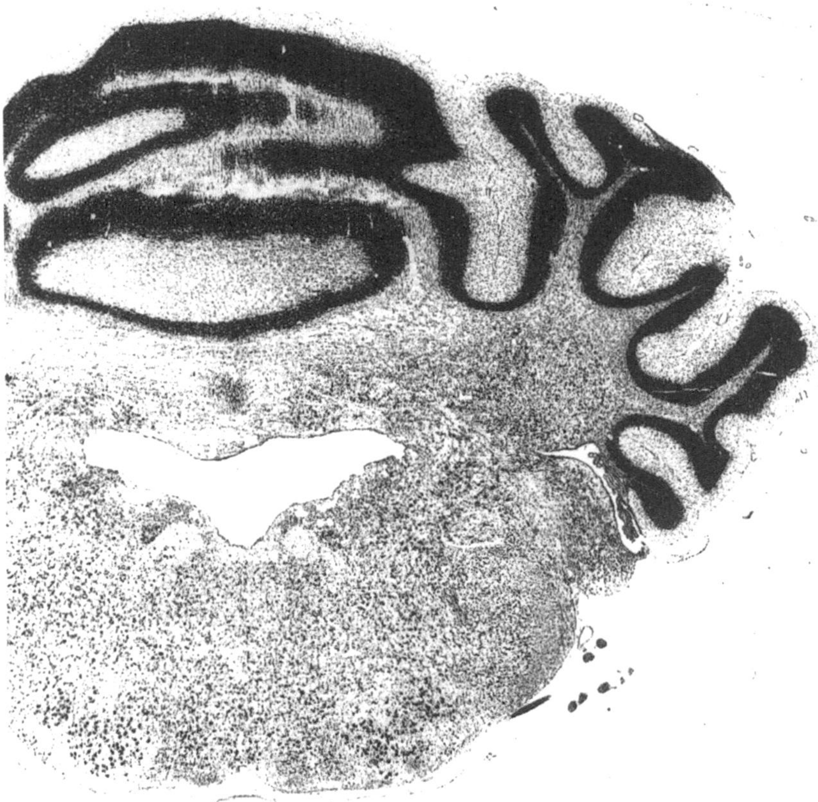

P 4

P 4

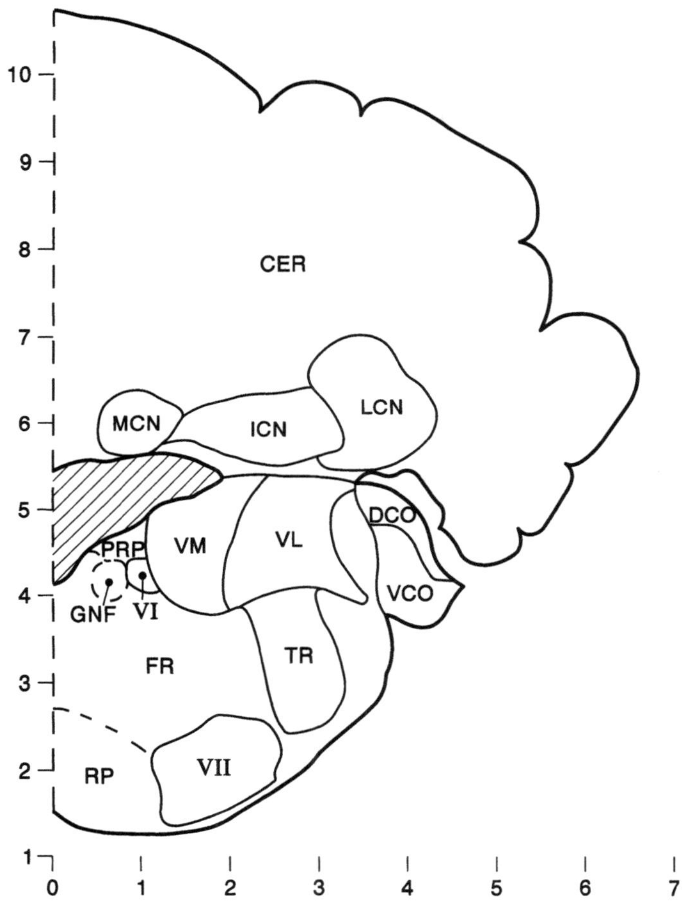

CER	cerebellum	RP	ncl. reticularis paramediani
DCO	ncl. cochlearis dorsalis	TR	complexus sensorius nervi trigemini
FR	formatio reticularis	VCO	ncl. cochlearis ventralis
GNF	genu nervi facialis	VL	ncl. vestibularis lateralis
ICN	nucleus interpositus cerebelli	VM	ncl. vestibularis medialis
LCN	nucleus lateralis cerebelli	VI	nucleus nervi abducentis
MCN	nucleus medialis cerebelli	VII	nucleus nervi facialis
PRP	ncl. prepositus hypoglossi		

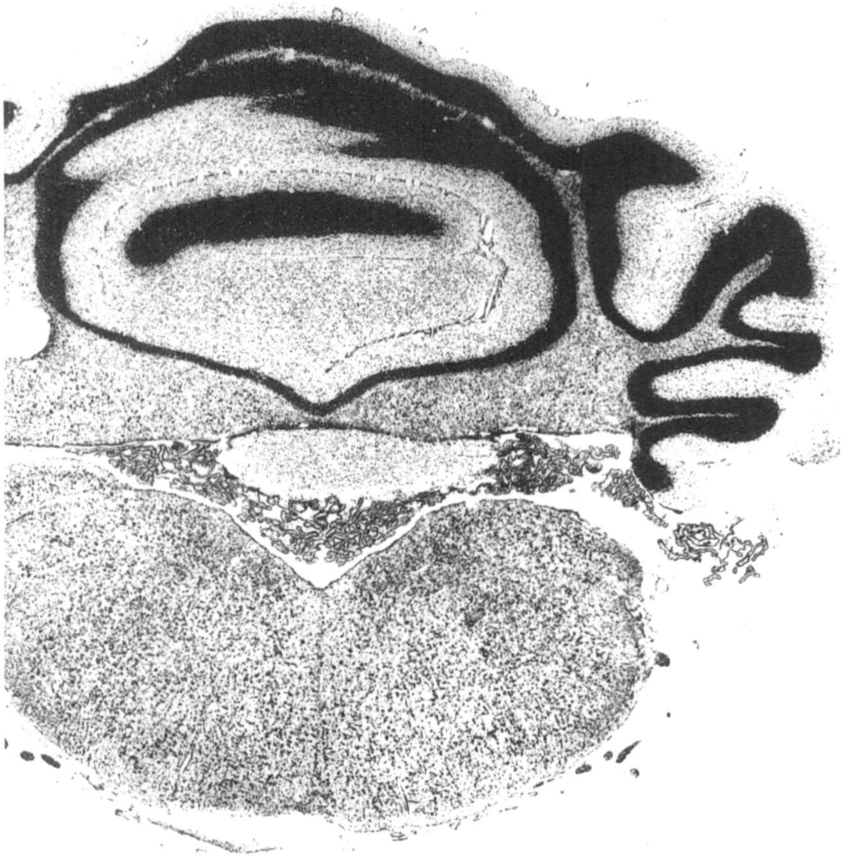

P 5

P 5

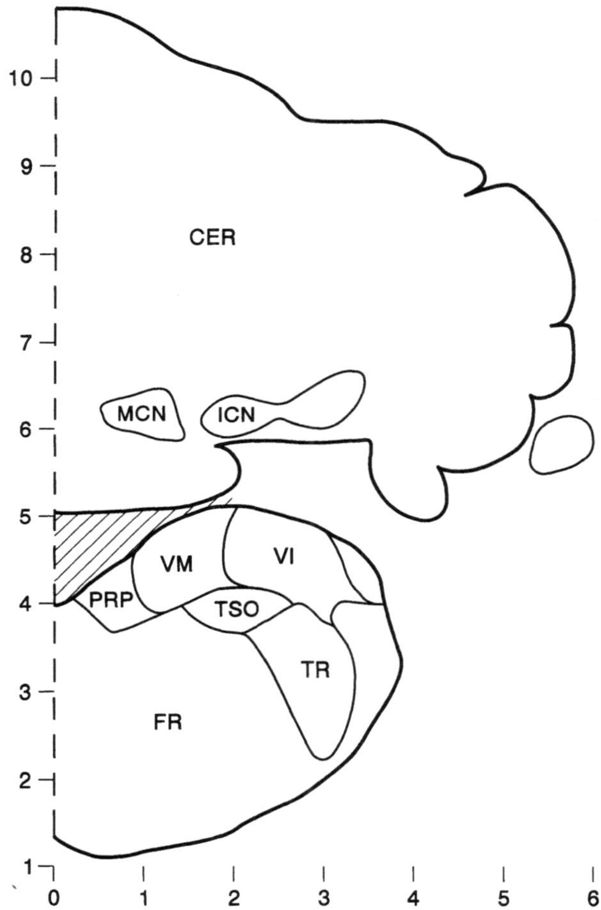

CER	cerebellum	TR	complexus sensorius nervi trigemini
FR	formatio reticularis	TSO	nucleus tractus solitarii
ICN	nucleus interpositus cerebelli	VI	ncl. vestibularis inferior
MCN	nucleus medialis cerebelli	VM	ncl. vestibularis medialis
PRP	ncl. prepositus hypoglossi		

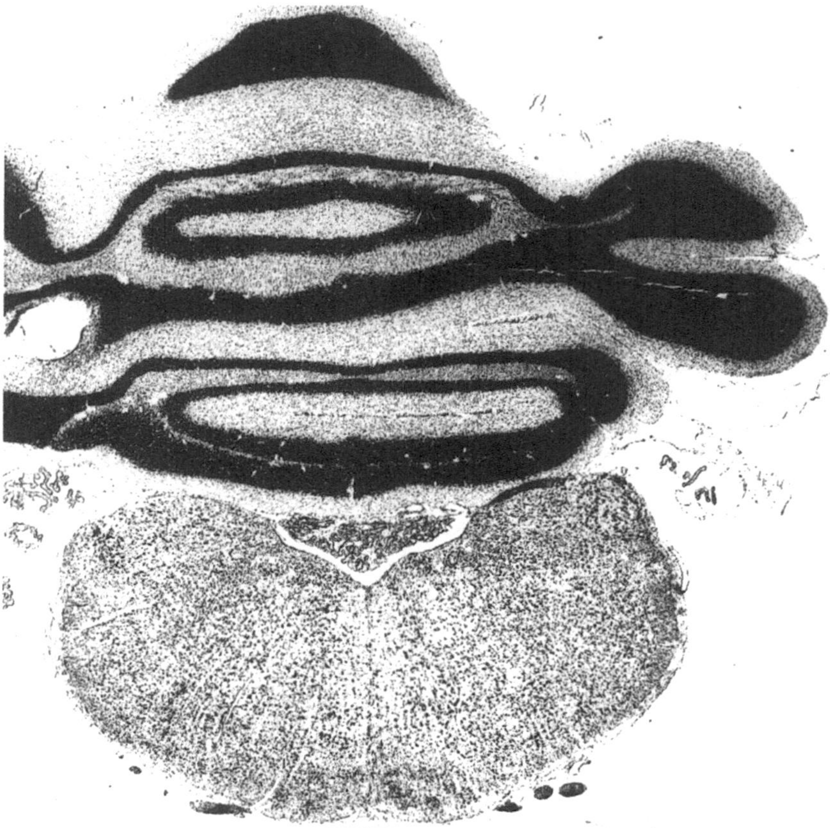

P 6

P 6

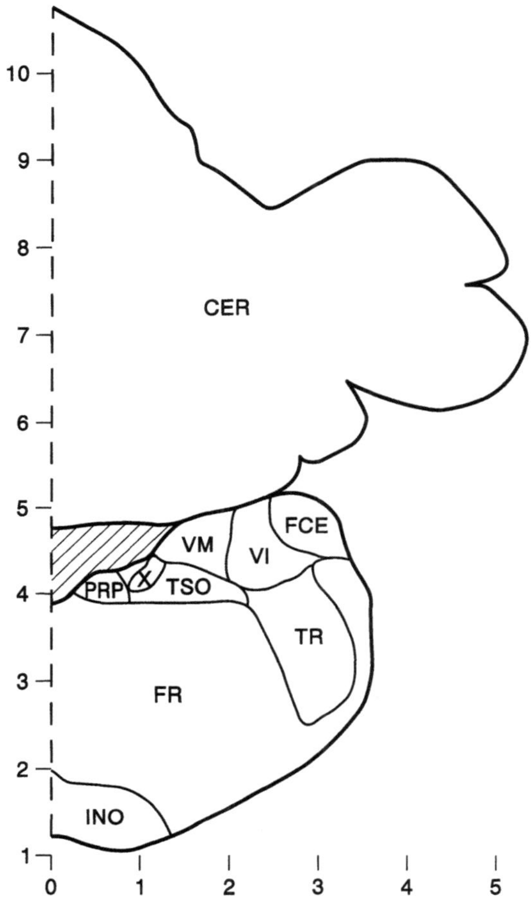

CER	cerebellum	TR	complexus sensorius nervi trigemini
FCE	ncl. fascicularis cuneatus externus	TSO	nucleus tractus solitarii
FR	formatio reticularis	VI	ncl. vestibularis inferior
INO	inferior olive	VM	ncl. vestibularis medialis
PRP	ncl. prepositus hypoglossi	X	ncl. dorsalis motorius nervi vagi

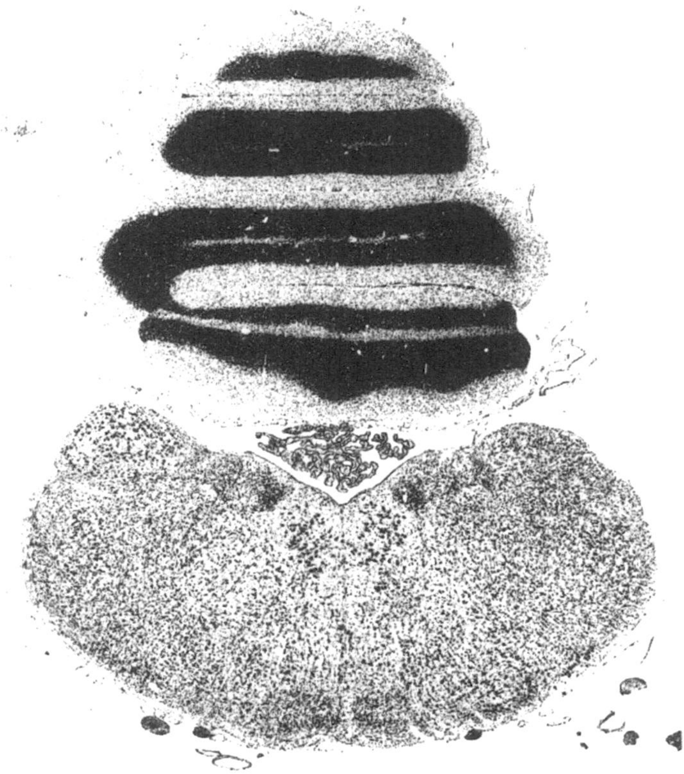

P 7

P 7

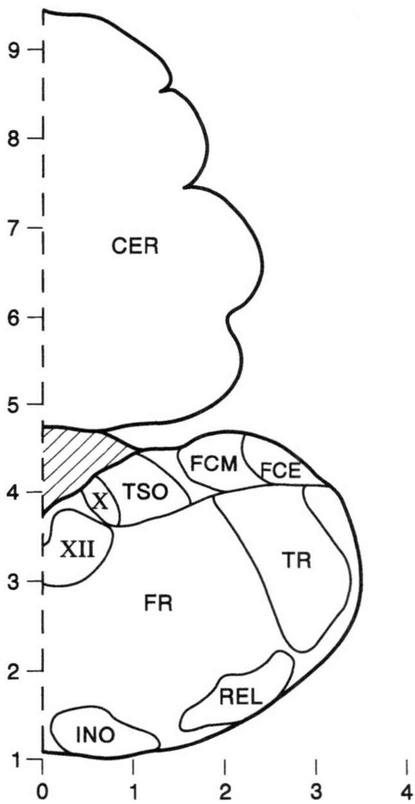

CER	cerebellum	REL	ncl. reticularis lateralis
FCE	ncl. fascicularis cuneatus externus	TR	complexus sensorius nervi trigemini
FCM	ncl. fascicularis cuneatus medialis	TSO	nucleus tractus solitarii
FR	formatio reticularis	X	ncl. dorsalis motorius nervi vagi
INO	inferior olive	XII	nucleus nervi hypoglossi

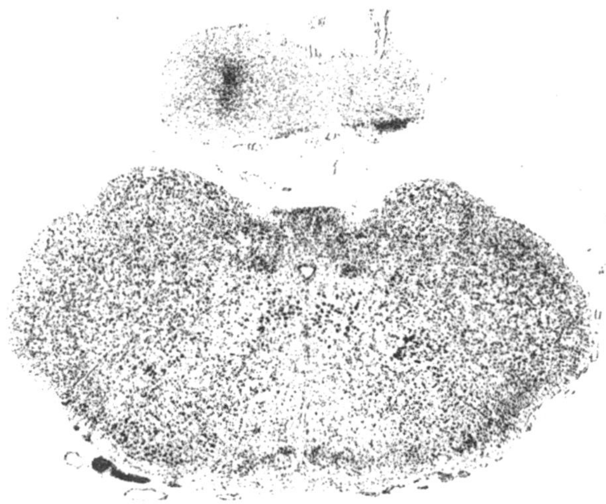

P 8

P 8

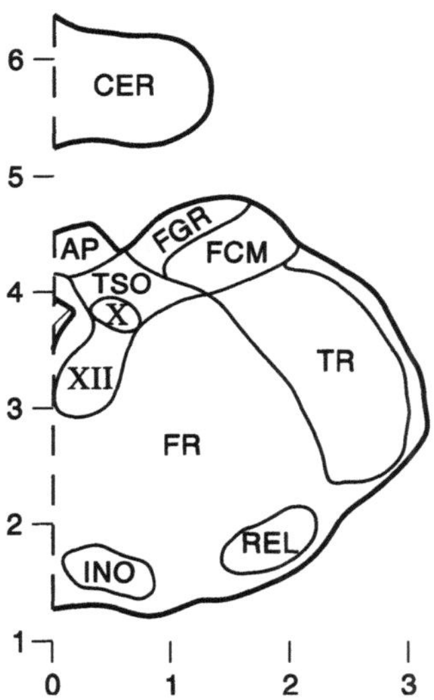

AP	area postrema	REL	ncl. reticularis lateralis
CER	cerebellum	TR	complexus sensorius nervi trigemini
FCM	ncl. fascicularis cuneatus medialis	TSO	nucleus tractus solitarii
FGR	ncl. fascicularis gracilis	X	ncl. dorsalis motorius nervi vagi
FR	formatio reticularis	XII	nucleus nervi hypoglossi
INO	inferior olive		

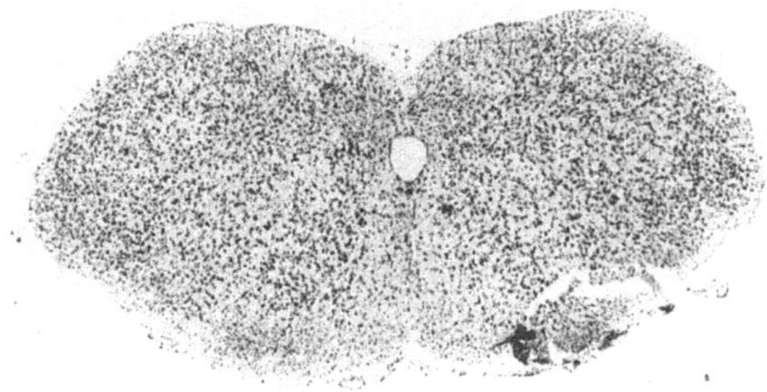

P 9

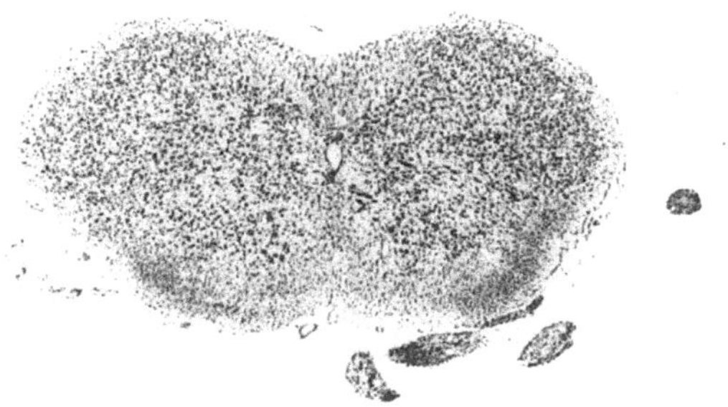

P 10

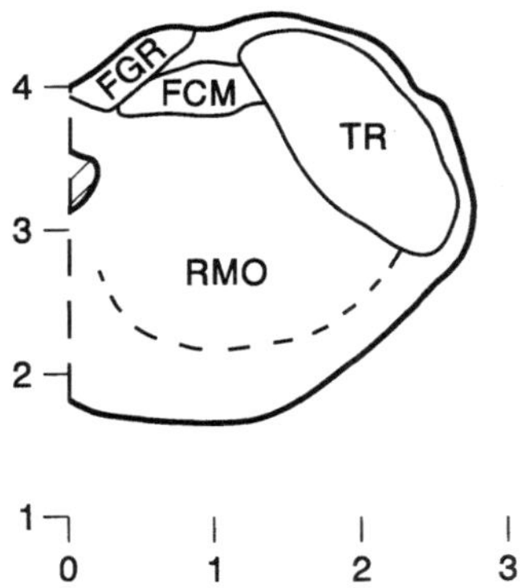

P 9

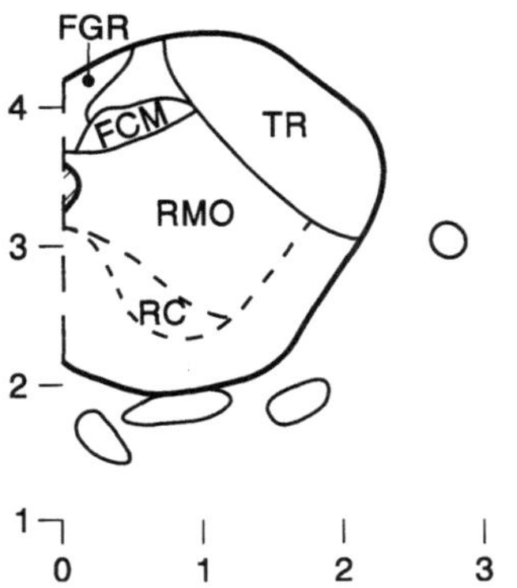

P 10

FCM	ncl. fascicularis cuneatus medialis
FGR	ncl. fascicularis gracilis
RC	nucleus residualis cornus ventralis
RMO	nucleus reticularis medullae oblongatae
TR	complexus sensorius nervi trigemini

12 Subject Index

bold faced: page numbers of Tables ***bold-faced italic:*** page numbers of Figures ***bold-faced italic, marked A:*** page numbers of Atlas Figures

bold faced: page numbers of Tables
bold-faced italic: page numbers of Figures
bold-faced italic, marked A: page numbers of Atlas Figures

bold faced: page numbers of Tables
bold-faced italic: page numbers of Figures
bold-faced italic, marked A: page numbers of Atlas Figures

bold faced: page numbers of Tables
bold-faced italic: page numbers of Figures
bold-faced italic, marked A: page numbers of Atlas Figures

bold faced: page numbers of Tables
bold-faced italic: page numbers of Figures
bold-faced italic, marked A: page numbers of Atlas Figures

bold faced: page numbers of Tables
bold-faced italic: page numbers of Figures
bold-faced italic, marked A: page numbers of Atlas Figures

bold faced: page numbers of Tables
bold-faced italic: page numbers of Figures
bold-faced italic, marked A: page numbers of Atlas Figures

bold faced: page numbers of Tables
bold-faced italic: page numbers of Figures
bold-faced italic, marked A: page numbers of Atlas Figures

bold faced: page numbers of Tables ***bold-faced italic:*** page numbers of Figures ***bold-faced italic, marked A:*** page numbers of Atlas Figures

MIX
Papier aus verantwortungsvollen Quellen
Paper from responsible sources
FSC® C105338

If you have any concerns about our products,
you can contact us on
ProductSafety@springernature.com

In case Publisher is established outside the EU,
the EU authorized representative is:
Springer Nature Customer Service Center GmbH
Europaplatz 3, 69115 Heidelberg, Germany

Printed by Libri Plureos GmbH
in Hamburg, Germany